NANOSCIENCE AND NANOTECHNOLOGY

NANOSCIENCE AND NANOTECHNOLOGY

ENVIRONMENTAL AND HEALTH IMPACTS

Edited by

Vicki H. Grassian

WILEY

A John Wiley & Sons, Inc., Publication

Library of Congress Cataloging-in-Publication Data:

ISBN 978-0-470-08103-7

Printed in the United States of America
10 9 8 7 6 5 4 3 2 1

To my family

CONTENTS

**7. Surface Oxides on Carbon Nanotubes (CNTs): Effects on
CNT Stability and Sorption Properties in Aquatic Environments** 133
*Howard Fairbrother, Billy Smith, Josh Wnuk, Kevin Wepasnick,
William P. Ball, Hyunhee Cho, and Fazlullah K. Bangash*

Nanoscience is the study of the fundamental principles of molecules and structures with at least one dimension between 1 and 100 nm. **Nanotechnology** is the application of these nanostructures into useful nanoscale devices.

BACKGROUND

The ability to manipulate and fabricate matter on the nanometer length scale is far better than ever before, with new capabilities to prepare and synthesize highly uniform nanoscale materials, and the advent of new instrumentation to investigate nanoscale materials. On the nanometer length scale, properties of matter can substantially differ and can exhibit size-dependent behavior. With these properties, a host of new materials have now become available for energy and environmental applications, and for improving human health (e.g., for disease detection, drug delivery, etc.).

The properties of matter depend on chemical composition and phase, and on the nanoscale the properties of matter also depend on size. On the nanoscale, electronic, optical, and magnetic properties can be size dependent. In addition, because of the high surface/volume ratio, surface properties, surface free energies, and surface coatings will also change the properties of nanomaterials. Given that on the nanoscale, the properties of matter depend on size, it can be asked, how well do we understand the environmental health and safety risks of nanomaterials? For example, can material safety data sheets developed for bulk graphite be used for nano-based carbon materials such as carbon nanotubes or buckyballs? What necessary measures are needed to ensure that the implications of nanomaterials are well understood as these materials are developed and used in a wide range of applications? Furthermore, what can be done to reduce the uncertainties in our understanding of the environmental health and safety implications of nanotechnology?

In the United States, the Environmental Protection Agency (EPA) provides funds not only for cleanup at superfund sites but also for superfund centers to provide a scientific basis for understanding toxicological impacts of these sites. A much better approach for future technologies is to fund centers for the study of environmental implications as technologies are developed. This new paradigm of doing business for the U.S. EPA is evident in the recent plan to partner with the National Science Foundation to fund a national center for the environmental implications of nanomaterials. In addition to a national center, there are several current funding initiatives that support the efforts of individual researchers and small research teams through several federal funding agencies in the United States, Canada, Europe, Australia, and other nations. In the United States, these include the Environmental Protection Agency and National Science Foundation, as noted above, as well as the National Institute of Environmental Health Science and the National Institute of Occupational Safety and Health. This represents a new paradigm for balancing technology development with environmental health and safety concerns.

The ultimate goal of the research activities discussed in each of the chapters of this book is to provide a strong scientific basis to the understanding of the environmental health and safety of nanomaterials. A high level of scientific understanding is essential so that sound environmental policies, if needed, can be developed and implemented with certainty while nanoscience and nanotechnology continue to grow. The avoidance of environmental health and safety problems, such as those that occurred with the development and use of chlorofluorocarbons (CFCs), polychlorinated biphenyls (PCBs), asbestos, and others, is imperative. In addition, from an economic standpoint even the avoidance of perceived risks is essential if nanotechnologies are to be accepted by the public. So from many perspectives—Occupational Health: Will this be the next asbestos? Environmental Health: Will this be the next PCB or CFC? Manufacturing/Marketing: Will this be the next genetically modified foods?—it is critical that issues related to the environmental and health impacts of nanomaterials be evaluated and understood.

OVERVIEW OF BOOK

Nanomaterials are of varying chemical complexity (bulk and surface), size, shape, and phase. Therefore, there exist large challenges in understanding the environmental health and safety of nanomaterials, and truly interdisciplinary efforts are needed. This book reflects the interdisciplinary nature of the research on the environmental and health impacts of nanoscience and nanotechnology. Chapter authors come from a variety of disciplines including chemical engineering, chemistry, civil and environmental engineering, environmental microbiology, geoscience, occupational and environmental health, pathology, pharmacology, and plant and soil science. The research described herein represents a compilation of some of the most recent studies and the current state of the science of the environmental and health impacts of nanoscience and nanotechnology. The book is divided into three parts as shown below.

PART I ENVIRONMENTAL AND HEALTH IMPACTS OF NANOMATERIALS: OVERVIEW AND CHALLENGES

Part I consists of three chapters: *Nanomaterials and the Environment* discusses the different types of nanomaterials that are being commercially produced and the potential for these materials to get into the environment; *Assessing the Life Cycle Environmental Implications of Nanomanufacturing: Opportunities and Challenges* discusses different approaches toward life cycle assessment that are being explored, as well as the difficulties; and the last chapter of this part, *An Integrated Approach Toward Understanding the Environmental Fate, Transport, Toxicity, and Health Hazards of Nanomaterials*, focuses on the importance and need to integrate high-quality nanomaterial characterization with studies related to the environmental health and safety of these materials and shows examples of different techniques that can be used for nanomaterials' physicochemical characterization.

PART II   FATE AND TRANSPORT OF NANOMATERIALS IN THE ENVIRONMENT

Part II focuses on what happens to nanomaterials once they get into the environment. Six chapters cover this topic with an emphasis on water and soil environments. Three of the chapters focus on metal and metal oxide nanomaterials. These include *Properties of Commercial Nanoparticles that Affect Their Removal During Water Treatment*; *Transport and Retention of Nanomaterials in Porous Media*; and *Transport of Nanomaterials in Unsaturated Porous Media*. Two chapters focus on carbon-based nanomaterials: *Surface Oxides on Carbon Nanotubes (CNTs): Effects on CNT Stability and Sorption Properties in Aquatic Environments* and *Chemical and Photochemical Reactivity of Fullerenes in the Aqueous Phase*. The last chapter of this part, *Bacterial Interactions with CdSe Quantum Dots and Environmental Implications*, looks at a different class of nanomaterials, crystalline semiconductor quantum dots, and their interactions with bacteria.

PART III   TOXICITY AND HEALTH HAZARDS OF NANOMATERIALS

Part III contains eight chapters on the toxicity and health hazards of nanomaterials. This part focuses on the impact nanomaterials have on the environment and its encompassing biota. Investigations described in Part III focus on living systems at all scales, from biological components to cells to simple organisms to fish to animals to humans. The first chapter in this part, *Potential Toxicity of Fullerenes and Molecular Modeling of Their Transport Across Lipid Membranes*, focuses on toxicity of fullerenes, C_{60}, and the transport of these nanomaterials across membranes. The next two chapters focus on *in vitro* studies and include *In Vitro Models for Nanoparticle Toxicology* and *Biological Activity of Mineral Fibers and Carbon Particulates: Implications for Nanoparticle Toxicity and the Role of Surface Chemistry*. The next two chapters in this part investigate environmental health and safety from an organism perspective. Chapter 13, *Growth and Some Enzymatic Responses of E. coli to Photocatalytic TiO_2*, examines the response of *E. coli* in the presence of illuminated TiO_2 nanoparticles and Chapter 14, *Bioavailability, Trophic Transfer, and Toxicity of Manufactured Metal and Metal Oxide Nanoparticles in Terrestrial Environments*, focuses on bioaccumulation and trophic transfer of metal and metal oxide nanoparticles.

The next two chapters of Part III focus on issues related to the toxicity of inhaled nanomaterials and the impact on human health. *Health Effects of Inhaled Engineered Nanoscale Materials* and *Neurotoxicity of Manufactured Nanoparticles* examine the multitude of potential health risks that result from inhalation of nanomaterials. Since there are concerns that inhalation exposure may be particularly problematic for those working in the industry, the last chapter, *Occupational Health Hazards of Nanoparticles*, focuses on this concern.

FUTURE OUTLOOK

The development of nanotechnology-based consumer products is predicted to grow substantially in the next 10 years and beyond. Along with this growth, it is clear that there will be many issues and questions that need to be addressed related to the potential impact this technology will have on the environment, living organisms, and human health. We hope that this book inspires some readers to rise to the challenges that are faced so that the environmental and health impacts of nanoscience and nanotechnology can be understood and therefore properly controlled as new commercial uses and applications emerge.

Director, Nanoscience and Nanotechnology   VICKI H. GRASSIAN
Institute at The University of Iowa,
Collegiate Fellow, College of Liberal Arts and Sciences
Professor, Departments of Chemistry
Chemical and Biochemical Engineering,
and Environmental and Occupational Health

■■■■ Contributors

Peter Aldous, Center for Health and the Environment, University of California, Davis, One Shields Avenue, Davis, CA 95616, USA. E-mail: paldous@byu.net

Bhavik R. Bakshi, Department of Chemical and Biomolecular Engineering, The Ohio State University, Columbus, OH 43210, USA. E-mail: bakshi.2@osu.edu

William P. Ball, Department of Geography and Environmental Engineering, Johns Hopkins University, 3400 North Charles Street, Baltimore, MD 21218, USA. E-mail: bball@jhu.edu

Young-Min Ban, Department of Chemical Engineering, University of Florida, Gainesville, FL 32611, USA. E-mail: half0min@ufl.edu

Fazlullah K. Bangash, Institute for Chemical Sciences, University of Peshawar, Peshawar 25120, Pakistan. E-mail: Fazlullah52@yahoo.com

Paul Bertsch, Department of Plant and Soil Sciences, University of Kentucky, Lexington, KY 40546, USA. E-mail: paul.bertsch@uky.edu

Gabriel Bitton, Department of Environmental Engineering Sciences, University of Florida, Gainesville, FL 32611, USA. E-mail: gbitton@ufl.edu

Jean-Claude Bonzongo, Department of Environmental Engineering Sciences, University of Florida, Gainesville, FL 32611, USA. E-mail: bonzongo@ufl.edu

Dan Cha, Department of Civil & Environmental Engineering, University of Delaware, Newark, DE 19716, USA. E-mail: cha@ce.udel.edu

Lixia Chen, School of Civil Engineering and Environmental Science, The University of Oklahoma, 202 W. Boyd Street, Rm 334, Norman, OK 73019, USA. E-mail: lxchen@ou.edu

Yongsheng Chen, Department of Civil and Environmental Engineering, Arizona State University, Box 5306, Tempe, AZ 85287, USA. E-mail: Yongsheng.chem@asu.edu

Hyunhee Cho, Department of Geography and Environmental Engineering, Johns Hopkins University, 3400 North Charles Street, Baltimore, MD 21218, USA. E-mail: Hcho25@jhu.edu

Jed Costanza, School of Civil and Environmental Engineering, Georgia Institute of Technology, 311 Ferst Drive, Atlanta, GA 30332, USA.
E-mail: jc394@mail.gatech.edu

John Crittenden, Department of Civil and Environmental Engineering, Arizona State University, Box 5306, Tempe, AZ 85287, USA. E-mail: J.Crittenden@asu.edu

Prabir K. Dutta, Department of Chemistry, The Ohio State University, Columbus, OH 43210, USA. E-mail: dutta@chemistry.ohio-state.edu

Sherrie Elzey, Department of Chemistry, University of Iowa, Iowa City, IA 52246, USA. E-mail: sherrie-elzey@uiowa.edu

Ayca Erdem, Department of Civil & Environmental Engineer, University of Delaware, Newark, DE 19716, USA. E-mail: ayca@ce.udel.edu

Howard Fairbrother, Department of Chemistry, Johns Hopkins University, 3400 North Charles Street, Baltimore, MD 21218, USA. E-mail: howardf@jhu.edu

John D. Fortner, School of Civil and Environmental Engineering, Georgia Institute of Technology, 200 Bobby Dodd Way, Atlanta, GA 30332, USA.
E-mail: f228@mail.gatech.edu

Jie Gao, Department of Environmental Engineering Sciences, University of Florida, Gainesville, FL 32611, USA. E-mail: dencyl@ufl.edu

Vicki H. Grassian, Department of Chemistry, University of Iowa, Iowa City, IA 52246, USA. E-mail: vicki-grassian@uiowa.edu

Geoffrey Grubb, Department of Chemical and Biomolecular Engineering, The Ohio State University, Columbus, OH 43210, USA. E-mail: grubb.284@osu.edu

Jaime M. Hatcher, Center for Neurodegenerative Disease, Department of Neurology, Emory University School of Medicine, 615 Michael Street, Atlanta, GA 30322, USA. E-mail: jmhatch@emory.edu

Patricia A. Holden, Donald Bren School of Environmental Science & Management, University of California, Santa Barbara, CA 93106-5131, USA. E-mail: Holden@bren.ucsb.edu

Chin Pao Huang, Department of Civil & Environmental Engineering, University of Delaware, Newark, DE 19716, USA. E-mail: huang@udel.edu

Joseph B. Hughes, School of Civil and Environmental Engineering, Georgia Institute of Technology, 200 Bobby Dodd Way, Atlanta, GA 30332, USA.
E-mail: joseph.hughes@ce.gatech.edu

Simona Hunyadi, Department of Plant and Soil Sciences, University of Kentucky, Lexington, KY 40546, USA. E-mail: simonamurph@gmail.com

Dean P. Jones, Department of Medicine, Emory University School of Medicine, 615 Michael Street, Atlanta, GA 30322, USA. E-mail: dpjones@emory.edu

Vikas Khanna, Department of Chemical and Biomolecular Engineering, The Ohio State University, Columbus, OH 43210, USA. E-mail: khanna.105@osu.edu

Tohren C.G. Kibbey, School of Civil Engineering and Environmental Science, The University of Oklahoma, 202 W. Boyd Street, Rm 334, Norman, OK 73019, USA. E-mail: kibbey@ou.edu

Jae-Hong Kim, School of Civil and Environmental Engineering, Georgia Institute of Technology, 200 Bobby Dodd Way, Atlanta, GA 30332, USA. E-mail: jaehong.kim@ce.gatec

Dmitry I. Kopelevich, Department of Chemical Engineering, University of Florida, Gainesville, FL 32611, USA. E-mail: dkopelevich@che.ufl.edu

Jaesang Lee, School of Civil and Environmental Engineering, Georgia Institute of Technology, 200 Bobby Dodd Way, Atlanta, GA 30332, USA. E-mail: jaesang.lee@ce.gatech.edu

John F. Long, Department of Veterinary Biosciences, The Ohio State University, Columbus, OH 43210, USA. E-mail: long15@osu.edu

Amy K. Madl, ChemRisk, Inc., 25 Jessie Street at Ecker Square, Suite 1800, San Francisco, CA 94105, USA. E-mail: amadl@chemrisk.com

Gary W. Miller, Center for Neurodegenerative Disease, Department of Neurology, Emory University School of Medicine, 615 Michael Street, Atlanta, GA 30322, USA. E-mail: gary.miller@emory.edu

Jay L. Nadeau, Department of Biomedical Engineering, McGill University, Montreal, Quebec, Canada H3A 2B4. E-mail: jay.nadeau@mcgill.ca

Mai A. Ngo, Center for Health and the Environment, University of California, Davis, One Shields Avenue, Davis, CA 95616, USA. E-mail: maingo@ucdavis.edu

Patrick T. O'Shaughnessy, College of Public Health, The University of Iowa, 100 Oakdale Campus, #126 IREH, Iowa City, IA 52242-5000, USA. E-mail: patrick-oshaughnessy@uiowa.edu

Kurt D. Pennell, School of Civil and Environmental Engineering, Georgia Institute of Technology, 311 Ferst Drive, Atlanta, GA 30332, USA. E-mails: kurt.pennell@ce.gatech and kpennel@emory.edu

John M. Pettibone, Department of Chemical and Biochemical Engineering, University of Iowa, Iowa City, IA 52246, USA. E-mail: john-pettibone@uiowa.edu

Kent E. Pinkerton, Center for Health and the Environment, University of California, Davis, One Shields Avenue, Davis, CA 95616, USA. E-mail: kepinkerton@ucdavis.edu

John H. Priester, Donald Bren School of Environmental Science & Management, University of California, Santa Barbara, CA 93106-5131, USA. E-mail: Priester@bren.ucsb.edu

Suzette Smiley-Jewell, Department of Anatomy, Physiology, and Cell Biology, University of California, Davis, One Shields Avenue, Davis, CA 95616, USA. E-mail: smsmiley@ucdavis.edu

Billy Smith, Department of Chemistry, Johns Hopkins University, 3400 North Charles Street, Baltimore, MD 21218, USA. E-mail: bsmith90@jhu.edu

Galen D. Stucky, Department of Chemistry and Biochemistry, University of California, Santa Barbara, CA 93106, USA. E-mail: stucky@chem.ucsb.edu

Ryan A. Tasseff, School of Chemical and Biomolecular Engineering, Cornell University Ithaca, NY 14853, USA. E-mail: rat44@cornell.edu

Jason Unrine, Department of Plant and Soil Sciences, University of Kentucky, Lexington, KY 40546, USA. E-mail: jason.unrine@uky.edu

John M. Veranth, Department of Pharmacology and Toxicology, University of Utah, 30 South 2000 East, Salt Lake City, UT 84112, USA. E-mail: John.Veranth@m.cc.utah.edu

W. James Waldman, Department of Pathology, The Ohio State University, Columbus, OH 43210, USA. E-mail: james.waldman@osuma.edu

Yonggang Wang, School of Civil and Environmental Engineering, Georgia Institute of Technology, 311 Ferst Drive, Atlanta, GA 30332, USA. E-mail: ywang32@gatech.edu

Kevin Wepasnick, Department of Chemistry, Johns Hopkins University, 3400 North Charles Street, Baltimore, MD 21218, USA. E-mail: kaw@jhu.edu

Paul Westerhoff, Department of Civil and Environmental Engineering, Arizona State University, Box 5306, Tempe, AZ 85287, USA. E-mail: p.westerhoff@asu.edu

Marshall V. Williams, Department of Immunology and Medical Genetics, The Ohio State University, Columbus, OH 43210, USA. E-mail: williams.70@osu.edu

Josh Wnuk, Department of Chemistry, Johns Hopkins University, 3400 North Charles Street, Baltimore, MD 21218, USA. E-mail: wunkjd@gmail.com

Yang Zhang, Department of Civil and Environmental Engineering, Arizona State University, Box 5306, Tempe, AZ 85287, USA. E-mail: yzhang16@yahoo.com

Yi Zhang, Department of Chemical and Biomolecular Engineering, The Ohio State University, Columbus, OH 43210, USA. E-mail: zhang.468@osu.edu

PART I

ENVIRONMENTAL AND HEALTH IMPACTS OF NANOMATERIALS: OVERVIEW AND CHALLENGES

CHAPTER 1

Nanomaterials and the Environment

MAI A. NGO[1], SUZETTE SMILEY-JEWELL[1], PETER ALDOUS[1], and KENT E. PINKERTON[1,2]

[1]Center for Health and the Environment, University of California, Davis, One Shields Avenue, Davis, CA 95616, USA
[2]Department of Anatomy, Physiology, and Cell Biology, University of California, Davis, One Shields Avenue, Davis, CA 95616, USA

1.1 INTRODUCTION

The use of nanomaterials, manufactured products having one or more dimensions 100 nm or less, has grown dramatically in the last decade and promises to continue to grow in the future. Currently, nanomaterials are used in medical devices, pharmaceuticals, environmental remediation, and in scores of consumer products ranging from cosmetics to electronics, with numerous forthcoming applications (1, 2). Forecasts predict nanotechnology, the science of using nanomaterials and nanodevices, to be a $10 billion industry by 2010, growing to $1 trillion by 2015 (3, 4). Figure 1.1 illustrates in graphic form the rapid growth in data found in the literature on nanoproducts via a search of life science journal articles using the terms "nanotechnology," "nanomaterial," and "nanoparticle."

Nanomaterials can be composed of many different base materials and have different structures. A few examples of nanomaterials are listed in Table 1.1 (adapted from Reference (5)). Typical nanomaterials include carbon-based fullerenes (buckminsterfullerene, buckyballs, and C_{60} fullerenes) and nanotubes, which have been used to selectively target and eliminate cancer cells (6); quantum dots, which are nanoscale semiconductor crystals used to track protein transport in biological systems; metal oxanes, such as titanium oxide (TiO_2) and zinc oxide, which are used in sunscreens for their ultraviolet (UV) reflecting capability in cosmetics, and in the formulation of membranes and films (7); and silver nanoparticles, which are widely incorporated into products such as antibacterial and antifungal elements (8–10).

Nanoscience and Nanotechnology Edited by Vicki H. Grassian
Copyright © 2008 John Wiley & Sons, Inc.

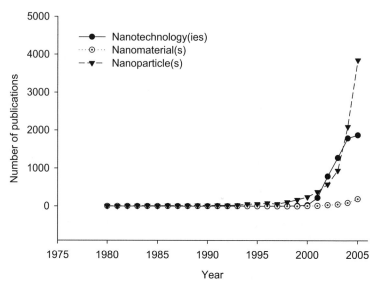

FIGURE 1.1 Growth in Nanotechnology publications: number of nanotechnology-related publications by year from 1990 to 2005 via literature search of life science journal articles on nanotechnology, nanomaterials, and nanoparticles.

What makes nanomaterials unique is their nanometer scale. At this very small size, the properties of materials can change. For example, quantum-mechanical effects can alter the behavior of small particles, such as their light-emitting color and electronic properties. Nanomaterials may have increased chemical reactivity as a consequence of increased surface area to volume ratio, compared to materials with the same chemical composition synthesized at the macroscale. This is because the chemical reactivity of a material depends on its surface area, which increases as the particle size decreases. For instance, gold in its bulk form has been regarded to be chemically inert, especially to

TABLE 1.1 Examples and Characteristics of Nanoproducts

Structure	Size	Base Material	Examples
Tubes	Diameter: 1–100 nm	Carbon	Single-walled carbon nanotubes (SWCNTs), multiwalled carbon nanotubes (MWCNTs)
Wires	Diameter: 1–100 nm	Metals, semiconductors, tellurium nanowires	Silicon nanowires, cadmium oxides, sulfides, nitrides
Crystals, clusters	Radius: 1–10 nm	Semiconductors, metals, metal oxides	Quantum dots, titanium dioxide, silicon dioxide
Spheres	Radius: <100 nm	Carbon	Fullerenes (C_{60}, buckyballs)

Adapted from Reference 5.

reactions with oxygen and hydrogen. However, at the nanoscale, gold has been found to be extremely reactive and is extensively being developed as a catalyst for a large range of chemical reactions (11–13).

Because nanomaterials have so many applications, with more to come, the prevalence of these materials in industry and society is ensured. It is evident from past experience that development of innovative products is not always associated with benign consequences in the long term, as illustrated by methyl tertiary butyl ether (MTBE), dichlorodiphenyltrichloroethane (DDT), and chlorofluorocarbons (CFCs), all of which have had significant adverse health and environmental impacts. While the novel properties that arise when a material is reduced to the nanoscale make them useful for a wide range of applications, little is known about how these properties will affect human health or how they will behave in the environment. These novel properties may also confer unusual mechanisms of toxicity and risks that cannot be predicted by existing knowledge or extrapolation from what is known regarding the behavior of these materials at the macroscale (14, 15). The degree of potential exposure to populations is also unknown and, at present, there is a lack of standard risk evaluation and safe handling guidelines for nanomaterials. The potential risks nanomaterial poses promises to be contentious until the uncertainties regarding toxicity and exposure are known.

1.2 NANOMATERIALS AND THE ENVIRONMENT

To address the potential influence of nanomaterials on the environment, assessment of their physiochemical properties will be needed to determine their fate, mobility, degradation, persistence, and bioavailability. The biological effects and the detection of nanomaterials in nature and in any environmentally exposed species, including humans, are also essential elements to consider. Nanomaterial risk assessment must evaluate the following: the toxicity of particular nanomaterials, the extent of their dispersion in the environment, environmental fate and transport, tranformations and modifications in the environment that may affect bioavailability, absorption, and toxicity upon exposure to biological systems, biological and ecological relevance of exposure, acute versus low level chronic exposure, and the ability to determine and measure exposure to the environment and to biological systems. In humans, the extent of exposure in several settings will need to be considered. These include exposure from occupational, commercial, and environmental sources. Given the bulk of knowledge needed for a proper assessment of the growth of nanoparticle use and manufacture, it seems that there will always be a lag in research as the properties of these nanoparticles are identified.

Recent studies indicate that some effects are not completely benign to biological and environmental targets (16–21). Concerns regarding potential health risks and the environmental impact of engineered nanomaterials should prompt a proactive approach to ensure that the rapidly growing nanotechnology industry is environmentally benign and sustainable. Since fate and transport of nanomaterials under a variety of conditions have not been systematically studied, potential risks and hazards of nanomaterials in the environment will be addressed in a general way using the few examples that are available.

1.2.1   Exposure

Release of nanomaterials into the environment will most likely arise from several points in the nanomaterial "life cycle," including release during the manufacturing process, use of products into which nanomaterials are incorporated, and, inevitably, with the disposal of these products. Yet, little is known about a nanomaterial's fate and behavior in the environment. Mobility and interaction with biological systems largely depend on the specific size, shape, and chemistry of a nanomaterial. Thus, different types of nanomaterials will have different types of impacts on the environment and human health.

1.2.2   Fate and Transport

Unlike free particles, nanoparticles contained in a matrix are not considered dangerous. Nevertheless, products into which nanomaterials are incorporated will be expected to breakdown with time and to release nanomaterials in some form, possibly as free particles, into the environment. As with chemical compounds leaching from landfills, nanoparticles may also enter and travel through the natural environment, potentially in different ways than larger particles. Once in the environment, a number of factors may sequester, modify, mobilize, or degrade nanomaterials, thereby influencing their bioavailability and potential toxicity. Again, the fate and behavior of nanoparticles in the environment may not necessarily be predicted from what is known of chemically similar material at a larger scale.

One example that underscores the unique and unpredictable behavior of nanoparticles is that of fullerenes. It would be predicted that an all-carbon molecule, such as a fullerene, would be quite hydrophobic and insoluble in water. Interestingly, researchers have found the hydrophobic fullerene molecules acquire charge, forming clustered structures [nano-C_{60} (nC_{60})] suspended in water (22–27). Such suspensions would permit greater distribution of fullerenes in the aqueous environment. Additionally, while salt-free solutions solubilized the nC_{60} fullerene crystals (28), high salt conditions neutralized the nC_{60} fullerene crystals, causing them to fall out of solution (29, 30) and setting up the potential for their accumulation in sediments. These studies and others demonstrate that the fate of fullerenes can change depending upon the properties of the aquatic ecosystem in which they are found.

Indeed, a principal factor that will affect the environmental fate of nanomaterials is water solubility. Water solubility correlates with the potential for dispersion over vast areas and increases the potential for exposure. It is unknown whether nanomaterials with low water solubility will bind with organic molecules in the aquatic environment to increase their affinity for the aqueous phase and potentially increase exposure to biota. As discussed below, interactions with bacterial communities, mineral surfaces, and pH gradients are also likely to influence the fate and potential risks of water-soluble nanomaterials.

Water solubility of a nanomaterial is an important attribute for many applications. For these purposes, unique surface coatings and derivatives are being developed to deliver nanomaterials for *in situ* groundwater remediation and delivery of drugs. For example, nanoiron is coated before being introduced into the environment for groundwater remediation (31, 32). Quantum dots consist of a metal complex core that is coated for biological activity (18). Hydroxylation of fullerenes increases their solubility (33) and permits their use in drug delivery and other therapeutic applications. The cytotoxicity of fullerenes derived for the purpose of water solubility was found to differ up to several orders of magnitude (34). Adding charged functional groups or surface coatings generally improves water solubility and suspension characteristics of nanoparticles. However, they can also impart selectivity in physical or chemical interactions, both with their intended targets and with indiscriminate entities in the natural environment. As a result, such modifications may influence the toxicological or ecotoxicological profile of a nanoparticle.

An additional factor that will affect the fate of nanomaterials in the environment is the ability to form large colloidal aggregates. Carbon nanotubes and nanoiron particles form aggregates of much larger dimensions than the individual nanosized particles (35–37). The large surface areas of smaller nanoparticles may allow for enhanced bioavailability and, consequently, greater potential for exposure and toxic effects. Evidence suggests that the adverse effects of fullerenes on bacterial populations are greater for smaller aggregates than for larger ones (38), and the antimicrobial properties of TiO_2 nanoparticles are inversely related to particle size (39). In contrast, the tendency of a nanomaterial to attach to surfaces or form aggregates can prevent its flow through porous media such as soil or reduce its bioavailability and exposure potential. Thus, factors that promote aggregation of nanomaterials in the environment can modify transport and subsequent toxicity. Depending on the specific nanomaterial, aggregation may mitigate both exposure and toxicity in some cases, while in others, it may increase the risk of toxic effects.

Apart from possibly being toxic in their own right, nanoparticles may provide surfaces that bind and transport toxic chemical pollutants. Possible nanoparticle association with naturally occurring compounds should be considered together with how this might affect their bioavailability and uptake into cells and organisms.

1.2.3 Transformation

Transformation of a nanoparticle in the environment will significantly influence its properties, including toxicity to living systems. Oxidation–reduction (redox) reactions are believed to be important for the transformation and environmental fate of nanomaterials. Chemical or biological oxidation is able to add, remove, or modify functional groups associated with mineral-based nanomaterials. Their high electron affinity and ability to participate in redox reactions (40) make fullerenes capable of producing reactive oxygen species (ROS) that can oxidize organic compounds in the environment and be key players in the induction and propagation of oxidative stress *in vivo*.

Microorganisms often mediate redox reactions in the environment. It is highly plausible that the interaction of nanomaterials with microorganisms normally present in soil and groundwater may alter nanomaterial transport, retention, bioavailability, and toxicological characteristics. Whether biotransformation of nanomaterials by microorganisms can occur to an appreciable extent in the natural environment and how such transformations affect nanomaterial toxicity is unclear. The contribution of bacteria and fungi in soil and water to the degradation of manufactured nanomaterials needs further examination. Conversely, the effects of nanomaterials on microbial populations in the natural environment are also important to consideration. Microorganisms are a principal component of many ecosystems, serving as a basis of food webs and as important contributors to soil health. As such, disturbing microbial homeostasis would ultimately affect biota in all parts of the ecosystem. The ability of photoactivated TiO_2 nanoparticles to effectively oxidize organic matter and inactivate microbes (41–43) is not only highly useful for treatment of water systems but also presents a possible risk to natural microbial populations if discharged without regulation into the environment.

As seen with TiO_2 nanoparticles, UV radiation can also change the characteristics of nanomaterials (10, 44). Fullerenes strongly absorb UV light and in a manner similar to TiO_2, photoexcitation can lead to the generation of cytotoxic reactive oxygen species (40). Quantum dots have been found to degrade under photolytic and oxidative conditions (18, 45), as occurs in the natural environment. Compromising the bioactive coat of quantum dots can expose the metalloid core, which is potentially toxic and may lead to undesirable or unanticipated reactions. Thus, chemodynamic reactions in the environment, such as photochemical changes and biodegradation, can contribute to the breakdown of nanomaterials.

In addition to the transformations that nanomaterials may undergo in the natural environment, it is also important to consider the ability of nanomaterials to affect elements with which they come in contact. TiO_2 and ZnO have long been used as fine powders in many sunscreens because of their ability to absorb or reflect UV radiation and reduce the amount of UV rays penetrating and damaging the deeper dermal layers of the skin. TiO_2 and ZnO have semiconductor and, consequently, photocatalytic activity that can promote the transformation of organic molecules upon absorption of radiation (46, 47). For this purpose, TiO_2 nanoparticles have been used to catalyze the photodegradation of environmental contaminants (48–50). A point of consideration may be the toxicity of TiO_2-initiated photodegradative reactions and any resulting products in the environment. It also remains to be seen whether the UV absorptive qualities of TiO_2 have the potential to disrupt food chains by competing for the UV energy on which marine and freshwater phytoplankton are dependent.

In summary, environmental fate depends on the chemistry of the specific nanomaterial and can be further influenced by environmental conditions such as physical binding or chemical reactivity with other compounds, water solubility, aggregation, and oxidation–reduction reactions. Modifications will greatly influence the behavior of nanomaterials, including their interaction with fauna and flora, accumulation in biological systems, and toxicity.

1.3 NANOMATERIALS AND BIOLOGICAL SYSTEMS

Although hundreds of tons of nanoparticles are released through emissions into the environment annually, little is known of their interactions with biological systems.

Once exposure occurs, the biological fate of nanomaterials depends on the balance of four processes: absorption, distribution, metabolism, and excretion. To determine the risk of nanomaterials, information is needed for each of these processes. Some studies have been done, but more are needed. Below is a discussion of the biological fate of nanomaterials—what is known and areas needed to be further explored.

1.3.1 Exposure and Absorption

Exposure to nanomaterials can occur via dermal, gastrointestinal, or inhalation routes. While exposure will depend on a number of factors, such as how well nanomaterials are contained during manufacture, how widespread their use becomes, and if they are biodegradable or recoverable, the extent of absorption into the biological system depends on the chemical and physical properties of the nanomaterial.

Nanomaterials are incorporated into many consumer products that are meant to be applied to the skin. Whether a nanomaterial enters the body through the skin can depend upon whether the skin is injured, to what degree the skin is flexed [51], and to what degree the nanomaterial is lipid soluble. The latter may depend upon the surface coating or carrier of the nanomaterial. For example, studies are currently being done to optimize carriers of nanomaterials for topically applied medicine [51, 52]. As mentioned above, current dermal exposure to nanomaterials occurs via sunscreen and cosmetics, such as by UV absorbing TiO_2 and ZnO. There have been no human clinical reports of toxicity from TiO_2 and ZnO, and these two nanomaterials have only been found toxic at very high doses in cell culture systems [53]. However, quantum dots topically applied to intact skin at occupationally relevant doses were found to penetrate and localize within the dermis in a few hours, which is a concern for occupational settings [54]. Synthesized ^{14}C-labeled C_{60} was also found to be taken up by human epidermal cells when administered as a fine aqueous suspension [55]. The C_{60} rapidly accumulated in human cells, although it did not cause acute toxicity or affect the proliferation of human keratinocytes until relatively high doses were administered [55, 56]. If nanomaterials contaminate the water, air, or soil, dermal exposure may become widespread for many organisms.

Gastrointestinal tract (GI) exposure may occur from the use of nanomaterial-containing cosmetics or drugs or as a result of the mucociliary escalator clearing nanomaterials from the respiratory tract [57]. Animals may also ingest nanomaterials on their skin when grooming. Once in the GI tract, nanomaterials can be absorbed, although the extent of absorption depends upon particle size, with smaller particles crossing the GI tract more readily than larger particles [58]. In fact, a challenge with designing nanomaterial drug delivery systems is that nanomaterials tend to aggregate in the gut, which increases their size and lowers their absorption [59]. Chemical property also influences absorption; for example, ingested iridium nanomaterials are not taken up by the GI tract very well [60], but C_{60} fullerenes are readily absorbed [61] due to their hydrophobic nature. The extent to which nanomaterials can enter the food chain and whether they will bioaccumulate is unknown.

Besides dermal and GI tract exposure, there is the potential for inhalation of nanomaterials. This is a primary concern in industrial settings. Widespread inhalation of nanomaterials may also occur if nanomaterials become airborne and enter the atmosphere. Due to their small size, nanomaterials can form light dusts that are easily distributed in the air, inhaled, and deposited in the lung. The size (single or aggregate) and shape of a nanomaterial help determine where it deposits within the lung. Because of the difficulty in conducting uniform, controlled inhalation experiments, many pulmonary studies of nanomaterials have been done via intratracheal instillation in which a fine dust of nanomaterials is placed in the trachea and the animal breathes the dust into the lung. Intratracheal instillation of nanotubes (62, 63), SiO_2 (64), and TiO_2 (65) has resulted in pulmonary inflammation, granulomas, and/or interstitial fibrosis. We do not know if breathing nanomaterials would result in acute toxicity in humans or whether toxicity would be latent, as occurs with asbestos. Even so, studies support the ability of TiO_2 to produce pulmonary inflammation in rodents (66, 67).

1.3.2   Distribution

It appears that once nanomaterials are absorbed into the biological system, they can distribute throughout the body. ^{14}C-labeled C_{60}, after being intravenously injected into female Sprague–Dawley rats, was found to be rapidly cleared from the circulation and accumulated in the liver (56). The ^{14}C-labeled C_{60} persisted in the liver for 120 h following the i.v. administration, suggesting long-term accumulation in this organ. Quantum dots injected into the dermis translocate to the lymph nodes (68). Macrophages and dendritic cells in the lymph nodes may take up these particles (69, 70), leading to perturbation of the immune system. Self-protein interactions with particles may change their antigenicity, initiating autoimmune responses. Nanomaterial–protein complexes have also been used to facilitate antigen uptake by dendritic cells leading to enhanced immune response (71).

Due to the cardiovascular effects of ultrafine particles, it is probable that nanomaterials can distribute throughout the circulatory system (57, 72). When rats were gavaged with I^{125}-labeled polystyrene particles (50 nm), a small percentage of the particles were found 8 days later in the liver, spleen, blood, and bone marrow (58). Research has shown that nanoparticles, such as TiO_2, can leave the lungs of exposed animals and distribute to other organs (73). Within the lung, nanomaterials can be taken up by alveolar macrophages (74), enter epithelium (57) or interstitium (75), or translocate across the alveolar epithelium (76, 77). Once through the alveolar epithelium, gold nanoparticles were found in the pulmonary capillaries (76) and blood (73). Silica-coated magnetic nanoparticles containing rhodamine B isothiocyanate given intraperitoneally to mice were able to cross the blood–brain barrier and be found in the brain (78). Nanomaterials are known to enter the brains of monkeys, rats, and fish via the olfactory bulb (79–81).

Taken together, these studies demonstrate that nanomaterials can distribute within the cells and fluids of the body and translocate from organ to organ. Nonetheless, this translocation is highly variable and, again, dependent on particle size, surface characteristics, and chemical composition. The extent to which a nanomaterial enters into and translocates within the body will have a significant impact on human health.

There is currently an incomplete understanding of what forms of nanomaterials are bioavailable or whether they will bioaccumulate. It is too early to know whether nanoparticles applied on the skin, inhaled, or ingested can find their way to organ systems distal from the site of absorption or what concentrations they will obtain in these organs. As always, effects seen for cultured cells might not apply to the human body.

1.3.3 Metabolism

Metabolism is a salient component in the clearance of a compound from the body. Again, metabolism will be dependent on the characteristics and chemical composition for a particular nanoparticle. Usually, polycyclic aromatic compounds are metabolized by the cytochrome P450 system. However, as mentioned above, after ^{14}C-labeled C_{60} was intravenously injected into female Sprague–Dawley rats, it was found to rapidly accumulate in the liver (56). Evidence suggested that the relatively long-term accumulation of ^{14}C-labeled C_{60} in the liver may have been the result of the C_{60} not being oxidized, as normally occurs with polycyclic aromatics.

Once nanomaterials distribute and reach a target, results can be either beneficial, as for a drug, or toxic. Metabolism plays a major role. To date, the primary cause of nanomaterial-induced toxicity is thought to be the generation of reactive oxygen species (15, 57). Quantum dots (82), single-walled nanotubes (83–85), and fullerenes (16) are associated with oxidative stress and/or the production of ROS. For example, lipid peroxidation was found in the brains of largemouth bass exposed to fullerenes (84), and ROS production, lipid peroxidation, oxidative stress, and mitochondrial dysfunction were found after keratinocytes and bronchial epithelial cells were incubated with single-walled nanotubes (15). The shape of nanomaterials (leading to electron instability), as well as the presence of surface metals and/or redox-cycling organic chemicals, is thought to be instrumental in producing free radicals (86).

ROS is harmful because it can damage cell membranes (membrane lipid peroxidation), leading to permeability problems; cross-link and fragment proteins; and cause lesions in DNA. Ultimately, this can lead to cell death or, if the system is overwhelmed, death of the organism. For instance, the ability of both fullerenes (87) and TiO_2 (7) to generate ROS confers antimicrobial properties. While bacteria are limited in their ability to take up particles >5 nm, it has been speculated that they die from nanomaterial-induced oxidative stress caused by their cell membrane, which houses their electron transport/ATP energy generating system, interacting with nanomaterial-induced ROS (17).

Veronesi and colleagues used commercially available titania nanoparticles approximately 30 nm in size, which they added to cultures of mouse microglia (61). They found that the titania nanoparticles engulfed by microglia triggered the release of ROS in a prolonged manner. If this was to occur in a real-life exposure situation, the prolonged release of ROS could subject the brain to oxidative stress, a mechanism underlying some neurodegenerative diseases, such as Parkinson's and Alzheimer's. Biological systems do have oxidative defense systems that can combat ROS. Currently, it is unclear whether nanomaterials can produce enough protein/DNA damage to cause cancer and at what levels or time of exposure this may occur.

Fullerenes and other nanomaterials are often derivatized for compatibility with biological systems. This means they will interact with cellular membranes and may even be endocytosed. Redox-sensitive nanomaterials, such as fullerenes, could participate in oxidation reactions to damage the cell membrane and affect cell permeability and fluidity, leaving cells more susceptible to osmotic stress or hindering nutrient uptake, electron transport, and energy transduction.

1.3.4   Excretion

Only a few studies have reported clearance mechanisms. For example, 98% of ingested C_{60} fullerenes was reported in feces of rats (57), and alveolar macrophages were found to phagocytize nanomaterials (88, 89). Surface chemistry and physical properties will influence to what extent nanomaterials are cleared from the body by urine/feces, dermis, hair, breast milk, and breath, or whether they will be sequestered in body tissue for many years. Ultimately, excretion influences the length of time that the body is exposed to nanomaterials and how nanomaterials are released into the environment.

In summary, there is evidence that nanomaterials can be absorbed, distributed, metabolized, and excreted by the biological system. At present, the scientific data from studies on ultrafine airborne particles are the best approximation available for potential health effects from exposure to nanoparticles. However, much more information is needed to adequately evaluate the biological toxicity of nanomaterials.

1.4   CONCLUSIONS AND DIRECTIONS FOR THE FUTURE

Whether or not applications for nanomaterials involve their direct introduction into the environment, as with the use of nanoiron in groundwater remediation and the widespread incorporation of nanomaterials into commercial products, it is clear that nanomaterials will eventually enter the environment. The great diversity of nanomaterials and types of applications, multiple potential points of release into the environment, and varied routes of exposure to populations make addressing the potential risks a complex task. Only recently have researchers begun to study the potential ecological risks and impacts of nanomaterial release into the environment (15, 88, 90, 91). To date, how much exposure to nanomaterials may adversely affect living organisms remains unknown, as do specific mechanisms of toxicity.

Rather than waiting for adverse effects or potential problems to appear before taking measures toward "damage control," a precautionary approach with evaluation of risk to human health and implications for the environment may be conducted concurrently with the development of new nanomaterials. As long as the potential risks of nanomaterials are unclear, measures may be taken to ensure that nanomaterial release into the environment and exposure to the public is not left unchecked.

If nanomaterials prove harmful after being widely distributed in commerce, then the consequences will not be limited to only adverse health and environmental consequences. Expensive remediation efforts may ensue. A negative public perception of nanomaterials, as illustrated by the case of genetically modified organisms, may thwart the many potential benefits nanotechnology can offer. Given the fact that there are suggestions that nanoparticles may affect human health and that commonly used sunscreens and many other products containing nanoparticles are on store shelves worldwide, consumers may demand labeling laws to inform the presence of nanoparticles in products.

There is a high level of uncertainty in terms of potential toxicity and latent, unforeseen impacts of nanomaterials in the environment. To appropriately address potential risks, better methods for the detection and quantification of nanomaterials in the workplace and environment (air, water, and soil) must be developed. Given that different types of nanomaterials will have different toxicological properties and types of impacts on human health and the environment, there is a need to establish a standardized set of criteria for assessing the most critical toxicological and ecotoxicological parameters and potential risks of nanomaterials on an individual basis. A standard of safety and suitable protective measures must be determined for those who are likely to be exposed to nanomaterials, especially in the occupational setting. Educational outreach services and information need to be made available to industrial hygienists, physicians, healthcare providers, veterinarians, and wildlife managers about possible exposure scenarios and proper precautionary measures to be taken to avoid undue exposure. Potential environmental, health, and safety concerns should be assessed early on in the researching and processing of manufactured nanomaterials. A system by which researchers and manufacturers may be encouraged to register, identify likely exposure scenarios, and provide at least a basic toxicological profile for their products should be established. Clearly, it is in the best interest of the public, regulatory agencies, and industry to integrate safety, toxicology, and environmental concerns into the research and manufacture of nanomaterials. Nanotechnology and its prospects should be promoted in a responsible and safe manner that acknowledges potential risks of nanomaterials being incorporated into widespread commercial production.

REFERENCES

1. Akerman, M.E., Chan, W.C., Laakkonen, P., Bhatia, S.N., Ruoslahti, E. Nanocrystal targeting *in vivo*. *Proc Natl Acad Sci USA* 2002;99(20): 12617–12621.

2. Harris, T.J., von Maltzahn, G., Derfus, A.M., Ruoslahti, E., Bhatia, S.N. Proteolytic actuation of nanoparticle self-assembly. *Angew Chem Int Ed Engl* 2006;45(19):3161–3165.

3. Roco, M.C. International strategy for nanotechnology research and development. *J Nanopart Res* 2001;3(5–6):353–360.

4. Roco, M.C., Bainbridge, W.S. Converging technologies for improving human performance: integrating from the nanoscale. *J Nanopart Res* 2002;4(4):281–295.

5. Jortner, J., Rao, C.N.R. Nanostructured advanced materials: perspectives and directions. *Pure Appl Chem* 2002;74(9):1491–1506.

6. Kam, N.W., Dai, H. Carbon nanotubes as intracellular protein transporters: generality and biological functionality. *J Am Chem Soc* 2005;127(16):6021–6026.

7. Wiesner, M.R., Lowry, G.V., Alvarez, P., Dionysiou, D., Biswas, P. Assessing the risks of manufactured nanomaterials. *Environ Sci Technol* 2006;40(14):4336–4345.

8. Morones, J.R., Elechiguerra, J.L., Camacho, A., Holt, K., Kouri, J.B., Ramirez, J.T., Yacaman, M.J. The bactericidal effect of silver nanoparticles. *Nanotechnology* 2005;16 (10):2346–2353.

9. Sondi, I., Salopek-Sondi, B. Silver nanoparticles as antimicrobial agent: a case study on *E-coli* as a model for Gram-negative bacteria. *J Colloid Interface Sci* 2004;275(1):177–182.

10. Maness, P.C., Smolinski, S., Blake, D.M., Huang, Z., Wolfrum, E.J., Jacoby, W.A. Bactericidal activity of photocatalytic TiO_2 reaction: toward an understanding of its killing mechanism. *Appl Environ Microbiol* 1999;65(9):4094–4098.

11. Haruta, M., Gold as a low-temperature oxidation catalyst: factors controlling activity and selectivity. *3rd World Congress on Oxidation Catalysis* 1997;110:123–134.

12. Bond, G.C. Gold: a relatively new catalyst. *Catal Today* 2002;72(1–2):5–9.

13. Hutchings, G.J. New directions in gold catalysis. *Gold Bull* 2004;37(1–2):3–11.

14. Donaldson, K., Stone, V., Tran, C.L., Kreyling, W., Borm, P.J.A. Nanotoxicology. *Occup Environ Med* 2004;61(9):727–728.

15. Nel, A., Xia, T., Madler, L., Li, N. Toxic potential of materials at the nanolevel. *Science* 2006;311(5761): 622–627.

16. Oberdorster, E. Manufactured nanomaterials (fullerenes, C-60) induce oxidative stress in the brain of juvenile largemouth bass. *Environ Health Perspect* 2004;112(10):1058–1062.

17. Long, T.C., Saleh, N., Tilton, R.D., Lowry, G.V., Veronesi, B. Titanium dioxide (P25) produces reactive oxygen species in immortalized brain microglia (BV2): implications for nanoparticle neurotoxicity. *Environ Sci Technol* 2006;40(14):4346–4352.

18. Hardman, R. A toxicologic review of quantum dots: toxicity depends on physicochemical and environmental factors. *Environ Health Perspect* 2006;114(2):165–172.

19. Sayes, C.M., Gobin, A.M., Ausman, K.D., Mendez, J., West, J.L., Colvin, V.L. Nano-C60 cytotoxicity is due to lipid peroxidation. *Biomaterials* 2005;26(36):7587–7595.

20. Cai, R.X., Kubota, Y., Shuin, T., Sakai, H., Hashimoto, K., Fujishima, A. Induction of cytotoxicity by photoexcited TiO_2 particles. *Cancer Res* 1992;52(8):2346–2348.

21. Kreyling, W.G., Semmler-Behnke, M., Moller, W. Health implications of nanoparticles. *J Nanopart Res* 2006;8(5):543–562.

22. Andrievsky, G.V., Klochkov, V.K., Karyakina, E.L., Mchedlov-Petrossyan, N.O. Studies of aqueous colloidal solutions of fullerene C-60 by electron microscopy. *Chem Phys Lett* 1999;300(3–4):392–396.

23. Simonin, J.P. Solvent effects on osmotic second virial coefficient studied using analytic molecular models: application to solutions of C-60 fullerene. *J Phys Chem B* 2001;105 (22):5262–5270.

24. Bensasson, R.V., Bienvenue, E., Dellinger, M., Leach, S., Seta, P. C60 in Model biological-systems: a visible-UV absorption study of solvent-dependent parameters and solute aggregation. *J Phys Chem* 1994;98(13):3492–3500.

25. Schuster, D.I., Cheng, P., Jarowski, P.D., Guldi, D.M., Luo, C.P., Echegoyen, L., Pyo, S., Holzwarth, A.R., Braslavsky, S.E., Williams, R.M., Klihm, G. Design, synthesis, and photophysical studies of a porphyrin-fullerene dyad with parachute topology; charge recombination in the Marcus inverted region. *J Am Chem Soc* 2004;126(23):7257–7270.

26. Andrievsky, G.V., Klochkov, V.K., Bordyuh, A.B., Dovbeshko, G.I. Comparative analysis of two aqueous-colloidal solutions of C-60 fullerene with help of FTIR reflectance and UV-Vis spectroscopy. *Chem Phys Lett* 2002;364(1–2):8–17.

27. Deguchi, S., Alargova, R.G., Tsujii, K. Stable dispersions of fullerenes, C-60 and C-70, in water: preparation and characterization. *Langmuir* 2001;17(19):6013–6017.

28. Li, H., Jia, X., Li, Y., Shi, X., J. H. A salt-free zero-charged aqueous onion-phase enhances the solubility of fullerene C60 in water. *J Phys Chem B* 2006;110(1):68–74.

29. Lyon, D.Y., Fortner, J.D., Sayes, C.M., Colvin, V.L., Hughe, J.B. Bacterial cell association and antimicrobial activity of a C60 water suspension. *Environ Toxicol Chem* 2005;24 (11):2757–2762.

30. Brant, J., Lecoanet, H., Wiesner, M.R. Aggregation and deposition characteristics of fullerene nanoparticles in aqueous systems. *J Nanopart Res* 2005;7:545–553.

31. Chang, M.C., Shu, H.Y., Hsieh, W.P., Wang, M.C. Using nanoscale zero-valent iron for the remediation of polycyclic aromatic hydrocarbons contaminated soil. *J Air Waste Manag Assoc* 2005;55(8):1200–1207.

32. Carter, S.R., Rimmer, S. Aqueous compatible polymers in bionanotechnology. *IEE Proc Nanobiotechnol* 2005;152(5):169–176.

33. Rodriguez-Zavala, J.G., Guirado-Lopez, R.A. Structure and energetics of polyhydroxy-lated carbon fullerenes. *Phys Rev B* 2004;69:075411.

34. Sayes, C.M., Fortner, J.D., Guo, W., Lyon, D., Boyd, A.M., Ausman, K.D., Tao, Y.J., Sitharaman, B., Wilson, L.J., Hughes, J.B., West, J.L., Colvin, V.L. The differential cytotoxicity of water-soluble fullerenes. *Nano Lett* 2004;4(10):1881–1887.

35. Schrick, B., Hydutsky, B.W., Blough, J.L., Mallouk, T.E. Delivery vehicles for zerovalent metal nanoparticles in soil and groundwater. *Chem Mater* 2004;16:2187–2193.

36. Zhang, W.-x., Elliott, D.W. Applications of iron nanoparticles for groundwater remediation. *Remediation* 2006;16(2):7–21.

37. Donaldson, K., Aitken, R., Tran, L., Stone, V., Duffin, R., Forrest, G., Alexander, A. Carbon nanotubes: a review of their properties in relation to pulmonary toxicology and workplace safety. *Toxicol Sci* 2006;92(1):5–22.

38. Lyon, D.Y., Adams, L.K., Falkner, J.C., Alvarez, P.J.J. Antibacterial activity of fullerene water suspensions: effects of preparation method and particle size. *Environ Sci Technol* 2006;40(14):4360–4366.

39. Verran, J., Sandoval, G., Allen, N.S., Edge, M., Stratton, J. Variables affecting, the antibacterial properties of nano and pigmentary titania particles in suspension. *Dyes Pigments* 2007;73(3):298–304.

40. Guldi, D.M., Prato, M. Excited-state properties of C(60) fullerene derivatives. *Acc Chem Res* 2000;33(10):695–703.

41. Agustina, T.E., Ang, H.M., Vareek, V.K. A review of synergistic effect of photocatalysis and ozonation on wastewater treatment. *J Photochem Photobiol C-Photochem Rev* 2005; 6(4):264–273.

42. Sunada, K., Kikuchi, Y., Hashimoto, K., Fujishima, A. Bactericidal and detoxification effects of TiO_2 thin film photocatalysts. *Environ Sci Technol* 1998;32(5):726–728.

43. Kashige, N., Kakita, Y., Nakashima, Y., Miake, F., Watanabe, K. Mechanism of the photocatalytic inactivation of Lactobacillus casei phage PL-1 by titania thin film. *Current Microbiol* 2001;42(3):184–189.

44. Wold, A. Photocatalytic properties of TiO_2. *Chemi Mater* 1993;5(3):280–283.

45. Derfus, A.M., Chan, W.C., Bhatia, S.N. Probing the cytotoxicity of semiconductor quantum dots. *Nano Lett* 2004;4(1):11–18.

46. Vione, D., Minero, C., Maurino, V., Carlotti, A.E., Picatonotto, T., Pelizzetti, E. Degradation of phenol and benzoic acid in the presence of a TiO_2-based heterogeneous photocatalyst. *Appl Catal B-Environ* 2005;58(1–2):79–88.

47. Picatonotto, T., Vione, D., Carlotti, M.E., Gallarate, M. Photocatalytic activity of inorganic sunscreens. *J Dispersion Sci Technol* 2001;22(4):381–386.

48. Bahnemann, D. Photocatalytic water treatment: solar energy applications. *Solar Energy* 2004;77(5):445–459.

49. Hoffmann, M.R., Martin, S.T., Choi, W.Y., Bahnemann, D.W. Environmental applications of semiconductor photocatalysis. *Chem Rev* 1995;95(1):69–96.

50. Hashimoto, K., Irie, H., Fujishima, A. TiO_2 photocatalysis: a historical overview and future prospects. *Jpn J Appl Phys* 2005;44(12):8269–8285.

51. Tinkle, S.S., Antonini, J.M., Rich, B.A., Roberts, J.R., Salmen, R., DePree, K., Adkins, E.J. Skin as a route of exposure and sensitization in chronic beryllium disease. *Environ Health Perspect* 2003;111(9):1202–1208.

52. Souto, E.B., Wissing, S.A., Barbosa, C.M., Muller, R.H. Comparative study between the viscoelastic behaviors of different lipid nanoparticle formulations. *J Cosmet Sci* 2004;55 (5):463–471.

53. Sayes, C.M., Wahi, R., Kurian, P.A., Liu, Y., West, J.L., Ausman, K.D., Warheit, D.B., Colvin, V.L. Correlating nanoscale titania structure with toxicity: a cytotoxicity and inflammatory response study with human dermal fibroblasts and human lung epithelial cells. *Toxicol Sci* 2006;92(1):174–185.

54. Ryman-Rasmussen, J.P., Riviere, J.E., Monteiro-Riviere, N.A. Penetration of intact skin by quantum dots with diverse physicochemical properties. *Toxicol Sci* 2006;91(1):159–165.

55. Scrivens, W.A., Tour, J.M., Creek, K.E., Pirisi, L. Synthesis of C-14-labeled C-60, its suspension in water, and its uptake by human keratinocytes. *J Am Chem Soc* 1994;116 (10):4517–4518.

56. BullardDillard, R., Creek, K.E., Scrivens, W.A., Tour, J.M. Tissue sites of uptake of C-14-labeled C-60. *Bioorg Chem* 1996; 24(4):376–385.

57. Oberdorster, G., Oberdorster, E., Oberdorster, J. Nanotoxicology: an emerging discipline evolving from studies of ultrafine particles. *Environ Health Perspect* 2005;113(7):823–839.

58. Jani, P., Halbert, G.W., Langridge, J., Florence, A.T. Nanoparticle uptake by the rat gastrointestinal mucosa: quantitation and particle size dependency. *J Pharm Pharmacol* 1990;42(12):821–826.

59. Florence, A.T. Issues in oral nanoparticle drug carrier uptake and targeting. *J Drug Target* 2004;12(2):65–70.

60. Kreyling, W.G., Semmler, M., Erbe, F., Mayer, P., Takenaka, S., Schulz, H., Oberdorster, G., Ziesenis, A. Translocation of ultrafine insoluble iridium particles from lung epithelium to extrapulmonary organs is size dependent but very low. *J Toxicol Environ Health A* 2002;65(20):1513–1530.

61. Yamago, S., Tokuyama, H., Nakamura, E., Kikuchi, K., Kananishi, S., Sueki, K., Nakahara, H., Enomoto, S., Ambe, F. *In vivo* biological behavior of a water-miscible fullerene: 14C labeling, absorption, distribution, excretion and acute toxicity. *Chem Biol* 1995;2(6):385–389.

62. Warheit, D.B., Laurence, B.R., Reed, K.L., Roach, D.H., Reynolds, G.A., Webb, T.R. Comparative pulmonary toxicity assessment of single-wall carbon nanotubes in rats. *Toxicol Sci* 2004;77(1):117–125.

63. Shvedova, A.A., Kisin, E.R., Mercer, R., Murray, A.R., Johnson, V.J., Potapovich, A.I., Tyurina, Y.Y., Gorelik, O., Arepalli, S., Schwegler-Berry, D., Hubbs, A.F., Antonini, J., Evans, D.E., Ku, B.K., Ramsey, D., Maynard, A., Kagan, V.E., Castranova, V., Baron, P. Unusual inflammatory and fibrogenic pulmonary responses to single-walled carbon nanotubes in mice. *Am J Physiol Lung Cell Mol Physiol* 2005;289(5):L698–708.

64. Chen, Y., Chen, J., Dong, J., Jin, Y. Comparing study of the effect of nanosized silicon dioxide and microsized silicon dioxide on fibrogenesis in rats. *Toxicol Ind Health* 2004;20 (1–5):21–27.

65. Warheit, D.B., Webb, T.R., Sayes, C.M., Colvin, V.L., Reed, K.L. Pulmonary instillation studies with nanoscale TiO$_2$ rods and dots in rats: toxicity is not dependent upon particle size and surface area. *Toxicol Sci* 2006;91(1):227–236.

66. Baggs, R.B., Ferin, J., Oberdorster, G. Regression of pulmonary lesions produced by inhaled titanium dioxide in rats. *Vet Pathol* 1997;34(6):592–597.

67. Rehn, B., Seiler, F., Rehn, S., Bruch, J., Maier, M. Investigations on the inflammatory and genotoxic lung effects of two types of titanium dioxide: untreated and surface treated. *Toxicol Appl Pharmacol* 2003;189(2):84–95.

68. Kim, S., Lim, Y.T., Soltesz, E.G., De Grand, A.M., Lee, J., Nakayama, A., Parker, J.A., Mihaljevic, T., Laurence, R.G., Dor, D.M., Cohn, L.H., Bawendi, M.G., Frangioni, J.V. Near-infrared fluorescent type II quantum dots for sentinel lymph node mapping. *Nat Biotechnol* 2004;22(1):93–97.

69. Reddy, S.T., Rehor, A., Schmoekel, H.G., Hubbell, J.A., Swartz, M.A. *In vivo* targeting of dendritic cells in lymph nodes with poly(propylene sulfide) nanoparticles. *J Control Release* 2006;112(1):26–34.

70. Lutsiak, M.E.C., Robinson, D.R., Coester, C., Kwon, G.S., Samuel, J. Analysis of poly(D,L-lactic-co-glycolic acid) nanosphere uptake by human dendritic cells and macrophages *in vitro. Pharm Res* 2002;19(10):1480–1487.

71. Fifis, T., Gamvrellis, A., Crimeen-Irwin, B., Pietersz, G.A., Li, J., Mottram, P.L., McKenzie, I.F.C., Plebanski, M. Size-dependent immunogenicity: therapeutic and protective properties of nano-vaccines against tumors. *J Immunol* 2004;173(5):3148–3154.

72. Kreyling, W.G., Semmler, M., Moller, W. Dosimetry and toxicology of ultrafine particles. *J Aerosol Med* 2004;17(2):140–152.

73. Takenaka, S., Karg, E., Kreyling, W.G., Lentner, B., Moller, W., Behnke-Semmler, M., Jennen, L., Walch, A., Michalke, B., Schramel, P., Heyder, J., Schulz, H. Distribution pattern of inhaled ultrafine gold particles in the rat lung. *Inhal Toxicol* 2006;18(10):733–740.

74. Ferin, J., Oberdorster, G., Soderholm, S.C., Gelein, R. Pulmonary tissue access of ultrafine particles. *J Aerosol Med* 1991;4(1):57–68.

75. Ferin, J., Oberdorster, G., Penney, D.P. Pulmonary retention of ultrafine and fine particles in rats. *Am J Respir Cell Mol Biol* 1992;6(5):535–542.

76. Berry, J.P., Arnoux, B., Stanislas, G., Galle, P., Chretien, J. A microanalytic study of particles transport across the alveoli: role of blood platelets. *Biomedicine* 1977;27(9–10):354–357.

77. Geiser, M., Rothen-Rutishauser, B., Kapp, N., Schurch, S., Kreyling, W., Schulz, H., Semmler, M., Hof, V.I., Heyder, J., Gehr, P. Ultrafine particles cross cellular membranes by nonphagocytic mechanisms in lungs and in cultured cells. *Environ Health Perspect* 2005;113(11):1555–1560.

78. Kim, J.S., Yoon, T.J., Yu, K.N., Kim, B.G., Park, S.J., Kim, H.W., Lee, K.H., Park, S.B., Lee, J.K., Cho, M.H. Toxicity and tissue distribution of magnetic nanoparticles in mice. *Toxicol Sci* 2006;89(1):338–347.

79. Oberdorster, E. Manufactured nanomaterials (fullerenes, C60) induce oxidative stress in the brain of juvenile largemouth bass. *Environ Health Perspect* 2004;112(10):1058–1062.

80. Elder, A., Gelein, R., Silva, V., Feikert, T., Opanashuk, L., Carter, J., Potter, R., Maynard, A., Ito, Y., Finkelstein, J., Oberdorster, G. Translocation of inhaled ultrafine manganese oxide particles to the central nervous system. *Environ Health Perspect* 2006;114(8):1172–1178.

81. Oberdorster, G., Sharp, Z., Atudorei, V., Elder, A., Gelein, R., Kreyling, W., Cox, C. Translocation of inhaled ultrafine particles to the brain. *Inhal Toxicol* 2004;16(6–7):437–445.

82. Tsay, J.M., Michalet, X. New light on quantum dot cytotoxicity. *Chem Biol* 2005;12 (11):1159–1161.

83. Manna, S.K., Sarkar, S., Barr, J., Wise, K., Barrera, E.V., Jejelowo, O., Rice-Ficht, A.C., Ramesh, G.T. Single-walled carbon nanotube induces oxidative stress and activates nuclear transcription factor-kappaB in human keratinocytes. *Nano Lett* 2005;5(9):1676–1684.

84. Shvedova, A.A., Castranova, V., Kisin, E.R., Schwegler-Berry, D., Murray, A.R., Gandelsman, V.Z., Maynard, A., Baron, P. Exposure to carbon nanotube material: assessment of nanotube cytotoxicity using human keratinocyte cells. *J Toxicol Environ Health A* 2003;66 (20):1909–1926.

85. Yamakoshi, Y., Umezawa, N., Ryu, A., Arakane, K., Miyata, N., Goda, Y., Masumizu, T., Nagano, T. Active oxygen species generated from photoexcited fullerene (C60) as potential medicines: O2-* versus 1O2. *J Am Chem Soc* 2003;125(42):12803–12809.

86. Fortner, J.D., Lyon, D.Y., Sayes, C.M., Boyd, A.M., Falkner, J.C., Hotze, E.M., Alemany, L.B., Tao, Y.J., Guo, W., Ausman, K.D., Colvin, V.L., Hughes, J.B. C60 in water: nanocrystal formation and microbial response. *Environ Sci Technol* 2005;39(11):4307–4316.

87. Wei, C., Lin, W.Y., Zainal, Z., Williams, N.E., Zhu, K., Kruzic, A.P., Smith, R.L., Rajeshwar, K. Bactericidal activity of TiO_2 photocatalyst in aqueous media: toward a solar-assisted water disinfection system. *Environ Sci Technol* 1994;28(5):934–938.

88. Colvin, V.L. The potential environmental impact of engineered nanomaterials. *Nat Biotechnol* 2003;21(10):1166–1170.

89. Liu, W.T. Nanoparticles and their biological and environmental applications. *J Biosci Bioeng* 2006;102(1):1–7.

90. Biswas, P., Wu, C.Y. Nanoparticles and the environment. *J Air Waste Manag Assoc* 2005;55(6):708–746.

91. Chow, J.C., Biswas, P., Eatough, D., McDade, C., Mueller, P., Overcamp, T., Watson, J., Wu, C.Y., Comm, C.R. Nanoparticles and the environment — introduction. *J Air Waste Manage Assoc* 2005;55(6):706–707.

Assessing the Life Cycle Environmental Implications of Nanomanufacturing: Opportunities and Challenges

VIKAS KHANNA, YI ZHANG, GEOFFREY GRUBB, and BHAVIK R. BAKSHI

Department of Chemical and Biomolecular Engineering, The Ohio State University, Columbus, OH 43210, USA

2.1 INTRODUCTION

A retrospective look at the history of scientific discoveries highlights several technologies that changed the world around us in unexpected ways. These discoveries greatly contributed to enhancing our way of living and making life easier. However, in many cases, the scientific community failed to address the broader impacts of several of these technologies before their widespread adoption, resulting in wide reaching unintended consequences in many instances. One of the classic examples is the development of chlorofluorocarbons (CFCs) that was considered as a great technological milestone. This class of compounds was found to be nontoxic, nonreactive, and widely used as refrigerants, propellants, and for several other industrial purposes. It was not until the 1970s when the destructive effects of CFCs on the stratospheric ozone layer started becoming apparent. The companies eventually started phasing out CFCs in the late 1970s under societal pressure and environmental concerns over their use, but the ozone hole resulting from their use is still with us. DDT, asbestos, and polychlorinated biphenyls are some of the other examples in this regard. The take-home message is that the failure to acknowledge the risks associated with the adoption of newer technologies can have far-reaching consequences. However, a systems approach encompassing a holistic understanding of products and processes can help address these concerns.

Nanotechnology is a fast emerging technology that deals with the art of manipulating matter at the atomic scale. Size and shape are attributes that are critical to the commercial utility of the engineered nanomaterials. By virtue of their altered physicochemical properties, nanomaterials are being explored for diverse applications. Some of these include the use of nanoparticles for drug delivery, nanoparticle-reinforced high-performance composite materials, antimicrobial nanomaterials, self-cleaning surfaces, electrical and electronic components, carbon-based nanomaterials for information technology and so on. However, the same properties that enable novel applications also open up the possibility of new health and environmental risks. Engineered nanomaterials may enter the environment through air, water, and/or soil. Release of nanomaterials into the environment can occur at either the manufacturing, processing, use, or the end phase along a product's life cycle. Little is known about the environmental fate, transport, and mechanisms of damage caused to ecosystems by nanomaterials. Most studies have focused on the human health hazard of nanoparticles. Traditional models used for estimating the environmental fate and exposure to chemicals are not valid for engineered nanomaterials (1), and new models will need to be developed for this purpose. Research is specifically required for studying the fate, exposure, and human and environmental effects of nanomaterials. Furthermore, it need not be obvious that a new technology is necessarily environmentally superior to traditional alternatives. Many studies have shown that technologies that may seem to be environment-friendly may simply shift the negative impacts outside the analysis boundary or impact categories. Examples include the use of electric vehicles potentially shifting CO_2 emissions from the tailpipe to the power plant, and the use of corn ethanol instead of gasoline reducing greenhouse gas emissions and fossil fuel consumption, but increasing nitrogen runoff, soil erosion, and consumption of most other resources. A holistic, systems view is essential to avoid such shifting of impact to avoid "unexpected" surprises.

Among the various systems analysis tools and techniques, life cycle assessment (LCA) offers great promise in the environmental evaluation of nanotechnology. This is especially important for nanotechnology applications intended to replace conventional products and technologies. The goal of this chapter is to discuss the needs, challenges, and benefits of life cycle analysis in the environmental evaluation of nanotechnology. The rest of the chapter is organized as follows. Section 2.2 provides a detailed description of the LCA approach and describes the unique challenges surrounding the LCA of nanotechnology. Section 2.3 discusses the benefits of the nanotechnology LCA and provides an overview of the major studies on the LCA of nanotechnology. A brief discussion along with some guidelines is provided for compiling a life cycle inventory of nanomanufacturing. A case study on life cycle analysis of carbon nanofibers (CNFs) is presented in Section 2.4. Section 2.5 provides a general discussion about the LCA of specific nanoproducts. The role of predictive approaches in the evaluation of nanotechnology is discussed in Section 2.6. Finally, Section 2.7 provides a summary of the overall discussion.

2.2 LIFE CYCLE ASSESSMENT AND CHALLENGES

2.2.1 LCA Approach

LCA is a methodology to quantify the environmental impact of a product or a process over its entire life cycle. The methodology is standardized via ISO 14000. In its

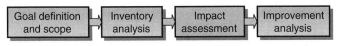

FIGURE 2.1   Phases of an LCA based on ISO 14040, 1997.

simplest form, the LCA methodology involves four basic steps (2, 3). Figure 2.1 gives a schematic description of these four phases. The four phases are briefly described below.

2.2.1.1 *Goal and Scope Definition*   The scope definition involves defining the boundaries of the life cycle study and specifies which major and minor processes are considered. In general, the scope of an LCA study comprises the resource extraction, inputs synthesis, product manufacture, waste treatment, packaging, and use phase, and ends when the product is finally disposed off or its constituents recycled back. Such an analysis is termed as "cradle-to-grave." The purpose of the LCA study should be stated clearly in this phase. This can be either comparative assessment of nanoproducts versus alternatives, evaluation of a nanoproduct with different synthesis routes, and/or a single nanoproduct for a new application. This is especially important to aid in defining the appropriate functional unit for the study to result in a fair comparison. As an example, consider the life cycle environmental comparison of plastic versus paper grocery bags. One paper bag might physically hold the same amount of groceries as two plastic bags. Thus, the appropriate functional unit in this case should be the comparison of one paper bag versus two plastic bags. Once the functional unit is defined, the process boundary should be specified clearly to define the scope of the LCA study. Since data collection and analysis can be very time consuming and expensive, it is tempting to narrow the inputs considered or make the boundary too narrow leading to invalid or flawed results. Also, since a process or product is connected to the rest of the economy through infinite links, attempting to get each and every input may make the study too expensive or time consuming. For these reasons, proper care and vigilance are crucial during this stage to yield meaningful and interpretable results. In addition, doing multiple studies with different boundaries to determine the effect of the boundary limits can be helpful in certain situations. The life cycle practitioner should always state the purpose of the study and this usually helps in defining the boundary.

2.2.1.2 *Inventory Analysis*   This is the most resource intensive phase of the LCA. It involves identifying and collecting input and output data for each of the processes that are included in the goal and scope definition phase. The quantity of data needed depends on the boundaries of the study. There are many ways of obtaining data for LCA. This includes public databases such as the National Renewable Energy Laboratory (NREL), or commercial databases such as Ecoinvent, often accessed via software such as SimaPro, Ecoinvent, GABI, and Umberto. Input data typically includes material and energy resources, while labor, capital, and equipment inputs are often not available. Output data consists of products, byproducts, and emissions of substances to air, water, and soil.

2.2.1.3 Impact Assessment This phase involves classifying and characterizing emissions into various impact categories to provide several indicators for analyzing the potential contributions of the extracted resources and emitted wastes in an inventory to a number of potential impacts (2). The individual results for the different impact categories can be further normalized and weighted based on valuation techniques. Since normalization and valuation techniques are subjective, these are often controversial, thus only classification and characterization steps are performed in this study. Several life cycle impact assessment (LCIA) approaches exist and these have been described and critically assessed in detail in the literature (4–6). The choice of which impact categories and hence impact assessment approach to use is a subjective one and at the discretion of the LCA practitioner. More often, the LCIA methodology to be used for the problem at hand is selected based on the goal of the study, that is, whether the goal is to make comparisons between alternative products serving the same utility or identifying areas for process improvement.

The emissions from the life cycle of a product are available at the aggregate level. These are classified and characterized based on their impact into various impact categories often described as midpoint indicators. Common impact categories include global warming potential (GWP), acidification potential, eutrophication potential, human toxicity potential, ozone layer depletion potential, and photochemical smog formation potential. Different chemicals can have different impacts. For example, carbon dioxide causes global warming and Hydrochlorofluorocarbons have an adverse effect on the ozone layer. Classification comprises combining different chemical impacts into a common metric. For example, 1 kg of methane has a global warming impact equivalent to 23 kg of carbon dioxide; similarly, 1 kg of methyl chloride has a global warming impact equivalent to 16 kg of carbon dioxide. Thus, the global warming potential characterization factors relative to carbon dioxide are 23 and 16 for methane and methyl chloride, respectively. These impacts are then expressed in terms of total equivalents of carbon dioxide to quantify the global warming potential. A similar approach is followed for other impact categories. Characterization factors are available in the literature for a large number of chemicals and their impacts in various categories.

Midpoint indicators can be further aggregated to obtain end-point indicators. Endpoints can be described as entities that are valuable to society. These can include buildings, natural resources, humans, and plant and animal species. End-point indicators are often combined together to provide more meaningful and interpretable damage indicators. For example, Eco-indicator 99 methodology is commonly used to obtain damage indicators (7). Under this approach, all impacts are ultimately aggregated into only three kinds of environmental damages, namely, damage to human health, ecosystem quality, and resources. Damage to human health is expressed in terms of disability adjusted life years (DALYs). The DALY scale, originally developed for the World Health Organization and the World Bank, incorporates damage to human health in the form of years of life lost (YLL) and years lived disabled (YLD) due to the impact of the selected emission. Damage to ecosystem quality refers to the damage caused to nonhuman species, that is, plant and animal species. Under Eco-indicator 99, damage to ecosystem quality is expressed in terms of PDF times area times year (PDF $\times$ m^2 $\times$ year), where PDF stands for potentially disappeared fraction and includes percentage of species that are threatened or that disappear from a given area in a given time. Finally, damage to resource quality is expressed in terms of megajoules (MJ) surplus. The underlying idea in quantifying damage to resources is that extraction of mineral resources and fossil fuels for the manufacture of a product will result in increasing expenditure of energy for future extraction. This is formulated in terms of MJ surplus per kg and represents the expected increase in extraction energy per kg of extracted material.

2.2.1.4   Interpretation   This is the step in LCA where the results of the impact assessment and other phases are put together in a form that can be used directly to draw conclusions and make recommendations. Before the interpretation is done, the LCA practitioner will have to verify the validity of the results by doing sensitivity and/or uncertainty analysis. Where possible, if enough data are available, statistical methods are of great utility at this stage. LCA is a data intensive approach that is finding wide use in industry. Many corporations are finding it advantageous to employ LCA techniques as a way of moving beyond compliance to create win–win business opportunities by improving the quality of their products while minimizing the associated environmental impact. LCA encompasses detailed information about the resource consumption and emissions across the entire life cycle of a product. Despite the fact that the LCA methodology is standardized via ISO 14000, and pertinent software and inventory databases are available, the methodology faces several challenges. Some of these include getting high-quality life cycle inventory data, combining data in disparate units and at multiple spatial and temporal scales, dealing with high-dimensionality data involving varying degree of uncertainty, and dealing with processes having a range of emissions (2). These challenges make it even more difficult to apply LCA to emerging technologies such as nanotechnology.

2.2.2   Nanotechnology LCA Challenges

The need for the LCA of potential nanoproducts has been identified and discussed by researchers and various agencies (8–10). However, LCA of nanotechnology poses several formidable challenges, some of which are listed below:

- Existing life cycle inventory databases are limited in scope and are useful for evaluating only common products and processes.
- Little is known about the inputs and outputs of nanomanufacturing processes since most of the data is proprietary or available at the laboratory scale.
- There is very little quantifiable data available on the human health and ecosystem impacts of products and byproducts of nanomanufacturing.
- Forecasting nanotechnology life cycle processes and activities is difficult since the technology is in its infancy and evolving rapidly.

In the light of these challenges, LCA of emerging nanotechnologies has received very little attention and there are only a handful of studies available addressing the life cycle issues and complexities of nanoproducts and nanoprocesses.

2.3   LIFE CYCLE ASSESSMENT OF NANOTECHNOLOGY

2.3.1   Expected Benefits

Although not a panacea, especially for emerging technologies, LCA can help address many critical concerns such as material, energy, and environmental impact intensity of products, which complemented with the toxicological information of nanomaterials will have the potential to inform so that the emerging field of nanotechnology can develop in a sustainable manner and contribute positively to the sustainability of human activities. Some of the expected benefits of nanotechnology LCA are listed below:

- Identify phases in a product's life cycle that have the maximum energy and environmental impact and thus potential for improvement.
- Quantify how much of the energy savings and environmental impact during the use phase of nanoproducts are offset during their production phase.
- Identification of end-of-life scenarios that might be specific to nanoproducts.
- LCA results can help in selection of various process alternatives during product design, which may positively influence the downstream phase of the product life cycle.
- Guide decision making by assessing the risk and trade-off between economic and environmental aspects of nanoproducts versus conventional ones.

Such insight may be useful for identifying and allaying public concerns about this emerging field. It can also be used for risk assessment and as a screening tool for evaluating competing technologies during the preliminary phases of a product design.

2.3.2   Existing Work

The challenges described in Section 2.2.2 have hindered the LCA of emerging nanotechnologies and there has been limited progress in this direction. Some of the major studies in this area are summarized here.

Lloyd and Lave (11) studied the life cycle energy and environmental implications of replacing automotive body panels made of steel with those of polymer nanocomposites and aluminum. The study employed economic input–output LCA (EIOLCA) framework (12) for quantifying savings in petroleum use and production and hence reductions in CO_2 emissions thereof by replacing conventional steel body panels with polymer nanocomposites. The nanocomposite considered was polypropylene reinforced with nanoscale montmorillonite clay particles. Polymer nanocomposites were found to be better than steel on the basis of their net life cycle energy use and environmental impact. These researchers have also studied the use of nanotechnology to stabilize platinum group metals in automobile catalytic converters (13). However, both studies utilize the EIOLCA model that consists of coarse and aggregated data for the different industrial sectors of the U.S. economy. Specific details about the different manufacturing and processing steps are missing, especially for the synthesis of nanoparticles.

Osterwalder and coworkers (14) compared the wet and dry synthesis methods for oxide nanoparticle production with respect to energy consumption and emissions impact as CO_2 equivalents. Titanium dioxide and zirconia were chosen in these studies for their high industrial application. The dry synthesis method was found to be energy intensive compared to the wet synthesis route owing to the high temperature requirements of the former one. Release and impact of nanoparticle and other emissions were not accounted for most likely due to the lack of detailed data. Besides, not all life cycle stages could be included in these studies especially the end-of-life and disposal phase of nanoproducts, again due to the lack of data.

Isaacs et al. (15) evaluated the economic and environmental trade-offs of single-wall carbon nanotubes (SWCNTs) production. Technical cost models were developed for three different manufacturing processes: arc ablation, chemical vapor deposition (CVD), and high-pressure carbon monoxide (HiPCO) process. A comparison of different synthesis routes revealed that the life cycle of SWCNTs is hugely energy intensive with numbers ranging from 1,440,000 to 2,800,000 MJ/kg of carbon nanotubes. Again, as in the previous cases, release and impact of nanotubes themselves along the different life cycle phases were not modeled due to the lack of toxicological data. More recently, Khanna et al. (16) studied the life cycle energy and environmental impact of CNF production. The results of this analysis revealed high life cycle energy and environmental impact of CNFs compared to traditional materials such as primary aluminum, steel, and polypropylene on an equal mass basis. The detailed case study on CNFs is described and explained in Section 2.4. The comparison of SWCNTs or CNFs on an equal mass basis with conventional materials might not be appropriate from a functional unit point of view but still reveals the energy intensive nature of their respective life cycles. Besides, studies such as these two can be especially useful since they will lay the basis for compiling the life cycle inventory for nanoparticle synthesis and hence pave the way for LCA of specific nanoproducts.

Any LCA study of nanoproducts will most likely suffer from high uncertainty at the early stages of nanotechnology research. As a starting step, efforts toward compiling a life cycle inventory of engineered nanomaterials can help pave the way for LCA of specific nanoproducts. The next subsection provides a brief overview of compiling life cycle inventory for nanomaterials and nanoproducts.

2.3.3 Inventory for LCA of Nanotechnology

The ISO methodology as outlined in ISO series 14040 and 14044, in general, is applicable for evaluating potential nanoproducts and nanoprocesses. However, as mentioned in Section 2.2.2, the major hindrance is the severe lack of life cycle data for nanomanufacturing. This subsection describes the general approach that can be followed for compiling the LCA of potential nanoproducts in the light of the challenges mentioned in the preceding sections.

Compiling life cycle inventory for nanomaterials and nanoproducts can be an excruciating and resource intensive task. This is because nanoprocesses are evolving rapidly and most of the industry data are proprietary. As a starting step, LCA practitioners can compile the life cycle inventory for nanoproducts based on laboratory experience and data available in the open literature. Wherever possible, plant-specific data should be used. In the absence of any information, parallels can be drawn with similar technologies to get data on material and energy consumption. For example, the CVD process used for the synthesis of carbon-based nanomaterials, such as SWCNTs and CNFs, is similar in nature to the CVD process used for chip manufacturing in the semiconductor industry. Missing data may be reconciled ensuring the satisfaction of conservation laws of mass and energy (17). Unlike traditional products and processes, additional information about the possible quantity and mode of release of nanoparticles is crucial for estimating the extent of human and ecological impact of nanoproducts. Various assumptions related to the boundary and selected processes should also be clearly stated. Since nanoproducts have only recently started penetrating the consumer market, information regarding their end-of-life is missing. Sensitivity analyses based on different scenarios can provide useful insight and extreme bounds for the end-of-life impact of nanoproducts. The next section illustrates the use of the life cycle approach in the environmental evaluation of one kind of engineered nanoparticles, carbon nanofibers.

2.4 CARBON NANOFIBERS: A CASE STUDY

Carbon nanofibers belong to a new class of carbon-based nanomaterials that have attracted the interest of the scientific community because of their many potential applications. CNFs consist of monomolecular carbon fibers with diameters ranging from tens of nanometers to 200 nm. CNFs are characterized by high tensile strength (12,000 MPa) and high Young's modulus (600 GPa) that is approximately 10 times that of steel (18). Besides mechanical strength, CNFs possess novel electrical properties such as high electrical conductivity. Potential applications of CNFs include their use for lighter and stronger polymer nanocomposites and in electronic components. In this case study, CNF synthesis from hydrocarbons is considered for LCA and quantifying the life cycle energy requirements. The system boundary employed for the LCA of CNFs is depicted in Fig. 2.2.

CNFs are produced by catalytic pyrolysis of hydrocarbons in the presence of a transition metal catalyst. Trace amounts of sulfur is added to the feed to promote the formation of carbon nanofibers. The hydrocarbon feed along with the catalyst, hydrogen gas, and a sulfur source are introduced into the reactor. Hydrogen serves as the carrier gas. The organometallic catalyst decomposes forming clusters of metallic particles that act as nuclei for the formation and further growth of CNFs. The fibers grow as they move along the reactor and the diameter of fibers increase. Catalytic pyrolysis of hydrocarbons occurs at a high temperature in the range of 1100–1200°C. The entire reactor assembly is enclosed within an electric furnace to supply the necessary heat. The characteristics of the fibers such as their thickness and length and hence their properties can be controlled by carefully manipulating the reactor-operating conditions.

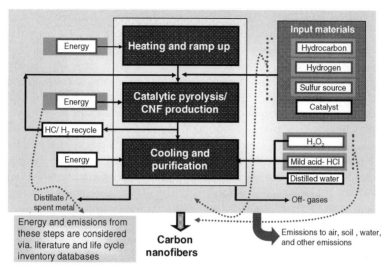

FIGURE 2.2   LCA of carbon nanofibers: System boundry.

2.4.1 Life Cycle Energy Analysis

Most of the inventory data for this case study is obtained directly from the published literature. Missing data are estimated by ensuring the satisfaction of mass and energy balances. Figure 2.3 presents a direct comparison of the life cycle energy requirements for CNF synthesis for different starting feedstocks with those of traditional materials: aluminum, steel, and polypropylene.

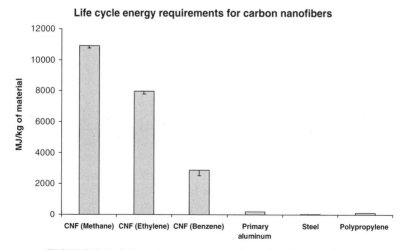

FIGURE 2.3   Life cycle energy comparison of carbon nanofibers.

Three different feedstocks, namely, methane, ethylene, and benzene, are examined for the life cycle energy analysis of VGCNFs. Different scenarios for unreacted hydrocarbon and carrier gas recycle rates are evaluated to address uncertainty and evaluate the life cycle energy consumption and emissions. The life cycle energy requirement for CNFs consists of three major components: process energy, indirect effects, and purification energy. Process energy is primarily electrical energy needed to maintain the high pyrolysis temperature. The indirect effects correspond to the life cycle energy requirements due to the consumption of material inputs such as hydrocarbon feedstock, carrier gas, sulfur source, and electricity. Figure 2.3 shows that the life cycle of CNFs is energy intensive compared to traditional materials with the life cycle energy requirements for CNF synthesis ranging from 2800 MJ/kg for benzene feedstock to around 10,900 MJ/kg for methane feedstock. The life cycle energy requirements for aluminum, steel, and polypropylene are obtained directly from SimaPro life cycle inventory database.

Process energy consumption constitutes a large fraction of the overall life cycle energy requirements ranging from 79% to 86% for benzene and methane feedstocks, respectively. As described earlier, this is primarily electrical energy and is required for the high-temperature CVD process.

The error bars in Fig. 2.3 show the effect of unreacted feedstock and carrier gas recycle rates on the life cycle energy consumption. Even with complete recycle of the unreacted hydrocarbons and a 90% recycle of the hydrogen stream that is only theoretical in nature, the total energy consumption decreases by only 2–12%. Thus, process energy requirements still outweigh the energy savings due to material recycling. It is noteworthy to mention here that the energy required for separating unreacted hydrocarbons and hydrogen from the off-gases coming out of the reactor is not accounted here. Therefore, the numbers presented in Fig. 2.3 represent an upper bound on the savings in energy consumption that can be achieved when a gas separation and recycle system is employed. The high life cycle energy requirements of the CNFs can be attributed to the high-temperature vapor-phase synthesis route with low efficiency that requires significant energy investment. Besides, production yields of carbon nanofibers are low even for continuous operations in the range of 10–30% by weight of the feedstock.

2.4.2   Environmental LCA of Carbon Nanofibers

This subsection illustrates the application of the basic LCA methodology described in Section 2.2.1 for carbon nanofibers. This study is a "cradle-to-gate" analysis of CNF synthesis instead of "cradle-to-grave" since no specific nanoproducts are investigated. In the case of CNF synthesis, since most of the data are either missing or proprietary information, certain assumptions are made to carefully define the process boundary. For example, emissions from the CNF synthesis step, that is, the process scale emissions are not included due to the lack of information. In addition, emissions corresponding to the catalyst life cycle are not included again due to the lack of data. Emissions corresponding to the production of all other inputs required for CNF synthesis are incorporated in the study.

The inventory data for the inputs required for CNF synthesis and electricity required for process heating during CNF synthesis are primarily obtained from SimaPro database, 6.0 demo version. SimaPro consists of "cradle-to-gate" life cycle inventory data for Western European technologies (SimaPro,[TM] http://www.pre.nl/simapro. Accessed 2007 Aug 9.). The database quantifies emissions from the materials, intermediate products, and products per kilogram (or 1 MJ for electricity) basis. Inventory data for traditional materials such as aluminum, steel, and polypropylene are also obtained directly from SimaPro.

Two base cases are evaluated for CNF synthesis, one with methane and the other with ethylene as the feedstock. Both cases are considered to have hydrogen as the carrier gas in accordance with the current industrial schemes. More details about this study are in Reference (16). Figure 2.4a shows a higher GWP for both methane and ethylene-based CNFs when compared with aluminum, steel, and polypropylene on an equal mass basis. A closer look at Fig. 2.4 reveals that the GWP potential of 1 kg methane-based CNF is equivalent to about 65 kg of primary aluminium, whereas 1 kg of ethylene-based CNF has a GWP equivalent of about 47 kg of steel. Similar trends are observed for other impact categories as displayed in Fig. 2.4. The impact numbers for CNFs represent only lower bounds whereas those for traditional materials are expected to be more precise since detailed and relatively more complete inventory data are available in the literature for traditional materials. Release and impact of CNFs on humans and ecosystem species during manufacturing are not taken into account in the analysis. Although the results presented in Figs. 2.3 and 2.4 are obtained based on conservative set of assumptions, they still represent important step toward the need and development of life cycle inventory modules for nanomaterials and nanoproducts.

Figure 2.5 presents the end-point indicators obtained from SimaPro based on Eco-indicator 99. Figure 2.5a indicates higher impact of CNFs in the category of human health measured as DALYs compared to traditional materials on an equal mass basis. Similar trends are observed for the damage to ecosystems and damage to resource quality as depicted in Fig. 2.5b and c, respectively. It is important to reiterate that human and ecosystem impact of CNFs is not accounted due to lack of information about their toxicological impact. Detailed knowledge and quantifiable data about possible fate, transport, and mechanism of damage of CNFs are not available. Thus, the impact numbers presented in Figs. 2.4 and 2.5 reflect only the material and energy use during the synthesis of CNFs.

2.5 DISCUSSION OF NANOTECHNOLOGY LCA

The case study on carbon nanofibers discussed in Section 2.4 and the study on SWCNTs described in Section 2.3 highlight the hugely energy intensive nature of the production processes of carbon-based nanomaterials, as well as their high life cycle environmental impact as compared to traditional materials. However, these results are on a per unit mass basis, making it difficult to reach any general conclusions about the environmental impact of specific nanoproducts based on such engineered nanomaterials without taking into account their complete life cycle and the possible

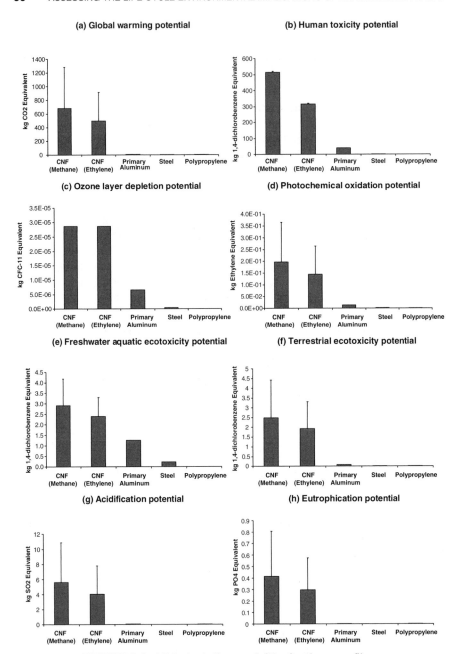

FIGURE 2.4 Midpoint indicators: LCA of carbon nanofibers.

(a) Damage to human health

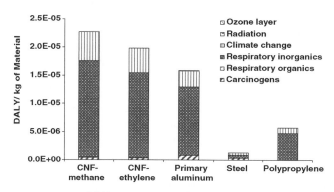

(b) Damage to ecosystems

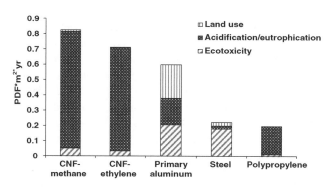

(c) Fossil fuels (MJ surplus)

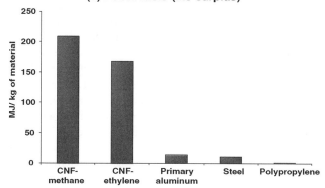

FIGURE 2.5   End-point indicators: LCA of carbon nanofibers.

small mass of nanoproduct required for a relatively large benefit. Thus, evaluation of potential nanoproducts on functional unit basis is essential. Such comparisons can help identify if the increased upstream production energy requirements are offset by savings during the use phase of the product or not. Besides, for certain products such as automotive or aerospace components, the use phase might dominate the energy use and emissions across the complete life cycle and hence may offer net life cycle energy and environmental savings. In other cases, the manufacturing phase might play a more significant role and hence offset any potential energy or environmental benefits achieved in other life cycle phases. Besides, end-of-life challenges such as recovery and recycle specific to engineered nanomaterials might be encountered for certain nanoproducts. In the absence of detailed environmental fate and transport, scenario analysis might help in screening different alternatives. Besides, scenario analysis coupled with LCA can also provide novel insights about the long-term impact of nanomaterial use across the economy.

2.6   FUTURE DIRECTIONS: PREDICTIVE APPROACHES FOR LCA OF NANOTECHNOLOGY

Although LCA of emerging technologies such as nanotechnology can be very informative, the traditional LCA approach described in Section 2.2 and illustrated in Section 2.4 suffers from several limitations. Traditional LCA is mainly an "output-side" method due to its focus on detailed data about emissions and their impact throughout the life cycle. As mentioned in Section 2.2.2, such data are often not available for nanotechnology and other emerging technologies due to a sheer lack of information or the slow pace of toxicological studies. In contrast to the lack of output information, data about material and energy inputs for even emerging technologies are more readily available from the scientific literature and laboratory or pilot plant studies. Since the use of materials and energy is connected to emissions from the process, it is likely that input-side metrics may be a proxy indicator of environmental impact. Such a relationship, if available, can serve as a way of doing quick or streamlined LCA of new technologies, such as nanotechnology, or at early stages of decision making.

Toxicological studies on quantifying the human and ecological impact of emissions such as nanoparticles are expected to take a few years, and it would be useful to have proxy indicators until detailed information is available. For nanoparticles, it seems that properties such as their size, surface area, reactivity, and so on may provide such indicators. Furthermore, thermodynamic measures such as the work that the environment needs to do to bring the particles in equilibrium with the environment may also correspond to their impact. This section summarizes current findings and ongoing research in these emerging areas of predicting the life cycle impact of emerging technologies and the environmental impact of nanoparticles based on proxy indicators.

2.6.1   Input Side Indicators of Life Cycle Environmental Impact

A variety of input-side methods have been developed for quantifying the life cycle environmental implications of products and processes. These include biophysical methods based on mass, energy, exergy (available energy), and emergy; economic methods based on monetary value; and other approaches such as ecological footprint analysis based on land area requirements. The focus of this chapter is on biophysical methods.

Material flow analysis (MFA) quantifies the mass of materials used in a life cycle. Aggregate indicators such as total mass or cumulative mass consumption (CMC) represent the overall material consumption in the life cycle of a product or service. Other indicators such as material input per service unit (MIPS) have also been defined as proxy indicators for a full LCA (19). This approach is appealing due to its simplicity, but cannot capture many resources including fuels. Furthermore, the mass of material used can be a poor indicator of environmental impact.

Energy analysis seeks to determine cumulative energy consumption (CEnC), that is, how much energy is consumed in all activities necessary to provide goods and services. This approach usually focuses only on fossil or nonrenewable fuels, although some efforts have combined energy content of renewable and nonrenewable resources (20). Unlike material flow analysis, energy analysis only accounts for fuels, and ignores nonfuel resources such as minerals. Furthermore, differences in the quality of energy resources such as their ability to do useful work are ignored. This is equivalent to ignoring the second law of thermodynamics, and can be particularly problematic when the energy content of renewable and nonrenewable resources is added. This is because, in general, a much smaller fraction of the energy contained in renewable resources, as compared to that in nonrenewable resources, can be converted to work. The objective of studying mass and energy flow is to reduce resource consumption and enhance process efficiency. But these methodologies potentially have inherent connections with environmental impact. Researchers have studied whether input-side metrics based on mass and energy can predict environmental impact. Van der Voet et al. (21) studied the relation of total mass/total energy and environmental impact for 28 European countries and found positive correlations. Two other independent studies (21, 22) used large life cycle inventory modules and found strong relation between cumulative energy demand (CEnD) and impact. The correlation can be higher than 90% for air pollution categories.

Although input-side metrics based on mass and energy show a certain correlation with impact, a thermodynamic view indicates that they may not be the best tools. Total mass only captures the property of weight. It is clear that a ton of iron is not equivalent to a ton of mercury in terms of environmental impact. Similarly, energy content lacks the ability to distinguish between the qualities of the energy. For instance, a Joule contained in solar energy cannot do the same amount or types of work as a Joule of electricity. Moreover, each of the methods considers only part of the total inputs. MFA includes materials, while energy analysis includes energy resources. The different units, gram versus Joule, make it difficult to compare them in a scientifically rigorous manner.

Principles of thermodynamics indicate that the second law may be better at capturing the relationship with environmental impact. This is because all industrial and ecological processes and their life cycles are networks of energy flows. Manufacturing involves reduction of entropy to make the desired products, which according to the second law, must result in an increase of entropy or disorder in the surroundings. Schematically, this is depicted in Fig. 2.6. This implies that among alternatives with similar utility, the process with a higher life cycle thermodynamic efficiency should have a smaller portion of useful work converted to entropy, and so a smaller life cycle environmental impact. Therefore, a working hypothesis could be as follows: life cycle thermodynamic efficiency is inversely proportional to life cycle environmental impact.

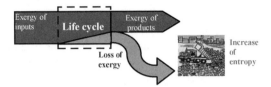

FIGURE 2.6   Second law and environmental impact.

Available energy or *exergy* is a concept that accounts for the first and second laws. It is defined as the maximum amount of work that can be extracted when a material or energy stream is brought to equilibrium with its environment (23). It captures information from mass, energy, concentration, velocity, location, and other properties of a material. Its unit is Joule, the same as energy, but has a different meaning. Besides the ability to indicate entropy formation by exergy loss, exergy can also compare and combine diverse material and energy flows, making it more comprehensive than mass or energy. When only one unitary measurement is desired, exergy is a more natural and scientifically rigorous choice for resource consumption than mass and energy. Therefore, exergy is also expected to be more attractive than mass or energy in connecting life cycle inputs with life cycle impacts (24).

Cumulative exergy consumption (CEC) analysis is a method that measures the exergy consumption and efficiency to make a product or service. Traditionally, exergy analysis focuses on single process to increase the exergy efficiency. But such an analysis ignores the life cycle information, and it is not able to help environmental decision making on a broader scale. Engineering methods have focused on including the exergy consumed in industrial processes to result in an approach that may be called *industrial cumulative exergy consumption* (ICEC) analysis (23, 25). ICEC takes a life cycle view, counting exergy consumed in all the steps from natural resources extraction to the final product. ICEC ignores the ecological processes that are required for producing or concentrating the resources. Different natural resources are made available by different ecological processes that vary considerably in their thermodynamic efficiencies. For instance, solar radiation is obtained freely from the sun, whereas coal is made available to the economic system by the geological cycle resulting in a much lower efficiency than solar energy. Accounting for the exergy consumption in the relevant ecosystem processes permits better representation of the quality differences between different types of natural resources.

Ecological cumulative exergy consumption (ECEC) analysis extends ICEC analysis by including ecological stages of the production network. These stages contain natural processes for producing, transporting, and concentrating natural resources, such as coal and rain; various pollution dissipation functions in ecological systems are also part of ECEC analysis. Take power generation from solar energy and coal as an example to compare ICEC and ECEC (25). Without considering ecological work, coal is more efficient for generating electricity than solar energy because coal is a higher quality natural resource. This result contradicts with traditional LCA, which shows that sunlight is a cleaner energy resource than coal. By including ecological stages, ECEC

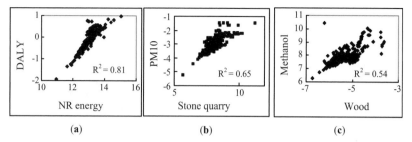

FIGURE 2.7   (a) Cumulative nonrenewable energy consumption versus overall impact (DALY); (b) Stone quarry versus PM10 emission; (c) Wood versus methanol emission.

of coal is much larger than solar energy, which reduces the life cycle thermodynamic efficiency of coal as compared to solar energy. This simple illustration indicates that life cycle thermodynamic efficiency defined with respect to ECEC, and not ICEC or cumulative energy, may be a better indicator of life cycle environmental impact.

Ongoing research is exploring whether useful relationships exist between the cumulative consumption of various biophysical quantities discussed above and life cycle impact. Zhang and Bakshi (26) studied the possibility of using life cycle input-side metrics to predict life cycle environmental impact. Resources are evaluated in different units: cumulative consumption of mass, energy, exergy without ecosystems, and exergy with ecosystems. The impact of emissions is quantified via DALYs.

Preliminary results are obtained based on a model for ecologically based LCA (EcoLCA) (27) that augments the 1997 U.S. Economic Input–Output model with information about flow of resources from nature to various sectors and emissions. This model provides data for almost 500 sectors and is well suited for statistical analysis. In addition, LCA based on process models have also been completed. Figure 2.7 shows the preliminary results from these studies. It shows that specific emissions do have a reasonably high correlation with the life cycle consumption of some resources. For example, CO_2 emissions are most correlated with the use of fossil fuels, emission of particulate matter (PM10) is most correlated with the use of stone, and methanol emissions are most correlated with the use of wood. If a single aggregated metric is desired, then ECEC seems to have the best correlation with life cycle environmental impact. This may indicate that ECEC, which accounts for the exergy consumed in industrial and ecological systems may be the most appropriate way of combining different types of resources. Although such a study is not comprehensive enough for validating the relation, this method seems to be worth further study, and with advanced statistical models. The results of such statistical analysis may be useful for guiding the development of nanotechnology and other emerging technologies until detailed LCA can be completed.

2.6.2   Predictive Toxicology and Applications for Nanotechnology

Although a powerful tool for assessing the environmental performance of products and processes, traditional LCA also falls short of estimating the toxicity impact of engineered nanomaterials and nanoproducts. The main hindrance in achieving this objective is the lack of sufficient data about the human and ecological health impact of nanoparticles. A complete understanding of the various characteristics of engineered nanomaterials that influence the biological activity is central for developing safe nanoproducts. The risk assessment approach currently followed by EPA is relevant in this regard. It consists of the hazard identification, evaluation of dose–response relationship, exposure routes, and extent of exposure. This is then quantified in terms of the risk characterization. In the case of nanotechnology, the biggest challenge is the evaluation of risks for a range of materials along their life cycle.

Assessing the correct exposure route of nanomaterials is critical to understanding the mechanism of damage. Primary exposure routes can be inhalation, dermal, ingestion, or implantation. Inhalation is more likely to be the exposure route for those involved in the manufacturing processes. Release of and exposure to nanoscale materials will also be highly dependent on whether the nanoscale particles involved are bound or unbound and hence the application involved. Higher levels of occupational exposures are expected for operations such as milling, spraying, and etching. Besides assessing the correct exposure route for nanomaterials, a crucial question is the selection of appropriate metric for quantifying the toxicity impact of nanomaterials. Traditionally, exposure in the context of toxicological assessments has been usually associated with the mass of material. However, in the case of nanomaterials, metrics such as particle size, surface area, and surface chemistry may play a more significant role. Studies exist in the literature that reveal how changing the shape alone of the zinc oxide nanomaterials alters the physicochemical properties that can be a direct function of the biological activity (28). Currently, there is little consensus, if any, in the scientific community on a common metric to quantify the toxicity impact of engineered nanomaterials, and is an area of active research.

Although not a replacement for detailed toxicological studies, a relatively new approach for assessing the toxicity of substances is predictive toxicology. The following subsection provides a brief description of predictive toxicology followed by its potential applications in the field of nanotechnology.

2.6.2.1 *Predictive Toxicology*   Toxicology is the study of chemicals and their effect on living systems or organisms. Throughout history, toxicology has become very important to the medical community. As more man-made chemicals find their way into our lives, there is a duty to find out what effect they have on animals, the environment, and especially human health. Toxicology is a very labor intensive field with thousands of researchers around the world dedicating their careers to experimenting and testing the toxicity of old and new compounds. In many ways, toxicology is also predictive or preventive. Inferences are made that similar chemicals will have similar adverse effects. Predictive toxicology formalizes these inferences to project the toxicity of a new or potential chemical from prior knowledge of similar chemicals. As with all models, a toxicological model is only useful when built upon large amounts of high-quality data. This data is used to create statistical models that allow for predictions and inferences. This process is outlined in Fig. 2.8. Helma (29) provides a good introduction to predictive toxicology.

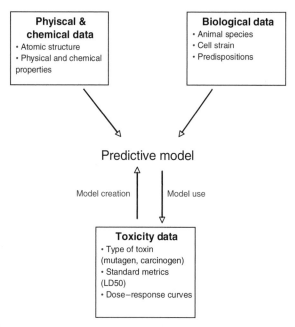

FIGURE 2.8   Model creation and use for predictive toxicity.

2.6.2.2 Methods   Several statistical methods are currently used in predictive toxicology both to create models and to interpret results. The methods we will consider are quantitative structure–activity relationships (QSAR), expert systems, regression methods, and data mining.

Quantitative Structure–Activity Relationships   Structure–activity relationships (SARs) identify links between specific atomic substructures in chemicals and their biological or chemical effects on cells or organisms. These relationships depend on the reasonable assumption that similar structures will provoke similar responses. For instance, if a specific substructure is common to several different chemicals all known to be mutagens, then it can be reasonably assumed that a new unknown chemical with this substructure is likely a mutagen. Quantitative structure–activity relationships work in the same way but also assign numerical values to the biological responses. QSARs are valuable in predictive toxicology because they allow for quantitative modeling and prediction.

Expert Systems   In a sense, expert systems attempt to mimic human problem solving in a virtual environment. The idea is to "train" the program using simple rules such as "if A, then B" or "if C and D, then not E." Experts in a given field build up complex networks of dependent and independent rules to form a model. When new information is introduced, the model uses the rules to arrive at conclusions. In predictive toxicology, this can be effective, but is limited by its exactitude. For example, if there is a study that connects chemical X with a relatively high incidence of brain cancer in rats, the rule might be "if chemical is X, then it causes brain cancer." But this rule is not exactly correct. All we know is that chemical X is somehow related to brain cancer in rats. Not all of the rats exposed developed cancer, and the mechanism that caused cancer in rats may effect humans in a different way or not at all. Modifications can be made to this type of model to make it more useful in predictive toxicology.

Data Mining and Regression   The use of very large bioassays in toxicological research necessitates the use of statistical methods such as regression and data mining to extract useful information. Data mining is simply the process of sorting through large sets of data to find patterns that were not otherwise obvious. Data mining algorithms are important to predictive toxicology for identifying relationships for use in QSAR-based prediction models. For these methods, the output or response variable may be some measure of toxicity such as LC50 or inflammatory response. The input or predictor variables can be numerous and could include factors such as atomic structure, physicochemical properties, thermodynamic properties, as well as information about the animal or cell strain used in the study.

As mentioned earlier, predictive models are only as good as the data they are built upon. For this reason, it is of utmost importance to use high-quality data in predictive toxicology. Therefore, the source of the data may be as important as the data itself. The experimental procedure and all relevant variables must be clearly defined, and careful characterization of chemicals is crucial to ensure that data are applicable. Also, data should be reported with uncertainty information. Often, it is desirable to design and execute experiments with the model in mind. This requires multidisciplinary cooperation. Statisticians, toxicologists, and chemists or other experts in a given field should work together to design relevant experiments for the development of an effective predictive model.

2.6.2.3   *Applications for Nanotechnology*   Little is known about the fate and effects of nanoscale elements. There is little data available about the toxicity of even common nanostructures such as carbon nanofibers and tubes, fullerenes, or nanoparticle metal oxides. Some research suggests that the toxicity of nanoscale materials can be much greater than their bulk phase counterparts due to higher surface to volume ratios leading to increased chemical activity. In addition, there are concerns that smaller particles can more easily penetrate biological systems and cross cellular boundaries.

Predictive toxicology could play a large part in filling the holes in toxicity data available for nanoscale structures and particles. As nanotechnology is still in its infancy, there is a great opportunity here to develop and tailor the methods of predictive toxicology to an emerging field. It could allow companies developing nanoproducts to be proactive about toxicity instead of reactive once their products are already on the market. Another advantage of an effective predictive model is cheaper and possibly more accurate estimation of toxic effects without fabricating new particles or products and testing them in the laboratory.

2.6.2.4 Challenges and Future Directions   There are some challenges that must be overcome before the full potential of predictive toxicology can be applied to nanotechnology. First of all, predictive toxicology was developed to deal with chemical species, not metal oxide particles or carbon fibers. Methods such as QSAR are tailored to chemical substructures and do not include parameters for the effects of size. Also, predictive toxicology is very data intensive and relies on large stores of historical toxicity data covering hundreds of thousands of different chemical compounds. Data on such a large scale are not available for nanoscale materials, and may not be available for several decades. The data that are currently available is new, incomplete, and untested. Toxicologists working with nanoparticles typically know very little about nanotechnology, and therefore do not characterize the particles they test correctly. This means a large portion of the data available is close to unusable for a predictive model. These obstacles are daunting, but there are ways they can be corrected or worked around.

To use predictive toxicology for nanotechnology, existing methods must be adapted. This will require starting from the beginning with simple methods such as multivariate linear regression to build simple models for individual compounds. Eventually, it should be possible to develop more complex methods for nanoscale elements. Structures common to differing nanostructures can be identified and classified for use in a QSAR model. Obviously, more and better data are required. Existing data for bulk materials might be used to project toxicity at the nanoscale. Also, a standard for evaluating toxic end points for nanotechnology needs to be established. Higher quality data in this field will only become available through more interdisciplinary cooperation. Toxicologists need to become experts in nanotechnology and vice versa. Characterization of nanoparticles also needs to be standardized before data from different studies in different labs can be used in the same model. In the short term, due to the current lack of data, statistical methods less reliant on data will have to be used. Methods such as Bayesian estimation may play a role in incorporating expert knowledge in the predictive models, and correlations between the toxicity of nanomaterials and their biophysical and thermodynamic properties such as those discussed in Section 2.4 may be found.

2.7 SUMMARY

It is beyond doubt that the inherent properties of nanoscale materials do pose several new challenges. One of the most outstanding challenges includes examining the large-scale impacts of potential nanoproducts and nanotechnologies before their widespread commercialization. Absent detailed data regarding the manufacturing processes and toxicity impact of nanoproducts, several system analysis approaches are discussed. These include the traditional LCA for assessing existing technologies, predictive LCA for emerging technologies, and predictive toxicology. All of these approaches have their own set of pros and cons. No single approach is comprehensive by itself and is meant to complement rather than replace each another. Traditional LCA can help in providing metrics for environmental evalua-tion of potential nanoproducts versus alternatives. LCA based on biophysical quantification of resource use can offer preliminary insight into the broader impact of technologies in the absence of detailed ecotoxicological data. Predictive toxi-cology could play a large part in providing insights about the toxicity of nano-scale structures and particles before detailed toxicological data become available. Few would argue that successful implementation of nanotechnology will require addressing and managing risks posed by this emerging technology to human health and environment. Nanotechnology is currently in its infancy with respect to its widespread adoption. Currently, there are no regulations as far as development of nanotechnology is concerned. Addressing environmental, health, and safety con-cerns in a timely manner will also help identify regulatory needs and address public skepticism about the emerging field of nanotechnology. It is inevitable that this powerful technology will have a pronounced impact on future science and technol-ogy. To ensure that the impact is positive, potential benefits and risks need to be balanced as the industry proceeds toward widespread commercialization.

It is often believed that increasing the efficiency of technological products and processes will translate into a smaller environmental impact. This is usually the basis of many claims of the environment friendliness of new technologies. Thus, new developments in nanotechnology will almost certainly reduce the material intensity and use of nanoparticles in individual nanomaterials. Significant improvements in the energy efficiency, recovery at the end-of-life, and cost are bound to occur over time. Historically, such improvements have also increased the popularity of the relevant products. Thus, even though the environmental impact per unit of a product has decreased, the overall impact the technology often increases due to its wide adoption and huge scale of production. Such a rebound effect could cancel the environmental benefits of the new technology. Addressing this challenge is typically beyond the realm of engineering and requires careful formulation of economic policies and socioeconomic aspects. The types of systems analysis methods discussed in this chapter can provide the basis for truly sustainable policies by cutting across disci-plinary boundaries.

REFERENCES

1. Available at http://www.epa.gov/osa/nanotech.htm. Accessed 2007 Aug 9.
2. Rebitzer, G., Ekvall, T., Frischknecht, R., Hunkeler, D., Norris, G., Rydberg, T., Schmidt, W.P., Suh, S., Weidema, B.P., Pennington, D.W. Life cycle assessment part 1: framework, goal and scope definition, inventory analysis, and applications. *Environ Int* 2004;30:701–720.
3. Sonnemann, G., Francesc, C., Schuhmacher, M. Integrated life-cycle and risk assessment for industrial processes. *Advanced Methods in Resource and Waste Management*. Boca Raton: Lewis Publishers; 2004.
4. Bare, J., Gloria, T.P. Critical analysis of the mathematical relationships and comprehen-siveness of life cycle impact assessment approaches. *Environ Sci Technol* 2006;40 (4):1104–1113.

5. Heijungs, R., Goedkoop, M., Struijs, J., Effting, S., Sevenster, M., Huppes, G. VROM Report, Leiden University Institute of Environmental Sciences (CML). Unpublished report. Leiden: CML; 2003. Available at http://www.leidenuniv.nl/cml/ssp/publications/.

6. Van Oers, L., De Koning, A., Guinee, J.B., Huppes, G. Abiotic resource depletion in LCA. Improving Characterization Factors for Abiotic Resource Depletion as Recommended in the New Dutch LCA Handbook. RWS-DWW Report. 2002.

7. Goedkoop, M., Spriensma, R. The Eco-indicator 99: a damage oriented method for life cycle impact assessment. Available at http://www.pre.nl/eco-indicator99/ei99-reports. htm. Accessed 2007 Nov 15.

8. The Royal Society and the Royal Academy of Engineering (RS RAEng). Nanoscience and nanotechnologies: Opportunities and Uncertainties. Available at www.nanotec.org.uk/finalReport.htm, Accessed 2007 Jan 20.

9. The National Nanotechnology Initiative. Report on Environmental, Health, and Safety Research Needs for Engineered Nanoscale Materials, Septemeber 2006. Available at www.nano.gov. Accessed 2007 Jan 20.

10. Karn, B.P. International Society for Industrial Ecology, Newsletter, vol. 4, issue 4, 2004.

11. Lloyd, S.M., Lave, L.B. Life cycle economic and environmental implications of using nanocomposites in automobiles. *Environ Sci Technol* 2003;37(15):3458–3466.

12. Carnegie Mellon Green Design Institute. Available at www.eiolca.net.

13. Lloyd, S.M., Lave, L.B., Matthews, H.S. Life cycle benefits of using nanotechnology to stabilize platinum-group metal particles in automotive catalysts. *Environ Sci Technol* 2005;39(17):1384–1392.

14. Osterwalder, N., Capello, C., Hungerbuhler, K., Stark, W.J. Energy consumption during nanoparticle production: how economic is dry synthesis? *J Nanopart Res* 2006; 8:1–9.

15. Isaacs, J.A., Tanwani, A., Healy, M.L. *Proceedings of the 2006 IEEE International Symposium on Electronics and the Environment;* 2007 May 8–11; San Francisco, CA; pp. 38–41.

16. Khanna, V., Bakshi, B.R., Lee, J.L. *Proceedings of the 2007 IEEE International Symposium on Electronics and the Environment;* 2007 May 7–10; Orlando, FL; pp. 128–131.

17. Hau, J.L., Yi, H.-S., Bakshi, B.R. Enhancing life cycle inventories via reconciliation with the laws of thermodynamics. *J Ind Ecol* 2007;11(4):1–21.

18. Mordkovich, V.Z. Carbon nanofibers: a new ultrahigh-strength material for chemical technology. *Theor Found Chem Eng* 2003;37(5):429–438.

19. Hinterberger, F., Kranendonk, S., Welfens, M.J., Schmidt-Bleek, F. Increasing resource productivity through eco-efficient services. Wuppertal Paper nr 13, Wuppertal Institute, Wuppertal, Germany.

20. Haberl, H., Weisz, H., Amann, C., Bondeau, A., Eisenmenger, N., Erb, K., Fischer-Kowalski, M., Krausmann, F. The energetic metabolism of the European Union and the United States: decadal energy input time-series with an emphasis on biomass. *J Ind Ecol* 2006;10(4):151–171.

21. Van der Voet, E., Van Oers, L., Moll, S., Schütz, H., Bringezu, S., de Bruyn, S., Sevenster, M., Warringa, G. Policy review on decoupling: development of indicators to assess decoupling of economic development and environmental pressure in the EU-25 & AC-3 countries. Available at http://www.leidenuniv.nl/cml/ssp/projects/dematerialisation/policy review on decoupling.pdf. Accessed 2007 Nov. 15.

22. Huijbregts, M.A.J., Rombouts, L.J.A., Hellweg, S., Frischknecht, R., Hendriks, A.J., Van de Meent, D., Ragas, A.M.J, Reijnders, L., Struijs, J. Is cumulative fossil energy demand a useful indicator for the environmental performance of products? *Environ Sci Technol* 2006;40(3):641–648.

23. Szargut, J., Morris, D.R., Steward, F.R. Exergy Analysis of Thermal, Chemical and Metallurgical Processes. New York: Hemisphere Publishers; 1988.

24. Ukidwe, N.U., Bakshi, B.R. Thermodynamic accounting of ecosystem contribution to economic sectors with application to 1992 U.S. Economy. *Environ Sci Technol* 2004; 38(18):4810–4827.

25. Hau, J.L., Bakshi, B.R. Expanding exergy analysis to account for ecosystem products and services. *Environ Sci Technol* 2004;38(13):3768–3777.

26. Zhang, Y., Bakshi, B.R. *Proceedings of the 2007 IEEE International Symposium on Electronics and the Environment;* 2007 May 7–10; Orlando, FL; pp. 117–122.

27. Zhang, Y., Baral, A., Bakshi, B.R. Ecologically based life cycle assessment, AIChE Annual Meeting; 2007 Nov 4–9; Salt Lake City, UT.

28. Wang, Z.L. Nanostructures of zinc oxide. Materials Today; 2004;7(6):26–33.

29. Helma, C., editor *Predictive Toxicology.* Boca Raton, Florida: Taylor & Francis Group; 2005.

CHAPTER 3

An Integrated Approach Toward Understanding the Environmental Fate, Transport, Toxicity, and Health Hazards of Nanomaterials

JOHN M. PETTIBONE[1], SHERRIE ELZEY[1], and VICKI H. GRASSIAN[1,2,3]

[1]Department of Chemical and Biochemical Engineering, University of Iowa, Iowa City, IA 52246, USA
[2]Department of Chemistry, University of Iowa, Iowa City, IA 52246, USA
[3]Nanoscience and Nanotechnology Institute at UI, Iowa City, IA 52246, USA

3.1 INTRODUCTION

Nanoscience and nanotechnology offer new opportunities for making superior materials for use in industrial, environmental, and health applications (1–6). As nanomaterials continue to develop and become more widespread, we can expect that these manufactured materials have the potential to get into the environment sometime during production, distribution, use or disposal, that is, sometime during the life cycle of these materials. There are now over 500 nano-based products commercially available and a number of toxicological studies are being done to ensure the safety of these materials (7). With the introduction of this new area in toxicology, nanotoxicology, databases have been created to catalog existing research in addition to keeping updated information of ongoing work in the field. The knowledge base on nanotoxicity and risks associated with nanomaterials has been growing due to an increase in the number of nanomaterials and nanomaterial-based consumer goods that are becoming commercially available. Databases that include the International Council on Nanotechnology (ICON) (Rice University), Project on Emerging Nano-technologies (Woodrow Wilson International Center), Nanoparticle Information Library (NIOSH), Nanoscale Science and Engineering Center (University of Wisconsin, Madison), SAFENANO (Institute of Occupational Medicine), and Nanomedicine Research Portal are being compiled to provide accessible information on the latest

Nanoscience and Nanotechnology Edited by Vicki H. Grassian

TABLE 3.1   Environmental Health and Safety Databases and Reports for Nanomaterials

International Council on Nanotechnology (ICON)
  http://icon.rice.edu/research.cfm
Project on Emerging Nanotechnologies (PEN): Woodrow Wilson International Center
  http://www.nanotechproject.com/index.php?id=18
NIOSH Nanoparticle Information Library (NIL)
  http://www.cdc.gov/niosh/topics/nanotech/nil.html
Nanoscale Science and Engineering Center University of Wisconsin, Madison:
  http://www.nsec.wisc.edu/NanoRisks/NS–NanoRisks.php
SAFENANO: Institute of Occupational Medicine
  http://www.safenano.org/AdvancedSearch.aspx
Nanomedicine Research Portal
  http://www.nano-biology.net/
EPA Nanotechnology White Paper
  http://www.epa.gov/OSA/nanotech.htm
Environmental Health and Safety Research Needs for Engineered Nanoscale Materials
  http://www.nano.gov/html/news/EHS_research_needs.html

scientific data. These databases, listed in Table 3.1 along with website information, are helpful tools for identifying the latest results on a number of environmental health, safety, and toxicity studies that have been conducted on a broad range of nanomaterials.

Nanoparticles, the primary building blocks of many nanomaterials, may become suspended in air or may get into water systems, for example, drinking water systems, groundwater systems, estuaries, and lakes (8–10). Therefore, manufactured nanoparticles can become a component of the air we breathe or the water we drink.

There has been one study thus far that has suggested that manufactured nanomaterials may become entrained in the atmosphere. In a field study done near Houston, Texas, 10 nm SiO_2 particles were detected with a single-particle mass spectrometer (11). Although the exact source of these particles is unknown, it appeared to be from an industrial source. Since nanoparticles and their aggregates are likely to be in the respirable size range, it is important to investigate the potential health effects of these particles suspended in air (12–14). Research on ultrafine particles has shown that extremely small particles (<100 nm) have greater adverse health effects than larger particles (13). This is because ultrafine particles have the ability to penetrate deep into the lungs (13–15). Given that manufactured nanoparticles are a specific subset of ultrafine particles, it is reasonable to investigate whether there are deleterious health effects associated with the inhalation of engineered nanoparticles (16).

Similarly in water systems, because of the high surface areas associated with them, nanoparticles in particular and nanomaterials in general may sequester heavy metals or other contaminants. Thus, nanomaterials associated with heavy metals or contaminants can potentially be toxic if ingested and there is a need to study the environmental and health impacts of manufactured nanomaterials in water systems.

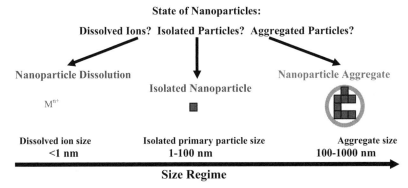

FIGURE 3.1   An important question when considering the environmental fate or toxicity of nanoparticles is what is the state of the nanoparticles? Nanoparticles can be present as isolated particles, can form aggregates up to a micron in size, or can dissolve into ions in solution.

One important issue in such studies is to understand the nature and state of nanoparticles in air or water. That is, will nanoparticles be present as isolated particles or in the form of aggregates? Under what conditions are nanoparticles present as isolated particles or aggregates? Furthermore, will nanoparticles dissolve in aqueous solution? Under what conditions will nanoparticles dissolve? As shown schematically in Figure 3.1, the size regime that needs to be considered or modeled, for example, in environmental transport or lung deposition models, will be very different depending on the state of the nanoparticles: as dissolved ions, isolated nanoparticles, or nanoparticle aggregates.

Studies directed toward understanding the environmental and biological fate of nanoparticles and the potential transformation of nanoparticles have just begun in recent years (17, 18). The environmental and biological fate, as well as the toxicity of nanoparticles, depends on nanoparticle size, shape, bulk composition and phase, and surface composition (19–28). From the most recent studies, it is becoming increasingly clear that nanoparticle aggregation is an important consideration in the fate of nanoparticles (29, 30) and their impact on living systems. There are a number of factors, including pH and surface chemistry, that control both the propensity of nanoparticles to aggregate and the size of the aggregate (31–33).

Many questions still remain pertaining to the environmental fate and transport of nanomaterials, their impact on environmental health and safety of living systems, and specifically to determine how physicochemical properties influence biological interactions. In fact, it is clear from some of the recent literature that the full impact, or even partial impact, of manufactured nanomaterials on human health and environment has yet to be fully explored (13, 17, 23, 24, 34–41). Thus, there are a number of questions and issues that need to be addressed so as to better understand the environmental fate and transport in addition to the toxicity of nanomaterials. There is a clear need for an integrative approach that combines extensive nanomaterial characterization with any investigation of the toxicity of nanomaterials or their impact on environmental health and safety, as discussed below.

3.2 IMPORTANCE OF AN INTEGRATED APPROACH TOWARD UNDERSTANDING THE ENVIRONMENTAL FATE, TRANSPORT, TOXICITY, AND HEALTH HAZARDS OF NANOMATERIALS

Because of the unique and size-dependent properties of nanomaterials, risks associated with these materials cannot be expected to be the same as their bulk counterparts. To better understand risks associated with nanomaterials, there needs to be a full understanding of the materials themselves. Studies devoted to the environmental fate and transport, biological fate, toxicity, and overall risk assessment of nanomaterials need to have nanomaterial characterization as a central component of the study design. There is a growing consensus that an integrated approach is necessary, as described in several workshop and agency reports. These reports providing information on suggested strategies, best practices, and important concepts of environmental health and nanomaterial toxicity are reviewed below.

3.2.1 Recommendations from Recent Workshop and Agency Reports

The recognition of the need for an integrative approach toward understanding the environmental and health effects of nanomaterials has been discussed in a number of workshop and agency reports. These include a "Hazard assessment for nanoparticles report from an interdisciplinary workshop" (42), "Testing strategies to establish the safety of nanomaterials: conclusions of an ECETOC workshop" (43), and a forum series titled "Research strategies for safety evaluation of nanomaterials" in *Toxicological Sciences* that consists of eight separate publications with each discussing a different issue related to the evaluation of nanomaterials (19, 44–50). The U.S. Environmental Protection Agency (EPA) has released the "Nanotechnology White Paper" (51) and the Woodrow Wilson International Center for Scholars released in 2006 "Nanotechnology: a research strategy for addressing risk" (7), both addressing the assessment and implementation of prioritized research strategies for current and emerging nanotechnologies. The issues discussed in these reports include biological fate, collection and types of data needed, risk assessment, characterization methods, international collaboration, and developing effective tools to evaluate these materials. The common theme that emerges from these different reports is the necessity for complete characterization of the nanomaterials, the effect of nanomaterials on living systems and ecosystems, the biological fate of nanomaterials, and importantly what physical and chemical properties of nanomaterials cause toxicity.

Traditionally in the field of toxicology, toxicity of materials is assessed in terms of composition and mass as the dose metric evaluation. The physical and chemical properties of materials can change on the nanoscale especially below 10 nm (52). Size is important to the physical and chemical properties of nanomaterials as it also affects the biological response. Several published articles have focused on the issue of dose metrics for nanomaterials, mass or surface area. These articles include "Characterization of the size, shape, and state of dispersions of nanoparticles for toxicological studies" (53), "Particokinetics *in vitro*: dosimetry considerations for *in vitro* nanoparticle toxicity assessments" (54). Other publications have given extensive discussions on the importance of primary size versus aggregate state and how the dispersion state may affect assessment (55, 56). Discussions on the appropriate dose metrics used for nanomaterials are ongoing and are not universal for the broad range of commercial materials available. As these discussions continue, nanomaterial characterization is accepted as an essential component of these studies.

3.2.2 Nanoparticle Characterization: Bulk and Surface Properties

As already emphasized above, correlating a biological response or lack of response with specific physicochemical properties can only be accomplished with full characterization of nanomaterials in environmental health and safety studies. Full characterization includes determining the bulk and surface properties of the nanomaterials because both of these will be important factors in biological response. Bulk characterization methods typically examine the shape, size, phase, electronic structure, and crystallinity of nanomaterials. Surface characterization methods typically include surface area, arrangement of surface atoms, surface electronic structure, and surface composition and functionality.

Examples of different characterization techniques are discussed in this section. These examples focus on the characterization of metal and metal oxide-based nanomaterials as recent inventories have suggested that these are some of the most common types of nanomaterials currently found in consumer-based products (7). Since metal and metal oxide-based nanomaterials represent about half the nanomaterials present in consumer products, with silver, zinc, titanium, and gold the most abundant, it is most likely that these nanomaterials will make their way into the environment and therefore should be of high priority in environmental health, safety, and toxicity studies. Select examples of the types of bulk and surface characterization data that can be used to better understand nanomaterials are given in this section.

The phase of a nanomaterial is an important characteristic and can be identified using powder X-ray diffraction (XRD) techniques (57). Because the total free energy of nanoparticles is the sum of bulk and surface contributions,

$$G = G_{surface} + G_{bulk} \tag{3.1}$$

the surface contribution becomes an increasingly important component of the total free energy as the particle decrease in size. Thus, the size of the particle can have a large influence on the most thermodynamically stable phase. For example, nearly 30–40% of the total atoms are at the surface for nanoparticles that are 4 nm in diameter (58). Therefore, differences in surface free energy for the lowest energy surface faces for different phases become increasingly important contributions to the energy as a function of nanoparticle size. For TiO_2, rutile is the more stable phase for the bulk material and, although anatase has a higher G_{bulk}, anatase has a lower $G_{surface}$ compared to rutile. Because of this, nanoparticles below 14 nm produced under thermodynamically controlled conditions at room temperature form 100% anatase (59).

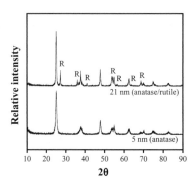

FIGURE 3.2   Powder X-ray diffraction patterns of 5 and 21 nm TiO$_2$ particles are shown and the diffraction lines due to the rutile phase are labeled with "R". The 5 nm particles are purely anatase but the 21 nm particles also contain rutile. Reproduced from Reference 73 with permission from *Informa Healthcare*.

This can be seen in the X-ray diffraction data shown in Fig. 3.2 for 5 and 21 nm TiO$_2$ particles. The 5 nm particles are pure anatase whereas the 21 nm particles are largely anatase but also contain rutile (peaks labelled with R). The amount of rutile from a powder XRD pattern that does not contain excessive broadening of peaks can be quantified by the ratio of intensities of the anatase and rutile peaks because the polymorphs have the same mass attenuation coefficient (60). More complex analysis is needed if the coefficients are not equivalent. From the simple analysis, it is estimated that $\sim$15% of the 21 nm particles is rutile and 85% is anatase.

In addition to phase, line analysis of XRD patterns can be used as a measure of crystallite size. According to Scherer's equation,

$$t = \frac{\kappa\lambda}{\beta\cos\theta} \tag{3.2}$$

where t is crystal size, κ is the shape factor, λ is the wavelength, θ is the diffraction angle and β is the full width at half maximum of the diffraction peaks, the crystalline size is inversely proportional to the width of the peaks. This broadening can be seen in Fig. 3.2 as the line width of the peak at $2\theta \sim 25$ in the XRD pattern for the 5 nm particles is larger than that observed for the 21 nm particles.

Electron microscopy, transmission (TEM) and scanning (SEM), is the most widely used method to determine nanoparticle size and shape. An example of a TEM image of commercially available silver nanoparticles is shown in Fig. 3.3a. By analyzing the size of many particles, a size distribution can be determined as shown in Fig. 3.3b. High-resolution TEM can also provide information about the crystallinity and crystalline phase of the nanoparticles (59). The size and shape of nanoparticles can also be used to derive surface area information based on simple geometrical considerations.

Besides microscopy, new characterization methods are being developed for determining nanoparticle size. These include scanning mobility particle sizers (SMPS)

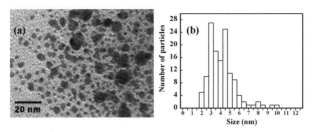

FIGURE 3.3 (a) TEM of silver nanoparticles taken from an aqueous solution. (b) Size distribution of the silver nanoparticles is shown.

and size exclusion chromatographs (SEC). These methods are emerging as new techniques with potential to determine single particle size in an ever-increasing variety of applications. These methods allow for *in situ*, rapid analysis of nanoparticle size. Compared to more established techniques such as TEM and SEM, these new methods offer economically a more feasible option for particle size measurements, and are often used in combination with traditional methods as part of a suite of single-particle sizing. In addition, these instruments can be used to determine the size of inorganic/organic composites that often consist of an inorganic core and organic coating. Because the organic coating has minimal contrast in electron microscopy, it is difficult to determine the coating thickness and usually only the size of the inorganic core can be measured. However, SMPS and SEC methods work on different principles and can be used to measure nanoparticle size including the organic coating.

Combining an electrospray aerosol generator with an SMPS shows considerable promise for nanoparticle characterization. A schematic of this type of instrument developed at the University of Iowa is shown in Fig. 3.4. The electrospray generates an aerosol of nanoparticles by drawing a liquid dispersion containing nanoparticles through a capillary tip by an applied voltage. Charged droplets containing the

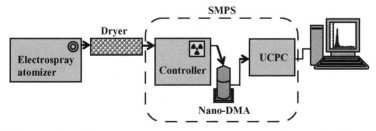

FIGURE 3.4 Schematic of an electrospray aerosol generator coupled to an SMPS. An aerosol containing isolated nanoparticles is generated in a ca. 150 nm droplet. The droplet is dried and the nanoparticles are sent through a diffusional dryer to evaporate the water. The polydispersed aerosol nanoparticles are radioactively charged and enter the Nano-DMA for classification based on electrical mobility. A monodisperse aerosol enters the UCPC where particles are counted and a particle size distribution is observed.

nanoparticles are sprayed from the capillary tip into a stream of dry air. The air mixture promotes liquid evaporation, and the remaining nanoparticles are neutralized with an ionizer. The initial droplet size is ~150 nm in diameter.

Nanoparticles exiting the electrospray are further dried before entering the SMPS. One drying technique is to flow the nanoparticles through the inner tube of a diffusion dryer where remaining liquid diffuses away from the particles and is captured by silica beads contained in an outer tube. Additional drying can be accomplished by mixing the nanoparticle stream with an excess of dry, filtered air, or by sending the nanoparticles through a heated zone. The nanoparticle aerosol stream then enters the SMPS for size determination.

The SMPS for nanoparticle characterization shown in Fig. 3.4 consists of a nano-differential mobility analyzer (Nano-DMA, TSI Inc.) and an ultrafine condensation particle counter (UCPC, TSI, Inc.). This instrument is capable of determining particle size distributions within the range of 2–80 nm. Particles entering the SMPS are charged using a radioactive source and their size is then classified based on electrical mobility as the particle stream flows down the high voltage column of the Nano-DMA. The particle trajectory down the column depends on electrical mobility, Z_p, of the particle in an applied electrical field and is defined by

$$Z_p = \frac{neC_c}{3\pi\eta D_p} \tag{3.3}$$

where n is the number of charges on the particle, e is the elementary charge, C_c is Cunningham slip correction, η is the dynamic viscosity of air, and D_p is the diameter of the particle. Therefore, only particles within a narrow mobility diameter range are allowed to exit the Nano-DMA and the polydisperse aerosol stream at the inlet is converted into a monodisperse aerosol stream at the outlet. The monodisperse particles then enter the UCPC where the particle concentration is determined. The UCPC uses butanol condensation to grow particles of known diameter to an optically active droplet size, and the droplet is counted using a laser beam and photodetector. The coupled particle sizing of the Nano-DMA and particle counting of the UCPC allow the SMPS to provide a single particle size distribution for a nanomaterial sample. For spherical particles, the mobility diameter and the volume equivalent diameters are the same. For nanoparticles with a different shape, these two diameters are related by a shape factor, χ.

The electrospray/SMPS as a method for determining single particle size has been successfully used to characterize surface coatings on manufactured gold nanoparticles (61). The surface coverage of thiol-modified single-stranded DNA (ssDNA) coated on the surface of gold nanoparticles was determined. A salt solution containing the ssDNA was used to coat the gold particles, and a particle size distribution was collected before and after surface coating. Based on changes in the particle size distribution, the coating thickness, which is dependent on the spatial configuration of the ssDNA, was determined. From the measured coating thickness, it was concluded that the strands exist in a random coil configuration on the gold nanoparticle surface. Identifying the spatial configuration allowed total surface coverage to be calculated using the known size of an ssDNA molecule. Such characterization of biological surface coatings may have applications in designing and controlling targeting agents for cancer treatments (62).

The primary advantages of the electrospray/SMPS method of characterizing particles are its excellent resolution, discrimination between changes in size as small as 0.2 nm and the ability to size particles as small as 2 nm. The electrospray aerosol generator also has the advantage of requiring very small volumes (<1 mL) of particle solutions or suspensions. One drawback of this method is that any particles or ions present in addition to the particles of interest will also be detected. Thus, salt peaks or "water impurity peaks" may be observed in the particle size distribution due to the formation of particles as liquid evaporates from droplets not containing the particles of interest (63, 64). Such impurities can also coat the surface of the particles of interest, adding to their measured size distribution. Therefore, this method is very sensitive to the nature of the solution in which the particles are suspended or dissolved. When the impurity size distributions are well resolved compared to the particle size distribution, this problem can be overcome by subtracting the impurity result from that of the particles of interest.

Another new method for determining single particle size is SEC that consists of a column packed with a porous polymer or gel material, through which particles flow down via a liquid referred to as the mobile phase. The packing material contains a variety of pore sizes, often on the order of a few hundred nanometers. As the particles travel through the column, those that are too large to fit through the pores will flow around the packing material and travel less distance before exiting the column than smaller particles. Particles small enough to enter the largest pores will pass through the porous material and travel a greater distance, while even smaller particles will travel through the smaller pores, and so on. Thus, the retention time for a particle to elute is size dependent, and a mathematical relationship between retention time and particle size can be determined for a material.

SEC as a method for determining single particle size has been successfully used to characterize gold nanoparticles prepared through seed-assisted synthesis (65). A polymer-based column with a pore size of 400 nm was used to separate and determine the size of the gold nanoparticles as they flowed down the column in a sodium dodecyl sulfate mobile phase at 1.0 mL/min. The mobile phase contained particles of 9.8, 40.1, 59.9, and 79.1 nm. The four particle sizes were separated and a linear relationship between retention time and particle size was found. Such studies indicate SEC can be used for rapid characterization of single particle size, monitoring of particle growth during synthesis, and particle separation based on size.

SEC has found extensive applications in the development, characterization, stability evaluation, and quality control of nanoparticle drug delivery systems (66). By monitoring the particle sizes present in a mobile phase, SEC has been used to track particle association/aggregation behavior and particle stability. Synthesis conditions have been optimized with SEC by studying the particle size dependence on synthesis temperature, pH, and reactant concentrations. Quality control has been verified with SEC by monitoring batch to batch conformity, and desired products were purified by SEC through particle size separation. Advantages of SEC are that a mobile phase containing particles of different sizes can flow through the column, each particle size can be measured, and the components can be collected separately. The analysis is *in situ*, rapid, and requires small sample volumes (submicroliter).

Besides nanoparticle size, surface characterization is of central importance to understanding the behavior and properties of nanomaterials. Surface characterization techniques for nanoparticles include surface area measurements. For powdered samples of nanoparticles, surface area measurements are typically done using the Brunner–Emmet–Teller (BET) method. Commercial BET instruments are available for routine surface area measurements of nanomaterials. Nitrogen is used as the adsorbate for surface area determination. Methods for investigating surface chemical composition and surface functional groups are quite important as a number of very important nanoparticle properties, such as solubility and biocompatibility, are determined by surface composition.

A powerful technique in probing the surface composition of nanomaterials is X-ray photoelectron spectroscopy (XPS). XPS is a well-established technique in surface science. An X-ray source (typically an Al anode) is used to eject core electrons from atoms in a solid. XPS uses X-rays to eject a core electron from an atom after which an electron from a higher energy level relaxes down to the lower energy state with the simultaneous release of a photoelectron. The photoelectron has a binding energy characteristic of a given element. The kinetic energy of the electron is typically measured and is related to the binding energy according to the relationship

$$E_B = h\nu - E_K - \phi \qquad (3.4)$$

where E_B is the binding energy referenced to vacuum level, $h\nu$ is the energy of the incoming X-ray photon, E_K is the kinetic energy of the electron, and ϕ is the work function. The surface sensitivity of this technique is derived from the short mean free path of the electrons being emitted from the solid. Atoms excited deep into the solid will not have electrons that will be ejected into the vacuum level. Because the kinetic energy is largely related to the binding energy of the electron in the atom, it is element specific since each element has well-defined core electron binding energies. The binding energy of the core electrons is a function of the nuclear charge and the shielding of other electrons from the nucleus; thus, XPS is inherently element specific. Furthermore, the observed binding energies can shift up to a 1–2 eVs depending on the chemical environment of the atom. Therefore, XPS yields information on the elements present in the near-surface region, the oxidation state of the elements, and surface functional groups present (e.g., hydroxyl groups on oxide nanoparticle surfaces or surface-modified nanomaterials that contain organic groups with carboxylic acids to increase water solubility). XPS measurements can be quantitative as the ratio of the peak areas can be related to atomic concentrations. One of the major drawbacks of the technique is that it is typically done in ultrahigh vacuum, which is not under the environmental conditions (e.g., air or water) of interest.

Examples of data acquired using some of the methods discussed above are shown in Fig. 3.5 for commercially manufactured iron nanoparticles. These metal

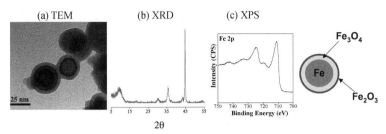

FIGURE 3.5   (a) TEM image, (b) XRD diffraction pattern, and (c) XPS measurements are shown for 25 nm iron particles. These data taken together provide the basis for the cartoon shown on the right.

nanoparticles, according to the manufacturer, have been partially passivated with oxygen. The TEM image shows the primary particle size of the nanoparticle is 25 nm in diameter with a $\sim$2 nm oxide coating clearly visible around the outer edge. This identity of this oxide coating is determined from the XRD data shown in Fig. 3.3b. Diffraction lines corresponding to metallic Fe, Fe_3O_4 and γ-Fe_2O_3 are present as determined by comparison to known standards of these materials. XPS measurements of the nanoparticles (and standard materials) in the Fe 2p region show that the outer layer is the more oxidized γ-Fe_2O_3. From these data, it appears that the structural makeup of the nanoparticle is the one shown in Fig. 3.3 in that the particle contains a metallic core, a partially oxidized Fe_3O_4 layer, and then is terminated with a fully oxidized layer, γ-Fe_2O_3.

3.2.3   Nanoparticle Characterization in Air and Water

Another tool for investigating the surface properties of nanomaterials especially in solution or in air is attenuated total reflectance Fourier transform infrared (ATR-FTIR) spectroscopy (67–69). In this method, the infrared beam is internally reflected through a high refractive index crystal creating an evanescent wave that decays exponentially away from the crystal surface and has a penetration depth proportionate to the index of refraction of the crystal and the sample. Because of the limited penetration depth of the beam, usually $\sim$1 μm, the signal mostly comprises the sample and not the solution above the sample. This technique allows *in situ* measurements of solution-phase adsorption studies or adsorption studies in air on the surface of thin films of nanoparticles coating the ATR crystal surface (70).

Surface reactivity of the nanoparticles such as adsorption or dissolution properties can also be a factor in the biological fate. If a particle can dissolve into ions after exposure, the soluble ions may cause a different biological response than the remaining solid or the combination may have an enhanced effect (20, 71). Furthermore, besides surface area the physical and chemical properties can change in the nanometer size regime especially $\leq$10 nm (52), and this will influence surface reactivity. For example, there are more edge and corner sites for chemical reactivity. A shift in bandgap energy for semiconductor nanoparticles can change the particle surface reactivity. Changes in the electronic bandgap, ΔE_g, can be estimated for a spherical particle by

$$\Delta E_g \approx \left(\frac{\pi^2 h^2}{2R^2}\frac{1}{\mu}\right) - \frac{1.8e^2}{\varepsilon R} \qquad (3.5)$$

where R is the particle radius, μ is the reduced mass of the exciton or the electron–hole pair, ε is the dielectric constant, and h is Planck's constant. As seen from Equation 3.5, the smaller the nanoparticle the larger the change in the bandgap energy.

Another consideration in air and water is that nanoparticles can form stable aggregates or agglomerates that are much larger in size. The American Society for Testing and Materials (ASTM) International has differentiated between agglomerates and aggregates in the following way. Particle agglomerates are defined as a group of particles held together by relatively weak forces, for example, van der Waals or capillary, that may break apart into smaller particles upon processing. In contrast, particle aggregates are defined as a group of particles in which the various individual components are not easily broken apart, such as in the case of primary particles that are strongly bonded together, for example, fused, sintered, or metallically bonded. (72). One tool for measuring the size of the nanoparticles that do form agglomerates/aggregates in aqueous environments including biological fluids is dynamic light scattering (DLS). DLS is a size measurement technique that uses scattered light to measure diffusion rates (Brownian motion) of particles in stable suspensions to determine a size based on Stokes–Einstein equation,

$$D_H = \frac{kT}{f} = \frac{kT}{3\pi\eta D} \qquad (3.6)$$

where D_H is the hydrodynamic diameter, k is the Boltzmann constant, T is absolute temperature, f is the particle frictional coefficient that consists of solvent viscosity, η, and the diffusion constant, D. Equation 3.6 was developed assuming hypothetical hard spheres that do not represent many nanoparticles, therefore, the size of nonspherical, hydrated, or solvated particles are related to the hypothetical hard sphere that would behave the same as the measured particles. The speed of the diffusion is measured by correlating the patterns of scattered light from time t to $(t + \Delta t)$, where Δt is on the order of nanoseconds. The smaller the particle the less correlation between the examined time points representing faster diffusion through the medium; analogously, the larger particles will diffuse much slower and the intensity pattern on the detector will change much slower. Using this technique, the hydrodynamic diameter of the nanoparticle aggregates/agglomerates can be measured in a relatively simple experimental system. An example of DLS measurements for TiO_2 nanoparticles is shown in Fig. 3.6 with increasing concentration (73). The DLS measurements show that for both 5 and 21 nm particles increasing concentration increases particle size (aggregation). Although the 21-nm particles at the lowest concentration appear to be fairly dispersed, this is not seen for the 5-nm particles and always seem to be aggregated in the solution.

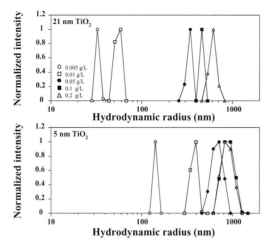

FIGURE 3.6   DLS measurements for 21 and 5 nm TiO$_2$ particles are shown with increasing solid concentrations in a saline solution. The data show that the nanoparticles are present as agglomerates in solution with the smaller nanoparticles forming larger agglomerates for each concentration. Reproduced from Reference 73 with permission from *Informa Healthcare*.

It is also noteworthy that the 5-nm particle sizes at all concentrations are larger than the 21-nm particle inferring that there is a greater propensity for the TiO$_2$ nanoparticles below 10 nm to aggregate.

DLS instruments can also measure the zeta potential of nanoparticles, which describes the charge on a particle. The zeta potential will also affect the distribution of the nanoparticles in solution as well as influence surface reactive properties that have been shown to affect biological responses (74). A common way of measuring the zeta potential of particles is applying an electric field to the particle suspension. The movement of the charged particles relative to the liquid suspension is electrophoresis. The velocity of the particle in the medium when the electric field is applied can be measured and a zeta potential can be calculated using the Henry equation,

$$U_E = \frac{2\varepsilon z f(ka)}{3\eta} \tag{3.7}$$

where U_E is electrophoretic mobility, z is the zeta potential, ε is the dielectric constant, η is the viscosity of the suspension, and $f(ka)$ is Henry's function, which is approximated as either 1.0 or 1.5 depending on particle size and ionic strength. The effect of applied electric fields can also be measured on liquids moving over stationary charged surfaces, electroosmosis, the forced flow of liquid past a charged surface generating an electric field, streaming potential, or when charged particles move relative to a stationary liquid generating an electric field, sedimentation potential.

Different techniques are needed to characterize nanoparticles or nanoparticle agglomerates in air as aerosols. This can be done with a variety of instruments depending on the size of the agglomerate. Characterizing the size of the generated aerosol is commonly done using a SMPS that consists of a DMA and particle counter and measures the mobility diameter of particles or agglomerates/aggregates below 700 nm. For these larger sizes, the Nano-DMA described earlier and shown in Fig. 3.4 is replaced with a long DMA (TSI, Inc.). If the nanoparticle agglomerates are large or the distribution of size contains both agglomerate sizes below and above ca. 700 nm, SMPS measurements are coupled to an aerodynamic particle sizer (APS) that is useful for detecting the larger particle agglomerates found in the micron size regime to obtain a more complete size distribution. Because the SMPS and APS instruments operate on very different principles, the measured diameters need to be related for comparison purposes.

For spherical particles, it is easy to relate the measured diameters from the different instruments mentioned above because no corrections need to be made for shape and volume, but diameters for nonspherical particles or agglomerates and aggregates that are irregularly shaped are not equivalent. The APS reports an aerodynamic diameter, D_a, for the irregular particle by comparing the settling velocity to a spherical particle with a density of 1 g/cm^3 to compute the particle size. As mentioned before, the SMPS measures a mobility diameter. The difference in measurements creates a need for a volume equivalent diameter, D_{ve}, which is defined as the volume of sphere with the same volume as a particle with an irregular shape. The relationships between D_m, D_a, and D_{ve} were outlined in Hudson et al. (75):

$$D_{ve} = D_a \sqrt{\chi \frac{\rho_o}{\rho_p} \frac{C_s(D_a)}{C_s(D_{ve})}} \tag{3.8}$$

$$D_{ve} = D_m \frac{C_s(D_{ve})}{\chi C_s(D_m)} \tag{3.9}$$

$$D_m = D_a \chi^{3/2} \sqrt{\frac{\rho_o}{\rho_p}} \frac{C_s(D_m)\sqrt{C_s(D_a)}}{C_s(D_{ve})^{3/2}} \tag{3.10}$$

where χ is the dynamic shape factor, ρ_o is the reference density (1 g/cm^3), ρ_p is the density of the particle, and $C_s(D_m)$, $C_s(D_a)$, and $C_s(D_{ve})$ are the Cunningham slip factors for the mobility, aerodynamic, and volume equivalent diameters, respectively. For spherical particles the dynamic shape factor, χ, is equal to one, and the volume equivalent diameter (D_{ve}) is equal to the measured mobility diameter (D_m).

Electron microscopy can also be used to confirm the mobility diameters from the SMPS measurements and aerodynamic diameters recorded using the APS. The microscopy techniques can not only give information on the size of the aggregates/agglomerates that are being introduced in the exposure but also give visual insight into the particle interactions. In a recent inhalation toxicology study, TEM was used to investigate the agglomeration of different sized TiO_2 nanoparticles, 5 and 21 nm. It was shown that although the aerosols that formed from the two different size nanoparticles had similar mobility diameters, the density and packing of the two different size nanoparticle agglomerates were in fact quite different (73). The images of the agglomerates are shown in Fig. 3.7 to highlight these results since differences in

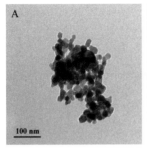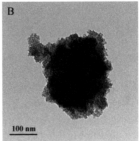

FIGURE 3.7   TEM images of 21 (a) and 5 (b) nm TiO$_2$ agglomerates generated in a whole-body exposure chamber are shown. The mobility diameters of the agglomerates are similar but there are distinct differences in the packing of the particle agglomerates for the two different size nanoparticles. Reproduced from Reference 73 with permission from *Informa Healthcare*.

the agglomeration states of the nanoparticles, whether tightly packed or consisting of void space, could have a significant effect on the inflammatory response.

Other techniques that can be useful in measuring nanoparticle agglomerates that are immobilized on a substrate include scanning probe microscopy techniques such as atomic force microscopy (AFM) and scanning tunneling microscopy (STM). Furthermore, for nanoparticles suspended in air, single particle mass spectrometry is unique as it is designed to measure particle size, typically the aerodynamic or agglomerated size, and is usually coupled with chemical composition data. However, only a few single-particle mass spectrometers (MS) are capable of measuring particles below 100 nm in diameter. The need for additional nanoparticle studies has been recently discussed in a review by Prather et al. (76).

A list of the instrumentation discussed in Sections 3.2.2 and 3.2.3 to characterize nanomaterials bulk, and surface properties as well as the agglomeration/aggregation states are summarized in Table 3.2. The information in the table can along with surface chemical studies, provide insight into the physicochemical properties of the nano-materials. By including these types of characterization methods in environmental health, safety, and toxicity studies of nanomaterials, there would potentially be a better understanding of nanoparticle toxicity as well as nanoparticle biocompatibility.

3.2.4   Testing Strategies and Commonly Used Markers for Inflammation and Response, the Need for Additional *In Vivo* Measurements of Nanoparticles

The continual growth of nanomaterial production has invoked the need for more standardized testing for effective comparisons. Many of the workshops, reports, and studies have highlighted specific testing strategies to quickly and efficiently evaluate the potential risk of new and existing nanomaterials (23, 27, 42, 77). Direct correlation of *in vitro* and *in vivo* studies have been unsuccessful (25, 54, 78). The different pathways for biological response can change as a function of the exposure method due

TABLE 3.2   Nanoparticle Characterization: Examples of Instrumentation and Methods

Bulk Characterization
  X-ray diffraction: crystallinity and phase
  Transmission electron microscopy: nanoparticle size and shape
  Scanning mobility particle sizer: size
  Size exclusion chromatography: size

Particle Aggregate/Agglomerate Size
  Dynamic light scattering: solution
  Scanning mobility particle sizer: aerosol
  Transmission and scanning electron microscopy: substrate deposited
  Atomic force microscopy: substrate deposited

Surface Properties
  X-ray photoelectron spectroscopy: surface composition—elemental and functional group characterization
  Attenuated total reflection-fourier transform infrared spectroscopy: surface composition—molecular nature of surface functional groups in air or water
  Braunner–Emmett–Teller analysis: surface area measurement)
  DLS surface potential: surface charge

Surface Reactivity
  Quantitative adsorption studies
  Dissolution measurements

to the specific properties of the nanomaterials. Currently, a consensus has been made to use tier-down testing approach (42, 43). The top tier consists of preliminary testing or screening of the targeted nanomaterial. *In vitro* and short-term *in vivo* exposures, consisting of both inhalation and instillation exposure methods, are recommended with the largest biological investigations on pulmonary effects. After the initial screening process, the remaining tiers would incorporate a more rigorous testing along with *in vitro* studies focused on specific cell lines that are targeted in the exposure models. The types of biological markers will depend on the nano-material that is being screened.

There are a large number of markers that can be used for monitoring biological response. The type of nanomaterial and response will dictate the experimental protocol used. One common technique used for recovery of pulmonary cells for cytologic analysis is the extraction of bronchoalveolar lavage (BAL) fluid. A saline wash is injected into the airways and alveolar region of the lungs for the recovery of inflammatory cells. Markers that are recovered from the BAL fluid include the total number of cells along with differentiation between macrophages, neutrophils, and lymphocytes, and the presence of increased number of cells is a possible sign of inflammation. Other markers found in the BAL fluid are total protein counts, which quantify the increased permeability of the BA-capillary barrier, and the activity of lactate dehydrogenase (LDH), which is an indicator of cytotoxicity. LDH is a cytoplasmic enzyme present in all cells of major organs and is rapidly released when the plasma membrane is damaged. There are also other markers for cytotoxicity that are not measured from the BAL fluid.

Cytokines are small secreted proteins that mediate and regulate immunity, inflammation, and hematopoiesis and are specific indicators of different types of cellular responses. The quantifying of cytokines must be limited to proteins produced *de novo* in response to an immune stimulus. Specific types of cytokines are categorized by the cell, which produces the protein, or their function; for example, lymphokines are produced by lymphocytes, monokines are produced by monocytes, chemokines are produced by cytokines with chemotactic activities, and interleukins are produced by one leukocyte and act on other leukocytes. Monitoring the cytokine levels in exposure studies can give valuable information as to what cells are targeted and what type of injury is occurring because of the specific function that each cytokine performs. Identifying specific cytokines can be difficult when their functions are synergistic or antagonistic in nature complicating the interpretation of the response. There are more than 18 cytokines that have been identified, and a list of the different types and the cells that secrete the proteins can be found elsewhere (80). Cytokines that are commonly examined in toxicity studies can be grouped into proinflammatory, anti-inflammatory, and cell proliferation and differentiation stimuli subsets. Measuring cytokine levels in conjunction with information from the BAL fluid leads to a better understanding of the types of cells being activated and the types of injuries induced by the introduction of the nanomaterial.

A study design for nanoparticle toxicity using the previously described characterization techniques as well as biological markers is shown in Fig. 3.8 as

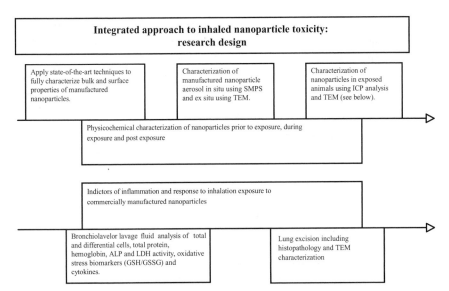

FIGURE 3.8 Research design for the integrated approach to determine toxicity of nanomaterials. Two parallel paths are shown: the physicochemical characterization of nanomaterials and biological assessment of the response induced by nanomaterials.

two parallel paths that need to be taken and are necessary for complete evaluation of the nanomaterials and their exposure response. The top section outlines the physicochemical characterization methods and the bottom section contains the toxicological evaluation methods that are needed for thorough examination. The physicochemical properties studied would ideally include bulk and surface characterization at three different time points: as received, during exposure, and postexposure (most difficult). Characterization of the nanomaterials as they are received is important, but understanding the particles may change during the aerosol generation or be processed in biological media possibly affecting the properties of the materials and consequently changing the cellular response is also important. The ability to characterize nanoparticles at these different endpoints can give valuable information about the processes that the materials may undergo in the model.

For the toxicological evaluation, combinations of different assays and markers are needed to gather enough information to accurately evaluate the cause of a cellular response. The nanoparticles' physicochemical properties as well as exposure route will dictate the biological fate after exposure forcing researchers to examine more than pulmonary effects. The deposition of nanoparticles in the lungs is not always the final destination of nanoparticles inhaled or instilled. Translocation of nanoparticles into the brain by passing through the blood brain barrier (BBB) (81, 82) has been found, and translocation to other organs can occur if they can transport into the blood stream (83, 84). The translocation of nanoparticles can be quantitatively measured in tissue analysis using an inductively coupled plasma (ICP) instrument (85). The information that is sought will direct the type of experimental studies conducted.

With the growing commercial use of new nanomaterials being developed, effective, fast, and affordable screening is needed on new and already available nanomaterials. One viable option that is getting considerable attention and has been mentioned in numerous reports is the development of *in vitro* studies to screen nanomaterials. The validity of these tests would create simpler, faster, and less expensive studies for screening compared to *in vivo* testing. The use of animals could be limited to more rigorous toxicity screening. Although *in vitro* tests can be useful in gathering important mechanistic information, correlation between *in vitro* and *in vivo* testing has not been obtained. Some of the major problems with comparing *in vitro* cell assays with complex systems were outlined by Sayes et al. (25) and include the particle dose issued, selection of cell types in the lung microenvironment related to single versus coculture systems, interactions of cells with particles in different biological media, time course effects (acute versus chronic), and end points for evaluation. It is obvious that there is a need for more extensive research on a broad range of nanomaterials to achieve ideal testing procedures, which many stem from problems associated with comparability between *in vitro* and *in vivo* testing, but *in vitro* testing will be necessary for the large amount of throughput screening that will be needed as the market for nanomaterials continues to grow.

3.2.5   Example of a Combined Characterization and Toxicological Study Design for Inhaled Nanomaterials and a Review of Some Recent Results

Because of the widespread use of TiO_2 in sunscreens, lotions, paints, pigments, and other coating applications (59) as well as the occupational hazard of inhaling suspended nanoparticles during production, transport or distribution, the inevitability for human exposure to these nanomaterials makes it necessary to study the biological effects caused by TiO_2 nanoparticles. Recently, the smallest commercially available titanium dioxide nanoparticles were used for inhalation and instillation exposures in murine models (73, 86, 87).

As mentioned in previous sections, full characterization of the TiO_2 nanoparticles were conducted determining size, surface properties, and dispersion of primary and aggregated nanoparticles for as received and as exposed to the model. A whole-body exposure chamber was used for the inhalation experiments. Experiments of different concentrations were run to examine the effects on the biological response and determine the appropriate dose metric. Gravimetric analysis on the outflow air from the chamber was also used to obtain mass concentrations. TEM images were taken of the aerosol inside the chamber for size and dispersion comparisons of the mobility and aerodynamic diameters from the SMPS and APS measurements.

Some of the characterization results for the reported average particle size (APS) 5 nm TiO_2 particles have been previously shown for XRD, DLS, and the TEM data. Summarizing the characterization, XRD was used to determine the phase of the particles to be purely anatase with a primary particle size of 3.5 ± 1 nm determined by counting 50 random nanoparticles using TEM images. TEM was also used to examine the aerosol generated in the whole-body chamber and showed tightly packed aerosols. ATR-FTIR and XPS spectra showed the TiO_2 nanoparticles had adsorbed water on the surface and were hydroxyl terminated, which is expected from metal oxide surfaces. Table 3.3 summarizes the characterization of the nanoparticles.

TABLE 3.3   Physicochemical Characterization Data of TiO_2 Nanoparticles

Property	Characterization
Crystalline or amorphous material	Crystalline
Phase	Anatase
Primary particle distribution	3.5 ± 1.0 nm
BET surface area	219 ± 3 m^2/g
Surface functionalization	O, O–H, and H_2O
Aerosol size distribution GM [GSD]	123 nm [1.6][a]
	120 nm [1.6][b]
	128 nm [1.7][c]

GM, geometric mean; GSD, geometric standard deviation. Reprinted from Ref. 86 with permission from *Environmental Health Perspectives*.

[a]Acute exposure, low concentration.

[b]Acute exposure, high concentration.

[c]Subacute exposure.

Two different inhalation studies, acute and subacute, were conducted on mice (86). The subacute study ran for two weeks exposing the mice 4 h a day. Mice were necropsied at 0, 1, 2, and 3 weeks postexposure, and the acute study was done for 1 day for 4 h and mice were necropsied immediately. In both studies, the 5 nm particles induced statistically significant but moderate responses in the murine models. The macrophages were the only cells measured in the BAL fluid to be enhanced by the 5 nm TiO_2 in the subacute study and their levels returned to baseline three weeks postexposure. Proinflammatory cytokines, IL-1 and IL-6, were monitored along with the cell proliferation cytokine INF-γ, which activates macrophages, in the subacute studies and all levels were elevated immediately postexposure but returned to baseline at 1 week postexposure. The 5 nm TiO_2 particles showed a moderate response in the murine model, but all specimens returned to baseline 3 weeks postexposure.

The moderate response of the 5 nm TiO_2 particles was compared with manufactured 21 nm TiO_2 particles under the same experimental conditions for acute inhalation studies (73). Results from acute inhalation studies done at two different concentrations and necropsied after 4 and 24 h showed very limited variation between the 5 and 21 nm particles. Mice necropsied 4 h after exposure did show significant increase in the number of macrophages present for the 5 and 21 nm particles but only at the high dose. Mice necropsied 24 h postexposure, which were exposed to similar high doses of nanoparticles, did not show statistically any significant increase in any of the markers examined in the study. The lack of a size-dependent effect was also seen in another acute pulmonary study of instilled 3 and 20 nm TiO_2 particles (87). Although there were not significant changes in biological response, there was a physical difference in the agglomeration states of the two different size nanoparticles that was shown in Fig. 3.7. The 5 nm particles form more dense agglomerates; in contrast, the 21 nm particles formed more loosely packed agglomerates with void space visible in the agglomerate, and individual particles were more easily discernable. This difference in agglomeration morphologies could lead to possible suppression of biological response of the 5 nm nanoparticles due to the stronger particle–particle interactions causing them to have less projected surface area and therefore a decrease in nanoparticle–cellular interactions. The complexity of exposure studies involving nanomaterials shows a necessity for the integrated approach as outlined in this chapter.

3.3 FUTURE ISSUES AND NEEDS

The continual need for risk assessment of manufactured nanomaterials is necessary to ensure safety of workers in occupational settings, the broader public, and ecosystems. With the introduction of large quantities of new nanomaterials into commercial markets, methods that are more effective at quickly screening nanomaterials need to be a priority along with more intensive studies of nanomaterials that have been identified as potentially being the most toxic. Studies on susceptible populations are also important as it is not clear how the response to exposure to nanomaterials may change for those with lowered immune systems. Because of the changes in physical and chemical properties of nanomaterials, there needs to be an assessment of the classical approaches of toxicology and the types of assays that are applicable for nanomaterial-based products. Continual advances in measuring and modeling clearance, deposition, and translocation of these particles in the lungs and into other primary organs are also important to more accurately determine the retained lung burden and nanoparticle processing that occurs in the body after exposure. Important correlations between physicochemical properties with biological responses can only be accomplished through collaboration of investigators in different scientific disciplines. This endeavor is interdisciplinary by its very nature and it is quite clear that a research team approach is needed to tackle and address these issues.

ACKNOWLEDGMENTS

Although the research described in this chapter has been funded in part by the Environmental Protection Agency through grant number EPA RD-83171701-0 to VHG, it has not been subjected to the Agency's required peer and policy review and therefore does not necessarily reflect the views of the Agency and no official endorsement should be inferred. This material is based upon work partially supported by the National Science Foundation under Grant CHE0639096 and the National Institutes of Health under Grant R01OH009448. Any opinions, findings, and conclusions or recommendations expressed in this chapter are those of the authors and do not necessarily reflect the views of the National Science Foundation. The authors would like to thank Professors Peter Thorne and Patrick O'Shaughnessy and Dr. Andrea Adamcakova-Dodd for their integral roles in the toxicity studies. Some of the research presented here was supported by the Department of Defense through the National Defense Science and Engineering Graduate Fellowship (NDSEG) Program to Sherrie Elzey.

REFERENCES

1. Anselmann, R. Nanoparticles and nanolayers in commercial applications. *J Nanopart Res* 2001;3:329.
2. Doumanidis, H. The Nanomanufacturing programme at the National Science Foundation. *Nanotechnology* 2002;13:248.
3. Emerich, D.F., Thanos, C.G. Nanotechnology and medicine. *Expert Opin Biol Ther* 2003;3:655.
4. Falkenhagen, D. Small Particles in Medicine. *Artif Organs* 1995;19:792.
5. Lowe, T. The revolution in nanometals. *Adv Mater Process* 2002;160:63.
6. McAllister, K., Sazani, P., Adam, M., Cho, M.J., Rubinstein, M., Samulski, R.J., DeSimone, J.M. Polymeric nanogels produced via inverse microemulsion polymerization as potential gene and antisense delivery agents. *J Am Chem Soc* 2002;124:15198.
7. Maynard, A.D. *Nanotechnology: A Research Strategy for Addressing risk.* Woodrow Wilson International Center for Scholars. 2006.

8. Alvarez, P.J. Nanotechnology in the environment - The good, the bad, and the ugly. *J Environ Eng Asce* 2006;132:1233.

9. Wiesner, M.R., Lowry, G.V., Alvarez, P., Dionysiou, D., Biswas, P. Assessing the risks of manufactured nanomaterials. *Environ Sci Technol* 2006;40:4336.

10. Richardson, S.D. Water analysis: Emerging contaminants and current issues. *Anal Chem* 2007;79:4295.

11. Phares, D.J., Rhoads, K.P., Johnston, M.V., Wexler, A.S. Size-resolved ultrafine particle composition analysis - 2. Houston. *J Geophys Res Atmos* 2003;108:8420.

12. Bang, J.J., Murr, L.E. Collecting and characterizing atmospheric nanoparticles. *J Miner Met Mat Soc* 2002;54:28.

13. Oberdorster, G., Oberdorster, E., Oberdorster, J. Nanotoxicology: An emerging discipline evolving from studies of ultrafine particles. *Environ Health Perspect* 2005;113:823.

14. Wilson, R., Spengler, J. Emissions, dispersion, and concentration of particles. In: Wilson, R., Spengler, J., editors. *Particles in Our Air: Concentrations and Health Effects.* Havard University Press; 1996, p. 41.

15. Daigle, C.C., Chalupa, D.C., Gibb, F.R., Morrow, P.E., Oberdorster, G., Utell, M.J., Frampton, M.W. Ultrafine particle deposition in humans during rest and exercise. *Inhal Toxicol* 2003;15:539.

16. Oberdorster, G., Stone, V., Donaldson, K. Toxicology of nanoparticles: A historical perspective. *Nanotoxicology* 2007;1:2.

17. Guzman, K.A.D., Taylor, M.R., Banfield, J.F. Environmental risks of nanotechnology: National nanotechnology initiative funding, 2000–2004. *Environ Sci Technol* 2006;40:1401.

18. Lecoanet, H.F., Bottero, J.Y., Wiesner, M.R. Laboratory assessment of the mobility of nanomaterials in porous media. *Environ Sci Technol* 2004;38:5164.

19. Borm, P., Klaessig, F.C., Landry, T.D., Moudgil, B., Pauluhn, J., Thomas, K., Trottier, R., Wood, S. Research strategies for safety evaluation of nanomaterials, Part V: Role of dissolution in biological fate and effects of nanoscale particles. *Toxicol Sci* 2006;90:23.

20. Brunner, T.J., Wick, P., Manser, P., Spohn, P., Grass, R.N., Limbach, L.K., Bruinink, A., Stark, W.J. In vitro cytotoxicity of oxide nanoparticles: Comparison to asbestos, silica, and the effect of particle solubility. *Environ Sci Technol* 2006;40:4374.

21. Chen, X., Tam, U.C., Czlapinski, J.L., Lee, G.S., Rabuka, D., Zettl, A., Bertozzi, C.R. Interfacing carbon nanotubes with living cells. *J Am Chem Soc* 2006;128:6292.

22. Karakoti, A.S., Hench, L.L., Seal, S. The potential toxicity of nanomaterials - The role of surfaces. *J Miner Met Mat Soc* 2006;58:77.

23. Maynard, A.D., Aitken, R.J., Butz, T., Colvin, V., Donaldson, K., Oberdorster, G., Philbert, M.A., Ryan, J., Seaton, A., Stone, V., Tinkle, S.S., Tran, L., Walker, N.J., Warheit, D.B. Safe handling of nanotechnology. *Nature* 2006;444:267.

24. Oberdorster, G., Maynard, A., Donaldson, K., Castranova, V., Fitzpatrick, J., Ausman, K., Carter, J., Karn, B., Kreyling, W., Lai, D., Olin, S., Monteiro-Riviere, N., Warheit, D., Yang, H. Principles for characterizing the potential human health effects from exposure to nanomaterials: Elements of a screening strategy. *Part Fibre Toxicol* 2005. Doi:10.1186/1743-8977-2-8.

25. Sayes, C.M., Wahi, R., Kurian, P.A., Liu, Y.P., West, J.L., Ausman, K.D., Warheit, D.B., Colvin, V.L. Correlating nanoscale titania structure with toxicity: A cytotoxicity and

inflammatory response study with human dermal fibroblasts and human lung epithelial cells. *Toxicol Sci* 2006;92:174.

26. Warheit, D.B., Brock, W.J., Lee, K.P., Webb, T.R., Reed, K.L. Comparative pulmonary toxicity inhalation and instillation studies with different TiO$_2$ particle formulations: Impact of surface treatments on particle toxicity. *Toxicol Sci* 2005;88:514.

27. Warheit, D.B., Webb, T.R., Colvin, V.L., Reed, K.L., Sayes, C.R. Pulmonary bioassay studies with nanoscale and fine-quartz particles in rats: Toxicity is not dependent upon particle size but on surface characteristics. *Toxicol Sci* 2007;95:270.

28. Warheit, D.B., Webb, T.R., Sayes, C.M., Colvin, V.L., Reed, K.L. Pulmonary instillation studies with nanoscale TiO$_2$ rods and dots in rats: Toxicity is not dependent upon particle size and surface area. *Toxicol Sci* 2006;91:227.

29. Chen, K.L., Elimelech, M. Aggregation and deposition kinetics of fullerene (C-60) nanoparticles. *Langmuir* 2006;22:10994.

30. Skebo, J.E., Grabinski, C.M., Schrand, A.M., Schlager, J.J., Hussain, S.M. Assessment of metal nanoparticle agglomeration, uptake, and interaction using high-illuminating system. *Int J Toxicol* 2007;26:135.

31. Chen, K.L., Elimelech, M. Influence of humic acid on the aggregation kinetics of fullerene (C-60) nanoparticles in monovalent and divalent electrolyte solutions. *J Colloid Interface Sci* 2007;309:126.

32. Huang, F., Gilbert, B., Zhang, H.H., Banfield, J.F. Reversible, surface-controlled structure transformation in nanoparticles induced by an aggregation state. *Phys Rev Lett* 2004;92: Article Number 155501.

33. Moskovits, M., Vlckova, B. Adsorbate-induced silver nanoparticle aggregation kinetics. *J Phys Chem B* 2005;109:14755.

34. Borm, P.J.A. Particle toxicology: From coal mining to nanotechnology. *Inhal Toxicol* 2002;14:311.

35. Colvin, V.L. The potential environmental impact of engineered nanomaterials. *Nature Biotechnol* 2003;21:1166.

36. Dagani, R. Nanomaterials: Safe or unsafe? *Chem Eng News* 2003;81:30.

37. Gogotsi, Y. How safe are nanotubes and other nanofilaments? *Mater Res Innov* 2003;7:192.

38. Hardman, R. A toxicologic review of quantum dots: Toxicity depends on physicochemical and environmental factors. *Environ Health Perspect* 2006;114:165.

39. Kleiner, K., Hogan, J. How safe is nanotech? *New Sci* 2003;177:14.

40. Masciangioli, T.M., Zhang, W.X. Environmental technologies at the nanoscale. *Abstr Pap Am Chem Soc* 2003;225:U952.

41. Nel, A., Xia, T., Madler, L., Li, N. Toxic potential of materials at the nanolevel. *Science* 2006;311:622.

42. Balbus, J., Maynard, A.D., Colvin, V.L., Castranova, V., Daston, G.P., Denison, R.A., Dreher, K.L., Goering, P.L., Goldberg, A.M., Kulinowski, K.M., Monteiro-Riviere, N.A., Oberdörster, G., Omenn, G.S., Pinkerton, G.E., Ramos, K.S., Rest, K.M., Sass, J.B., Silbergeld, E.K., Wong, B.A. Hazard Assessment for Nanoparticles—Report from an Interdisciplinary Workshop. *Environ Health Perspect* 2007;115:1654.

43. Warheit, D.B., Borm, P.J.A., Hennes, C., Lademann, J. Testing strategies to establish the safety of nanomaterials: Conclusions of an ECETOC workshop. *Inhal Toxicol* 2007;19:631.

44. Powers, K.W., Brown, S.C., Krishna, V.B., Wasdo, S.C., Moudgil, B.M., Roberts, S.M. Research strategies for safety evaluation of nanomaterials. Part VI. Characterization of nanoscale particles for toxicological evaluation. *Toxicol Sci* 2006;90:296.

45. Thomas, K., Aguar, P., Kawasaki, H., Morris, J., Nakanishi, J., Savage, N. Research strategies for safety evaluation of nanomaterials, Part VIII: International efforts to develop risk-based safety evaluations for nanomaterials. *Toxicol Sci* 2006;92:23.

46. Thomas, T., Thomas, K., Sadrieh, N., Savage, N., Adair, P., Bronaugh, R. Research strategies for safety evaluation of nanomaterials, part VII: Evaluating consumer exposure to nanoscale materials. *Toxicol Sci* 2006;91:14.

47. Tsuji, J.S., Maynard, A.D., Howard, P.C., James, J.T., Lam, C.W., Warheit, D.B., Santamaria, A.B. Research strategies for safety evaluation of nanomaterials, part IV: Risk assessment of nanoparticles. *Toxicol Sci* 2006;89:42.

48. Balshaw, D.M., Philbert, M., Suk, W.A. Research strategies for safety evaluation of nanomaterials, part III: Nanoscale technologies for assessing risk and improving public health. *Toxicol Sci* 2005;88:298.

49. Holsapple, M.P., Farland, W.H., Landry, T.D., Monteiro-Riviere, N.A., Carter, J.M., Walker, N.J., Thomas, K.V. Research strategies for safety evaluation of nanomaterials, part II: Toxicological and safety evaluation of nanomaterials, current challenges and data needs. *Toxicol Sci* 2005;88:12.

50. Thomas, K., Sayre, P. Research strategies for safety evaluation of nanomaterials, part I: Evaluating the human health implications of exposure to nanoscale materials. *Toxicol Sci* 2005;87:316.

51. U.S. Environmental Protection Agency. *Nanotechnology White Paper.* 2007.

52. Lucas, E., Decker, S., Khaleel, A., Seitz, A., Fultz, S., Ponce, A., Li, W.F., Carnes, C., Klabunde, K.J. Nanocrystalline metal oxides as unique chemical reagents/sorbents. *Chem Eur J* 2001;7:2505.

53. Powers, K.W., Palazuelos, M., Moudgil, B.J., Roberts, S.M. Characterization of the size, shape, and state of dispersion of nanoparticles for toxicological studies. *Nanotoxicology* 2007;1:42.

54. Teeguarden, J.G., Hinderliter, P.M., Orr, G., Thrall, B.D., Pounds, J.G. Particokinetics in vitro: Dosimetry considerations for in vitro nanoparticle toxicity assessments. *Toxicol Sci* 2007;95:300.

55. Moss, O.R., Wong, V.A. When nanoparticles get in the way: Impact of projected area on in vivo and in vitro macrophage function. *Inhal Toxicol* 2006;18:711.

56. Sager, T.M., Porter, D.W., Robinson, V.A., Lindsley, W.G., Schwegler-Berry, D.E., Castranova, V. Improved method to disperse nanoparticles for in vitro and in vivo investigation of toxicity. *Nanotoxicology* 2007;1:118.

57. Atkins, P., de Paula, J. *Physical Chemistry,* 7th ed. New York: W.H. Freeman and Company; 2002.

58. Kakkar, R., Kapoor, P.N., Klabunde, K.J. Theoretical study of the adsorption of formaldehyde on magnesium oxide nanosurfaces: Size effects and the role of low-coordinated and defect sites. *J Phys Chem B* 2004;108:18140.

59. Chen, X., Mao, S.S. Titanium dioxide nanomaterials: Synthesis, properties, modifications, and applications. *Chem Rev* 2007;107:2891.

60. Jenkins, R., Snyder, R.L. *Introduction to X-ray Powder Diffractometry.* New York: John Wiley & Sons, Inc.; 1996.

61. Pease, L.F., Tsai, D.H., Zangmeister, R.A., Zachariah, M.R., Tarlov, M.J. Quantifying the surface coverage of conjugate molecules on functionalized nanoparticles. *J Phys Chem C* 2007;111:17155.

62. Ito, A., Kuga, Y., Honda, H., Kikkawa, H., Horiuchi, A., Watanabe, Y., Kobayashi, T. Magnetite nanoparticle-loaded anti-HER2 immunoliposomes for combination of antibody therapy with hyperthermia. *Cancer Lett* 2004;212:167.

63. Ho, J., Kournikakis, B., Gunning, A., Fildes, J. Sub-Micron Aerosol Characterization of Water by a Differential Mobility Particle Sizer. *J Aerosol Sci* 1988;19:1425.

64. Krames, J., Buttner, H., Ebert, F. Submicron Particle Generation by Evaporation of Water Droplets. *J Aerosol Sci* 1991;22:S15.

65. Liu, F.K. SEC characterization of Au nanoparticles prepared through seed-assisted synthesis. *Chromatographia* 2007;66:791.

66. Otto, D.P., Vosloo, H.C.M., de Villiers, M.M. Application of size exclusion chromatography in the development and characterization of nanoparticulate drug delivery systems. *J Liq Chromatogr Relat Technol* 2007;30:2489.

67. Hug, S.J., Sulzberger, B. In-Situ Fourier-Transform Infrared Spectroscopic Evidence for the Formation of Several Different Surface Complexes of Oxalate on Tio2 in the Aqueous-Phase. *Langmuir* 1994;10:3587.

68. Dobson, K.D., McQuillan, A.J. In situ infrared spectroscopic analysis of the adsorption of aliphatic carboxylic acids to TiO_2, ZrO_2, Al_2O_3, and Ta_2O_5 from aqueous solutions. *Spectrochim Acta A* 1999;55:1395.

69. Yoon, T.H., Johnson, S.B., Musgrave, C.B., Brown, G.E. Adsorption of organic matter at mineral/water interfaces: I. ATR-FTIR spectroscopic and quantum chemical study of oxalate adsorbed at boehmite/water and corundum/water interfaces. *Geochim Cosmochim Acta* 2004;68:4505.

70. Pettibone, J.M., Baltrusaitis, J., Grassian, V.H. Chemical properties of oxide nanoparticles: surface adsorption studies from gas- and liquid-phase environments. In: Rodriguez, J.A., Fernandez-Garcia, M., editors. *Synthesis, Properties, and Applications of Oxide Nano-materials*. Hoboken, New Jersey: John Wiley and Sons, Inc., 2007. pp. 335.

71. Knaapen, A.M., Shi, T.M., Borm, P.J.A., Schins, R.P.F. Soluble metals as well as the insoluble particle fraction are involved in cellular DNA damage induced by particulate matter. *Mol Cell Biochem* 2002;234:317.

72. ASTM International Committee E56 on Nanotechnology. ASTM E2456-06 Standard Terminology for Nanotechnology is available at www.astm.org

73. Grassian, V.H., Adamcakova-Dodd, A., Pettibone, J.M., O'Shaughnessy, P.T., Thorne, P.S. Inflammatory response of mice to manufactured titanium dioxide nanoparticles: comparison of size effects through different exposure routes. *Nanotoxicology* 2007;1:211.

74. Schins, R.P.F., Duffin, R., Hohr, D., Knaapen, A.M., Shi, T.M., Weishaupt, C., Stone, V., Donaldson, K., Borm, P.J.A. Surface modification of quartz inhibits toxicity, particle uptake, and oxidative DNA damage in human lung epithelial cells. *Chem Res Toxicol* 2002;15:1166.

75. Hudson, P.K., Gibson, E.R., Young, M.A., Kleiber, P.D., Grassian, V.H. A newly designed and constructed instrument for coupled infrared extinction and size distribution measurements of aerosols. *Aerosol Sci Technol* 2007;41:701.

76. Prather, K.A., Hatch, C.D., Grassian, V.H. (July 2008) Analysis of Atmospheric Aerosols. *Annu Rev Anal Chem* 2008.

77. Royal Society. Nanoscience and nanotechnologies: opportunities and uncertainties; RS Policy Document 19/04, 2004.

78. Seagrave, J., McDonald, J.D., Mauderly, J.L. In vitro versus in vivo exposure to combustion emissions. *Experiment Toxicol Pathol* 2005;57:233.

79. Osier, M., Oberdorster, G. Intratracheal inhalation vs intratracheal instillation: Differences in particle effects. *Fundam Appl Toxicol* 1997;40:220.

80. Decker, J.M. *Immunology.* University of Arizona; 2007. Available at http://microvet. arizona.edu/Courses/MIC419/Tutorials/cytokines.html.

81. Oberdorster, G., Sharp, Z., Atudorei, V., Elder, A., Gelein, R., Kreyling, W., Cox, C. Translocation of inhaled ultrafine particles to the brain. *Inhal Toxicol* 2004;16:437.

82. Elder, A., Gelein, R., Silva, V., Feikert, T., Opanashuk, L., Carter, J., Potter, R., Maynard, A., Ito, Y., Finkelstein, J., Oberdorster, G. Translocation of inhaled ultrafine manganese oxide particles to the central nervous system. *Environ Health Perspect* 2006;114:1172.

83. Chen, Z., Meng, H.A., Xing, G.M., Chen, C.Y., Zhao, Y.L., Jia, G.A., Wang, T.C., Yuan, H., Ye, C., Zhao, F., Chai, Z.F., Zhu, C.F., Fang, X.H., Ma, B.C., Wan, L.J. Acute toxicological effects of copper nanoparticles in vivo. *Toxicol Lett* 2006;163:109.

84. Yu, L.E., Yung, L.L., Ong, C., Tan, Y., Balasubramaniam, K.S., Hartono, D., Shui, G., Wenk, M.R., Ong, W. Translocation and effects of gold nanoparticles after inhalation exposure in rats. *Nanotoxicology* 2007;1:235.

85. Chen, Z., Meng, H., Yuan, H., Xing, G.M., Chen, C.Y., Zhao, F., Wang, Y., Zhang, C.C., Zhao, Y.L. Identification of target organs of copper nanoparticles with ICP-MS technique. *J Radioanal Nucl Chem* 2007;272:599.

86. Grassian, V.H., O'Shaughnessy, P.T., Adamcakova-Dodd, A., Pettibone, J.M., Thorne, P.S. Inhalation exposure study of titanium dioxide nanoparticles with a primary particle size of 2 to 5 nm. *Environ Health Perspect* 2007;115:397.

87. Li, J.A., Li, Q.N., Xu, J.Y., Li, J., Cai, X.Q., Liu, R.L., Li, Y.J., Ma, J.F., Li, W.X. Comparative study on the acute pulmonary toxicity induced by 3 and 20 nm TiO_2 primary particles in mice. *Environ Toxicol Pharmacol* 2007;24:239.

PART II

FATE AND TRANSPORT OF NANOMATERIALS IN THE ENVIRONMENT

Properties of Commercial Nanoparticles that Affect Their Removal During Water Treatment

PAUL WESTERHOFF, YANG ZHANG, JOHN CRITTENDEN, and YONGSHENG CHEN

Department of Civil and Environmental Engineering, Arizona State University, Box 5306, Tempe, AZ 85287-5306, USA

4.1 INTRODUCTION

The fate of engineered nanoparticles in aquatic (natural or engineered) systems depends upon their size, number concentration, surface charge, and type of material. Characteristics of natural water systems, including pH, ion composition, and ionic strength, as well as constituents such as natural organic matter (NOM), affect the surface charge of nanoparticles and thus their fate in aquatic systems [1]. Natural nanoparticles are already present in surface and groundwaters [2]. For example, freshwaters contain approximately 10^{13} particles with a diameter of 10 nm/L. [3]. Engineered nanoparticles may enter aquatic systems via direct discharge, wastewater effluents, runoff, atmospheric deposition, or other processes. This chapter examines relationships between the characteristics of nanoparticles and their behavior (i.e., aggregation) in aquatic systems. Specifically, this chapter focuses on nanoparticle properties that affect their removal during potable water treatment. The chapter is primarily a literature review, but includes some of the authors' experimental results to demonstrate key points.

4.2 NANOPARTICLE PROPERTIES

4.2.1 Types of Nanoparticles

This chapter considers three types of engineered nanoparticles: (1) metals and metal oxides, (2) metal-core quantum dots (QDs), and (3) carbonaceous fullerenes

Nanoscience and Nanotechnology Edited by Vicki H. Grassian

TABLE 4.1 Representative Properties of Three Classes of Engineered Nanoparticles

	Metal or Metal Oxide Nanoparticles	Quantum Dots	Carbonaceous Nanoparticles
Examples	Silver, iron, gold, iron oxide, titanium dioxide	Cadmium–telluride, gold–silica with carboxyl functional groups	Fullerenes, fullerols
Primary particle size reported by manufacturers	10 to >100 nm	3 to 10 nm	~0.5 nm (water soluble aggregates ~75 nm)
Net surface charge at pH 7	Negative, neutral, or positive	Usually negative	Negative

(Table 4.1). These nanoparticles are discussed because their properties vary widely and because their commercial applications are such that the particles have the potential to enter waterways. First, metal oxide nanoparticles (e.g., titanium dioxide, zinc oxide, and iron oxide) are widely used in paints, pigments, sunscreens, and other products. Some metal nanoparticles (e.g., silver) serve as antibacterial agents in clothing products, while elemental iron is used to dehalogenate pollutants in an environmental treatment process. Metal and metal oxide nanoparticles are closely related to natural colloids (e.g., hematite) that have been well studied as "model" materials during water treatment research (4, 5). Second, quantum dots are being developed for biomedical applications, pigments, sensor platforms, and other applications. Quantum dots contain a metal core and/or metal coating; organic capping ligands are chemically bound to the metal surface via thiols or other functional groups. The capping ligands ensure that the quantum dots become stable colloidal particles. Carbonaceous nanoparticles, including C_{60} (i.e., fullerene) and functionalized C_{60} (i.e., fullerol), have diverse uses including in cosmetics, biomedical applications, and electronic applications.

4.2.2 Particle Size

Nanoparticles are often defined as having at least one dimension less than 100 nm. The materials identified in Table 4.1 are roughly spherical, so particles with diameters less than 100 nm (0.1 mm) will operationally be defined in this chapter as nanoparticles. While transmission (TEM) or scanning (SEM) electron microscopy can image materials less than 100 nm in solid samples, accurate determination of nanoparticle size and concentration in aqueous samples can be complicated. Optical and electrical sensing systems for particle counting have traditionally required certain particle sizes (e.g., detection above 500 nm only) or small numbers of particles in solution (e.g., <10,000 particles per liter are needed for detection) that are suitable only for ultrapure water applications (e.g., semiconductor manufacturing) and not natural aquatic systems that contain trillions of particles per milliliter smaller than a few hundred nanometers.

Due to their small sizes, colloidal particles exhibit two well-known behaviors, the Tyndall effect and Brownian motion. For a colloidal system, because the wavelength of light that passes through the system is larger than the sizes of the colloidal particles, light scattering occurs. This is called the Tyndall effect (6). Generally, the Tyndall effect can be used to identify the presence of colloidal particles in a system. In addition, colloidal particles can move randomly and do not settle quickly under gravity, a phenomenon discovered by Robert Brown in 1827 and named Brownian motion (7, 8). According to Brownian motion, colloidal particles can diffuse without consumption of energy. The movement rates of colloidal particles in "Brownian motion" are related to their sizes, the temperature, and the viscosity of water. Dynamic light scattering (DLS) uses Brownian motion and the Tyndall effect to determine the sizes of colloids suspended in liquid. DLS is also often used to establish the size of particles in solution, although it tends to be better suited for homogeneous than heterogeneously sized particle solutions.

Figure 4.1 presents DLS results and electron micrographs of representative natural nanoparticles. The hematite appears as discrete nanoparticles in the SEM and has a

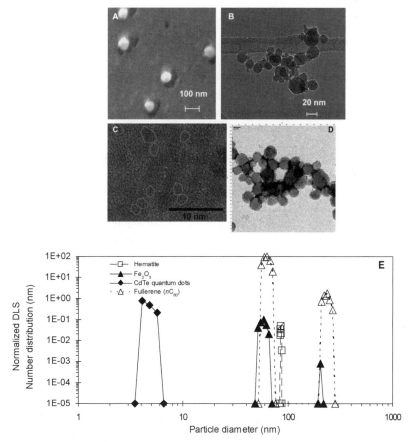

FIGURE 4.1 Electron micrograph of (a) lab-synthesized hematite, (b) commercially available iron oxide nanoparticles, (c) quantum dots, and (d) aggregated C_{60}. DLS for the four nanoparticles are shown in the bottom plot (e).

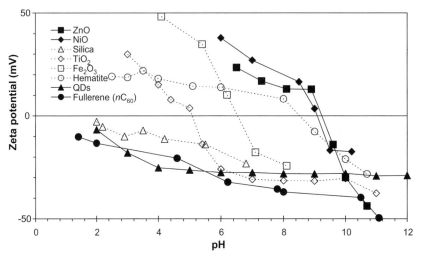

FIGURE 4.2   Zeta potentials of nanoparticles in 10 mM KNO_3 solution as a function of pH. *Note*: TiO_2 represents two types of TiO_2 (powder and 5 wt% suspension) that have similar zeta potentials as a function of pH.

DLS weighted average diameter of 85 nm, which is consistent with literature reports (4, 5, 9). The commercial iron oxide is typical of most commercial metal and metal oxide nanoparticles. Commercial nanoparticles purchased as either dry powders or in liquid solutions contain aggregated nanoparticles. Upon addition of these commercial nanoparticles to water, the nanoparticles remain aggregated. Sonication cannot disaggregate the commercial nanoparticles in water. For the commercial iron oxide illustrated in Fig. 4.2, the primary particle size is nanoscale, on the order of 5–25 nm, but the aggregated form is far greater in size and may not be considered a nanoparticle by the operational definition cutoff of 100 nm. Similar phenomenon is observed with commercial TiO_2 nanoparticles (10) and other commercial nanoparticles (11–16), which showed that such so-called nanoparticles were aggregates composed of primary nanoparticles with sizes less than 100 nm. We believe that commercial metal and metal oxide nanoparticles aggregate during their manufacturing processes, but that they can be manufactured as discrete nanoparticles by adding functional groups to the surface (e.g., citrate).

Quantum dots are semiconductor nanocrystals that can be as small as 1–10 nm (17, 18) and are generally composed of 1000–100,000 atoms. Quantum dots usually have organically functionalized capping ligands (e.g., thioglycolate ($HS\text{-}CH_2\text{-}COO^-$)), resulting in stable colloidal particles with metal cores (e.g., CdTe).

Single fullerenes are on the order of 0.5 nm in diameter. Although the hydrophobicity of C_{60} fullerene is well known (solubility $\approx 1.3 \times 10^{-13}$ μg/L (19)), its low water solubility can be overcome by proper derivatization with hydrophilic functional groups, by oxidation or by the formation of stable aqueous nC_{60} through stirring or sonication of C_{60} in a mixture of organic solvents and water (20–23). Stable colloidal C_{60} aggregates (nC_{60}) are on the order of 50–100 nm in diameter.

4.2.3 Surface Charge

The net charge on nanoparticles can be estimated by measuring particle movement in an electric field to determine the zeta potential, the electrokinetic charge on the shear plane around a particle. Zeta potential values are usually less than the actual surface potential (i.e., surface charge), but can be used to assess the energy of interactions between particles in solution. Figure 4.2 presents the zeta potentials for several engineered nanoparticles as measured in our laboratories (ZetaPlus Analyzer by Brookhaven Instruments Corp.). Over the pH range of most environmental waters ($5.5 < pH < 8.5$), engineered nanoparticles may be positive, neutral, or negatively charged. However, water also contains natural organic matter that affects surface charge, which usually results in particles having negative zeta potentials. For example, hematite has a zeta potential of $+19$ mV in a 10-mM KCl solution at pH 8.0, but the addition of only 0.5 mg/L dissolved organic carbon changes its zeta potential at the same pH to -37 mV.

4.2.4 Quantification of Nanoparticles in Water

Organic and inorganic colloids are ubiquitous in natural waters, and many different techniques have been employed to quantify their presence. Common separation methods include filtration/ultrafiltration (24, 25), centrifugation/ultracentrifugation (26), electrophoresis/capillary electrophoresis (27), and size exclusion chromatography (24, 25). Recently, additional technologies such as field flow fractionation (28) and split flow thin (SPLITT) fractionation (2) have been developed to fractionate colloidal particles in natural water without the potential aggregation problems associated with filtration and centrifugation (29).

Published research often used a combination of different techniques to determine the characteristics of colloidal particle, such as colloid concentration, size distribution, mineral composition, and surface properties (25, 27, 28). For instance, UV detection, fluorescence spectroscopy, and gravimetric analysis are used to determine colloid concentration. Light scattering, photon correlation spectroscopy, and fractionation procedures such as ultrafiltration and size exclusion chromatography can be utilized for size distribution measurements. Atomic force microscopy, scanning electron microscopy, transmission electron microscopy, small angle neutron scattering, X-ray diffraction, infrared spectroscopy, and inductively coupled plasma mass spectrometry provide information about mineral constituents and general surface properties (30). While these techniques are mature for the detection of natural nanoparticles, they have not been applied to engineered nanoparticles. Most studies on engineered nanoparticles have occurred in controlled solutions and at concentrations significantly higher than would be expected in the environment. Consequently, these studies measure the primary metal by either atomic absorption or ion-coupled plasma analysis methods (for metal oxide, metal, and quantum dot nanoparticles) or UV–Vis light scattering (for fullerenes and carbon nanotubes). Recently our group demonstrated that solid-phase extraction of nC_{60} was possible and that fullerenes could be quantified to submicrogram per liter levels using liquid chromatography mass spectroscopy (31).

4.3 NANOPARTICLE REMOVAL MECHANISMS DURING WATER TREATMENT

In general, water treatment plants (WTPs) are designed to effectively remove particulates using a series of chemical and physical processes (Fig. 4.3). Metal salt coagulants (e.g., hydrous aluminum sulfate, or alum) are added to destabilize negatively charged particles in water and precipitate new particles (e.g., amorphous aluminum hydroxide) that sorb or enmesh pollutants, thus facilitating their removal during sedimentation or filtration. The process of flocculation involves mixing chemically destabilized water to promote aggregation of smaller particles into larger particles that can settle out of solution. Particles that do not settle out can be removed during granular or membrane filtration. In granular filtration, chemically destabilized particles (those having near-neutral net surface charge) collide and chemically attach to sand grains rather than being removed by "straining" mechanisms at the top of the filter. The collision mechanisms between the particles and granular media involve interception, gravity separation, or Brownian diffusion for transport of particles from the fluid stream to the surface of the granular media. For this reason, sand filters are usually 1–2 m deep to promote a large number of collisions to occur between particles and granular media. In contrast, membrane filters (e.g., 0.2 μm microfilters) "strain" out particles larger than the pore size and do not require chemical particle destabilization. However, destabilized particles can improve membrane performance by reducing fouling or by agglomerating smaller particles approaching the membrane. Biological particles (e.g., bacteria, protozoa, and viruses) that are not removed during sedimentation or filtration are inactivated by the addition of chemical disinfectants (e.g., chlorine) or ultraviolet irradiation. Clearwells provide adequate contact time for chemical disinfectants to kill infectious, pathogenic organisms.

Traditionally, WTPs have optimized their processes to remove microbial particulates the size of bacteria (0.5 to 5 μm), oocysts (e.g., Giardia is 8–12 μm, Cryptosporidium is 3–5 μm), and viruses (10–100 nm). The particle counters used by WTPs are capable only of detecting particles larger than 1 μm because larger sized oocysts are generally more difficult to inactivate by chemical disinfectants than are viruses. Monitoring smaller, nanometer size particles has not been a primary objective, although mechanistic models describe how such particles should behave based upon their physicochemical properties. Using established literature reports, this chapter

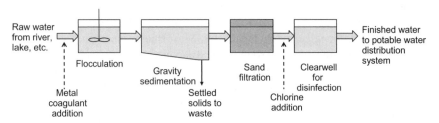

FIGURE 4.3   Schematic diagram of conventional water treatment processes and locations of chemical addition.

describes the relationship between physicochemical properties and likely removal mechanisms for engineered nanoparticles. The chapter concludes with potential research that will enhance our understanding of the fate of engineered nanoparticles during water treatment.

4.3.1 Coagulation

The presence of salts or NOM in water affects the zeta potentials and thus stability of nanoparticles. Stable nanoparticles are operationally defined as those that stay in suspension; collisions between stable nanoparticles do not lead to aggregation into larger, easier to settle out particles. Destabilized nanoparticles are operationally defined as those that tend to aggregate into larger particles. For example, Brant and coworkers (32, 33) reported that 168 nm nC_{60} nanoparticles were not stable in relatively weak electrolyte solutions (1 mM). Others have observed that both monovalent electrolytes and polyelectrolytes destabilized hematite nanoparticles through compression of their electrical double layers (EDLs), charge neutralization, and polymer bridging (5, 34–36). NOM affects the stability of particles by interacting with their surfaces. Fulvic acids and some NOMs can stabilize inorganic colloids (e.g., hematite nanoparticles) by increasing their absolute surface potentials, while alginic organic matter (AOM, chain-structure organic carbon) causes the aggregation of colloids by interparticle bridging (34, 37–39).

Derjaguin–Landau–Verwey–Overbeek (DLVO) theory can be used to evaluate the stability of colloidal particles in water, which depends on two major forces between particles, van der Waals attraction (Φ_{vdW}) and electrical double layer repulsion (Φ_{EDL}). The EDL interaction originates from the surface charges of particles; it occurs when particles approach each other and their electrical double layers overlay. A nanoparticle EDL can be considered spherical because the thickness of the double layer is comparable to the size of a nanoparticle, and thus the assumption of a flat double layer is invalid. The EDL repulsive interaction energy for two identical spherical nanoparticles is expressed by the following equation (18):

$$\Phi_{EDL} = \frac{4\pi\varepsilon r^2\zeta^2}{H + 2r}\exp\left[-\kappa r\cdot\left(\frac{H + 2r}{r} - 2\right)\right] \tag{4.1}$$

where r is the particle radius, ε is the dielectric constant, ζ is the zeta potential, κ is the inverse of the Debye length, and H denotes the shortest distance between two particle surfaces. In addition, London–van der Waals interactions (Φ_{vdW}) occur due to molecular dispersion forces. The expression for the van der Waals attractive interaction energy between two identical spherical particles can be written as follows (40–42):

$$\Phi_{vdW} = -\frac{A}{6}\left[\frac{2r^2}{H^2 + 4rH} + \frac{2r^2}{H^2 + 4Hr + 4r^2} + \ln\left(\frac{H^2 + 4Hr}{H^2 + 4Hr + 4r^2}\right)\right] \approx -\frac{rA}{12H} \tag{4.2}$$

where A is the London–Hamaker constant of particles in water. The sum of the above two forces determines whether the net interaction between two particles will be repulsive (i.e., nanoparticles are stable) or attractive (i.e., nanoparticles are destabe):

$$\Phi_{Total} = \Phi_{EDL} + \Phi_{vdW} \tag{4.3}$$

The graph of the interaction energy between two particles versus separation distance in Fig. 4.4 has an energy barrier and two energy minima, one primary and one secondary. Depending on the particles' Brownian kinetic energy, particles may aggregate at the primary minima by overcoming the energy barrier or remain bound at the secondary minima. Consequently, two critical parameters—the height of the energy barrier and the depth of the secondary minima—determine the stability of colloidal particles in a system. The stability ratio W_t, derived from the above two parameters, is an index representing the stability of particles (41, 43):

$$W_t^{-1} = \left[\frac{2a}{y_m^2}\left(\frac{2kT\pi}{\Phi''(y_m)}\right)^{1/2} \exp\left(\frac{\Phi_{MAX}}{kT}\right)\right]^{-1} + \left[\frac{1}{1-\exp(\Phi_{MIN}/kT)}\right]^{-1} \tag{4.4}$$

where k is Boltzmann's constant, T is absolute temperature, Φ_{MAX} and Φ_{MIN} are the respective net energies of the energy barrier and secondary minimum, y_m is the center-to-center distance of separation at the maximum in the net energy, and $\Phi''(y_m) = -d^2\Phi_{TOTAL}/dy^2$ is evaluated at $y = y_m$. A higher stability ratio indicates that the colloidal particles are more stable.

In DLVO theory, the size and surface charge are the two critical factors impacting the stability of colloidal particles in water. Small particle sizes result in low net energy

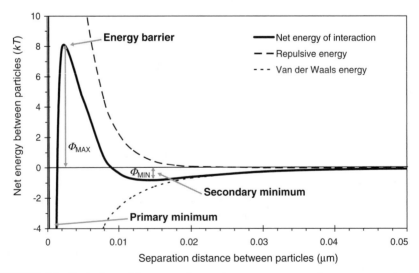

FIGURE 4.4 Typical repulsive, attractive, and net energy curves between two spherical colloidal particles under the following conditions: $r = 40\,nm$, $I = 0.01$, $\zeta = -40\,mV$, $A = 9.0 \times 10^{-20}\,J$, $T = 298K$.

barriers such that small particles are more unstable than large ones. Previous studies also showed that the values of secondary minima in net interaction energy profiles decreased with decreasing particle sizes and that the aggregation of nanosized particles (which was irreversible) occurred mainly in the primary minima (43–46). In addition, electrolytes and NOM in water can also affect the stability of colloidal particles by changing their surface charge and compressing their EDLs (37–39, 47–50).

Classical DLVO theory provides a means of quantifying nanoparticle interactions (i.e., stability) in water. However, previous research has documented that for many cases, especially in environmental systems, classical DLVO theory does not provide an adequate explanation for colloid stability (51–56). Accordingly, some non-DLVO interactions, such as hydrophilic, hydrophobic, and steric interactions, have been incorporated into DLVO to create extended DLVO theory. For example, hydrophobic interactions are likely to be related to the special arrangement of water molecules around hydrophobic particle surfaces, which results in an increase in free energy and a decrease in entropy (55, 57–60). Hydrophobic interactions arise from the removal of water molecules from the vicinity of hydrophobic particle surfaces (61). Van Oss and coworkers developed a polar interfacial interaction theory based on electron acceptor–electron donor (Lewis acid–base) theory to describe hydrophobic interaction energy (62, 63). Thus, as engineered nanoparticles are increasingly used in society and enter waterways, it would be useful to have extended DLVO parameters available to understand their potential stability in water.

In coagulation, WTPs destabilize particulates suspended in water by adding a chemical coagulant so that the particles form flocs during subsequent mixing (i.e., flocculation processes). Coagulation mechanisms that destabilize particulates include (1) compression of the electrical double layer, (2) adsorption and charge neutralization, (3) adsorption and interparticle bridging, and (4) sweep floc (64).

Electrical double layers can be compressed by increasing the electrolyte concentration (ionic strength) in a water sample (41). By decreasing the EDL thickness, the repulsive electrostatic force between particulates is reduced and they can approach each other due to Brownian motion. When the electrical double layer thickness is smaller than the distance that van der Waals forces extend, particle collisions and thus aggregation can occur.

Solutions of nanoparticles with negative zeta potentials are unlikely to rapidly aggregate because particles of like-charge will repel each other, so such a solution is referred to as stable. As addition of countercharged ions or compression of EDLs upon ion addition reduces a zeta potential to near zero (± 5 mV), aggregation can occur when particles colloid (65). Thus, the zeta potential emerges as a useful parameter for understanding the ability of particles to aggregate during water treatment. Increasing ionic strength decreases the absolute value of the zeta potential (i.e., less positive or less negative zeta potential), so that it is closer to zero. Consequently, the nanoparticles become destabilized and form aggregates of increasing diameter. Such aggregates are easier to remove using sedimentation and filtration.

Table 4.2 summarizes experimental work conducted in the authors' laboratory using lab-synthesized hematite and seven commercial, engineered nanoparticles. Representative results show the effect of salt type and concentration on the zeta potential and average particle size in solution after standard mixing conditions. For most

TABLE 4.2   Summary of Experiments Addressing the Impact of Salt Concentration and Salt Type on Zeta Potential, Aggregate Size, and Ability to Remove Nanoparticles

Type of Nanoparticle	Salt Concentration and Type	Zeta Potential (mV)	Average Diameter by DLS (nm)	Fraction Remaining After Coagulation, Sedimentation and Filtration
Hematite	0	+14	90	>80%
	100 mM KCl	+6	>1000	NA
TiO_2	0	−22	550	10%
	20 mM KCl	−12	>800	NA
	100 mM KCl	−4	>1000	NA
	100 mM $MgCl_2$	+2	>1200	5%
ZnO	0	+24	300	95% (sed. only)[a]
	20 mM KCl	+14	>800	NA
	100 mM KCl	+8	>900	NA
	100 mM $MgCl_2$	+8	>1100	70% (sed. only)[a]
Silica	0	−25	700	60% (sed. only)[a]
	40 mM KCl	−25	700	NA
	100 mM KCl	−10	>1000	NA
	100 mM $MgCl_2$	−6	>1300	50% (sed. only)[a]
Fe_2O_3	0	+16	200	70% (sed. only)[a]
	20 mM KCl	+9	>1500	NA
	100 mM KCl	+7	>1500	NA
	100 mM $MgCl_2$	+6	>1800	65% (sed. only)[a]
NiO	0	+38	750	70% (sed. only)[a]
	20 mM KCl	+22	>800	NA
	100 mM KCl	+14	>1000	NA
	100 mM $MgCl_2$	+6	>1300	70% (sed. only)[a]
CdTe Quantum Dot	0	−32	8	100%
	100 mM KCl	−20	8	NA
	0.5 mM $CaCl_2$	−26	>800	NA
	1 mM $MgCl_2$	−24	>3000	NA
Fullerene (nC_{60})	0	NA	NA	100%
	10 mM NaCl			60%
	100 mM NaCl			5%

Jar tests were used with 1 min of rapid mixing (100 rpm), 30 min of flocculation (30 rpm), 60 min of sedimentation (no mixing) using a Phipps & Bird Inc. (PB-700) jar test apparatus. Filtration was achieved using a 0.45-μm filtration (Nylaflo™ Membrane Disc Filter, Pall Corp., NY).

[a]Percentage remaining is after sedimentation only; additional removal would occur had 0.45 μm filter membranes also been used.

nanoparticles, increasing ionic strength decreases the absolute value of zeta potentials so that they become closer to zero (i.e., destabilizes the particles). As a result, the average diameter increases (i.e., aggregation occurs). Without salt addition, many of the nanoparticles do not settle out of solution (right-hand column in Table 4.2). Upon salt addition, particle destabilization occurs, large aggregates form, and these settle out of solution. In the case of fullerenes, where 0.45 μm membrane filtration was also used, the percentage of nC_{60} in solution changed from 100% without salt addition to 60% and 5% upon addition of 10 mM and 100 mM NaCl, respectively. However, carboxylated quantum dots behave differently from the other nanoparticles. They remain stable in the presence of 100 mM KCl, but rapidly aggregate in the presence of only 0.5 or 1 mM of divalent salts ($CaCl_2$ or $MgCl_2$). The divalent salts form complexes with the carboxyl functional groups on the quantum dot capping ligands, leading to charge neutralization and intraparticle bridging, and thus aggregation into larger particles.

4.3.2 Flocculation and Sedimentation

Flocculation is the process by which destabilized particles aggregate rapidly to form large flocs that can be removed by sedimentation or filtration. Depending upon the size of the particles, different mechanisms affect the rate of aggregation (Table 4.3). Perikinetic flocculation occurs due to Brownian motion of particles; their diffusion coefficients determine the flocculation rate. Perikinetic flocculation is the dominant mechanism of aggregation for particles smaller than 0.1 μm (64). Orthokinetic flocculation arises from collisions between particles that have different fluid velocities. Gentle mixing of water causes the velocity gradients. Orthokinetic flocculation is the dominant aggregation mechanism for particles 0.1–30 μm in diameter. When the

TABLE 4.3   Key Equations for Flocculation, Sedimentation, and Filtration

Relationship	Equation
Perikinetic flocculation (for a uniform particle size distribution)	$\dfrac{dN}{dt} = -\alpha \dfrac{4kT}{3\mu} N^2$
Orthokinetic flocculation (for a uniform particle size distribution)	$\dfrac{dN}{dt} = -\dfrac{4}{\pi} \alpha \Phi G N$
Differential settling	$\dfrac{dN}{dt} = -N_1 N_2 \dfrac{\pi}{4}(d_1 + d_2)^2(v_1 - v_2)$
Stokes settling velocity (v)	$v = \dfrac{g(\rho_P - \rho_L)d^2}{18\mu}$
Filtration for "clean-bed removal"	$\dfrac{N}{N_0} = \exp\left[\dfrac{-3(1-f)\alpha\eta L}{2D_C}\right]$

Definition of symbols: α = collision frequency factor; k = Boltzmann's constant; T = absolute temperature; μ = viscosity of water; N_i = number of particles of a size "i"; Φ = floc volume; G = Camp mixing intensity; v = Stokes settling velocity; g = gravitational acceleration constant; ρ_P = density of the particle; ρ_L = density of water; d = diameter of the particle; f = porosity of filter bed; η = single filter collection efficiency function; L = length of filter bed; D_c = diameter of filter collector media.

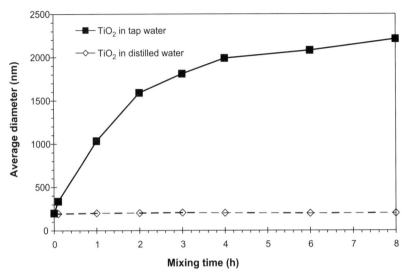

FIGURE 4.5 Change in average particle diameter (measured by dynamic light scattering) over time for TiO_2 nanoparticles (prepared from a 5% suspension; initial concentration is 5 mg/L as titanium) in ultrapure water and tap water.

particles are very large, differential settling can lead to collisions and aggregation. Differential settling is an important aggregation mechanism in systems containing particles that have a wide size range.

The flocculation models in Table 4.3 are a function of the size and number of particles in solution and can be used to estimate the initial rate of aggregation. A critical term is the stickiness factor, "α," the value of which ranges from 0 to 1, with $\alpha = 0$ representing no aggregation after a collision occurs and $\alpha = 1$ indicating that all collisions lead to aggregation. The value of α depends upon surface charge and chemistry, as seen in Fig. 4.4. Adding alum neutralizes surface charge, which increases α values. Figure 4.5 illustrates this effect with the change in TiO_2 particle size as a function of mixing time. In ultrapure water (i.e., low salt content), the zeta potential is -30 mV at pH 8, and aggregation does not occur over 8 h of mixing. However, in a local tap water with a conductance of ~ 800 μS/cm, aggregation occurs rapidly as smaller TiO_2 nanoparticles form large particles. The salts present in the tap water destabilize the TiO_2 nanoparticles (the zeta potential is -17 mV at pH 8), which essentially increases the α value and leads to a faster rate of aggregation (dN/dt).

While compression of the electrical double layer due to increasing salinity is important in many aquatic systems (e.g., estuaries, blends of two water sources), it is not a practical mechanism for particle destabilization during water treatment because the electrolyte concentration required to destabilize particulates completely tends to be much higher than the acceptable level in drinking water. In the United States, aluminum sulfate (alum) is the most common water treatment coagulant, used by 72% of water treatment utilities (64). When alum salts are added to water, aluminum ions form various hydrated species. At low pH values and alum dosages, charge neutralization is the dominant coagulation mechanism. Current water treatment protocols typically call for alum dosages of 10–150 mg/L and pH ranges from 5 to 8, depending on the raw water quality. Rapid mixing is very important immediately following the addition of alum (66, 67) because the hydrolysis of aluminum ions is quite rapid; amorphous aluminum hydroxide precipitates have formed within 1 s (65, 68). In most water treatment plants, the mixing time tends to be 10–30 s.

Alum addition forms new aluminum hydroxide particles in solution that aid in removal of "target" particles in water (e.g., virus, protozoa, bacteria, or sediment that causes turbidity). These new particles may "enmesh" or aggregate with existing particles in solution. Alum addition also increases the number of particles present in solution, thereby increasing the rate of aggregation. Larger particles have higher settling velocities (see the Stokes law relationship in Table 4.3). Furthermore, aggregated, larger particles are easier to remove by membrane filtration with a fixed pore size (e.g., 0.2 μm).

Figure 4.6 illustrates the effect of alum addition on two types of engineered nanoparticles, TiO_2 and CdTe quantum dots. Without alum addition, less than 20% of

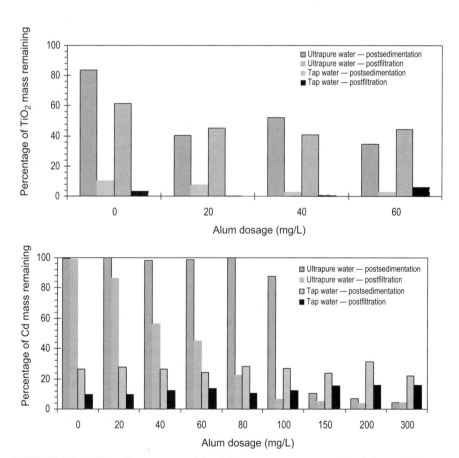

FIGURE 4.6 Effect of water composition (ultrapure versus tap water) and alum addition on removal of TiO_2 nanoparticles (top) and CdTe quantum dots (bottom).

the TiO_2 nanoparticles in ultrapure water settle out of solution, but nearly 90% can be filtered out using a 0.45 µm membrane filter. Similar experiments (no alum) conducted in tap water resulted in nearly 40% removal of TiO_2 nanoparticles by sedimentation alone, due to the higher α value as a result of salts present in solution, and nearly 95% removal after 0.45 µm membrane filtration. In both water types, adding alum slightly improves the removal of TiO_2 nanoparticles using sedimentation, but leads to a greater than 95% removal with 0.45 µm membrane filtration. Alum addition resulted in visual formation of new white flocs (i.e., amorphous aluminum hydroxide). In all cases, some titanium remained in the water after 0.45 µm membrane filtration. Since titanium is fairly insoluble, the presence of ionic titanium cannot account for the detected titanium after 0.45 µm membrane filtration, and thus nanosized titanium must have remained in the solution.

The effect of alum addition on the removal of CdTe quantum dots functionalized with carboxylic capping ligands is slightly different than that on TiO_2 nanoparticles. Without alum addition, QDs in ultrapure water are stable and not removed after sedimentation or 0.45 µm membrane filtration, while in tap water without alum addition, nearly 70% of the QD mass is removed by sedimentation alone (90% after 0.45 µm membrane filtration). In tap water, divalent cations (e.g., calcium) presumably formed complexes with the carboxylic capping ligands, thus destabilizing the QDs (e.g., increased α values) (69). The complexation may have "bridged" functional groups between different QDs, because the zeta potential decreased by less than 5 mV upon divalent cation addition (e.g., Table 4.2). Low alum dosages did not result in white floc precipitates in ultrapure water because the carboxylic capping ligands were complexing (i.e., binding) with cationic aluminum species, thereby preventing aluminum hydroxide precipitation. This complexation neutralized the QD zeta potential (not shown) and facilitated QD removal. At high alum dosages, white precipitates were formed, indicating complete complexation of the carboxylic capping ligands by cationic aluminum, and removal of QDs proceeded along mechanisms similar to those described for TiO_2 nanoparticles. As with TiO_2, a small percentage of metal (i.e., cadmium) was present even after 0.45 µm membrane filtration, indicating that some QDs remain in solution. Nanoparticles not removed by membrane filtration could pose an exposure risk in drinking water.

4.3.3   Filtration

The above experiments utilized membrane filtration instead of packed bed porous media filtration. Membrane filtration (0.1–0.4 µm) is increasingly being used in the United States, but granular filters remain the most widely used technology worldwide. Granular filters remove particles from water through several mechanisms that lead to collisions between particles and sand filter media. For most particles, the mechanisms of interception, diffusion, or gravity separation dominate. Collisions will only lead to permanent attachment when particles are properly destabilized. Therefore, models for particle removal in clean packed bed filters (Table 4.3) include terms for the "stickiness" of the particles to the filter media (α value) as well as a collector efficiency term (η) that is used to calculate the importance of different removal mechanisms. The

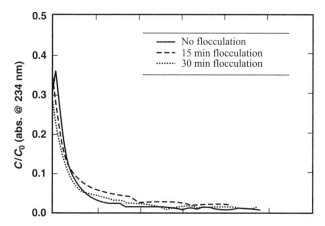

FIGURE 4.7   Fractional removal versus filter run time for 10 mg/L of 270 nm diameter latex particles with 0.05 M $CaCl_2$. *Source*: Reference 1.

"clean-bed removal" predicted by the equations in Table 4.3 accurately predicts the removal of nanoparticles. For example, a packed bed of glass beads removes ~65% of 270 nm latex particles that were chemically destabilized by adding 0.05 M $CaCl_2$ (70) (Fig. 4.7). However, filter ripening occurred as the latex particles were removed. In filter ripening, improved particle removal over time occurs because the previously retained particles aid in subsequent particle removal. This filter ripening is more difficult to mechanistically model. However, the main point is that packed bed filtration can efficiently remove particles to levels greater than that predicted by clean-bed filtration models (Table 4.3).

Research has examined the mobility in granular media of eight types of nanoparticles: anatase, ferroxane, alumoxane, fullerol, silica of two different sizes, nC_{60}, and single-wall carbon nanotubes. Ferroxane had the highest attachment efficiency, which was close to unity (12, 71). Chapter 7 in this book also addresses the fate of nanoparticles during filtration. The key point made here is that packed bed filtration is very effective at removing nanometer-sized particles (e.g., viruses) to levels greater than 99.99%. However, complete removal is never attainable. Thus, a small fraction of nanoparticles may pass through filters and pose an exposure risk in drinking water.

4.4   CONCLUSIONS

Based on decades of research on "colloids in water" (41) the fate of engineered nanoparticles in water can be inferred. Analogies between natural and engineered nanoparticles provide a context for understanding that the stability (ability to remain in water) of engineered nanoparticles in rivers, lakes, groundwaters, estuaries, and engineered treatment systems is a function of their size, surface charge, surface properties (e.g., hydrophilicity), number, and ability to interact with other constituents in water (through, e.g., ion complexation, EDL compression due to ionic strength, or complexation by NOM). Many engineered nanoparticles, particularly metal oxides, have direct analogies to natural nanoparticles or colloids. Other engineered nano-particles may initially seem "new," but can be represented by existing colloid models. For example, a CdTe quantum dot with carboxylic functionalized capping ligands is very stable and analogous to a natural metal colloid coated with polymeric NOM. Thus, the ability to *a priori* predict the fate of engineered nanoparticles necessitates measuring their zeta potentials (or surface charges) and ability to complex with ions or NOM. In all our tests to date, complete (100%) removal of the engineered nano-particles was never achieved in simulated conventional drinking water treatment processes, suggesting that a very small percentage of engineered nanoparticles (as is undoubtedly the current case with natural nanoparticles) will be present in potable drinking water and may pose an exposure pathway for humans.

ACKNOWLEDGMENTS

The authors gratefully acknowledge research funding (EPA Grant # RD831713) from the U.S. Environmental Protection Agency. Contributions from Ayla Kiser, Troy Benn, and Kiril Hristovski (graduate students) and Dr Zhou Chen are appreciated. Any opinions, findings, conclusions, or recommendations expressed in this chapter are those of the authors and do not necessarily reflect the view of the supporting organizations.

REFERENCES

1. Westerhoff, P. A study of the effects of flocculation on direct filtration performance. *Civil and Environmental Engineering*. Amherst, MA: University of Massachusetts;1991.
2. De Momi, A., Lead, J.R. Size fractionation and characterisation of fresh water colloids and particles: split-flow thin-cell and electron microscopy analyses. *Environ Sci Technol* 2006;40(21):6738–6743.
3. Buffle, J., van Leeuwen, H.P. *Environmental Particles,* vol. 1. John Wiley and Sons; 1992.
4. Penners, N.H.G., Koopal, L.K. Preparation optical properties of homodisperse hematite hydrosols. *Colloids Surf* 1986;19:337–349.
5. Zhang, J., Buffle, J. Kinetics of hematite aggregation by polyacrylic acid: importance of charge neutralization. *J Colloid Interface Sci* 1995;174:500–509.
6. Tyndall, J. On the blue color of the sky, the polarization of skylight and the polarization of light by cloudy matter generally. *Philos Mag* 1869;37:384–394.
7. Berg, H.C. *Random Walks in Biology*, expanded ed. Princeton: Princeton University Press; 1993.
8. Einstein, A. *Investigations on the Theory of the Brownian Movement*. New York: Dover; 1956.
9. Matijevic, E., Scheiner, P. Ferric hydrous oxide sol. 3. preparation of uniform particles by hydrolysis of Fe(III)-chloride, Fe(III)-nitrate, and Fe(III)-perchlorate solution. *J Colloid Interface Sci* 1978;63(3):509.

10. Yaremko, Z.M., Nikipanchuk, D.M., Fedushinskaya, L.B., Uspenskaya, I.G. Redispersion of highly disperse powder of titanium dioxide in aqueous medium. *Colloid J* 2001;63 (2):280–285.

11. Brunner, T.J., Wick, P., Manser, P., Spohn, P., Grass, R.N., Limbach, L.K., Bruinink, A., Stark, W.J. *In vitro* cytotoxicity of oxide nanoparticles: comparison to asbestos, silica, and the effect of particle solubility. *Environ Sci Technol* 2006;40:4374–4381.

12. Lecoanet, H.F., Bottero, J.Y., Wiesner, M.R. Laboratory assessment of the mobility of nanomaterials in porous media. *Environ Sci Technol* 2004;38:5164–5169.

13. Leong, Y.K., Ong, B.C. Critical zeta potential and the Hamaker constant of oxides in water. *Powder Technol* 2003;134:249–254.

14. Nurmi, J.T., Tratnyek, P.G., Sarathy, V., Baer, D.R., Amonette, J.E., Pecher, K., Wang, C., Linehan, J.C., Matson, D.W., Penn, R.L., Driessen, M.D. Characterization and properties of metallic iron nanoparticles: spectroscopy, electrochemistry, and kinetics. *Environ Sci Technol* 2005;39:1221–1230.

15. Park, K., Kittelson, D.B., Zachariah, M.R., McMurry, P.H. Measurement of inherent material density of nanoparticle agglomerates. *J Nanopart Res* 2004;6:267–272.

16. Soto, K.F., Carrasco, A., Powell, T.G., Garza, K.M., Murr, L.E. Comparative *in vitro* cytotoxicity assessment of some manufactured nanoparticulate materials characterized by transmission electron microscopy. *J. Nanopart Res* 2005;7:145–169.

17. Justin, C. Quantum dot conjugated gold/silica nanoshells for optical targeting and therapyof cancer. Summer Research Program at IBBME, 2004.

18. Klabunde, K.J. editor. *Nanoscale Materials in Chemistry.* New York: Wiley Interscience; 2001.

19. Heymann, D. Solubility of fullerenes C-60 and C-70 in seven normal alcohols and their deduced solubility in water. *Fullerene Sci Technol* 1996;4(3):509–515.

20. Bolskar, R.D., Alford, J.M. Chemical oxidation of endohedral metallofullerenes: identification and separation of distinct classes. *Chem Commun* 2003(11):1292–1293.

21. Deguchi, S., Alargova, R.G., Tsujii, K. Stable dispersions of fullerenes, C-60 and C-70, in water. Preparation and characterization. *Langmuir* 2001;17(19):6013–6017.

22. Labille, J., Brant, J.A., Villiéras, F., Pelletier, M., Wiesner, M.R., Bottero, J.-Y. Affinity of C-60 fullerenes with water. *Fullerenes Nanotubes Carbon Nanostruct* 2006;14 (2–3):307–314.

23. Tseluikin, V.N., Chubenko, I.S., Gun'kin, I.F., Pankst'yanov, A.Yu. Colloidal dispersion of fullerene C-60 free of organic solvents. *Russ J Appl Chem* 2006;79(2):325–326.

24. Doucet, F.J., Maguire, L., Lead, J.R. Size fractionation of aquatic colloids and particles by cross-flow filtration: analysis by scanning electron and atomic force microscopy. *Anal Chim Acta* 2004;522(1):59–71.

25. Ledin, A., Karlsson, S., Duker, A., Allard, B. Characterization of the submicrometer phase in surface waters: a review. *Analyst* 1995;120(3):603–608.

26. Kim, J.P., Lemmon, J., Hunter, K.A. Size-distribution analysis of submicron colloidal particles in river water. *Environ Technol* 1995;16(9):861–868.

27. Rodriguez, M.A., Armstrong, D.W. Separation and analysis of colloidal/nano-particles including microorganisms by capillary electrophoresis: a fundamental review. *J Chromatogr B* 2004;800(1–2):7–25.

28. Moon, J., Kim, S.H., Cho, J. Characterizations of natural organic matter as nano particle using flow field-flow fractionation. *Colloids Surf A* 2006;287(1–2):232–236.

29. Martin, S. Caldwell, K. Giddings, J.C. editors. *Field-Flow Fractionation Handbook.* New York: John Wiley & Sons, Inc.; 2000. p. 592.

30. Diallo, M.S., Glinka, C.J., Goddard, W.A., Johnson, J.H. Characterization of nanoparticles and colloids in aquatic systems 1. Small angle neutron scattering investigations of Suwannee River fulvic acid aggregates in aqueous solutions. *J Nanopart Res* 2005;7 (4–5):435–448.

31. Chen, Z., Westerhoff, P., Herckes, P. Quantification of C_{60} Fullerene Concentrations in Water. *Environ Toxic Chem* (in press Feb 2008).

32. Brant, J., Lecoanet, H., Hotze, M., Wiesner, M.R. Comparison of electrokinetic properties of colloidal fullerenes (n-C-60) formed using two procedures. *Environ Sci Technol* 2005;39 (17):6343–6351.

33. Brant, J., Lecoanet, H., Wiesner, M.R. Aggregation and deposition characteristics of fullerene nanoparticles in aqueous systems. *J. Nanopart Res* 2005;7:545–553.

34. Amal, R., Coury, J.R., Raper, J.A., Walsh,W.P., Waite, T.D. Structure and kinetics of aggregating colloidal haematite. *Colloids Surf* 1990;46:1–19.

35. Ferretti, R., Zhang, J., Buflte, J. Kinetics of hematite aggregation by polyacrylic acid: effect of polymer molecular weights. *Colloids Surf A* 1997;121:203–215.

36. Schudel, M., Behrens, S.H., Holthoff, H., Kretzschmar, R., Borkovec, M. Absolute aggregation rate constants of hematite particles in aqueous suspensions: a comparison of two different surface morphologies. *J Colloid Interface Sci* 1997;196:241–253.

37. Tiller, C.L., O'melia, C.R. Natural organic matter and colloidal stability: models and measurements. *Colloids Surf A* 1993;73:89–102.

38. Wilkinson, K.J., Joz-roland, A., Buffle, J. Different roles of pedogenic fulvic acids and aquagenic biopolymers on colloid aggregation and stability in freshwaters. *Limnol. Oceanogr.* 1997;42:1714–1724.

39. Wilkinson, K.J., Negre, J.C., Buffie, J. Coagulation of colloidal material in surface waters: the role of natural organic matter. *J Contam Hydrol* 1997 (26):229–243.

40. Snoswell, D.R.E., Duan, J., Fornasiero, D., Ralston, J. Colloid stability of synthetic titania and the influence of surface roughness. *J Colloid Interface Sci* 2005;286:526–535.

41. Hiemenz, P.C., Rajagopalan, R. *Principles of Colloid and Surface Chemistry,* 3rd ed. New York: Marcel Dekker; 1997.

42. Verwey, E.J.W., Overbeek, J.T.G. *Theory of the Stability of Lyophobic Colloids.* New York: Elsevier; 1948. p. 49.

43. Marmur, A. A kinetic theory approach to primary and secondary minimum coagulations and their combination. *J Colloid Interface Sci* 1979;72(1):41–48.

44. Joseph Petit, A.M., Dumont, F., Watillon, A. Effect of particle size on stability of monodisperse selenium hydrosols. *J Colloid Interface Sci* 1973;43:649–661.

45. Parfitt, G.D., Picton, N.H. Stability of dispersions of graphitized carbon blacks in aqueous solutions of sodium dodecyl sulfate. *Trans Faraday Soc* 1968;64:1955–1964.

46. Weise, G.R., Healy, T.W. Effect of particle size on colloid stability. *Trans Faraday Soc* 1970;66:490–499.

47. Filella, M., Buffle, J. Factors controlling the stability of submicron colloids in natural waters. *Colloids Surf A* 1993;73:255–273.

48. Filella, M., Buffle, J., Leppard, G.G. Characterization of submicrotre colloids in freshwaters; evidence for their bridging by organic structures. *Water Sci Technol* 1993;27:91–102.

49. Findlay, A.D., Thompson, D.W., Tipping, E. The aggregation of silica and haematite particles dispersed in natural water samples. *Colloids Surf A* 1996;118:97–105.

50. Pizarro, J., Belzile, N., Filella, M., Leppard, G.G., Negre, J.C., Perret, D., Buffle, J. Coagulation/sedimentation of submicron iron particles in a eutrophic lake. *Water Res* 1995;29:617–632.

51. Chheda, P., Grasso, D. Surface thermodynamics of ozone-induced particle destabilization. *Langmuir* 1994;10:1044–1053.

52. Grasso, D., Carrington, J.C., Chheda, P., Kim, B. Nitrocellulose particle stability: coagulation thermodynamics. *Water Res* 1995;29:49–59.

53. Grasso, D., Subramaniam, K., Butkus, M., Strevett, K., Bergendahl, J. A review of non-DLVO interactions in environmental colloidal systems. *Rev Environ Sci Biotechnol* 2002;1:17–38.

54. Hu, Y., Dai, J. Hydrophobic aggregation of alumina in surfactant solution. *Miner Eng* 2003;16:1167–1172.

55. Pashley, R.M. Hydration forces between mica surfaces in electrolyte solutions. *Adv Colloid Interface Sci* 1982;82:52–57.

56. Zhou, Z., Wu, P., Ma, C. Hydrophobic interaction and stability of collidal silica. *Colloids Surf* 1990;50:177–188.

57. Israelachvili, J.N. *Intermolecular and Surface Forces*. London: Academic Press; 1987.

58. Pashley, R.M., McGuiggan, P.M., Ninham, B.V. Attractive forces between uncharged hydrophobic surfaces: direct measurements in aqueous solution. *Science* 1985;229:1088–1089.

59. Sadowski, Z. A study on hydrophobic aggregation suspensions of calcite aqueous. *Powder Technol* 1994;80:93–98.

60. Xu, Z., Yoon, R.H. The role of hydrophobic interactions in coagulation. *J Colloid Interface Sci* 1989;132(2):532–541.

61. Claesson, P.M., Christenson, H.K. Very long range attractive forces between uncharged hydrocarbon and fluorocarbon surface in water. *J Phys Chem* 1988;92(6):1650–1664.

62. Van Oss, C.J., Chaudhury, M.K., Good, R.J. Interfacial Lifshitz–van der Waals and polar interactions in macroscopic systems. *Chem Rev* 1988;88:927–941.

63. Van Oss, C.J., Giese, R.F., Costanzo, P.M. DLVO and non-DLVO interactions in hectorite. *Clays Clay Miner* 1990;38(2):151–159.

64. Crittenden, J., Trussell, R., Hand, D., Howe, K., Tchobanoglous, G. *Water Treatment: Principles and Design*, 2nd ed. New York: John Wiley & Sons, Inc.; 2005.

65. Stumm, W. *Chemistry of the Solid–Water Interface*. New York: John Wiley & Sons, Inc.; 1992. pp. 269–286.

66. Amirtharajah, A., Mills, K.M. Rapid-Mix Design for Mechanisms of Alum Coagulation. *J Am Water Works Assoc* 1982;74(4):210–216.

67. Vrale, L., Jorden, R.M. Rapid Mixing in Water Treatment. *J Am Water Works Assoc* 1971;63(1):52–58.

68. Base, C.F., Mesmer, R.E. *The hydrolysis of Cations*. New York: John Wiley & Sons, Inc.; 1976.

69. Zhang, Y., Chen, Y., Hristovski, K., Westerhoff, P., Crittenden, J.C., Stability of Commercial Metal Oxide Nanoparticles in Water. *Water Res* 2008;42(8–9):2204–2212.

70. Tobiason, J.E., Edzwald, J.K., Reckhow, D.A., Switzenbaum, M.S. Effect of Pre-Ozonation On Organics Removal By in-Line Direct-Filtration. *Water Sci Technol* 1993;27(11):81–90.

71. Lecoanet, H.F., Wiesner, M.R. Velocity effects on fullerene and oxide nanoparticle deposition in porous media. *Environ Sci Technol* 2004;38(16):4377–4382.

Transport and Retention of Nanomaterials in Porous Media

KURT D. PENNELL[1,2], JED COSTANZA[1], and YONGGANG WANG[1]

[1]School of Civil and Environmental Engineering, Georgia Institute of Technology, 311 Ferst Drive, Atlanta, GA 30332-0512, USA
[2]Center for Neurodegenerative Disease, Department of Neurology, Emory University School of Medicine, 615 Michael Street, Atlanta, GA 30322-3090, USA

5.1 INTRODUCTION

Commercial production and use of manufactured nanomaterials is anticipated to increase dramatically over the next several decades (1), which will inevitably lead to the release of these materials into the environment during their manufacture, transport, application, and disposal. While the potential human health effects, aquatic toxicity, and antimicrobial properties of nanoscale particles have been investigated (2–5), far less attention has been directed toward understanding the processes that govern the fate and transport of manufactured nanomaterials in the environment. This lack of knowledge hinders current efforts to evaluate manufactured nanoparticle exposure risks, conduct life cycle analysis, and develop effective waste management strategies.

Here, we focus on one aspect of the environmental fate of manufactured nano-materials: transport and retention in the subsurface environment. Nanoparticles may reach the land surface from a number of sources including atmospheric deposition, accidental spills, and discharges, while direct releases into the subsurface may occur via injection wells, leaking underground storage tanks and sewer lines, and municipal or industrial landfills. The subsequent migration of manufactured nanomaterials will be governed primarily by the flow of water, in which nanoparticles can form stable or unstable aqueous suspensions. In many subsurface systems, the aqueous nanoparticle suspension will migrate through unconsolidated porous media, consisting of soil or aquifer materials. During transport, nanoparticles may be deposited and retained by the porous medium, which could limit their migration from a source area (e.g., spill or landfill). However, the possibility exists that a portion of the released nanoparticles may reach the water table and could be transported laterally to drinking water wells and surface water bodies including streams, lakes, and wetlands. Thus, transport and retention of manufactured nanomaterials in water-saturated porous media represents a potentially important human and environmental exposure pathway. To further explore this topic, this chapter provides a basic description of the subsurface environment and properties of aqueous nanoparticle suspensions, followed by a discussion of particle migration in porous media, which encompasses transport and deposition of several representative manufactured nanomaterials, including fullerene (C_{60}), titanium dioxide (TiO_2), and zero-valent iron (Fe^0).

5.2 THE SUBSURFACE ENVIRONMENT

The outer layer of the Earth's terrestrial surface consists primarily of unconsolidated (i.e., weathered and fragmented) mineral and organic matter, commonly referred to as soil. Beneath the surface soil horizons, one frequently encounters unconfined aquifer systems, which are particularly vulnerable to terrestrial contaminant releases. The subsurface can be divided into several zones based on the water content, flow, or distinct changes in soil properties (Fig. 5.1). In this context, the subsurface can be envisioned as a three-phase system, consisting of a solid phase (e.g., mineral grains), a liquid phase (typically water), and a gas phase (soil air). Depending upon the properties and arrangement of the solid particles, the void or pore space (soil porosity) occupies between 25% and 50% of the total soil volume. These soil pores are interconnected, allowing for both liquid and gas transport through the subsurface. During a rainfall or irrigation event, water will enter the soil profile and may completely fill or saturate soil pores (water-saturation = 1). Following this initial infiltration process, water existing within the root zone can be removed from the soil profile and reenter the atmosphere as a result of evaporation and/or plant transpiration (evapotranspiration). In contrast, water present below the root zone will tend to migrate downward, primarily due to gravitational forces, a process commonly referred to as gravity drainage or percolation. In an unconfined aquifer formation, as shown in Fig. 5.1, water may then enter (recharge) an underlying aquifer and be transported laterally with the existing groundwater flow. Thus, nanoparticles that enter the subsurface may encounter water saturations that vary in both space and time, ranging from 1.0 (i.e., groundwater and capillary fringe) to an irreducible or residual saturation in the root zone, which is typically less than 0.1 or 10% of the pore space. In this chapter, we focus on transport processes in water-saturated soils and aquifer materials, while the following chapter addresses unsaturated conditions (i.e., gas phase present, water saturation <1).

5.3 NANOMATERIAL TRANSPORT AND RETENTION IN POROUS MEDIA

Manufactured or "engineered" nanomaterials are generally defined as possessing at least one dimension that is in the range of 1–100 nm in length. Thus, nanomaterials are typically smaller than most bacteria (200–3000 nm) and equivalent to or slightly

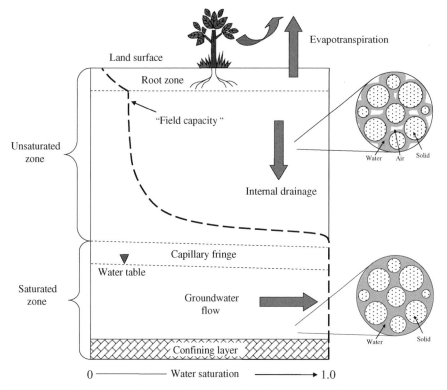

FIGURE 5.1   Schematic diagram of the subsurface illustrating key features, water distribution profile, and water flow pathways. *Root zone*: region immediately below the land surface occupied by plant roots, typically on the order of 0.5–1.0 m in depth. *Unsaturated zone*: region extending from the land surface to the top of the capillary fringe; void space usually contains a gas phase and is therefore "unsaturated" with respect to water. *Saturated zone*: region extending from the top of the capillary fringe to a lower confining layer; void space is completely filled or "saturated" with water. *Capillary fringe*: region immediately above the water table; void space is completely saturated with water, but the water is under suction or negative pressure. *Confining layer*: a geologic unit that does not transmit significant quantities of water (low permeability). The upper inset shows a three-phase system containing soil grains, water and air, while the lower inset shows a two-phase system containing water and soil grains.

smaller than viruses (10–400 nm), which are considered to be the smallest noncellular organisms. Due to their relatively small size, one might assume that nanomaterials would be readily transported through the subsurface since the effective pore diameter of soil is typically greater than 1000 nm (Fig. 5.2). However, several laboratory-scale studies have shown that nanomaterial transport can be quite limited, even at relatively high flow rates. For example, Lecoanet et al. (6) measured the transport of several nanomaterials (alumoxane, anatase, ferroxane, fullerene, fullerol, silica, and single-walled nanotubes) in water-saturated columns (2.5 cm inside diameter × 9.25 cm length) packed with 40–50 mesh ($d_{50} = 0.355$ mm) spherical glass beads (Fig. 5.3). Nanoparticles appeared in the column effluent after injecting less than one pore

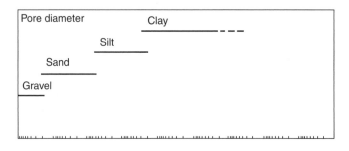

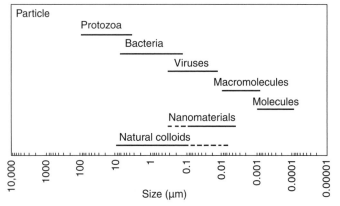

FIGURE 5.2   Comparison of the diameter of representative particles and the effective pore diameter of natural soil materials.

volume (PV) of each nanoparticle suspension and approached steady-state relative effluent concentrations (C/C_0, where C_0 is the applied concentration and C is the measured effluent concentration) that ranged from approximately 1.0 (fullerol, 1.2 nm diameter) down to 0.3 (ferroxane, 300 nm diameter). Thus, the observed steady-state effluent concentrations decreased with increasing nanoparticle diameter, indicating substantial retention of nanoparticles with diameters greater than 100 nm. In related laboratory-scale column experiments, steady-state relative effluent concentrations of nanoscale fullerene aggregates (nC_{60}, 100 and 168 nm diameter) ranged from 0.3 to 0.8 and increased with increasing flow rate (7, 8). To interpret their nano-material transport data, the authors employed clean-bed filtration theory, which has been used extensively to describe colloid removal in filter media. Based on this approach, Lecoanet et al. (6) estimated that the distance or bed length required for a 3-log reduction (99.9% or $C/C_0 = 0.1\%$) in concentration ranged from 10 to 20 cm for nanoparticles with diameters greater than 100 nm and up to 14 m for fullerol (1.2 nm diameter). Similarly, Cheng et al. (8) reported that distances of 0.31–1.32 m were required for 3-log removal of nanoscale fullerene aggregates (ca. 100 nm diameter) during steady-state transport through water-saturated aquifer material. The theoretical basis and mathematical development of the equations used for these calculations are discussed in detail below.

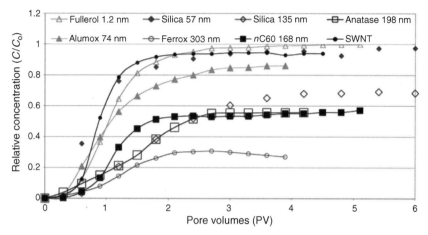

FIGURE 5.3 Effluent breakthrough curves obtained for nanomaterial suspensions (10 mg/L) introduced at a flow rate (Q) of 12 mL/min in columns (2.5 cm inside diameter × 9.25 cm length) packed with spherical glass beads (diameter = 0.355 mm) (Reproduced with permission from *Environ. Sci. Technol.* 2004, 38, 5164–5169. Copyright 2004 American Chemical Society).

5.3.1 Nanoparticle Transport and Filtration

The transport and retention of particles in a water-saturated porous medium can be described mathematically by considering the processes of advection, hydrodynamic dispersion, and particle deposition. Assuming conditions of homogeneity, laminar flow, absence of particle–particle interactions on the surface (i.e., clean bed at early time), and negligible particle release or decay, the one-dimensional advection–dispersion equation for a particle suspension can be written as (9):

$$\frac{\partial C}{\partial t} + \frac{\rho_b}{\theta_w}\frac{\partial S}{\partial t} = D_H \frac{\partial^2 C}{\partial x^2} - v\frac{\partial C}{\partial x} \qquad (5.1)$$

$$\frac{\rho_b}{\theta_w}\frac{\partial S}{\partial t} = k_{att}C \qquad (5.2)$$

where C is the particle concentration in the aqueous phase, t is the time, ρ_b is the solid bulk density, θ_w is the volumetric water content which is equivalent to the soil porosity in a water-saturated medium, S is the solid-phase particle concentration, D_H is the hydrodynamic dispersion coefficient, x is the distance, v is the interstitial or pore-water velocity, and k_{att} is the first-order particle attachment rate coefficient. If one further assumes that hydrodynamic dispersion is negligible and that the concentration in the aqueous phase has reached a steady-state condition, substitution of Equation 5.2 into Equation 5.1 yields:

$$v\frac{dC}{dx} + k_{att}C = 0 \qquad (5.3)$$

For a continuous injection of a particle suspension at the inlet boundary ($C = C_0$ at $x = 0$) applied for time t, the following analytical solutions can be obtained for the aqueous- and solid-phase concentrations as a function of distance from the column inlet:

$$C(x) = C_0 e^{\left[-\frac{k_{att}}{v}x\right]} \tag{5.4}$$

$$S(x) = \frac{t\theta_w k_{att} C_0}{\rho_b} e^{\left[-\frac{k_{att}}{v}x\right]} \tag{5.5}$$

If the underlying assumptions are valid, these equations indicate that both the aqueous- and solid-phase particle concentrations will decrease exponentially with distance from the column inlet. Furthermore, the attachment rate and pore-water velocity will govern the rate at which the aqueous- and solid-phase concentrations decline as a function of distance from the column inlet or injection point.

The particle attachment coefficient, k_{att}, can be related to the single collector efficiency, η, which is defined as the rate at which particles are deposited onto a collector divided by rate at which the particles are flowing toward the collector (9, 10):

$$k_{att} = \frac{3(1-\theta_w)v}{2d_c}\eta \tag{5.6}$$

where d_c is the diameter of collector grains or porous medium. Substitution of Equation 5.6 into Equation 5.4 yields the steady-state relative effluent concentration expression that is commonly used to obtain values of η from experimental column data as shown in Fig. 5.3:

$$\ln\left(\frac{C_L}{C_0}\right) = -\frac{3(1-\theta_w)L}{2d_c}\eta \tag{5.7}$$

where L is the column length. The experimentally determined collector efficiency, η, is commonly related to the theoretical (maximum) collector efficiency, η_0, as $\eta = \alpha\eta_0$, where α is the attachment or collision efficiency. Rearrangement of Equation 5.7 yields the collision efficiency expression:

$$\alpha = -\frac{2d_c}{3(1-\theta_w)L\eta_0}\ln\left(\frac{C_L}{C_0}\right) \tag{5.8}$$

that was used by (6–8) to calculate the travel distance, L, required for 3-log removal ($C_L/C_0 = 0.1\%$) of nanoparticles.

As nanoparticles migrate through a porous medium, several mechanisms may contribute to retention and deposition, including straining (sieving), sedimentation, diffusion, inertial impaction (interception), and hydrodynamic forces. Under most conditions, diffusion (Brownian motion) and sedimentation (gravity settling) are the dominant mechanisms responsible for particle deposition. Particle trajectory theory

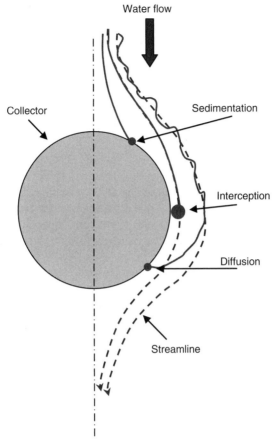

FIGURE 5.4   Schematic diagram illustrating the major processes governing particle deposition on a single collector based on trajectory theory (adapted from Yao et al. (10)).

was originally developed to account for the contributions of diffusion, interception, and sedimentation to the theoretical collector efficiency, where $\eta_0 = \eta_D + \eta_I + \eta_G$ (10) (Fig. 5.4). Here, the theoretical collector efficiencies attributed to diffusion (η_D), interception (η_I), and sedimentation or gravity (η_G) can be expressed as:

$$\eta_D = 4\left(\frac{qd_c}{D_p}\right)^{-2/3} = 4N_{Pe}^{-2/3}, \quad \eta_I = 1.5\left(\frac{d_p}{d_c}\right)^2 = 1.5N_R^2, \quad \eta_G = \frac{(\rho_p - \rho)gd_p^2}{18\mu q} = N_G$$

$$(5.9)$$

where D_p is the bulk particle diffusion coefficient, q is the superficial or Darcy velocity, d_c is the collector diameter, d_p is the particle diameter, ρ_p is the particle density, ρ is the fluid density, g is the gravitational constant, and μ is the dynamic viscosity of the suspension. To simplify these expressions, three dimensionless terms are incorporated: the Peclet number (N_{Pe}), the interception number (N_R), and the gravitation number (N_G). The diffusion term given above, η_D, was refined by Yao et al. (10) to account for collisions generated by diffusion using the Happel (11) flow field factor: $A_s = 2(1 - p^5)/(2 - 3p + 3p^5 - 2p^6)$, where $p = (1 - \theta_w)^{1/3}$. This model was subsequently corrected and refined by Rajagopalan and Tien (12) so that all terms represent fluid flows in the concentric spherical space surrounding the collector particles:

$$\eta_0 = 4A_s^{1/3}N_{Pe}^{-2/3} + A_sN_{Lo}^{1/8}N_R^{15/8} + 0.00338A_sN_G^{1.2}N_R^{-0.4} \tag{5.10}$$

where N_{Lo} represents the contribution of van der Waals attractive forces to particle removal, defined as $N_{Lo} = 4H/9\pi\mu d_p q$, and H is the Hamaker constant. As noted by Logan et al. (13), this form of the trajectory theory equation is consistent with the particle filtration transport equations derived using a mass balance approach (i.e., Eqs. 5.1–5.8).

Representative theoretical collector efficiency curves are shown in Fig. 5.5 as a function of particle diameter. Here, the corrected (RT) and uncorrected (Y) trajectory models are plotted for two particle densities, 1.002 and 1.05 g/cm³. The porous medium was assumed to be 20–30 mesh Ottawa sand ($d_{50} = 0.71$ mm), subject to a pore-water (interstitial) velocity of 0.78 cm/min. The minimum theoretical collector efficiency occurs between particle diameters of 1 and 2 μm for the RT model with a

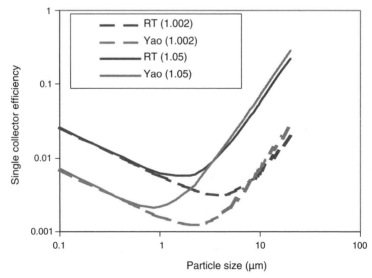

FIGURE 5.5 Theoretical single collector efficiency (η_0) as a function of particle diameter, calculated using the corrected (RT) model of Rajagopalan and Tien (12) and the uncorrected (Y) model of Yao et al. (10) for spherical particles with a density of either 1.002 and 1.05 g/cm³.

density of 1.05. For particles with diameters greater than 2 µm, removal efficiency increases rapidly with particle size, indicative of sedimentation and interception mechanisms. For nanoscale particles (diameter <1 µm), the theoretical collector efficiency increases with decreasing particle diameter, which results primarily from the contribution of diffusion to particle deposition.

The mathematical equations and underlying theories described above provide a framework from which to interpret and predict the transport of manufactured nanoparticles in water-saturated porous media. In effect, the basic principles developed for the removal of naturally occurring colloidal particles from drinking water have been extended to nanoparticles. While this is a reasonable first approach, almost no research has been undertaken to critically evaluate the applicability of clean-bed filtration theory to nanoparticle transport and retention. In the remaining sections of this chapter, we focus on several areas of nanoparticle transport research, including ionic strength, aggregation, and particle retention effects, which may require modification to existing filtration approaches.

5.3.2 Nanoparticle Aggregation

In the absence of surface coatings, nanoparticles frequently exhibit a strong tendency to aggregate or agglomerate into larger particles, particularly when the aqueous suspension contains a background electrolyte (14, 15). Thus, particles that were initially nanoscale in size may aggregate to form micron-sized particles in experimental systems that contain high salt concentrations, such as cell culture media. For example, Long et al. (16) reported that the mean diameter of TiO_2 (Degussa-P25) nanoparticles in Hanks basic salt solution (HBSS) nearly doubled after 30 min. Similarly, Hussain et al. (17) found that MnO_2 nanoparticles formed agglomerates that were greater than 1000 nm in diameter when dispersed in cell culture media. The aggregation and deposition kinetics on nanoscale fullerene (nC_{60}) aggregates in suspensions containing both monovalent and divalent cations was investigated by Chen and Elimelech (18). They found that the rate of nC_{60} nanoparticle aggregation was diffusion limited in the presence of NaCl or $CaCl_2$ and that nC_{60} nanoparticles become less negatively charged when the electrolyte concentration is increased.

Derjaguin–Landau–Verwey–Overbeek (DLVO) theory can be used to evaluate the effects of electrolyte species and concentration on electrostatic interactions between nanoparticles. Based on DLVO theory, the interaction energy (E_i) between two particles is composed of the electrical double layer repulsion energy (E_{edl}) and the van der Waals attraction energy (E_V). The value E_{edl} can be calculated using the equation from Gregory (19):

$$E_{edl} = \frac{64\pi nkTa}{\kappa^2} \tanh^2\left(\frac{ze\psi}{4kT}\right) e^{(-\kappa d)} \tag{5.11}$$

where n is the number of cations in solution, k is the Boltzmann constant, T is the absolute temperature, a is the particle radius, κ is the Debye–Huckel reciprocal length parameter, z is the charge number, e is the electron charge, ψ is the surface potential of the particle, and d is the surface to surface distance. The value of E_V can be computed as (20):

$$E_V = \frac{Ha}{12d}\left[1 - \frac{bd}{\lambda}\ln\left(1 + \frac{\lambda}{bd}\right)\right] \tag{5.12}$$

where b is a constant with a value of 5.32 and λ is the characteristic wavelength of the interaction. Assuming a fullerene-water-fullerene Hamaker constant, H, of 6.7×10^{-21} J (18) and a characteristic wavelength, λ, of 100 nm (21), the resulting interaction energy profiles between two representative nC_{60} aggregates are presented in Fig. 5.6 for suspensions containing either NaCl or $CaCl_2$. Here, negative values represent a net attractive force, while positive values correspond to a net repulsive force. In the presence of $CaCl_2$ (Fig. 5.6a), the energy minimum becomes more negative and approaches the surface as the electrolyte concentration is increased, consistent with strong suppression of the electrical double layer by divalent cations. Under these conditions, van der Waals attractive forces become dominant, which has been shown to result in the formation of nC_{60} agglomerates consisting of multiple-scale aggregates (22). In the presence of NaCl, a large positive interaction energy (repulsive force) is observed at all but the highest concentration considered (100 mM), where a small secondary minimum (attractive force) was present at a separation distance of approximately 10 nm (Fig. 5.6b). These findings suggest that a relatively strong repulsive force exists between nC_{60} aggregates, consistent with the observed stability of nC_{60} suspensions containing NaCl (15, 18, 22).

To further explore the role of surface charge on nC_{60} suspension stability, the zeta potential of nC_{60} aggregates has been determined as a function of electrolyte species and concentration. As the concentration of both NaCl and $CaCl_2$ was reduced, the surface charge (electrophoretic mobility) of nC_{60} aggregates became more negative, from slightly less than zero to -30 to -50 mV (23). Although the zeta potential of nC_{60} generally decreases (became more negative) as the concentration of $CaCl_2$ is decreased, a charge reversal of approximately $+18$ mV was observed at a $CaCl_2$ concentration of 0.1 mM (23). This effect was attributed to an excess of counterions surrounding the charged aggregate surface (23). In the presence of NaCl, the zeta potential of nC_{60} aggregates was slightly more negative compared to $CaCl_2$ over the same range of ionic strength. This difference is attributed to the stronger screening effect of Ca^{2+} ions (15, 23) and may also indicate specific adsorption of Ca^{2+} ions on the surface of nC_{60} aggregates (23).

Magnetic nanoparticles, such as zero-valent iron, experience magnetic interactions in addition to electrical double layer repulsion and van der Waals energies experienced by diamagnetic fullerene (C_{60}) particles. These magnetic interactions cause particles with diameters of greater than 10 nm to agglomerate into larger aggregates with diameters of greater than 1000 nm. Phenrat et al. (24) reported the formation of aggregates with hydrodynamic radius of up to 4000 nm within 10 min after adding zero-valent iron particles with initial radius of 109 nm to a 1 mM $NaHCO_3$ solution. Once these magnetic aggregates form, their magnetic interaction can only be overcome by thermal energy and fluid shear forces (25). Thus, describing

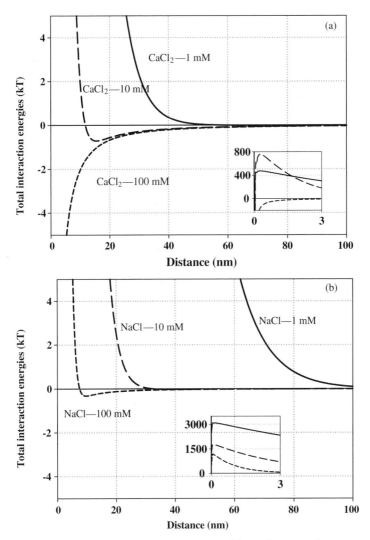

FIGURE 5.6   Interaction energies (E_i) between two 100 nm diameter nC_{60} aggregates as a function of (a) $CaCl_2$ and (b) NaCl concentrations at pH 7. Inset show energy barrier near the surface (Reproduced with permission from Wang et al. (27), *Environmental Toxicology & Chemistry*, Allen Press Publishing Services).

the interaction of ferromagnetic nanoparticles requires adding a magnetic term to the DLVO sum of interaction forces. For example, the 7 kT repulsion force predicted by classical DLVO theory changed into a strong attractive energy of greater than −15 kT when Phenrat et al. (24) incorporated magnetic interactions in their extended DLVO theory calculations. Thus, predicting the transport of nanoscale ferromagnetic particles will require developing numerical approaches beyond the clean-bed filtration model, which assumes negligible particle–particle interactions.

5.3.3 Nanoparticle–Solid Interactions

Several studies have demonstrated the importance of ionic strength in the transport of nC_{60} aggregates in water-saturated porous media. Espinasse et al. (26) reported that as the ionic strength of an nC_{60} suspension containing NaCl was increased from 0.01 to 0.6 M, the steady-state relative effluent concentration (C/C_0) decreased from approximately 0.94 to 0.15, while the corresponding collision efficient, α, increased from 0.03 to 0.73 (Eq. 5.8). Similarly, Wang et al. (27) reported that nC_{60} aggregates were readily transported through a water-saturated column packed with either 40–50 mesh ($d_{50} = 0.355$ mm) or 100–140 mesh ($d_{50} = 0.125$ mm) Ottawa sand at an ionic strength of 3.05 mM (NaCl) (Fig. 5.7a). However, when the ionic strength was increased to 30.05 mM, more than 95% of the applied nC_{60} aggregates were retained within the column, consistent with the observed steady-state relative effluent concentration of approximately 0.05 (Fig. 5.7a). These findings indicate that electrostatic interactions between the solid phase and nC_{60} aggregates strongly influences the migration and corresponding deposition processes. To explore these effects in more detail, DLVO theory can be used to calculate electrostatic interaction energy profiles as a function of electrolyte species and concentration. The double layer energy (E_{edl}) and van der Waals attraction energy (E_V) were calculated using the approach of Guzman et al. (28), which accounts for the small size of nanoparticles:

$$E_{edl} = \pi\varepsilon_0\varepsilon_r\kappa(\psi_p^2+\psi_s^2)$$
$$\times \int_0^a \left(-\coth\left[\kappa\left(d+a-a\sqrt{1-(r/a)^2}\right)\right]+\coth\left[\kappa\left(d+a+a\sqrt{1-(r/a)^2}\right)\right]\right)rdr$$
$$+ \int_0^a \frac{2\psi_s\psi_p}{\psi_s+\psi_p}\left(\csc h\left[\kappa\left(d+a-a\sqrt{1-(r/a)^2}\right)\right]-\csc h\left[\kappa\left(d+a+a\sqrt{1-(r/a)^2}\right)\right]\right)rdr$$

$$(5.13)$$

$$E_V = -\frac{A}{6}\left[\frac{a}{d}+\frac{a}{d+2a}+\ln\left(\frac{d}{d+2a}\right)\right] \tag{5.14}$$

where ε_0 is the permittivity of a vacuum, ε_r is the relative dielectric constant, and ψ_s and ψ_p are the surface potentials of the solid surface and the nanoparticle, respectively. In Fig. 5.7b, energy profiles are shown for a system consisting of quartz sand and nC_{60} aggregates, assuming a sand zeta potential of -30 mV and -22 mV in 3.05 mM and 30.05 mM NaCl solutions, respectively (29) with a fullerene-silica-water Hamaker constant of 4.71×10^{-21} J (23). In the presence of NaCl at 3.05 mM, a large primary repulsive force (145 kT) exists near the surface, (Fig. 5.7b). This repulsive force is consistent with the minimal retention of nC_{60} aggregate observed in column experiments conducted at an ionic strength of 3.05 mM (NaCl). At the higher ionic strength (30.05 mM), the primary repulsive force is reduced to approximately 125 kT, and the secondary attractive region became stronger (-1 kT) and moved closer to the surface, consistent with the much greater retention of nC_{60} aggregates in column experiment conducted at an ionic strength of 30.05 mM (Fig. 5.7a).

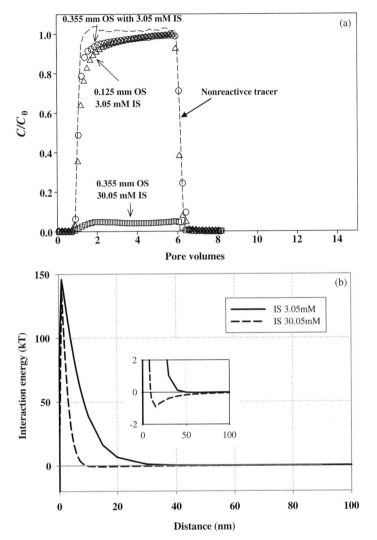

FIGURE 5.7   Effluent concentration profiles for a 3-PV pulse of nC_{60} introduced into water-saturated columns packed with either 40–50 mesh or 100–140 mesh Ottawa sand (a), and interaction energies between an nC_{60} aggregate and quartz sand at ionic strengths of 3.05 and 30.05 mM (NaCl) (b). The inset shows the secondary minimum attractive region (Reproduced with permission from Wang et al.(27), *Environmental Toxicology & Chemistry*, Allen Press Publishing Services).

5.3.4 Nanoparticle Retention

Based on clean-bed filtration theory, the solid-phase concentration of retained nanoparticles in a water-saturated porous medium will decrease exponentially with distance from the column inlet or release point, as shown in Equation 5.5. Interestingly, nearly all of the nanoparticle transport data reported to date lack measured retention data (i.e., only effluent concentrations are presented), which precludes mass balance calculations and a critical evaluation of applicability of filtration theory to describe nanoparticle retention profiles (e.g., 6–8, 23, 26). In a recent study, however, Wang et al. (30) reported nC_{60} aggregate (ca. 100 nm diameter) retention profiles in columns packed with water-saturated 40–50 mesh Ottawa operated at a Darcy velocity of 2.8 m/day. The resulting nanoparticle profiles are nearly constant (flat) over the column length and could not be accurately simulated using a mathematical model that incorporated clean-bed filtration theory. To address this discrepancy, Wang et al. (30) incorporated a blocking function expression, ψ_b, into Equation 5.1 (31):

$$\psi_b = \frac{S_{max} - S}{S_{max}} \qquad (5.15)$$

where S_{max} is the maximum particle retention capacity. The modified nanoparticle transport simulations closely matched both the measured retention profiles and the effluent concentration data. These findings suggest that Ottawa sand exhibits a limiting or maximum retention capacity for nC_{60} aggregates. Additional research is needed to determine if the observed limiting retention capacity applies to other nanomaterials and to other types of porous media, including natural soils containing significant quantities of organic matter and clay minerals.

5.4 SUMMARY

The transport of manufactured nanoparticles through the subsurface is a potentially important pathway for human and environmental exposure. To date, a limited number of experimental studies have been conducted to quantify the transport of nanoparticles in water-saturated porous media. Using clean-bed filtration theory, travel distances corresponding to 3-log removal have been estimated to range from 0.1 m for nanoparticles with diameters greater than 100 nm up to 14 m for fullerol nanoparticles (1.2 nm diameter). Subsequent studies have demonstrated the importance of ionic strength on nanoparticle transport, which readily occurred at low ionic strength but decreased substantially when the ionic strength was raised. These findings were consistent with electrostatic energy profiles calculated from DLVO theory, which indicate a large energy barrier at low ionic strength. As the ionic strength was increased, the primary energy barrier was either completely or substantially reduced and coincided with the appearance of secondary minimum attractive region. While the current body of literature generally supports the application of clean filtration theory to describe manufactured nanomaterial transport in water-saturated porous media, research is needed to describe nanoparticle transport and deposition processes in natural soils that contain relevant amounts of organic matter and clay minerals. In addition, the effects of surface coatings and stabilizing agents (e.g., natural organic matter, macromolecules, and surfactants) on nanoparticle transport must be carefully evaluated as their presence may greatly enhance transport in the subsurface, thereby increasing the potential for nanomaterials to migrate from a point of release to a drinking water supply or surface water body.

ACKNOWLEDGMENTS

This work has been supported by grant R-832535 from the U.S. Environmental Protection Agency's Science to Achieve Results (STAR) program. This work has not been subjected to any EPA review and therefore does not necessarily reflect the views of the Agency and no official endorsement should be inferred.

REFERENCES

1. Maynard, A.D., Aitken, R.J., Butz, T., Colvin, V., Donaldson, K., Oberdörster, G., Philbert, M.A., Ryan, J., Seaton, A., Stone, V., Tinkle, S.S., Tran, L., Walker, N.J., Warheit, D.B. Safe handling of nanotechnology. *Nature* 2006;444:267–269.
2. Block, M.L., Zecca, L., Hong, J.-S. Microglia-mediated neurotoxicity: uncovering the molecular mechanisms. *Nat Rev Neurosci* 2007;8:57–69.
3. Lyon, D.Y., Fortner, J.D., Sayes, C.M., Colvin, V.L., Hughes, J.B. Bacterial cell association and antimicrobial activity of a C60 water suspension. *Environ Toxicol Chem* 2005; 24:2757–2762.
4. Nel, A., Xia, T., Madler, L., Li, N. Toxic potential of materials at the nanolevel. *Science* 2006;311:622–627.
5. Oberdörster, G., Oberdörster, E., Oberdörster, J. Nanotoxicology: an emerging discipline evolving from studies of ultrafine particles. *Environ Health Perspect* 2005;113: 823–839.
6. Lecoanet, H.F., Bottera, J.-Y., Weisner, M.R. Laboratory assessment of the mobility of nanoparticles in porous media. *Environ Sci Technol* 2004;38:5164–5169.
7. Lecoanet, H.F., Weisner, M.R. Velocity effects on fullerene and oxide nanoparticle deposition in porous media. *Environ Sci Technol* 2004;38:4377–4382.
8. Cheng, X., Kan, A.T., Thomson, M.B. Study of C60 transport in porous media and the effect of sorbed C60 on naphthalene transport. *J Mater Res* 2005;20:3244–3254.
9. Tufenkji, N., Redman, J.A., Elimelech, M. Intepreting deposition patterns of microbial particles in laboratory-scale column experiments. *Environ Sci Technol* 2003;37:616–623.
10. Yao, K.-M., Habibian, M.T., O'Melia, C.R. Water and waste water filtration: concepts and applications. *Environ Sci Technol* 1971;5:1105–1112.
11. Happel, J. Viscous flow in multiparticle systems: slow motion of fluids relative to beds of spherical particles. *AIChE J* 1958;4:197–201.
12. Rajagopalan, R., Tien, C. Trajectory analysis of deep-bed filtration with the sphere-in-a-cell porous media model. *AIChE J* 1976;22:523–533.
13. Logan, B.E., Jewett, D.G., Arnold, R.G., Bouwer, E.J., O'Melia, C.R. Clarification of clean-bed filtration models. *ASCE J Environ Eng* 1995;121:869–873.
14. Deguchi, S., Alargova, R.G., Tsujii, K. Stable dispersions of fullerenes, C-60 and C-70, in water: preparation and characterization. *Langmuir* 2001;17:6013–6017.

15. Fortner, J.D., Lyon, D.Y., Sayes, C.M., Boyd, A.M., Falkner, J.C., Hotze, E.M., Alemany, L.B., Tao, Y.J., Guo, W., Ausman, K.D., Colvin, V.L., Hughes, J.B. C_{60} in water: nanocrystal formation and microbial response. *Environ Sci Technol* 2005;39:4307–4316.

16. Long, T.C., Saleh, N., Tilton, R.D., Lowry, G.V., Veronesi, B. Titanium dioxide (P25) produces reactive oxygen species in immortalized brain microglia (BV2): implications for nanoparticle neurotoxicity. *Environ Sci Technol* 2006;40:4346–4352.

17. Hussain, S.M., Javorina, A.K., Schrand, A.M., Duhart, H.M., Ali, S.F., Schlager, J.J. The interaction of manganese nanoparticles with PC-12 cells induces dopamine depletion. *Toxicol Sci* 2006;92:456–463.

18. Chen, K.L., Elimelech, M. Aggregation and deposition kinetics of fullerene (C-60) nanoparticles. *Langmuir* 2006;22:10994–11001.

19. Gregory, J. Interaction of unequal double-layers at constant charge. *J Colloid Interface Sci* 1975;51:44–51.

20. Gregory, J. Approximate expressions for retarded van der Waals interaction. *J Colloid Interface Sci* 1981;83:138–145.

21. Schenkel, J.H., Kitchener, J.A. A test of the Derjaguin–Verwey–Overbeek theory with a colloidal suspension. *Trans Faraday Soc* 1960;56:161–173.

22. Brant, J., Lecoanet, H., Wiesner, M.R. Aggregation and deposition characteristics of fullerene nanoparticles in aqueous systems. *J Nanopart Res* 2005;7:545–553.

23. Brant, J., Lecoanet, H., Hotze, M., Wiesner, M. Comparison of electrokinetic properties of colloidal fullerenes (n-C60) formed using tow procedures. *Environ Sci Technol* 2005;39:6343–6351.

24. Phenrat, T., Saleh, N., Sirk, K., Tilton, R.D., Lowry, G.V. Aggregation and sedimentation of aqueous nanoscale zerovalent iron dispersions. *Environ Sci Technol* 2007;41:284–290.

25. Rosensweig, R.E., *Ferrohydrodynamics*. Cambridge, UK: Cambridge University Press; 1985, p. 34.

26. Espinasse, B., Hotze, E.M., Wiesner, M.R. Transport and retention of colloidal aggregates of C60 in porous media: effects of organic macromolecules, ionic composition and preparation method. *Environ Sci Technol* 2007;41:7369–7402.

27. Wang, Y., Li, Y., Pennell, K.D. Influence of electrolyte species and concentration on the aggregation and transport of fullerene nanoparticles in quartz sands. *Environ Toxicol Chem* 2008; 27:Doi: 10.1897/08-039.1.

28. Guzman, K.A.D., Finnegan, M.P., Banfield, J.F. Influence of surface potential on aggregation and transport of titania nanoparticles. *Environ Sci Technol* 2006;40:7688–7693.

29. Kaya, A., Yukselen, Y. Zeta potential of clay minerals and quartz contaminated by heavy metals. *Can Geotech J* 2005;42:1280–1289.

30. Wang, Y., Li, Y., Fortner, J.D., Hughes, J.B., Abriola, L.M., Pennell, K.D. Transport and retention of nanoscale C60 aggregates in water-saturated porous media. *Environ Sci Technol* 2008;42:3588–3594.

31. Johnson, P.R., Elimelech, M. Dynamics of colloid deposition in porous media–blocking based on random sequential adsorption. *Langmuir* 1995;11:801–812.

CHAPTER 6

Transport of Nanomaterials in Unsaturated Porous Media

LIXIA CHEN and TOHREN C.G. KIBBEY

School of Civil Engineering and Environmental Science, The University of Oklahoma, 202 W. Boyd Street, Rm 334, Norman, OK 73019, USA

6.1 INTRODUCTION

With the significant increase in the use of nanomaterials in practical applications, manufactured nanomaterials will inevitably enter the subsurface through pathways such as infiltration of atmospheric dispersions, wet deposition, landfill leachate, or direct injection (e.g., in cases where they are used for groundwater remediation). As nanomaterials enter the subsurface, the unsaturated (or vadose) zone is likely to play an important role in their transport and fate. As such, this chapter focuses on the transport behavior of nanomaterials in unsaturated porous media, with an emphasis on mechanisms likely to influence their fate and transport.

The unsaturated zone is the portion of the subsurface above the water table. Porous media in the unsaturated zone are not completely saturated with water, but contain significant quantities of air. Nanomaterials are specialized materials with sizes between 1 and 100 nm in at least one dimension. Although nanomaterials are considered to be colloids (particles with sizes between 1 nm and 1 μm), manufactured nanomaterials may behave differently from many natural colloids due to their small sizes and specialized surface properties. To understand the transport behavior of nanomaterials in the unsaturated porous media, it is necessary to understand (1) the mechanisms of transport most important for colloidal particles, to provide a basis for interpreting nanomaterial behavior, and (2) the properties of unsaturated porous media that influence unsaturated transport.

As such, this chapter provides an overview of both the important mechanisms influencing the transport of colloidal particles in saturated and unsaturated porous media, and key properties of unsaturated porous media. Experimental data for the

Nanoscience and Nanotechnology Edited by Vicki H. Grassian

dynamic unsaturated transport of tin oxide (SnO_2) and latex nanoparticles are discussed in the context of transport mechanisms and unsaturated porous media properties.

6.2   MAJOR MECHANISMS INFLUENCING SATURATED AND UNSATURATED TRANSPORT OF COLLOIDS AND NANOMATERIALS

To understand unsaturated transport of colloids and nanomaterials, it is important to first understand saturated transport as all of the mechanisms of saturated transport also apply to unsaturated transport. The following Subsections describe key mechanisms in saturated (Section 6.2.1) and unsaturated (Section 6.2.2) transport.

6.2.1   Transport of Colloids and Nanomaterials in Saturated Porous Media

Saturated transport of colloidal particles has been extensively studied to date, and considerable information exists in the literature. In saturated porous media, colloidal particles can be present in the water phase or at the solid–water interface. The mobility of colloids in the water phase is influenced by particle–particle and particle–solid interactions that may remove colloids from the water phase. The removal of colloids from the water phase, which is often referred to as filtration in the literature, involves two major mechanisms: solid–water interface attachment and pore straining. Key processes related to these mechanisms are illustrated in Fig. 6.1 and discussed below.

Solid–water interface attachment, which is considered to be the primary mechanism accounting for the saturated filtration of colloidal particles, is influenced by physicochemical processes that cause colloids to collide with and adhere to solid surfaces. The processes responsible for solid–water interface attachment are related to both colloidal movement, which brings colloids into contact with the solid–water interface (interception, sedimentation, and Brownian diffusion), and particle–solid

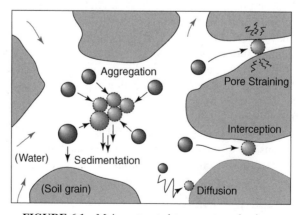

FIGURE 6.1   Major saturated transport mechanisms.

interactions, which cause colloids to remain at the solid–water interface (electrostatic, van der Waals, and hydrophobic interactions) (1–3).

For colloids larger than 1 μm, colloid movement is dominated by the interception and sedimentation mechanisms; movement of smaller colloids tends to be dominated by Brownian diffusion (3, 4). Interception is a result of the contact between the colloids suspended in flow streamlines and the stationary porous media. Sedimentation is the vertical transport of suspended colloids out of streamlines and toward porous media surfaces when the density of colloids is larger than the density of water. Brownian diffusion is a random motion of small colloids through thermal effects. For a given porous medium, colloid density, and set of environmental conditions, there exists a range of suspended colloid sizes for which the amount of filtration resulting from the three main mechanisms (interception, sedimentation, and diffusion) is at a minimum; that is, a size range where colloids exhibit maximum mobility (4). The existence of this size range means that small colloids will not necessarily be able to move more freely in porous media than larger colloids. For colloids with densities close to that of water, the colloid size corresponding to minimum filtration removal (maximum mobility) has been reported to be approximately 1 μm (4, 5). However, for iron nanoparticles with a density of 7.8 g/cm^3, minimum filtration removal (maximum mobility) has been reported to occur in the 100–200 nm particle size range (6).

In most cases, both colloids and porous media have charged surfaces in water, creating the potential for solid–water interface attachment due to electrostatic interactions between colloids and the solid surface if the two are oppositely charged. The charge on the colloid (or the solid) surface is balanced by an equal and opposite charge carried by counterions in the surrounding liquid. The counterions are attracted by the charged surface and form a double layer (Stern layer and diffuse layer) ion cloud (7). The Stern layer is considered to be strongly associated with the charged surface, but the diffuse layer is not (7). The electrical potential (relative to any point beyond the double layer) at the boundary between the Stern and diffuse layers is called zeta potential. Zeta potential has a strong influence on the stability and mobility of colloids in solution and porous media. Both the nature (positive or negative) and the magnitude of zeta potential of both colloids and solid surfaces are influenced by chemical conditions of the system, such as ionic strength and pH.

The increase in ionic strength (either by a high solution ion concentration or valence) decreases the double layer thickness and consequently the absolute value of zeta potential. The pH value influences the zeta potential mainly by changing the surface charge (density and sign) of ionizable surface sites. For example, Fig. 6.2 shows the zeta potential curve of SnO$_2$ nanoparticles (Sigma–Aldrich, St. Louis, MO) measured as a function of pH at 0.01 M ionic strength at a temperature of 22°C. The pH value where zeta potential is zero is referred to as point of zero charge (PZC; in this case approximately 4.3); the surface is positively charged at lower pH values and negatively charged at higher pH values. Figure 6.3 illustrates the effect of pH on colloid mobility. When the pH of the system falls between the PZCs of the colloid and porous media (Fig. 6.3), the colloid and porous media are oppositely charged and low colloid mobility (i.e., high attachment or filtration) will be observed due to electrostatic attraction between colloids and porous media surfaces. In contrast, when the pH is

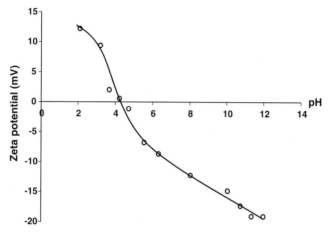

FIGURE 6.2   Zeta potential of aqueous tin oxide (SnO_2) suspension as a function of pH ($T = 22°C$, $I = 0.01$ M).

higher or lower than the PZCs of both colloid and porous media (Fig. 6.3), relatively high colloid mobility (i.e., low attachment or filtration) is typically observed due to electrostatic repulsion between colloids and porous media surfaces. When the electrostatic repulsion force exceeds the van der Waals attraction force between

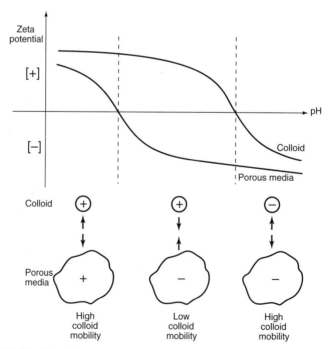

FIGURE 6.3   The effect of zeta potential on colloid–porous medium interactions.

colloids and porous media, there may exist an energy barrier to attachment in the Derjaguin–Landau–Verwey–Overbeek (DLVO) net interaction energy curve (8), and the system is considered to be unfavorable for colloid attachment. Note that not all colloids or solids have pH-dependent surface charges. For example, many clay minerals have strong fixed charges due to mineral composition; although some surface sites may have charges that are pH-dependent, they play a small role in determining the total surface charge, so surface charge is relatively independent of pH. Nevertheless, the same electrostatic forces still apply, so a colloid with the opposite charge to the porous medium will typically have low mobility, regardless of whether the charges are pH-dependent or not.

Considering Fig. 6.3 again, it should be noted that quartz sand, one of the most common mineral surfaces in environmental porous media, has a PZC of approximately pH 2. That means that under most environmentally-relevant conditions, quartz sand has a negative surface charge. As such, positively charged colloids would generally be expected to have a low mobility in the environment, although positively charged colloids with pH-dependent surface charge could be mobilized by changing pH conditions.

Under conditions unfavorable for attachment, pore straining is another major mechanism that may lead to colloid filtration during transport in saturated porous media (9). Pore straining (Fig. 6.1) is the trapping of colloids in down-gradient pores that are too small to allow them to pass, or trapping at the junction of two or more solid grains. Pore straining is controlled by particle size and pore size distributions (10, 11), and becomes significant in retarding colloids only when the ratio between the colloidal particle size and the mean porous media grain size is greater than a critical value. This critical colloid-to-media size ratio was found to be 0.05 by Herzig et al. (12), but more recent studies suggest it may be as low as 0.008 (13), 0.005 (10), or 0.0017 (14). It has also been proposed that pore straining is influenced by solution chemistry and hydrodynamics in the system (9), and published experiments examining pore straining have indicated greater colloid retention at the column inlet than would be predicted by classical attachment theory (9, 10, 14).

An important factor influencing the transport of colloids is the stability of colloidal suspensions. If a colloidal suspension is unstable, then smaller colloids will aggregate to form larger aggregates, increasing the importance of the interception, sedimentation, and pore straining mechanisms. Two basic mechanisms that influence the stability of colloids are electrostatic stabilization and steric stabilization. Electrostatic stabilization results from repulsive interactions between colloid surface electric double layers that prevent colloids from colliding and aggregating, while steric stabilization results from the presence of adsorbed macromolecular layers that physically prevent colloids from colliding (3). In the absence of significant quantities of adsorbed macromolecules, electrostatic stabilization is predominant. Any changes in solution chemistry, such as pH and ionic strength, which may increase the absolute value of colloid zeta potential will also increase the stability of colloids and nanomaterials (15, 16). Under the condition that electrostatic interaction controls, the colloid stability can be evaluated by DLVO theory, which considers the combined effects of electrostatic repulsion and van der Waals attraction energies (8).

According to DLVO theory, favorable aggregation or deposition (i.e., unfavorable stabilization or mobilization) occurs irreversibly at a deep primary potential energy minimum present at small interparticle distances, and reversibly at a shallow secondary minimum present at larger interparticle distances (17). The repulsive region at distances between the primary and secondary minima is called the potential energy barrier, the height of which is an indicator of the stability of the system. Both the primary and secondary minima play important roles in colloid transport. However, for colloids or nanomaterials that are sufficiently small, the secondary energy minimum may play a relatively small role, and the primary energy minimum may not be as deep as for larger colloids (18). As such, aggregation and deposition for some nanomaterials may be more reversible than for larger colloids (18). Further, for extremely small nanoparticles (<10 nm), complete overlap of the diffuse layers surrounding the nanoparticles is typical and electrostatic repulsion does not occur, so DLVO theory is not applicable (19).

Although saturated transport of colloids has been extensively studied, available information on saturated behaviors of manufactured nanomaterials is very limited. Recent studies have examined the aggregation of selected nanomaterials in aqueous dispersion and their mobility in saturated porous media by investigating the influence of ionic strength, pH values, flow rates, or concentrations (15, 16, 20–23). It has been found that the stability and the saturated mobility of nanoparticle suspensions vary with different nanomaterials (20, 21), and decrease significantly with the increase of electrolyte (15) and nanoparticle (23) concentrations and with the decrease of flow rate (20, 22). It has also been found that pH values influence the stability and mobility of TiO_2 nanoparticles in saturated porous media by influencing their surface potentials (16). In general, all of the above findings with nanomaterials are consistent with common colloid filtration theories. However, numerical simulations by Kallay and Zalac (19) suggest that nanoparticle systems may be less stable than similar colloid systems due to a greater influence of pH and ionic strength and the higher number concentration of nanoparticles, a result which may imply lower mobility for nanomaterials than similar colloids.

6.2.2 Transport of Colloids and Nanomaterials in Unsaturated Porous Media

Compared to the transport in saturated porous media, unsaturated transport is further complicated by the additional air–water interface created by the presence of the air phase. In addition to the mechanisms important in saturated transport, several additional mechanisms are also important in unsaturated transport, such as attachment to the air–water interface and film straining. Furthermore, the presence of air can enhance pore straining by redirecting flow through smaller pores as large pores are drained and filled with air. These mechanisms can cause the retention of colloids to be more significant in unsaturated porous media than in saturated porous media. Figure 6.4 illustrates the additional mechanisms important in unsaturated transport.

The importance of the air–water interface in colloid transport has been observed through the increased retention of colloids in unsaturated porous media as water

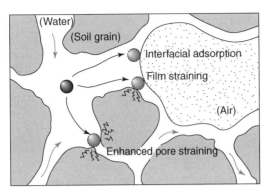

FIGURE 6.4 Additional transport mechanisms introduced by the presence of the air phase in unsaturated porous media.

content decreases (24–28). Attachment to the air–water interface has been observed for both hydrophobic and hydrophilic colloids with either positive or negative surface charges (24, 25, 29–32), but hydrophobic colloids have been reported to have a greater affinity for the air–water interface than hydrophilic colloids (24, 29–31, 33–35). Major forces controlling colloid attachment to the air–water interface include electrostatic, van der Waals, hydrophobic, and capillary interactions (25, 27, 31, 32). As has been observed for solid–water interface attachment, electrostatic interaction has also been found to be significant in air–water interface attachment, especially for hydrophilic colloids (31, 36, 37). Experimental measurements have shown that the air–water interface is negatively charged with surface potential ranging from -15 to $-65\,\mathrm{mV}$ depending on pH and ionic strength (38, 39). As such, positively charged colloids usually show greater adsorption to the air–water interface than negatively charged colloids (29, 37, 40). Hydrophobic interactions between colloids and the air–water interface are believed to be much stronger than DLVO forces (i.e., electrostatic and van der Waals forces), so are able to overcome the energy barrier for the air–water interface attachment, allowing even negatively charged colloids to attach to the air–water interface in some cases (31, 32, 41, 42). As is the case with attachment to the solid–water interface, attachment to the air–water interface also strongly depends on surface properties of colloids (hydrophobicity, zeta potential), solution chemistry (pH, ionic strength), and pore water flow rate. In addition, studies with clay colloids have suggested a strong influence of colloid shape on the retention by the air–water interface in unsaturated porous media (40).

Chen and Flury (43) reported that the repulsive interaction between colloids and the air–water interface was larger than that between colloids and the solid–water interface, causing colloids to be preferentially retained at the solid–water interface during unsaturated transport once the separation distance was small enough to overcome the energy barrier. However, some researchers have suggested that decreased solid–water interface attachment in unsaturated transport is likely, partly due to the decreased amount of solid–water interface that is accessible to colloids as water content decreases (36). In fact, many studies have found higher affinity of colloids for the air–water interface than the solid–water interface (24–26, 32). It is believed that colloid attachment to the air–water interface is irreversible due to the attractive capillary forces that hold colloids at the air–water interface once attachment occurs (29, 42, 44). For example, the capillary energy retaining carboxylate polystyrene latex spheres ($d = 0.05$–3 mm) at the air–water interface has been calculated to be approximately two to three orders of magnitude higher than the energy barrier obtained from DLVO theory for the same latex colloids (42). Under transient water flow, colloids attached to the air–water interface can be released into pore solution only when the quantity of air–water interface is diminished due to the resaturation of porous media (37, 42). If bubbles of air are mobilized by increased water flow, attached colloids may be transported by the migrating air bubbles (29, 30, 42).

Straining by the thin water films surrounding solid grains or by the contact line between the air–water interface and the solid phase (referred to as the air–water–solid contact line) (Fig. 6.4) is collectively referred to here as film straining. Water films surrounding solid grains may be as thin as tens of nanometers (36, 45), close to or smaller than the size of most of colloids and nanomaterials. As such, film straining is believed to be at least as important as air–water interface attachment in unsaturated transport (27, 32, 34, 35, 45–50). Studies with bacteria (34, 35), silica colloids (27), and latex microspheres (47–51) have suggested that film straining may be the dominant mechanism in immobilizing colloids during unsaturated transport in many systems. Through microscope visualization in thin sand chambers, negatively charged hydrophilic latex has been reported to be mainly retained by the air–water–solid contact line, but not by thin water films or the air–water interface (47–50). For hydrophobic latex, in addition to the primary retention at the contact line, attachment to the air–water interface and the solid–water interface has also been found to be significant (47, 50). Similar to the influence of pore straining on saturated retention, film straining in unsaturated transport has been reported to result in more significant retention at the column inlet than would be predicted by classical attachment theory (34, 35).

The significance of film straining primarily depends on the ratio between the particle size and the thickness of water films, as well as the pore water flow rate (45). As such, the stability (or the aggregation) of colloids and the water content in porous media play an essential role in film straining. Particle size has also been found to be important in determining the straining sites (air–water–solid contact line or thin water film) (47). In general, film straining is considered reversible, which means that the strained colloids can be quickly remobilized by the increase of water content (and the corresponding increase in film thickness), caused by transient water flow (46, 52, 53) or resaturation (43).

In addition to unsaturated filtration processes (air–water interface attachment and film straining), some saturated filtration processes may be enhanced in retaining colloids in unsaturated porous media compared to saturated porous media, and the dominant mechanisms for filtration may change as a function of saturation and system properties. For example, increased virus deposition to the solid–water interface has been observed at low saturations compared to more saturated conditions when the sand surface contains metal and metal oxides, but not when the surface is cleaned (54). Other researchers found the dominant mechanism for silica colloid retention in steady unsaturated flow changed from straining to air–water interface attachment, and finally to solid–water interface attachment as solution ionic strength increased from 2×10^{-4} M to 0.2 M (36).

Although there is little direct experimental evidence, pore straining has been suggested to be more significant under unsaturated conditions than saturated conditions for the same porous media, because water flow mainly occurs in increasingly smaller pores and the fraction of pores smaller than colloid size increases as saturation decreases (9). Under saturated conditions, water flow occurs preferentially in the larger, more conductive pores. As a porous medium drains and becomes unsaturated, the largest pores drain first, forcing flow through the smaller pores that are more likely to strain colloids. The likelihood of enhanced pore straining can be inferred from the pore size distribution. In unsaturated porous media, water distribution is controlled by capillary forces, and the effective pore size at a given saturation can be estimated from the measured capillary pressure–saturation curve using the Young–Laplace equation of capillarity (Eq. 6.1) (55):

$$r = \frac{2\gamma\cos\theta}{P_c} \qquad (6.1)$$

where, γ is the surface tension of the pore solution, θ is the contact angle between the air–water interface and the solid surface, P_c is the capillary pressure, and r is the effective pore radius of the largest saturated pore at a given P_c. From Equation 6.1, it is apparent that at higher capillary pressure (and correspondingly lower saturation) the effective radius of pores remaining saturated is smaller. Several authors have used calculations related to Equation 6.1 to assess the effects of straining in full (9) and partially filled (36) pores.

6.3 CAPILLARY PRESSURE–SATURATION (P_C–S) RELATIONSHIPS AND AIR–WATER INTERFACIAL AREA IN THE UNSATURATED ZONE

The magnitude and distribution of air–water interface in an unsaturated porous medium are expected to influence the significance of both air–water interface attachment and film straining. The quantity and spatial distribution of air–water interfacial area depend on the wetting–drying path followed to reach a particular water content. To fully understand the relationship between interfacial area and saturation, it is necessary to first understand capillary pressure–saturation (P_c–S) relationships.

6.3.1 Capillary Pressure–Saturation (P_c–S) Relationships of Porous Media

The capillary pressure–saturation (P_c–S) relationship, which is also known as the water retention function or moisture characteristic curve, is one of the fundamental hydraulic properties of porous media controlling unsaturated flow in the subsurface.

Saturation is defined as the ratio of wetting phase volume (V_w) to total void volume (V_v) in the medium $(S = V_w/V_v)$. At equilibrium, P_c is defined as the pressure difference between the nonwetting phase (P_n) and the wetting phase (P_w). In the unsaturated zone in the subsurface, the pressure of the nonwetting phase, air, is typically atmospheric pressure (zero gage); the pressure of wetting phase, water, is smaller than the air pressure due to capillarity. As such, capillary pressure in natural systems is often expressed as the soil matric suction, ψ $(\psi = -P_w)$.

Capillary pores in the unsaturated zone draw water up from the saturated zone below the water table to form the capillary fringe, which is a saturated region of soil above the water table. Because of irregular pore sizes in natural systems, the capillary fringe is also often irregular. The smaller the pores (or the finer the soil) the higher the capillary water rises above the water table. In consequence, soil saturation in the unsaturated zone typically increases with depth as the water table is approached. In fact, the natural soil moisture profile at equilibrium is equivalent to the P_c–S curve (Fig. 6.5).

Although P_c–S relationships are of practical use in a wide range of applications, their use is complicated by their hysteretic nature. Hysteresis in the P_c–S relationship means that at a given capillary pressure, a porous medium being drained has a higher saturation than the same medium undergoing imbibition. Figure 6.6 shows a detailed hysteretic P_c–S relationship of U.S. Silica F-95 sand measured by Chen et al. (56) using a rapid pseudostatic method, starting from full water saturation $(S = 1.0)$. As shown in Fig. 6.6, as P_c is increased from zero, the observed behavior follows the primary drainage (drying) curve until the residual wetting phase saturation (S_{wr}) is reached. The P_c value above which porous media start to drain significantly is called the displacement pressure. If P_c is then decreased to zero following primary drainage, the observed behavior follows the main imbibition (wetting) curve until the natural wetting phase residual saturation (S_{nwr}) is reached. Secondary drainage is then obtained when P_c is increased again. Scanning curves define points bounded by the secondary drainage and main imbibition curves, and can be obtained by reversing the applied capillary pressure at any point along the P_c–S relationship. The fact that P_c–S relationships are hysteretic indicates that the configuration and distribution of water within soil pores at a particular saturation,

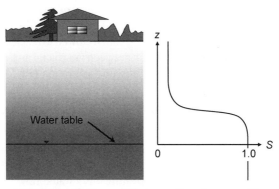

FIGURE 6.5 The equilibrium soil moisture profile.

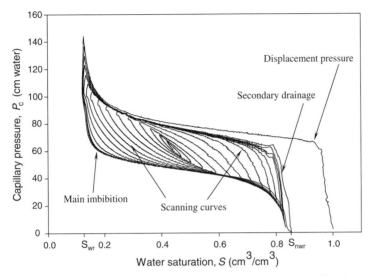

FIGURE 6.6   The measured air–water capillary pressure–saturation (P_c–S) relationship for U.S. silica F-95 fine sand. (Reprinted with permission from Chen et al. (56), copyright ASTM International.)

and the corresponding area of air–water interface, depend on the wetting/drying history the soil has experienced.

6.3.2   Air–Water Interfacial Areas in Porous Media

Because of the important role of the air–water interface in retarding the transport of colloidal particles, the quantification of air–water interfacial area (A) is essential to better understand the retention of colloids in the unsaturated porous media. The air–water interfacial area depends on both the distribution and the content of water phase in the porous media. Studies have found that the air–water interfacial area increases monotonically with decreasing water saturation, with the maximum interfacial area occurring at residual water saturation (S_{wr}) (57–61). Because the P_c–S curve is hysteretic, it is reasonable to expect that interfacial area itself may depend on wetting/drying history. As such, how the wetting/drying history influences the dependence of interfacial area on saturation has practical implications for both qualitative and quantitative understanding of colloid transport in unsaturated porous media.

   Using a new method developed to measure fluid–fluid interfacial areas of porous media during multiple drainages, Chen and Kibbey (61) measured the air–water interfacial area of a fine sand (U.S. Silica F-110) as a function of capillary pressure and saturation during primary, secondary, and one scanning drainages to explore the influence of wetting/drying history on interfacial area (Fig. 6.7). Figure 6.7 shows that over most of the saturation range studied, higher air–water interfacial area was observed for the higher level drainages (i.e., secondary and scanning drainages) at a given saturation. This implies that colloid retention during unsaturated transport may also depend on the wetting/drying history.

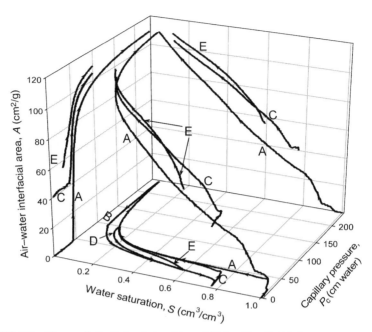

FIGURE 6.7   Measured P_c–S–A relationships of U.S. silica F-110 fine sand showing the influence of drying/wetting history. A: primary drainage; B: main imbibition; C: secondary drainage; D: scanning imbibition; E: scanning drainage. Note that interfacial areas were only measured during drainages. (Reprinted with permission from Chen and Kibbey (61), copyright American Chemical Society.)

6.4   MOBILIZATION AND TRANSPORT OF COLLOIDS DURING DRAINAGE AND IMBIBITION

A few studies have been reported relating the transport of colloids during drainage and imbibition, with most of them focusing on the mobilization of natural colloids during simulated infiltration events (27, 52, 62–64). Most studies have observed pulses of colloid release during consecutive multiple infiltration events with the peak of the release observed at the beginning of each infiltration event, and the height of the peak decreasing with each consecutive infiltration (52, 62–64). Some researchers have also found a second peak in each pulse after the infiltration event stops, with the height either smaller than (64) or equal to (63) the height of the peak released at the beginning of infiltration. This observation indicates that the shifting between imbibition and drainage may cause the redistribution of both water and air phases with the consequent movement of air–water interfaces, which release colloids into pore water. This conjecture was supported by the finding that multiple short infiltration cycles can mobilize more colloids than a single infiltration with the same effluent volume (64). It was further found that, similar to steady-state transport, the mobilization of colloids by infiltration is reduced by the increased ionic strength of the infiltration solution (64).

Lenhart and Saiers (27) investigated the dependence of colloid transport on the wetting/drying history by injecting a colloid suspension through a sand column of a particular saturation near water residual saturation ($S = 0.27$; $S_{wr} = 0.13$) reached by either drainage or imbibition. They found that at steady state, more colloids were retained when initial saturation had been reached by imbibition. When the saturation of the sand was subsequently increased, a higher percent of colloids that had been retained following imbibition were released compared to the colloids that had been retained following drainage.

Most studies of unsaturated transport published to date have focused on the transport of colloidal particles. However, little information about unsaturated transport of engineered nanomaterials is available. Furthermore, most reported work has emphasized steady-state unsaturated transport. However, unsteady conditions prevail in the vadose zone as a result of precipitation events and changing water table levels. In the subsequent sections, we examine the dynamic unsaturated transport of nanomaterials in porous media as saturation continuously decreases during drainage.

6.5 EXPERIMENTAL MATERIALS AND METHODS

6.5.1 Materials

Two kinds of nanoparticles are studied in this chapter: tin oxide nanopowder (SnO_2, Sigma–Aldrich, St. Louis, MO), with a nominal particle size of 18 nm and PZC of 4.3 (Fig. 6.2), and polystyrene latex nanospheres with a nominal size of 240 nm (Duke Scientific, Palo Alto, CA). Stable SnO_2 water dispersions were obtained by sonicating for 5 min (Ultrasonic Processor, Cole–Parmer, Vernon Hills, IL). Latex nanospheres, which are sold in highly concentrated suspensions stabilized by trace amounts of surfactant, were diluted to the desired concentration and used without further treatment.

For both latex and SnO_2, a concentration of 25 mg/L was used in unsaturated transport experiments. To minimize the attachment of nanoparticles to the solid–water interface, nanoparticle suspensions were adjusted to pH 10 using 0.1 M NaOH creating conditions under which both nanoparticles and porous media are negatively charged. By minimizing interactions with the solid surface, the experiments allowed independent examination of the influence of the air phase on the retention of nanoparticles. Ionic strength (I) of particle suspensions was maintained at 0.2 mM by adding NaCl. Under these conditions, the SnO_2 suspension is stable for over 24 h (eight times the duration needed for the unsaturated experiment, including preparation time). The latex suspension is stable indefinitely.

The zeta potential and particle size of SnO_2 and latex water suspensions were measured under experimental conditions (pH = 10, $I = 0.2$ mM) using a dynamic light scattering zeta potential/particle sizer (NICOMP 380 ZLS, Particle Sizing Systems, Santa Barbara, CA). The data are shown in Table 6.1. The number-weighed mean particle size of SnO_2 in water was found to be 100 nm, which is approximately six times the nominal size (18 nm) designated by the manufacturer. This indicates that, as observed by many other researchers with other metal oxide nanomaterials (20, 21),

TABLE 6.1   Characteristics of SnO$_2$ and Latex Nanoparticles

Nanomaterial	Nominal Size (nm)	Mean Size in Suspension (nm)	Point of Zero Charge	Zeta Potential (mV) (pH = 10, I = 0.2 mM)
SnO$_2$	18	100	4.3	−26.8
Latex	240	210	ND^a	−36.4

^{a}Not determined.

SnO$_2$ nanoparticles exist in suspension as stable clusters rather than as individual primary particles.

pherical glass beads with 0.5 mm grain size (Scientific Industries Inc., Bohemia, NY) were used as porous media. Prior to use, fines were removed by sieving using a #45 mesh screen and then glass beads were cleaned with a sequence of acid and base washes, as described by Lenhart and Saiers (27), to remove any metal oxides or other impurities. The cleaned glass beads were conditioned in pH 10 water to minimize the positively charged sites at the surface before packing. The packed porosity was measured to be 0.37. The geometric surface area is calculated to be 120 cm^2/cm^3 dry glass beads based on the assumption that glass beads are smooth and spherical.

6.5.2   Air–Water Interfacial Area Measurement

Air–water interfacial area for the 0.5 mm glass beads was measured as a part of previous work (65) using a method developed earlier (61). In the method, an isomerically pure anionic surfactant, sodium octylbenzene sulfonate (SOBS) (Sigma–Aldrich, St. Louis, MO), is used at a concentration of 1.7 mM to saturate the porous medium, which is subsequently drained. As drainage proceeds, the SOBS concentration in the effluent decreases due to adsorption to the newly generated air–water interfaces. The decrease of the aqueous SOBS concentration is detected by an in-line UV spectrometer (Model: SD2000, Ocean Optics, Inc., Dunedin, FL) with a flow-through optical cell. Based on a continuous mole balance, combined with the Szyszkowski and Langmuir equations (66), the air–water interfacial area can be calculated as a function of saturation. In this research, air–water interfacial area was measured as a function of saturation for primary drainage, although the method is capable of measuring area during drainages throughout the entire hysteretic P_c–S relationship.

6.5.3   Unsaturated Transport of SnO$_2$ and Latex Nanoparticles During Primary Drainage

The method used here to study the unsaturated transport of SnO$_2$ and latex nanoparticles is the same as previously described (65). The retention of nanoparticles in unsaturated porous media was studied simultaneously with the measurement of the capillary pressure–saturation (P_c–S) relationship for primary drainage using the same device and method as used for air–water interfacial area measurement (61), but replacing the SOBS tracer with a nanomaterial suspension. Glass beads conditioned in pH 10 water were wet-packed in a custom soil cell (2.54 cm ID × 1.27 cm HT) and were saturated with the nanoparticle suspension before drainage started. A hydrophilic nylon net filter with 30 μm pore size (Millipore, Billerica, MA) was used as the gas-phase capillary barrier and a hydrophobic Teflon® membrane with 0.22 μm pore size (Osmonics Inc., Minnetonka, MN) was used as a water-phase capillary barrier. During drainage, the decrease of nanoparticle concentration was continuously measured by the UV spectrometer as the suspension drained out of the cell. The full UV absorbance spectrum was used for particle quantification (61), and a continuous mass balance was used to determine the mass of nanoparticle retained by unsaturated porous media at any given saturation. By relating the retained mass to the air–water interfacial area measured at the same saturation, the role of air–water interface on the unsaturated retention of SnO_2 nanoparticles was evaluated.

To quickly saturate the porous media with nanoparticle suspensions and to minimize the possibility of saturated deposition during drainage, two different flushing rates were applied in the saturation process. At a flow rate of 1.0 mL/min (Darcy velocity 0.20 cm/min), the wet-packed cell was first flushed with pH 10 water to remove glass components dissolved during conditioning and was then flushed with nanoparticle suspension to replace the pore solution with a suspension of nanoparticles. After the breakthrough of nanoparticles, the flushing rate was decreased to the predominant drainage flow rate. Predominant drainage flow rate is defined here as the flow rate observed in the near-horizontal region of the P_c–S curve (between the saturations of 0.85 and 0.20), where the majority of drainage occurs. As the saturated flushing rate is reduced to the predominant drainage flow rate, flow rates in the system before and during drainage are relatively constant, which minimizes the possibility of saturated filtration during drainage due to significant flow rate decrease.

After glass beads were saturated with nanoparticle suspension and a stable effluent concentration was observed at the predominant drainage flow rate, the flow was stopped and a hydrophobic Teflon membrane with 0.22 μm pore size (Osmonics Inc., Minnetonka, MN) was placed at the top of the porous medium to prevent water phase from being pulled into the sintered stainless steel frit at the top of the porous medium chamber. The pressurized gas phase (nitrogen) was applied from the top of the cell and the pressure was automatically increased at the predetermined rate to initiate primary drainage. When the gas phase pressure is high enough, capillary pressure (P_c) exceeds the displacement pressure and drainage occurs (Fig. 6.6). Details on experimental setup, procedure, operation, and calculation have been previously described by Chen and Kibbey (61) and Chen et al. (56).

The dependence of unsaturated retention of nanomaterials on drainage rate has been investigated with TiO_2 nanoparticles in a previous study (65). In this chapter, the dynamic unsaturated transport of SnO_2 and latex during primary drainage is studied at a single gas pressure rate (the rate at which gas pressure is ramped to produce drainage). The gas pressure rate was controlled at 0.026 cm water per second after the displacement pressure had been exceeded. This gas pressure rate produces a predominant drainage flow rate of approximately 0.047 mL/min for the porous medium studied. Below the displacement pressure, faster rates were used to minimize the

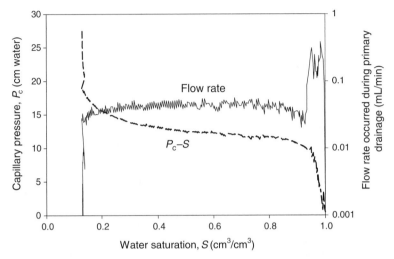

FIGURE 6.8   Air–water capillary pressure–saturation (P_c–S) relationship (dashed line) and flow rate as a function of saturation (solid line) for primary drainage of 0.5 mm glass beads. (Data reproduced with permission from Chen, et al. (65), copyright American Chemical Society.)

duration prior to the onset of drainage flow, to reduce the potential for filtration prior to drainage, which can occur in saturated porous media under slow flow conditions (20, 22). The duration of the entire primary drainage was approximately 54 min for each experiment. The measured P_c–S curve and corresponding flow rate as a function of saturation are shown in Fig. 6.8. Previous work has shown that the drainage curve produced at the applied gas pressure rate (0.026 cm water per second) is close to the static equilibrium curve (56, 65).

6.5.4 Saturated Transport of SnO$_2$ and Latex Nanoparticles

To rule out the possibility of saturated filtration due to flow rate fluctuations observed during unsaturated transport, saturated transport of SnO$_2$, and latex nanoparticles was also studied. Methods for the packing of glass beads and nanomaterial detection were the same as for unsaturated transport. The flow rates observed in unsaturated transport and applied in saturated flushing prior to drainage, as well as the duration at each rate, were duplicated in saturated transport experiments by a digital syringe pump (KDS 230, KD Scientific Inc., Holliston, MA).

6.6 RESULTS AND DISCUSSION

6.6.1 Saturated Transport of SnO$_2$ and Latex Nanoparticles

Figure 6.9 shows the saturated transport of SnO$_2$ and latex nanoparticles at flow rates equivalent to those observed in the drainage experiment (Fig. 6.8). From the figure, it is apparent that after breakthrough at the high flow rate (1.0 mL/min) both SnO$_2$ and latex concentrations remain relatively constant when flow rate is decreased to the

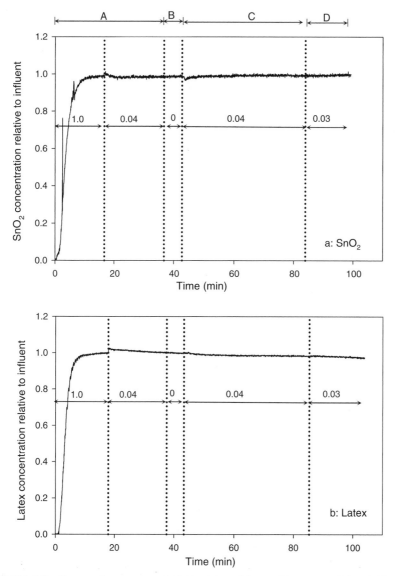

FIGURE 6.9   Saturated transport of SnO$_2$ (a) and latex (b) nanoparticles (pH = 10, $I = 0.2$ mM) in 0.5 mm glass beads at varying flow rates as observed in the primary drainage. Numbers shown in the figure are flow rates (mL/min) applied over different time intervals, as indicated by dotted vertical lines. Regions indicated at the top of the figure correspond to the following flow rates in an unsaturated drainage experiment: A. Initial cell saturation; B. Top membrane changes; C. Drainage — predominant flow rate; D. Drainage — decreased flow near residual saturation.

predominant drainage flow rate (0.04 mL/min). The constant nanoparticle concentrations continue upon further flow rate changes. The flow rate of zero represents the stopped flow when placing the hydrophobic Teflon membrane on top of the porous medium. The flow resumes at 0.04 mL/min for approximately 40 min, simulating the major part of drainage before approaching the residual saturation. The flow rate is then reduced to 0.03 mL/min, simulating the part of drainage close to residual saturation. The constant nanoparticle concentrations observed over the entire saturated transport experiment in Fig. 6.9 suggest that the saturated filtration of nanoparticles by the glass beads is negligible. Thus, any concentration decrease observed during drainage must result from the filtration induced by the introduction of the air phase.

6.6.2 Dynamic Unsaturated Retention of SnO$_2$ and Latex Nanoparticles During Primary Drainage

The mass of SnO$_2$ and latex nanoparticles (M) retained in porous media during primary drainage is plotted in Fig. 6.10 as a function of saturation (S). Individual replicate experiments for each nanomaterial are displayed in the 3D domain, and the averaged M–S relationships are shown on the M–S plane. In general, for both SnO$_2$ and latex, the retained mass increases with decreasing saturation, with a more dramatic increase observed as saturation drops below 0.4. The M–S trend observed here for SnO$_2$ and latex is the same as that observed for TiO$_2$ at four different drainage rates in the

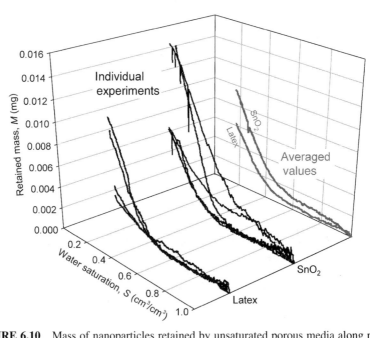

FIGURE 6.10   Mass of nanoparticles retained by unsaturated porous media along primary drainages. Black lines in the 3D domain display four replicates for latex and five replicates for SnO$_2$. Gray lines shown on the M–S plane are average retained mass as a function of saturation.

previous work (65). Although slightly higher mass retention is observed for SnO_2 at all saturations compared to latex, given the scatter in the data it would be difficult to say the difference is significant. The masses of SnO_2 and latex retained as a function of saturation are also on the same order of magnitude of the previous results for TiO_2 obtained under the same experimental conditions (65).

Recall that the results of saturated transport (Fig. 6.9) indicate that filtration due to saturated mechanisms is minimal at the flow rates observed in unsaturated transport (Fig. 6.9). Further, pore straining is likely negligible because the nanoparticle-to-porous media size ratios are approximately 2.0×10^{-4} for SnO_2 and 4.1×10^{-4} for latex — 8 and 4 times smaller, respectively, than the smallest critical value (0.0017) reported in the literature (14). Also, at the residual saturation (the end of drainage flow) shown in the measured P_c–S curve (Fig. 6.8), the capillary pressure is approximately 20 cm water, which corresponds to an effective saturated pore size of approximately 140 μm in diameter (Eq. 6.1). This saturated pore size is three orders of magnitude larger than the latex (205 nm or 0.2 μm) and SnO_2 (100 nm or 0.1 μm) average particle sizes (Table 6.1). As such, it is unlikely that increased pore straining plays a major role in the dynamic unsaturated transport experiments described here. As a consequence, the retention of SnO_2 and latex nanoparticles observed during primary drainage (Fig. 6.10) likely results from typical unsaturated phenomena, such as air–water interfacial attachment and film straining. As drainage proceeds and saturation decreases, more air–water interface is formed resulting in higher potential for air–water interfacial attachment. In addition, the water films around the grains become thinner as saturation decreases, increasing the likelihood of film straining. Both of these unsaturated processes could account for the unsaturated retention observed in Fig. 6.10.

6.6.3 The Dependence of the Dynamic Unsaturated Retention of SnO₂ and Latex Nanoparticles on the Air–Water Interfacial Area During Primary Drainage

The air–water interfacial area (A) measured for the 0.5 mm glass beads during primary drainage has been reported in previous work (65). In general, air–water interfacial area increases monotonically as saturation decreases. The A–S relationship for glass beads is similar to that of the fine sand shown in Fig. 6.7, except that a lower area is observed for the glass beads at a given saturation. At the residual saturation, the measured air–water interfacial area of the glass beads is approximately 58 cm^2/cm^3 dry grain volume (65), which is half that of F-110 fine sand shown in Fig. 6.7.

Comparison between Figs 6.10 and 6.7 shows that the measured air–water interfacial area shares the same trend as the mass of retained nanoparticles, in terms of their dependence on saturation. That is, both air–water interfacial area and mass of retained nanoparticles increase monotonically with decreasing saturation. To better understand the likely mechanisms responsible for the observed retention, Fig. 6.11 shows the area-normalized nanomaterial mass retention based on the averaged mass retention values shown in Fig. 6.10. If adsorption to air water interfaces was the primary mechanism causing retention in the unsaturated porous medium, the area-normalized curves shown in Fig. 6.11 might be expected to be near horizontal,

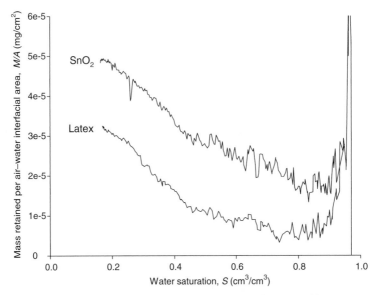

FIGURE 6.11   Air–water interfacial area-normalized mass of SnO_2 and latex nanomaterials retained by unsaturated porous media during primary drainage.

as newly formed interfacial area is occupied with a similar adsorption density of nanomaterials. Examining the curves for both nanomaterials in Fig. 6.11, it is apparent that the behavior at higher saturations (e.g., above $S \approx 0.4$) is not inconsistent with adsorption to air–water interfaces. That is, although there is considerable scatter in the data, the mass per area for each nanomaterial is approximately constant above a saturation of 0.4. (It should be noted that values very close to full saturation, $S = 1.0$, show substantial spikes due to normalization by near-zero area). As saturation drops below approximately 0.4, the amount retained per area begins to increase more significantly with decreasing saturation, suggesting that adsorption to air–water interfaces is not the major mechanism at low saturations. This observation is consistent with the increasing presence of thin water films as saturation decreases, and the increased likelihood of film straining. These observations are similar to the results obtained for TiO_2 nanoparticles in previous work (65).

Although the extent of scatter in the data makes firm conclusions difficult, the data in Fig. 6.11 also show that the SnO_2 nanoparticles may be retained to a greater extent than the latex nanoparticles over much of the saturation range. More work is needed to assess the reasons for this difference, but it may result at least partly from the different densities of the nanoparticles and the correspondingly different number concentrations. Furthermore, more work is needed to understand the effect of the different nanoparticle size distributions on retention behavior.

6.7   CONCLUSIONS

A number of mechanisms are responsible for the transport of colloids and nanomaterials in saturated and unsaturated porous media. In saturated porous media, the movement of colloids and nanomaterials is influenced by interactions with solid–water interfaces, suspension stability, as well as straining of colloids/nanoparticles and aggregates by saturated pores. The presence of an air phase in unsaturated porous media adds the potential for additional mechanisms, including adsorption to air–water interfaces, straining by water films, and enhanced straining by saturated pores as water flow is directed away from larger pores that are filled with air. The work described in this chapter was designed to examine dynamic unsaturated transport under conditions that minimize saturated transport mechanisms. Results showed that nanoparticle retention at high saturations is consistent with adsorption to air–water interfaces, with increasing contributions from film straining likely at lower saturations. Future work examining the transport under conditions that permit both saturated and unsaturated mechanisms would be useful for assessing the relative importance of unsaturated mechanisms under real-world conditions.

ACKNOWLEDGMENTS

Although the research described in this article has been funded by the United States Environmental Protection Agency through grant R832529, it has not been subjected to the Agency's required peer and policy review and therefore, does not necessarily reflect the views of the Agency and no official endorsement should be inferred.

REFERENCES

1. O'Melia, C.R. Aquasols: the behavior of small particles in aquatic systems. *Environ Sci Technol* 1980;14:1052–1060.
2. Elimelech, M., O'Melia, C.R. Kinetics of deposition of colloidal particles in porous media. *Environ Sci Technol* 1990;24:1528–1536.
3. O'Melia, C.R., Tiller, C.L. Physicochemical aggregation and deposition in aquatic environments. In: Buffle, J. van Leeuwen, H.P., editors. *Environmental Particles.* Chelsea, MI: Lewis Publishers; 1993. pp. 353–386.
4. Yao, K.-M., Habibian, M.T., O'Melia, C.R. Water and waste water filtration: concepts and applications. *Environ Sci Technol* 1971;5:1105–1112.
5. Tufenkji, N., Elimelech, M. Correlation equation for predicting single-collector efficiency in physicochemical filtration in saturated porous media. *Environ Sci Technol* 2004;38: 529–536.
6. Elliott, D.W., Zhang, W.-X. Field assessment of nanoscale bimetallic particles for groundwater treatment. *Environ Sci Technol* 2001;35:4922–4926.
7. Stumm, W., Morgan, J.J. *Aquatic Chemistry,* 3rd ed. New York: John-Willey & Sons, Inc.; 1996. Chapter 9, pp. 516–613.
8. Verwey, E.J.W., Overbeek, J.Th.G. *Theory of the Stability of Lyophobic Colloids* Amsterdam: Elsevier; 1948.
9. Bradford, S.A., Simunek, J., Bettahar, M., van Genuchten, M.T., Yates, S.R. Significance of straining in colloid deposition: evidence and implications. *Water Resour Res* 2006;42: W12S15. DOI: 10.1029/2005WR004791.

10. Bradford, S.A., Simunek, J., Bettahar, M., van Genuchten, M. Th., Yates, S.R. Modeling colloid attachment, straining, and exclusion in saturated porous media. *Environ Sci Technol* 2003;37:2242–2250.

11. Bradford, S.A., Simunek, J., Walker, S.L. Transport and straining of *E. coli* O157:H7 in saturated porous media. *Water Resour Res* 2006;42:W12S12. DOI: 10.1029/ 2005WR004805.

12. Herzig, J.P., Leclerc, D.M., Goff, P.L. Flow of suspensions through porous media — application to deep filtration. *Ind Eng Chem* 1970;62(5): 8–35.

13. Xu, S., Gao, B., Saiers, J.E. Straining of colloidal particles in saturated porous media. *Water Resour Res* 2006;42:W12S16. DOI: 10.1029/2006WR004948.

14. Bradford, S.A., Yates, S.R., Bettahar, M., Simunek, J. Physical factors affecting the transport and fate of colloids in saturated porous media. *Water Resour Res* 2002;38(12): 1327. DOI: 10.1029/2002WR001340.

15. Brant, J., Lecoanet, H., Wiesner, M.R. Aggregation and deposition characteristics of fullerene nanoparticles in aqueous systems. *J Nanopart Res* 2005;7:545–553.

16. Dunphy Guzman, K.A., Finnegan, M.P., Banfield, J.F. Influence of surface potential on aggregation and transport of titania nanoparticles. *Environ Sci Technol* 2006;40:7688–7693.

17. Franchi, A., O'Melia, C.R. Effects of natural organic matter and solution chemistry on the deposition and reentrainment of colloids in porous media. *Environ Sci Technol* 2003;37:1122–1129.

18. Mulvaney, P. Metal nanoparticles: double layers, optical properties, and electrochemistry. In: Klabunde, K.J. editor. *Nanoscale Materials in Chemistry*, John Wiley & Sons; 2001. pp. 121–168.

19. Kallay, N., Žalac, S. Stability of nanodispersions: a model for kinetics of aggregation of nanoparticles. *J Colloid Interface Sci* 2002;253:70–76.

20. Lecoanet, H.F., Wiesner, M.R. Velocity effects on fullerene and oxide nanoparticle deposition in porous media. *Environ Sci Technol* 2004;38:4377–4382.

21. Lecoanet, H.F., Bottero, J.-Y., Wiesner, M.R. Laboratory assessment of the mobility of nanomaterials in porous media. *Environ Sci Technol* 2004;38:5164–5169.

22. Cheng, X., Kan, A.T., Tomson, M.B. Study of C_{60} transport in porous media and the effect of sorbed C_{60} on naphthalene transport. *J Mater Res* 2005;20:3244–3254.

23. Phenrat, T., Saleh, N., Sirk, K., Tilton, R.D., Lowry, G.V. Aggregation and sedimentation of aqueous nanoscale zerovalent iron dispersions. *Environ Sci Technol* 2007;41:284–290.

24. Wan, J., Wilson, J.L. Colloid transport in unsaturated porous media. *Water Resour Res* 1994;30:857–864.

25. Schäfer, A., Ustohal, P., Harms, H., Stauffer, F., Dracos, T., Zehnder, A.J.B. Transport of bacteria in unsaturated porous media. *J Contam Hydrol* 1998;33:149–169.

26. Sim, Y., Chrysikopoulos, C.V. Virus transport in unsaturated porous media. *Water Resour Res* 2000;36:173–179.

27. Lenhart, J.J., Saiers, J.E. Transport of silica colloids through unsaturated porous media: experimental results and model comparisons. *Environ Sci Technol* 2002;36:769–777.

28. Cherrey, K.D., Flury, M., Harsh, J.B. Nitrate and colloid transport through coarse Hanford sediments under steady state, variably saturated flow. *Water Resour Res* 2003;39(6): 1165. DOI: 10.1029/2002WR001944.

29. Wan, J., Wilson, J.L. Visualization of the role of the gas-water interface on the fate and transport of colloids in porous media. *Water Resour Res* 1994;30:11–23.

30. Wan, J., Wilson, J.L., Kieft, T.L. Influence of the gas–water interface on transport of micro-organisms through unsaturated porous media. *Appl Environ Microbiol* 1994;60:509–516.

31. Schäfer, A., Harms, H., Zehnder, A.J.B. Bacterial accumulation at air–water interface. *Environ Sci Technol* 1998;32:3704–3712.

32. Lazouskaya, V., Jin, Y., Or, D. Interfacial interactions and colloid retention under steady flows in a capillary channel. *J Colloid Interface Sci* 2006;33:171–184.

33. Corapcioglu, M.Y., Choi, H. Modeling colloid transport in unsaturated porous media and validation with laboratory column data. *Water Resour Res* 1996;32:3437–3449.

34. Gargiulo, G., Bradford, S.A., Simunek, J., Ustonal, P., Vereecken, H., Klumpp, E. Transport and deposition of metabolically active and stationary phase *Deinococcus radiodurans* in unsaturated porous media. *Environ Sci Technol* 2007;41:1265–1271.

35. Gargiulo, G., Bradford, S.A., Simunek, J., Ustonal, P., Vereecken, H., Klumpp, E. Bacteria transport and deposition under unsaturated conditions: the role of the matrix grain size and the bacteria surface protein. *J Contam Hydrol* 2007;92:255–273.

36. Saiers, J.E., Lenhart, J.J. Ionic-strength effects on colloid transport and interfacial reactions in partially saturated porous media. *Water Resour Res* 2003;39:1256. DOI: 10.1029/2002WR001887.

37. Torkzaban, S., Hassanizadeh, S.M., Schijven, J.F., van den Berg, H.H.J.L. Role of air–water interfaces on retention of viruses under unsaturated conditions. *Water Resour Res* 2006;42: W12S14. DOI: 10.1029/2006WR004904.

38. Li, C., Somasundaran, P. Reversal of bubble charge in multivalent inorganic salt solutions — effect of magnesium. *J Colloid Interface Sci* 1991;146(1): 215–218.

39. Graciaa, A., Morel, G., Saulner, P., Lachaise, J., Schechter, R.S. The ζ–potential of gas bubbles. *J Colloid Interface Sci* 1995;172:131–136.

40. Wan, J., Tokunaga, T.K. Partitioning of clay colloids at air–water interfaces. *J Colloid Interface Sci* 2002;247:54–61.

41. Ducker, W.A., Xu, Z., Israelachvili, J.N. Measurements of hydrophobic and DLVO forces in bubble-surface interactions in aqueous solutions. *Langmuir* 1994;10:3279–3289.

42. Sirivithayapakorn, S., Keller, A. Transport of colloids in unsaturated porous media: a pore-scale observation of processes during the dissolution of air–water interface. *Water Resour Res* 2003;39(12): 1346. DOI: 10.1029/2003WR002487.

43. Chen, G., Flury, M. Retention of mineral colloids in unsaturated porous media as related to their surface properties. *Colloid Surface A* 2005;256:207–216.

44. Williams, D.F., Gerg, J.C. The aggregation of colloidal particles at the air–water interface. *J Colloid Interface Sci* 1992;152:218–229.

45. Wan, J., Tokunaga, T.K. Film straining of colloids in unsaturated porous media: conceptual model and experimental testing. *Environ Sci Technol* 1997;31:2413–2420.

46. Gao, B., Saiers, J.E., Ryan, J.N. Deposition and mobilization of clay colloids in unsaturated porous media. *Water Resour Res* 2004;40:W08602. DOI: 10.1029/2004WR003189.

47. Zevi, Y., Dathe, A., McCarthy, J.F., Richards, B.K., Steenhuis, T.S. Distribution of colloid particles onto interfaces in partially saturated sand. *Environ Sci Technol* 2005;39: 7055–7064.

48. Zevi, Y., Dathe, A., Gao, B., Richards, B.K., Steenhuis, T.S. Quantifying colloid retention in partially saturated porous media. *Water Resour Res* 2006;42:W12S03. DOI : 10.1029/2006WR004929.

49. Crist, J.T., McCarthy, J.F., Zevi, Y., Baveye, P., Throop, J.A., Steenhuis, T.S. Pore-scale visualization of colloid transport and retention in partly saturated porous media. *Vadose Zone J* 2004;3:444–450.

50. Crist, J.T., Zevi, Y., McCarthy, J.F., Throop, J.A., Steenhuis, T.S. Transport and retention mechanisms of colloids in partially saturated porous media. *Vadose Zone J* 2005;4:184–195.

51. Chen, G., Abichou, T., Tawfiq, K., Subramaniam, P.K. Impact of surface charge density on colloid deposition in unsaturated porous media. *Colloid Surface A* 2007;302:342–348.

52. Saiers, J.E., Lenhart, J.J. Colloid mobilization and transport within unsaturated porous media under transient-flow conditions. *Water Resour Res* 2003;39:1019. DOI: 10.1029/2002WR001370.

53. Keller, A., Sirivithayapakorn, S. Transport of colloids in unsaturated porous media: explaining large-scale behavior based on pore-scale mechanisms. *Water Resour Res* 2004;40:W12403. DOI: 10.1029/2004WR003315.

54. Chu, Y., Jin, Y., Flury, M., Yates, M.V. Mechanisms of virus removal during transport in unsaturated porous media. *Water Resour Res* 2001;37(2): 253–263.

55. Hiemenz, P.C., Rajagopalan, R. *Principles of Colloid and Surface Chemistry,* 3rd ed. New York: Marcel Dekker; 1997.

56. Chen, L., Miller, G., Kibbey, T.C.G. Rapid pseudo-static measurement of hysteretic capillary pressure-saturation relationships in unconsolidated porous media. *Geotech Test J* 2007;30(6):474–483.

57. Kim, H., Rao, P.S.C., Annable, M.D. Determination of effective air–water interfacial area in partially saturated porous media using surfactant adsorption. *Water Resour Res* 1997;33:2705–2711.

58. Saripalli, K.P., Kim, H., Rao, P.S.C., Annable, M.D. Measurement of specific fluid-fluid interfacial areas of immiscible fluids in porous media. *Environ Sci Technol* 1997;31:932–936.

59. Anwar, A.H.M.F., Bettahar, M., Matsubayashi, U.A. Method for determining air–water interfacial area in variably saturated porous media. *J Contam Hydrol* 2000;43:129–146.

60. Schaefer, C.E., DiCarlo, D.A., Blunt, M.J. Experimental measurement of air–water interfacial area during gravity drainage and secondary imbibition in porous media. *Water Resour Res* 2000;36:885–890.

61. Chen, L., Kibbey, T.C.G. Measurement of air–water interfacial area for multiple hysteretic drainage curves in an unsaturated fine sand. *Langmuir* 2006;22:6874–6880.

62. Ryan, J.N., Illangasekare, T.H., Litaor, M.J., Shannon, R. Particle and plutonium mobilization in macroporous soils during rainfall simulations. *Environ Sci Technol* 1998;32:476–482.

63. El-Farhan, Y.H., Denovio, N.M., Herman, J.S., Hornberger, G.M. Mobilization and transport of soil particles during infiltration experiments in an agricultural field. *Environ Sci Technol* 2000;34:3555–3559.

64. Zhuang, J., McCarthy, J.F., Tyner, J.S., Perfect, E., Flury, M. *In situ* colloid mobilization in Hanford sediments under unsaturated transient flow conditions: effect of irrigation pattern. *Environ Sci Technol* 2007;41:3199–3204.

65. Chen, L., Sabatini, D.A., Kibbey, T.C.G. 2008. The role of the air–water interface in the retention of TiO_2 nanoparticles in porous media during primary drainage. *Environ Sci Technol* 2008;42:1916–1921.

66. Rosen, M.J. *Surfactants and Interfacial Phenomena*, 2nd ed. New York: John-Wiley & Sons, Inc.; 1988. pp. 83 and 84.

CHAPTER 7

Surface Oxides on Carbon Nanotubes (CNTs): Effects on CNT Stability and Sorption Properties in Aquatic Environments

HOWARD FAIRBROTHER[1], BILLY SMITH[1], JOSH WNUK[1], KEVIN WEPASNICK[1], WILLIAM P. BALL[2], HYUNHEE CHO[2], and FAZLULLAH K. BANGASH[3]

[1]Department of Chemistry, Johns Hopkins University, 3400 North Charles Street, Baltimore, MD 21218, USA
[2]Department of Geography and Environmental Engineering, Johns Hopkins University, 3400 North Charles Street, Baltimore, MD 21218, USA
[3]Institute for Chemical Sciences, University of Peshawar, Peshawar 25120, Pakistan

7.1 OVERVIEW

The expanding commercial applications of carbon nanotubes (CNTs) have been responsible for rapidly increasing production rates of these engineered nanomaterials. This will inevitably result in greater quantities of CNTs being released into atmospheric, terrestrial, and aquatic environments (1–6), where their fate and impact remain unclear (3, 7–10). One systematic and scientifically rigorous approach to understanding the behavior of nanomaterials in the environment is to correlate measurable material characteristics (e.g., size, shape, aggregation state, surface area, chemical composition, and surface chemistry) with phenomenological properties (e.g., sorption affinities and colloidal stability) that will have a more direct bearing on the nanomaterial's environmental fate and effect. Results obtained from such studies will not only facilitate the development of models capable of predicting the behavior and impact of engineered nanomaterials in different environments, but may also be used to improve the design and environmental sustainability of engineered nanomaterials before more widespread commercialization occurs.

Nanoscience and Nanotechnology   Edited by Vicki H. Grassian

7.2 BACKGROUND

Carbon nanotubes consist of one or more graphene sheets rolled into a long, thin, and hollow cylinder. Two structural forms of CNTs exist: single-walled (SWCNTs) and multiwalled (MWCNTs) carbon nanotubes. A SWCNT is a single sheet of carbon atoms, normally capped at the ends, while an MWCNT is composed of two or more concentric cylinders of carbon held together by van der Waals forces (see Fig. 7.1) (11, 12). SWCNTs and MWCNTs exhibit extremely high surface area to volume ratios and large aspect ratios, with lengths between 1 and 100 μm and internal diameters typically between 1 and 25 nm (13, 14).

The unique physical and chemical properties of CNTs give rise to some remarkable material properties, including tensile strengths 100 times greater than steel (14, 15), an electrical conductivity as high as copper, and a thermal conductivity comparable to diamond (15, 16). This combination of highly desirable properties means that CNTs are attractive components in numerous real-world applications. For example, some product developers are using MWCNTs in polymer composites to add mechanical strength and electrical conductivity (17). CNTs are also attractive materials for hydrogen storage devices and catalytic supports as a consequence of their high surface area to volume ratios (18–21). In other applications, CNTs are being considered for incorporation into bullet proof vests, for use in a variety of microelectronic devices (22), as potential electron field emitters in flat-panel displays (23), as imaging agents in gene therapy (24), as vehicles for targeted drug delivery (25, 26), and as sorbents to remove harmful contaminants from water and wastewater (27–30).

The commercial market for CNTs is still in its infancy but growing rapidly. In 2004, the CNT marketplace was estimated to be worth $6 million and is projected to exceed $215 million by 2009 (31). This rapid growth in the projected applications of CNTs has

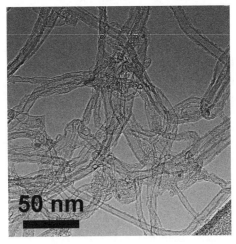

FIGURE 7.1   TEM image of pristine (as received) MWCNTs.

also been responsible for a dramatic increase in annual production rates: in 2004 the global annual production rate for MWCNTs was 100 tons and by 2008 this is expected to increase to 350 tons (32, 33).

7.2.1 Oxidation of Carbon Nanotubes

A growing number of potential applications, especially those in the biomedical sector, will require CNTs that can form stable aqueous suspensions. For the pristine nanomaterials, this requirement is precluded by the hydrophobic graphene surfaces that effectively prevents dispersion (suspension as individual particles) in polar liquids such as water and many polymeric resins (34–36). To render CNTs more hydrophilic and to facilitate the formation of well-dispersed suspensions, it is necessary to functionalize the exterior surfaces of CNTs (37) by incorporating hydrophilic substituents or adsorbing surfactants or other polymeric additives (37–39). Among the various methods available for CNT surface modification, the grafting of hydrophilic oxides onto the surface (Fig. 7.2) has emerged as a popular and versatile method for creating CNTs that more readily disperse in water (40–58). Various synthetic strategies have also been developed to deliberately introduce surface oxides into CNT surfaces, providing points of attachment in the creation of more complex nanostructures (59), such as metal-bearing nanoparticles for catalysis (20, 21) or biologically modified sensors (17).

The introduction of surface oxides onto CNTs is not always intentional. For example, a degree of surface oxidation always accompanies the use of aggressive oxidizing treatments (e.g., HNO_3, HNO_3/H_2SO_4, or H_2O_2) used in cleaning/purification processes designed to remove amorphous carbon and metallic impurities from pristine (as received) CNTs (40, 50, 60, 61). Surface oxidation can also occur when CNTs are released into the atmosphere and subsequently exposed to oxidizing agents such as ozone and hydroxyl radicals (46, 62–64). In aquatic environments, oxidation of CNTs is expected to occur in surface waters subjected to photolysis and some common water treatment processes such as UV irradiation, permanganate oxidation, and ozonolysis (46, 64). The various ways in which surface oxides can become incorporated into CNTs are summarized in Fig. 7.2.

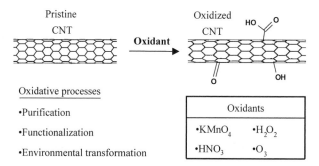

FIGURE 7.2 Oxidation of carbon nanotubes.

7.2.2 Influence of Surface Chemistry on the Environmental Impact of Carbon Nanotubes

Surface functionality regulates many important environmental characteristics and interfacial properties of black carbon (BC) particulates in aquatic environments, including their surface potential (zeta potential) at ambient pH, their point-of-zero charge (pH of surface neutrality), their reactivity in water (acid/base character), and their cohesive and adhesive properties (65). For CNTs, and other nanomaterials, the influence of surface chemistry on these properties will be accentuated since most of the atoms are located at or near the surface. Consequently, the introduction of surface oxides is expected to profoundly impact the aggregation state of CNTs in water as well as their interactions with other aqueous species, each other, and other solid phases in the environment, including their deposition properties during flow through porous media (65, 66). Thus, the interaction and uptake of nanomaterials by living organisms and the ability of nanomaterials to bind and transport pollutants are expected to depend critically upon surface chemistry, with corresponding implications for health and environmental safety. In fact, studies suggest that surface chemistry is the principal factor governing the biological activity and toxicity of nanoparticles (5, 67, 68).

In aquatic environments, carbon-based nanomaterials such as CNTs and fullerenes typically form colloidal aggregates due to van der Waals forces between the hydrophobic graphene surfaces (69, 70) and the lack of energetically favorable polar interactions with water. The absence of hydrophilic forces limits water adsorption. In fact, the entropic effects of water structuring around pristine CNTs add a substantial entropic contribution to the forces of aggregation—the so-called hydrophobic effect. Surface oxides, of which some will be negatively charged under environmentally relevant aquatic conditions, will influence the forces between individual CNTs as well as the hydration forces with surrounding water molecules. Changes in these intra- and intermolecular forces will profoundly influence the colloidal stability, aggregation state, and transport properties of CNTs in aquatic environments (69).

Depending on the extent of surface oxidation and the solution chemistry (e.g., pH, ionic strength), CNT aggregates may either form stable suspensions in water or rapidly settle as larger aggregates (71, 72). Similarly, CNT transport properties within the environment will be strongly influenced by the surface chemistry of the moving particles, the surface chemistry of the solids with which they collide (the collectors), and the solution chemistry (65, 73–75), owing to changes in forces of interaction between particles and "collectors." The importance of surface chemistry and the presence of surface oxides in particular is supported by the observation that fullerenes and fullerols (hydroxylated fullerenes) exhibit pronounced differences in their aquatic stability, aggregation state, transport properties, and mobility (76, 77).

The sorption properties of CNTs toward toxic chemicals in aquatic systems are also of environmental interest and concern not only because CNTs may find use as commercial sorbents but also because sorption can affect the behavior and/or the impact of CNTs that are deliberately or inadvertently introduced into the environment. This is of particular concern in regard to human and ecological toxicity, since the high surface area to volume ratios of CNTs means that even small concentrations of these nanomaterials can provoke profound changes in the distribution, availability, and transport properties of toxic chemicals (67, 68). Theory predicts, and prior studies have shown, that the presence of polar hydrophilic surface oxides will influence the sorption properties of carbonaceous materials toward both organic and inorganic contaminants (78, 79). For example, surface oxidation decreases the sorption capacity of soot particulates toward hydrophobic organic chemicals (HOCs) (80) and similar results have recently been observed for CNTs (81). In regard to metal contaminants (e. g., Cd^{2+}, Pb^{2+}, Cu^{2+}), studies have shown that CNTs (78, 79, 82) and activated carbons (83) that have been subjected to surface oxidation display higher adsorption capacities than the pristine materials.

With the ultimate goal of developing a better understanding of the environmental impact of CNTs, we have focused on characterizing the surface oxides present on CNTs and their influence on the behavior of these nanomaterials in aquatic environments. Our studies highlight the fact that even comparatively small changes in the surface composition provoke profound changes on the aquatic stability and sorption properties of CNTs.

7.3 PREPARATION OF OXIDIZED MWCNTs

7.3.1 Source of CNTs

In our studies, we have focused on the effects that oxidative processes exert on the properties of MWCNTs (diameter: 15 $\pm$ 5 nm; length: 1–5 μm and 5–20 μm; purity: 95%). The decision to focus on MWCNTs rather than SWCNTs was motivated by the fact that both the current annual production rate and the commercial market for MWCNTs were substantially greater than for SWCNTs (32, 33): in 2004, the worldwide market demand for MWCNTs and SWCNTs was estimated to be valued at $6 million and less than $1 million, respectively (31). Due to their greater production rates and more widespread commercial applications, it seems reasonable to expect that greater quantities of MWCNTs will be entering aquatic environments. Since the physical properties, particularly length, diameter, and purity of CNTs, can vary significantly between different manufacturers, we have restricted our studies to one commercial source of nanomaterials, Nanolab Inc.

7.3.2 Oxidative Treatment

Pristine (as received) MWCNTs were oxidized by refluxing in HNO_3; an oxidant commonly used to functionalize and purify CNTs (43, 84–89). Oxidizing conditions were varied by using solutions that contained different concentrations of HNO_3 ranging between 10% and 70% w/w. This range of HNO_3 concentrations was selected deliberately to replicate oxidative conditions that are frequently used for CNT purification and surface functionalization strategies (43, 90–92). In each oxidative treatment, the CNT mass to HNO_3 solution volume ratio was maintained at 0.4 mg/mL. Prior to oxidation, pristine MWCNTs were sonicated (Sonicator: Branson 1510, operating at 70 W) for 1 h in the selected concentration of HNO_3 to help break apart larger CNT aggregates. After sonication, the mixture of CNTs and HNO_3 was refluxed for 1.5 h at 140°C while being stirred vigorously. Once the oxidation process was complete, the HNO_3 solution, carbonaceous byproducts, residual oxidants, and metallic impurities were removed by repeatedly diluting the solution with milli-Q water, centrifuging, and decanting until the pH of the supernatant was ~5. MWCNTs were subsequently dried overnight in an oven at 100°C.

7.4   CHARACTERIZATION OF OXIDIZED CARBON NANOTUBES

Several analytical techniques, summarized in Table 7.1, have been used to determine the physicochemical effects of oxidation on MWCNTs.

7.4.1   Effects of Oxidation on the Physical Characteristics of MWCNTs

Transmission Electron Microscopy (TEM). The structural integrity of oxidized MWCNTs was examined using TEM. To prepare samples for analysis, MWCNTs were dispersed in water and a drop of the suspension placed on a holey-carbon TEM grid. A representative TEM image of MWCNTs that had been oxidized with 70% w/w HNO_3 is shown in Fig. 7.3. A comparison of Figs. 7.1 and 7.3 reveals that the MWCNTs are structurally unchanged by oxidation.

Atomic Force Microscopy (AFM). The influence of HNO_3 oxidation on the length distribution of MWCNTs was evaluated using an AFM (Pico SPM LE; Agilent Technologies) operating in tapping mode, using magnetically coated probes oscillating at 75 kHz (NSC 18/Co–Cr; MikroMasch). Prior to AFM analysis, both pristine and oxidized CNTs were sonicated in ethanol for 2 min to exfoliate the CNTs. Afterward, two drops of the dispersed CNTs were flash evaporated to limit the extent of reaggregation during the drying process onto an atomically smooth silicon substrate.

TABLE 7.1   Analytical Techniques Used to Characterize the Effects of Oxidation on the Physical and Chemical Properties of MWCNTs

Analytical Technique	Information Obtained on MWCNTs
Transmission electron microscopy (TEM)	Structural integrity of MWCNTs after oxidation
Atomic force microscopy (AFM)	Length distribution of MWCNTs before and after oxidation
BET isotherm	Surface area of oxidized MWCNTs
Infrared spectroscopy	Functional group identification
Elemental analysis	Overall chemical composition of oxidized MWCNTs
X-ray photoelectron spectroscopy	Surface chemical composition of oxidized MWCNTs
Chemical derivatization in conjunction with XPS	Concentration of surface hydroxyl (C—OH), carbonyl (C=O), and carboxylic acid (COOH) groups

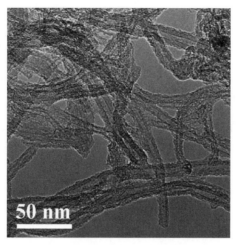

FIGURE 7.3   TEM of oxidized MWCNT samples were imaged using a Philips CM 300 field emission gun transmission electron microscope operating at 297 keV.

A sufficient number of $5.0\,\mu m^2$ AFM images were acquired and analyzed until a histogram of ~300 individual MWCNTs could be constructed.

The AFM data shown in Fig. 7.4 revealed that the pristine and oxidized MWCNTs exhibited length distributions of 1.9 ± 1.7 and $1.7 \pm 1.4\,\mu m$, respectively. More than 50% of the pristine and oxidized MWCNTs had lengths between 0.4 and $1.6\,\mu m$, and less than 7% exhibited lengths $<0.4\,\mu m$. Thus, while oxidation may shorten some of the MWCNTs, it does not appear to produce a significant change in the overall length distribution (81).

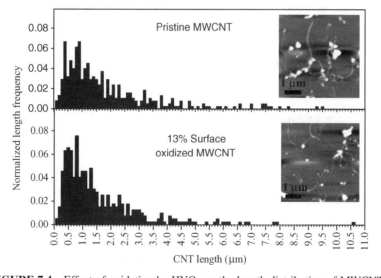

FIGURE 7.4   Effect of oxidation by HNO_3 on the length distribution of MWCNTs.

BET Isotherm. To determine the effect of oxidation on the surface area of MWCNTs, N_2 adsorption data were obtained on pristine and oxidized MWCNTs at 77 K using a high-resolution gas adsorption analyzer with high vacuum capacity (5.0×10^{-7} Pa) (ASAP 2010, Micromeritics, Norcross, GA). Following previously established protocols for carbonaceous materials (93), MWCNTs were outgassed at 300°C for 5 h prior to N_2 adsorption. From these measurements, the surface area of pristine and oxidized MWCNTs was found to be 283 and 287 m^2/g, respectively. Despite uncertainties in using BET measurements to quantitatively determine the surface area of nanomaterials, these results indicate that oxidation does not change the surface area of the MWCNTs (81).

In summary, TEM, AFM, and BET measurements all support the idea that *oxidation does not produce any significant changes in the physical/structural characteristics of MWCNTs.*

7.4.2 Effects of Oxidation on the Chemical Composition of MWCNTs

Infrared Spectroscopy. After oxidation, the introduction of oxygen-containing functional groups onto MWCNTs is observed experimentally by IR spectroscopy (Fig. 7.5), which shows vibrational modes associated with C=O, O—H and C—O—C. Although analytical techniques such as infrared (IR) spectroscopy (94, 95) can identify the presence of surface functional groups, they can only provide qualitative information on the concentration and distribution of different oxygen functional groups on BC materials, such as MWCNTs. This limitation is due to difficulties in unambiguously assigning different IR bands and an inability to quantitatively convert absorption peaks into functional group concentrations.

Elemental Analysis (Bulk and Surface). The bulk and surface oxygen concentration for pristine and oxidized MWCNTs are compared and contrasted in Fig. 7.6. The bulk oxygen concentration was determined by elemental analysis (Huffman Laboratories,

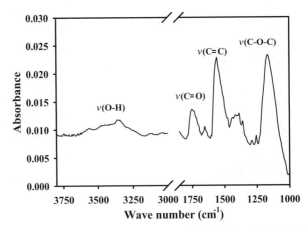

FIGURE 7.5 Transmission IR spectrum of MWCNTs oxidized by HNO_3.

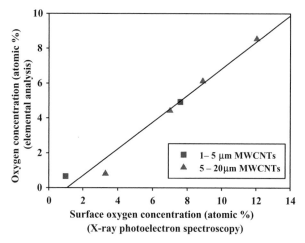

FIGURE 7.6 Correlation between the extent of bulk and surface oxidation for pristine and oxidized MWCNTs.

Inc., Golden, CO, USA), while the surface composition of oxidized MWCNTs was quantified by analysis of the X-ray photoelectron spectroscopy (XPS) survey spectrum obtained using a Φ 5400 system. Results from these studies as follows:

1. The bulk and surface oxygen concentration are strongly correlated with one another ($r = 0.98$) and independent of the MWCNT length or the extent of oxidation.

2. The concentration of oxygen atoms is always greater at the surface compared to the bulk, consistent with the idea that the effects of oxidation should occur preferentially at the surface of the MWCNT (96). The fact that the levels of bulk and surface oxidation are comparable (within a few atomic %) for any MWCNT is a reflection of the fact that in any nanomaterial, most of the atoms are at or in close proximity to the surface.

The extent of surface oxidation also exhibits a systematic dependence upon the oxidizing conditions, which were varied by refluxing MWCNTs with different HNO_3 concentrations, ranging from 10–70% w/w. Specifically, the surface oxygen concentration increase from ≈2% for pristine MWCNTs to a maximum value of ≈10% for MWCNTs refluxed with 70% HNO_3. Since less than 0.1% of the carbon atoms are located at the open ends of the MWCNTs, the fact that as much as 10% of the surface is composed of oxygen atoms after oxidation indicates that surface oxides are distributed predominantly along the sidewalls of the MWCNTs, likely localized around defect sites on the graphene surface.

Chemical Derivatization in Conjunction with XPS. As a result of oxidation, different surface oxides can become incorporated into the sidewalls of MWCNTs (Fig. 7.7) (97). In addition to the overall level of oxidation (i.e., oxygen concentration), the

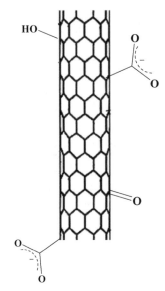

FIGURE 7.7 Surface oxides on CNTs.

distribution of these surface oxides (e.g., hydroxyl, carbonyl, and carboxylic acid groups) is also expected to influence the properties of BC materials in aquatic environments (77). For example, carboxylic acid groups that will be negatively charged under most environmentally relevant conditions are expected to play a particularly important role in removing aqueous metal cations from solution (98). Furthermore, different hydration forces will be associated with hydroxyl, carbonyl, and carboxylic acid groups (99), and solvation effects associated with hydroxyl and carboxylic acid groups can be further moderated by deprotonation. All of these effects will have important consequences for colloidal stability and interfacial properties of CNTs.

Most currently available analytical techniques (e.g., infrared spectroscopy (94, 95) and Boehm titrations (100)) provide only qualitative information on the concentration and distribution of surface oxides on CNTs. In principle, the C(1s) XPS region of BC materials, such as CNTs, can be used to obtain detailed information on the distribution of oxygen-containing functional groups by fitting the C(1s) spectral envelope to include contributions from distinct local carbon environments (e.g., C–C, C–O, C=O), each of which give rise to different C(1s) binding energies (Fig. 7.8). Accurate and reliable spectral deconvolution of the C(1s) region is, however, extremely difficult due to the close proximity of the C(1s) binding energies associated with the different oxides coupled with the lack of monochromatic X-rays and limited energy resolution characteristic of typical XP spectrometers (100–102). In the case of CNTs, spectral deconvolution is further complicated by the asymmetric C(1s) peak profile. This phenomenon is a consequence of a π–π^* "shake-up" feature to the higher binding energy side of the

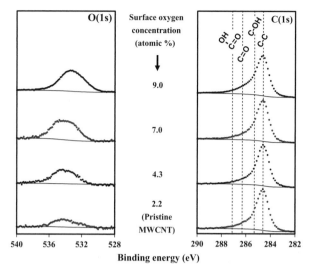

FIGURE 7.8   O(1s) and C(1s) XPS regions of pristine (as received) and oxidized MWCNTs. Dotted lines shown in the C(1s) region indicate the approximate binding energies for various surface oxides.

dominant C−C peak in any BC material, within the same spectral window as XPS peaks associated with surface oxides.

To overcome this limitation, we have developed a suite of chemical derivatization reactions that specifically target hydroxyl, carbonyl, or carboxylic acid functional groups present in BC materials, as shown in Fig. 7.9 (103–108). The surface product of each derivatization reaction contains three fluorine atoms. For CNTs, fluorine atoms are an ideal chemical tag because they are rarely present as a naturally occurring elemental component of BC materials and possess a high cross section for detection by XPS. As a result, the F(1s) signal recorded on a pristine or oxidized MWCNT after a derivatization reaction can be used to accurately quantify the concentration of a specific surface oxide. As a necessary component of this method development, we have quantified the selectivity and efficiency of each derivatization reaction shown in Fig. 7.9. This was accomplished by conducting control studies on a suite of selected polymers, each of which contained one type of oxygen-containing functional group. For example, using this approach we have demonstrated that trifluoroacetic anhydride reacts selectively and with ≈100% efficiency toward hydroxyl groups on carbonaceous surfaces. Analogous results have been obtained for the reactions of trifluoroethanol and trifluoroethyl hydrazine toward carboxylic acid and carbonyl groups, respectively.

An example of the results that can be obtained from these derivatization reactions, carried out on a series of pristine and oxidized MWCNTs, is shown in Fig. 7.10. The concentration of hydroxyl, carbonyl, and carboxylic acid groups on each MWCNT (obtained by separate derivatization reactions carried out in conjunction with XPS) is

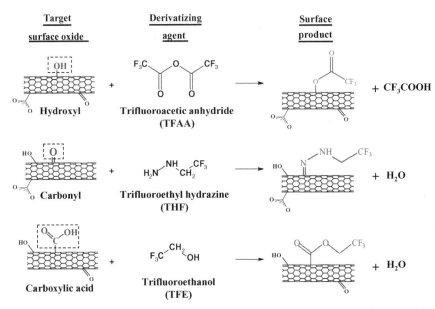

FIGURE 7.9   Surface oxides targeted by chemical derivatization reactions, the derivatizing agents, and the surface product of each derivatization reaction.

plotted for MWCNTs that exhibit different levels of overall surface oxygen concentration (shown as a dotted line for each MWCNT). Figure 7.10 shows that

1. The majority (>75%) of the oxygen-containing functional groups present in MWCNTs oxidized by HNO_3 are present as either hydroxyl, carbonyl, or

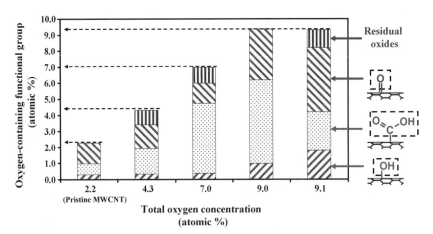

FIGURE 7.10   Distribution of oxygen containing functional groups (as determined by chemical derivatization for hydroxyl, carbonyl, and carboxylic acid groups) for pristine and oxidized MWCNTs.

carboxylic acid groups. This is in contrast to previous studies we have carried out on more traditional BC materials (e.g., soots and chars), in which hydroxyl, carbonyl, or carboxylic acid groups typically accounted for only ≈50% of the surface oxides present (103).

2. On MWCNTs oxidized with HNO_3, carboxylic acids are the dominant surface oxide. There are exceptions, however, as shown by the MWCNTs that exhibit 9.1% surface oxygen (data on the far right of Fig. 7.10). In this case, the carbonyl group concentration is greater than that of the carboxylic acids.

Thus, in summary *the effect of oxidation on MWCNTs is restricted to the incorporation of chemically distinct surface oxides onto the sidewalls without producing any significant changes in the physical or structural characteristics of the MWCNTs.* As a result, changes in the properties and behavior of oxidized MWCNTs are a reflection of the effects of these various surface oxides.

7.5 INFLUENCE OF SURFACE OXIDES ON THE AQUATIC STABILITY OF MWCNTs

To determine the aquatic stability of colloidal MWCNT particulates, we have quantified the concentration of MWCNTs that are stable in solution under different aquatic conditions (defined by pH and ionic strength) by measuring the UV–Vis absorbance at 500 nm (109). In these experiments, the UV–Vis extinction coefficient of oxidized MWCNTs was determined by dispersing a known mass of the nanomaterials into milli-Q water by sonication at 75 W overnight, or until macroscopic particles could no longer be observed with the naked eye. The UV–Vis extinction coefficient (ε) for the oxidized CNTs was calculated by measuring the absorbance at 500 nm (Abs_{500}) for a set of serial dilutions (109, 110). Analysis of the MWCNTs oxidized by HNO_3 has shown that $\varepsilon_{average} = 41.2$ L/mg/cm and is independent of the level of surface oxidation. This value is typical for the extinction coefficient of well-dispersed MWCNTs (110). Information obtained from dynamic light scattering (DLS) measurements showed that the CNT colloids exhibit a narrow particle size distribution (effective hydrodynamic radii $= 91 \pm 12$ nm) that is essentially independent of the level of surface oxidation.

The influence of surface oxides on the colloidal stability of MWCNTs is summarized in Fig. 7.11. Figure 7.11a shows three colloidal suspensions of MWCNTs, each of which is characterized by a different level of surface oxygen (shown on the cap of each vial) obtained by using varying strengths of HNO_3 during oxidation. Appropriate dilutions were made to ensure that in each vial the absorbance at 500 nm was the same. Since the average particle size was independent of oxidation level, this means that the initial concentration of MWCNT particulates was also the same. Once a MWCNT suspension had been allowed to settle, the NaCl concentration was increased to 0.07 M, the vial was vortexed for a few seconds thoroughly mix the different species, and the solution was then allowed to equilibrate unperturbed, under conditions of perikinetic flocculation for 5 h. Figure 7.11a

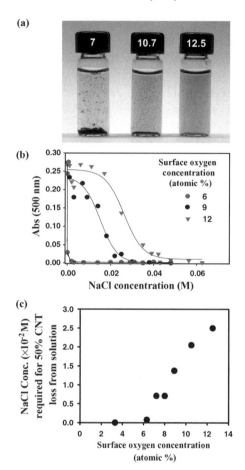

FIGURE 7.11 Influence of surface oxygen concentration on MWCNT colloidal stability: (a) MWCNTs with different levels of surface oxidation (shown on the cap of each vial) were allowed to settle unperturbed in a 0.07-M NaCl solution for 5 h. (b) the concentration of MWCNTs remaining in solution after 2 months of settling time plotted as a function of increasing NaCl concentration. In (a) and (b) the solution pH is ≈5.8. (c) The NaCl concentration required to remove 50% of the MWCNTs from solution after 2 months of settling time plotted as a function of the surface oxygen concentration.

shows that the relative rate at which MWCNTs aggregate to produce settleable aggregates under these conditions increases systematically as the extent of MWCNT surface oxidation decreases. The experiments thus provide clear evidence that the aquatic stability of MWCNTs is influenced by the extent of surface oxidation that is, in turn, a strong function of the oxidizing conditions used to purify and/or functionalize the CNTs.

In Fig. 7.11b, the resistance of MWCNT colloids toward coagulation and the formation of settleable aggregates in the presence of NaCl (the so-called salting out effect) is described more mathematically for three MWCNTs that contain different levels of surface oxygen. Consistent with Fig. 7.11a, MWCNT colloids that contain higher concentrations of surface oxygen remain stable in solution over a wider range of aquatic conditions. More detailed analysis of the results of Fig. 7.11b is shown in Fig. 7.11c. In this figure, the NaCl concentration required to remove 50% of the oxidized MWCNTs from solution has been plotted as a function of the MWCNT surface oxygen concentration, as measured by XPS. The monotonic increase in this required concentration with increasing surface oxygen is believed to be related to increasing negative surface charge and the associated increase in particle–particle repulsion as the oxygen atom concentration increases. The data of Fig. 7.11c also suggest that a critical oxygen concentration is necessary to stabilize MWCNTs in solution and that mathematical relationships can be established to link the MWCNTs' colloidal stability to their surface chemistry.

In regard to the hypothesized interparticle electrostatic effects, the extent to which oxidized MWCNT particulates follow the precepts of classical DLVO theory (111, 112) is explored in Fig. 7.12. In this figure, the effect of different electrolytes on the aquatic stability of oxidized MWCNTs is illustrated. Figure 7.12 shows the colloidal stability of the oxidized MWCNTs is greatly diminished in the presence of Mg^{2+} and Ca^{2+} compared to Na^+. This is a consequence of the more effective screening of the surface charge on the nanoparticles by these divalent cations, which mitigate surface charge more effectively than a monovalent cation such as sodium. This increased screening effectiveness allows a lower electrolyte concentration to produce the same compression of the double layer (72, 113). Thus, divalent cations require lower salt concentration to form settleable aggregates. From Fig. 7.12 we can estimate the minimum salt concentrations required to remove all of the MWCNT particulates from solution and thereby obtain a measure of the critical coagulation concentration (CCC). From this analysis the ratio of the CCC for $MgCl_2$, $CaCl_2$, and

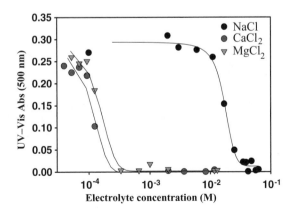

FIGURE 7.12 Aquatic stability of oxidized MWCNTs (level of surface oxidation = 13.5%) toward different inert electrolytes, measured by UV–Vis after 48 d of settling under the influence of perikentic flocculation.

NaCl was proportional to $z^{-4.2}$, where z is the counterion valence (i.e., $z = 2$ for $MgCl_2$ and $CaCl_2$ and $z = 1$ for NaCl). This is reasonable for a colloidal system that at least qualitatively follows the precepts of DLVO theory since the Schulze–Hardy rule predicts a proportionality of z^{-6} for particles of large zeta potential (72, 113), although a much weaker dependence on counterion charge (z^{-2}) has been observed for particles with smaller zeta potentials (114).

Figure 7.11 illustrates that for MWCNTs oxidized by HNO_3 the aquatic stability is governed principally by the overall level of surface oxidation. This is probably a reflection of the fact that the distribution of oxygen-containing functional groups remains essentially constant among all oxidized MWCNTs, irrespective of the oxygen concentration. In this regard, the data of Fig. 7.10 show that the relative distribution of oxygen among functional groups typically increases in the order: carboxylic acid (COOH) > carbonyl (C=O) > hydroxyl (C–OH) groups. However, oxidation by HNO_3 did on occasion generate MWCNTs when the concentration of carbonyl groups was greater than that of carboxylic acid groups (data on right hand side of Fig. 7.10).

An examination of aquatic stability where surface oxide distribution varied from the norm provided some insight into the role of different types of surface oxides. In Fig. 7.13, the vial on the far right contains oxidized MWCNTs with 9% overall surface oxygen concentration but with a higher concentration of carbonyl groups than carboxylic acid groups. Interestingly, the aquatic stability of these MWCNTs is visibly less than that of MWCNTs that contain slightly less total surface oxygen (8%) but which exhibited the more typical distribution of oxide functional groups (i.e., a higher concentration of carboxylic acid groups compared to carbonyl groups). This observation suggests that carboxylic acid groups, which are expected to be negatively charged under environmentally relevant conditions (pH = 5–7), are likely to play a major role in controlling the aquatic stability of CNTs. Further studies to quantify the influence of both oxygen concentration and oxide distribution on aquatic stability are currently being carried out in our laboratories.

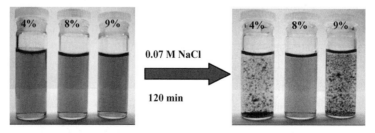

FIGURE 7.13 Influence of oxygen functional group distribution on the aquatic stability of oxidized MWCNTs. The solution chemistry of three vials, each containing an equal concentration of oxidized MWCNT particulates (overall oxygen concentration shown on the cap of each vial) at pH ≈ 7, has been modified by the addition of 0.07 M NaCl, vortexed and allowed to settle under the influence of perikinetic flocculation for 120 min.

7.6 INFLUENCE OF SURFACE OXIDES ON THE SORPTION PROPERTIES OF MWCNTs

Oxygen-containing functional groups are expected to play a pivotal role in moderating the sorption properties of CNTs and other carbon-based nanomaterials. To better understand the interplay between oxidation and CNT sorption properties, we have studied the adsorption of representative organic and inorganic contaminants encountered in aquatic environments with pristine and oxidized MWCNTs. A priori, we anticipated that the adsorption of hydrophobic organic chemicals (HOCs) and divalent metal cations onto oxidized MWCNTs would occur by two different mechanisms, as illustrated in Fig. 7.14. Apolar organic chemicals such as naphthalene (NAP) were expected to adsorb to the hydrophobic graphene sheets of CNTs, while divalent cations such as Zn^{2+} will interact most strongly with negatively charged surface oxides, particularly deprotonated carboxylic acid groups (see Fig. 7.14). With these precepts in mind, the addition of surface oxides, either as a result of intentional or incidental oxidation, should produce substantial changes in the sorption properties of CNTs toward both inorganic and organic contaminants.

7.6.1 Effects of Surface Oxides on Naphthalene Sorption with MWCNTs

The sorption of naphthalene to MWCNTs was examined as a function of surface oxygen concentration. In these experiments, aqueous concentrations of dissolved naphthalene were exposed onto MWCNTs in a simple batch system, using flame-sealed ampules and ^{14}C-labeled naphthalene to provide good precision and accuracy. NAP sorption onto MWCNTs was determined by scintillation counting of the ^{14}C-labeled NAP remaining in equilibrated solutions after centrifugation (3000 g, 30 min). In these sorption experiments, MWCNTs were in a macroscopic particulate form. Full

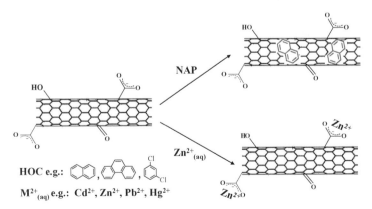

FIGURE 7.14 Schematic illustration of potential sorption interactions of hydrophobic organic chemicals and divalent metal cations (M^{2+}) with CNT graphene surfaces and CNT surface oxides, respectively.

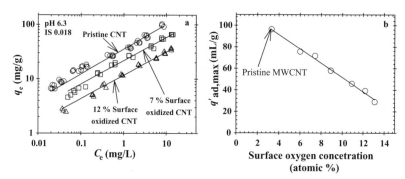

FIGURE 7.15 (a) Freundlich adsorption isotherms for naphthalene with pristine MWCNTs and oxidized MWCNTs of varying surface oxygen concentration. (b) Maximum adsorption capacity ($q'_{ad,max}$) for naphthalene plotted as a function of surface oxide concentration using the Polanyi-based Dubinin–Astikov adsorption model.

methods and results of this investigation are presented elsewhere (81); selected results are shown in Fig. 7.15. Here, C_e is the aqueous concentration of naphthalene at equilibrium and q_e is the concentration of naphthalene adsorbed onto MWCNTs at equilibrium. These results, shown in Fig. 7.15, demonstrate several aspects of naphthalene adsorption by MWCNTs: (1) adsorption is well described by a Freundlich isotherm with a slope substantially less than one (Fig. 7.15a), which implies a large amount of heterogeneity in the adsorption sites energies; (2) the extent of surface oxidation exerts a strong impact on the extent of NAP sorption, with decreasing NAP sorption on more highly oxidized MWCNTs; (3) q_e values (e.g., the $q_{e,max}$ that occurs at $C_e = $ solubility) followed a quantifiable linear decrease as a function of increasing oxygen content; and (4) slopes of the isotherms (the exponent n in the Freundlich relation, $q_e = K_f C_e^n$) are reasonably constant among the various sorbents. The latter result implies that the relative distribution of adsorption site energies is unaffected by oxidation (Fig. 7.10). Thus, the primary impact of oxidation appears to be to decrease the amount of adsorbed NAP concentration (q_e) for any given C_e. A polanyi-type analysis suggests that a 10% increase in oxygen concentration produces a 71% decrease in sorption capacity (Fig. 7.15b). It remains to be seen, however, as to whether such a relationship will exist when the relative distribution of surface oxides changes. Investigations of this type are ongoing.

7.6.2 Effects of Surface Oxides on Zinc Sorption with MWCNTs

Adsorption isotherms with MWCNTs in aqueous systems were also obtained for Zn^{2+}, using a similar batch technique, but with plastic sorption vessels in lieu of sealed glass ampules, and using atomic absorption spectrophotometry (AAS) to analyze aqueous phase metal ion concentrations. Sorption experiments to date have been carried out at pH = 6.5–6.8 and at Zn^{2+} concentrations well below solubility with respect to hydroxide or carbonate species.

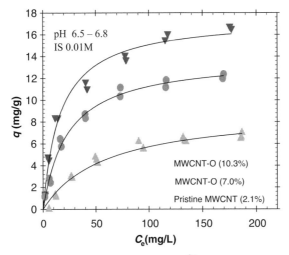

FIGURE 7.16   Langmuir adsorption isotherms for Zn^{2+} with MWCNTs of varying surface oxygen concentration.

Although these experiments are ongoing at the time of this writing, Fig. 7.16 illustrates the typical adsorption isotherms observed when adsorbing Zn^{2+} to a series of oxidized CNTs. The isotherm results (Fig. 7.16) are fit to a Langmuir rather than Freundlich isotherm, because the Langmuir formulation gave better statistical fits to the data with the same number of fitting parameters (i.e., two). Adsorption behavior that can be described by a Langmuir isotherm is typical of metal ions adsorbing onto BC sorbents, exhibiting a linear behavior at low metal ion concentrations and a plateau at high equilibrium concentrations due to a finite number of adsorption sites.

The variation in sorption capacity toward Zn^{2+} as a function of oxygen concentration shown in Fig. 7.16 is opposite to the trend observed with NAP, consistent with the expectations illustrated in Fig. 7.14. A comparison of the sorption isotherm data shown in Fig. 7.16 reveals a linear trend of increasing sorption capacity as a function of the surface oxygen concentration measured by XPS. More experiments are needed to confirm the implied quantitative relationship between $q_{ad(max)}$ and oxygen content (analogous to Fig. 7.15b for NAP) and, as with NAP, it also remains to be seen whether such a single relationship will exist when the relative distribution of oxygen-containing functional groups changes. In the case of Zn^{2+}, we expect that the concentration of carboxylic acid groups will play a determinant role in Zn^{2+} sorption owing to the presence of electrostatic interactions of the type illustrated in Fig. 7.14.

7.7 SUMMARY

The key finding from these studies is that *the concentration and distribution of surface oxides exert a profound influence on the aquatic stability and sorption properties of multiwalled carbon nanotubes*. Furthermore, even comparatively small changes in the surface chemistry are responsible for significant changes to the behavior of MWCNTs in aquatic environments. We anticipate that these results will also be qualitatively similar for single-walled carbon nanotubes.

In general, more aggressive oxidizing conditions lead to larger changes in the surface oxygen content without altering the physical characteristics of individual MWCNTs (e.g., length and structural integrity). The overall oxygen concentration present on MWCNTs can be accurately quantified by XPS, although an accurate determination of the distribution of different oxygen functional groups present mandates the application of chemical derivatization prior to XPS analysis. In terms of both aquatic stability and sorption properties, mathematical expressions are able to describe the relationships between materials properties and the effects of surface oxidation on MWCNTs.

By combining detailed information on the physicochemical characteristics of MWCNTs with environmentally relevant phenomenological properties, these studies significantly advance our understanding of the structure–activity relationships that describe the behavior of MWCNTs in aquatic environments. For example, by identifying the interplay between the MWCNT's surface chemistry, colloidal stability, and sorption properties, aqueous conditions can be identified where MWCNTs could act as "Trojan horses," facilitating the unwanted dispersion of toxic chemicals in water with implications for transport in waterways, aquifers, and biological media (including humans). To provide a more complete understanding of the environmental impact of oxidative processes on the impact of MWCNTs, however, it will also be necessary to better understand the mechanistic aspects of CNT aggregation and deposition and to conduct transport studies to measure the mobility of oxidized MWCNTs through porous media with and without the presence of potentially sorbing contaminant species. Quantitative data on aquatic stability, sorption, and transport properties are urgently needed in response to the increasing quantity of CNTs being produced. This information will also help to identify and prevent the potentially negative environmental consequences of surface functionalized and well-dispersed CNTs.

REFERENCES

1. Dreher, K.L. Health and environmental impact of nanotechnology: toxicological assessment of manufactured nanoparticles. *Toxicol Sci* 2004;77:3–5.

2. Dionysiou, D.D. Environmental applications and implications of nanotechnology and nanomaterials. *J Environ Eng* 2004;130:723–724.

3. Guzman, K.A.D., Taylor, M.R., Banfield, J.F. Environmental risks of nanotechnology: national nanotechnology initiative funding 2000–2004. *Environ Sci Technol* 2006;40:1401–1407.

4. Robichaud, C.O., Tanzil, D., Weilenmann, U., Wiesner, M.R. Relative risk analysis of several manufactured nanomaterials: an insurance industry context. *Environ Sci Technol* 2005;39:8985–8994.

5. Wiesner, M.R., Lowry, G.V., Alvarez, P., Dionysiou, D., Biswas, P. Assessing the risks of manufactured nanomaterials. *Environ Sci Technol* 2006;40(14):4336–4345.

6. Maynard, D.A. Safe handling of nanotechnology. *Nature* 2006;444:267–269.

7. Wiesner, M.R. Responsible development of nanotechnologies for water and wastewater treatment. *Water Sci Technol* 2006;53:45–51.

8. Muller, J., Huaux, F., Lison, D. Respiratory toxicity of carbon nanotubes: how worried should we be? *Carbon* 2006;44:1048–1056.

9. Lam, C.-W., James, J.T., McCluskey, R., Holian, A., Hunter, R.L. *Toxicity of Carbon Naonotubes and its Implications for Occupational and Environmental Health, vol. 5.* Wiley-VCH; 2006.

10. EPA White Paper on Nanotechnology, 2007.

11. He, X., Kitipornchai, S., Wang, C.M., Leiw, K.M. Modeling of van der Waals force for infinitesimal deformation of multi-walled carbon nanotubes treated as cylindrical shells. *Int J Solids Struct* 2005;42:6032–6047.

12. Terrones, M., Hsu, W.K., Kroto, H.W., Walton, D.R.M. *Nanotubes: A Revolution in Materials Science and Electronics,* vol. 199. Berlin: Springer;1999.

13. Iijima, S. Helical microtubules of graphitic carbon. *Nature* 1991;354:56–58.

14. Masciangioli, T., Zhang, W.-X. Environmental technologies at the nanoscale. Nanotechnology could substantially enhance environmental quality and sustainability through pollution prevention, treatment, and remediation. *Environ Sci Technol* 2003;102A–108A.

15. Ebbesen, T.W., Lezec, H.J., Hiura, H., Bennett, J.W., Ghaemi, H.F., Thio, T. Electrical conductivity of individual carbon nanotubes. *Nature* 1996;382:54–56.

16. Gibson, J.M., Ebbensen, T.W., Treacy, M.M.J. Exceptionally high Young's modulus observed for individual carbon nanotubes. *Nature* 1996;381:678–680.

17. Penza, M., Cassano, G., Aversa, P., Antolini, F., Cusano, A., Cutolo, A., Giordano, M., Nicolasis, L. Alcohol detection using carbon nanotubes acoustic and optical sensors. *Appl Phys Lett* 2004;85:2379–2381.

18. Li, L., Xing, Y. Pt-Ru nanoparticles supported on carbon nanotubes as methanol fuel cell catalysts. *J Phys Chem C* 2007;111:2803–2808.

19. Xing, Y. Synthesis and electrochemical characterization of uniformly-dispersed high loading Pt nanoparticles on sonochemically-treated carbon nanotubes. *J Phys Chem B* 2004;108:19255–19259.

20. Li, L., Xing, Y. Pt-Ru nanoparticles supported on carbon nanotubes as methanol fuel cell catalysts. *J Phys Chem C* 2007;111:2803–2808.

21. Xing, Y. Synthesis and electrochemical characterization of uniformly-dispersed high loading Pt nanoparticles on sonochemically-treated carbon nanotubes. *J Phys Chem B* 2004;108:19255–19259.

22. Rao, C.N.R., Satishkumar, B.C., Govindaraj, A., Nath, M. Nanotubes. *Chem Phys Chem* 2001;2:78–105.

23. Robertson, J. Practical applications of carbon nanotubes. *Mater Today* 2004;7:46–52.

24. Pastorin, G., Kostarelose, K., Prato, M., Bianco, A. Functionalized carbon nanotubes: towards the delivery of therapeutic molecules. *J Biomed Nanotechnol* 2005;1:133–142.

25. Bianco, A., Kostarelos, K., Prato, M. Applications of carbon nanotubes in drug delivery. *Curr Opin Chem Biol* 2005;9:674–679.

26. Alper, J., *NCI Alliance for Nanotechnology in Cancer;* 2006. pp. 1–4. Available at www.cancer.gov.

27. Peng, X., Li, Y., Luan, Z., Di, Z., Wang, H., Tian, B., Jia, Z. Adsorption of 1,2-dichlorobenzene from water to carbon nanotubes. *Chem Phys Lett* 2003;376:154–158.

28. Lu, C., Chung, Y.-L., Chang, K.-F. Adsorption of trihalomethanes from water with carbon nanotubes. *Water Res* 2005;39:1183–1189.

29. Li, Q.-L., Yuan, D.-X., Lin, Q.-M. Evaluation of multi-walled carbon nanotubes as an adsorbent for trapping volatile organic compounds from environmental samples. *J Chromatogr A* 2004;1026:283–288.

30. Li, Y.-H., Wang, S., Luan, Z., Ding, J., Xu, C., Wu, D. Adsorption of cadmium(II) from aqueous solution by surface oxidized carbon nanotubes. *Carbon* 2003;41: 1057–1062.

31. Thayer, A.M. *Chem Eng News* 2007;85:29–35.

32. Cientifica, Nanotubes, 2004.

33. Templeton, R.C., Ferguson, P.L., Washburn, K.M., Scrivens, W.A., Chandler, G.T. Life-cycle effects of single-walled carbon nanotubes (SWNTs) on an estuarine meiobenthic copepod. *Environ Sci Technol* 2006;40:7387–7393.

34. Marrs, B., Andrews, R., Pienkowski, D. Multiwall carbon nanotubes enhance the fatigue performance of physiologically maintained methyl methacrylate-styrene copolymer. *Carbon* 2007;45:2098–2104.

35. Breuer, O., Sundararaj, U. Big returns from small fibers: a review of polymer/carbon nanotube composites. *Polym Compos* 2004;25(6):630–645.

36. Vaisman, L., Marom, G., Wagner, H.D. Dispersions of surface-modified carbon nanotubes in water-soluble and water-insoluble polymers. *Adv Funct Mater* 2006;16:357–363.

37. Bourlinos, A.B., Georgakilas, V., Zboril, R., Dallas, P. Preparation of a water-dispersible carbon nanotube-silica hybrid. *Carbon* 2007;45:2126–2139.

38. Sluzarenko, N., Heurtefeu, B., Maugey, M., Zakri, C., Poulin, P., Lecommandoux, S. Diblock copolymer stabilization of multi-wall carbon nanotubes in organic solvents and their use in composites. *Carbon* 2006;44:3207–3212.

39. Liu, P. Modifications of carbon nanotubes with polymers. *Eur Polym J* 2005;41: 2693–2703.

40. Peng, Y., Liu, H. Effects of oxidation by hydrogen peroxide on the structures of multiwalled carbon nanotubes. *Ind Eng Chem Res* 2006;45:6483–6488.

41. Parekh, B., Debbies, T., Knight, P., Santhanam, K.S.V., Takacs, G.A. Surface functionalization of multiwalled carbon nanotubes with UV and vacuum UV photo-oxidation. *J Adhes Sci Technol* 2006;20:1833–1846.

42. Najafi, E., Kim, J.-Y., Han, S.-H., Shin, K. UV-ozone treatment of multi-walled carbon nanotubes for enhanced organic solvent dispersion. *Colloids Surf A* 2006;284–285: 373–378.

43. Rosca, I.D., Watari, F., Uo, M., Akasaka, T. Oxidation of multiwalled carbon nanotubes by nitric acid. *Carbon* 2005;43:3124–3131.

44. Zhang, N., Xie, J., Varadan, V.K. Functionalization of carbon nanotubes by potassium permanganate assisted with phase transfer catalyst. *Smart Mater Struct* 2002; 11:962–965.

45. Yang, D.-Q., Rochette, J.-F., Sacher, E. Functionalization of multiwalled carbon nanotubes by mild aqueous sonication. *J Phys Chem B* 2005;109:7788–7794.

46. Savage, T., Bhattacharya, S., Sadanadan, B., Gaillard, J., Tritt, T.M., Sun, Y.-P., Wu, Y., Nayak, Y.W., Car, R., Marzari, N., Ajayan, P.M., Rao, A.M. Photoinduced oxidation of carbon nanotubes. *J Phys Condens Matter* 2003;15:5915–5921.

47. Wang, Y., Iqbal, Z., Mitra, S. Rapidly functionalized, water-dispersed carbon nanotubes at high concentration. *J Am Chem Soc* 2006;128:95–99.

48. Rasheed, A., Howe, J.Y., Dadmun, M.D., Britt, P.F. The efficiency of the oxidation of carbon nanofibers with various oxidizing agents. *Carbon* 2007;45:1072–1080.

49. Grujicic, M., Cao, G., Rao, A.M., Tritt, T.M., Nayak, S. UV-light enhanced oxidation of carbon nanotubes. *Appl Surf Sci* 2003;214:289–303.

50. Banerjee, S., Hemraj-Benny, T., Balasubramanian, M., Fischer, D.A., Misewich, J.A., Wong, S.S. Surface chemistry and structure of purified, ozonized, multiwalled carbon nanotubes probed by nexafs and vibrational spectroscopies. *Chem Phys Chem* 2004;5:1416–1422.

51. Banerjee, S., Hemraj-Benny, T., Balasubramanian, M., Fisher, D.A., Misewich, J.A., Wong, S.S. Ozonized single-walled carbon nanotubes investigated using NEXAFS spectroscopy. *Chem Commun* 2004;7:772–773.

52. Simmons, J.M., Nichols, B.M., Baker, S.E., Marcus, M.S., Castellini, O.M., Lee, C.-S., Hamers, R.J., Eriksson, M.A. Effect of ozone oxidation on single-walled carbon nanotubes. *J Phys Chem B* 2006;110:7113–7118.

53. Jung, A., Graupner, R., Ley, L., Hirsch, A. Quantitative determination of oxidative defects on single walled carbon nanotubes. *Phys Status Solidi B* 2006;243:3217–3220.

54. Zhang, J., Zou, H., Qing, Q., Yang, Y., Li, Q., Liu, Z., Guo, X., Du, Z. Effect of chemical oxidation on the structure of single-walled carbon nanotubes. *J Phys Chem B* 2003;107:3712–3718.

55. Banerjee, S., Kahn, M.G.C., Wong, S.S. Rational chemical strategies for carbon nanotube functionalization. *Chem Eur J* 2003;9:1898–1908.

56. Ye, J., Lui, X., Cui, H.F., Zhang, W., F., Lim, T. Electrochemical oxidation of multi-walled carbon nanotubes and its application to electrochemical double layer capacitors. *Electrochem Commun* 2005;7:249–255.

57. Chen, Z., Ziegler, K.J., Shaver, J., Hauge, R.H., Smalley, R.E. Cutting of single-walled carbon nanotubes by ozonolysis. *J Phys Chem B* 2006;110:11624–11627.

58. Lim, S.C., Jo, C.S., Jeong, H.J., Shin, Y.M., Lee, Y.H., Samayoa, I.A., Choi, J. Effect of oxidation on electronic and geometric properties of carbon nanotubes. *Jpn J Appl Phys* 2002;41:5635–5639.

59. Peng, H., Alemany, L.B., Margrave, J.L., Khabashesku, V.N. Sidewall carboxylic acid functionalization of single-walled carbon nanotubes. *J Am Chem Soc* 2003;125:15174–15182.

60. Xing, Y., Li, L., Chusuei, C.C., Hull, R.V. Sonochemical oxidation of multiwalled carbon nanotubes. *Langmuir* 2005;21:4185–4190.

61. Kim, U.J., Furtado, C.A., Liu, X.-M., Chen, G., Eklund, P. Raman and IR spectroscopy of chemically processed single-walled carbon nanotubes. *J Am Chem Soc* 2005;127:15437–15445.

62. Vione, D., Maurino, V., Minero, C., Pelizzetti, E., Harrison, M.A.J., Olariu, R.I., Arsene, C. Photochemical reactions in the tropospheric aqueous phase and on particulate matter. *Chem Soc Rev* 2006;35:441–453.

63. Esteve, W., Budzinski, H., Villenave, E. Relative rate constants for the heterogeneous reactions of OH, NO_2 and NO radicals with polycyclic aromatic hydrocarbons adsorbed on carbonaceous particles. Part 1: PAHs adsorbed on 1–2 μm calibrated graphite particles. *Atmos Environ* 2004;38:6063–6072.

64. Song, W., Ravindran, V., Koel, B., Pirbazari, M. Nanofiltration of natural organic matter with H_2O_2/UV pretreatment: fouling mitigation and membrane surface characterization. *J Membr Sci* 2004;241:143–160.

65. Franchi, A., O'Melia, C.R. Effects of natural organic matter and solution chemistry on the deposition and reentrainment of colloids in porous media. *Environ Sci Technol* 2003;37:1122–1129.

66. Morel, F.M.M., Gschwend, P.M. The role of colloids in the partitioning of solutes in natural waters. In: Stumm, W., editor. *aquatic surface chemistry*, New York: John Wiley & Sons;1987.

67. Oberdorster, G., Oberdorster, E., Oberdorster, J. Nanotoxicology: an emerging discipline evolving from studies of ultrafine particles. *Environ Health Perspect* 2005; 13:823–840.

68. Sayes, C.M., Liang, F., Hudson, J.L., Mendez, J., Guo, W., Beach, J.M., Moore, V.C., Doyle, C.D., West, J.L., Billups, W.E., Ausman, K.D., Colvin, V.L. Functionalization density dependence of single-walled carbon nanotubes cytotoxicity *in vitro*. *Toxicol Lett* 2006;161:135–142.

69. Chen, Q., Saltiel, C., Manickavasagam, S., Schadler, L.S., Siegel, R.W., Yang, H. Aggregation behavior of single-walled carbon nanotubes in dilute aqueous suspension. *J Colloid Interface Sci* 2004;280:91–97.

70. Brant, J., Lecoanet, H., Wiesner, M.R. Aggregation and deposition characteristics of fullerene nanoparticles in aqueous systems. *J Nanopart Res* 2005;7:545–553.

71. Saltiel, C., Manickavasagam, S. Light-scattering and dispersion behavior of multiwalled carbon nanotubes. *J Opt Soc Am A* 2005;22:1546–1554.

72. Chen, K.L., Elimelech, M. Aggregation and deposition kinetics of fullerene (C_{60}) nanoparticles. *Langmuir* 2006;22:10994–11001.

73. O'Melia, C.R., Hahn, M.W., Chen, C.-T. Some effects of particle size in separation processes involving colloids. *Water Sci Technol* 1997;36:119–126.

74. Elimelech, M., O'Melia, C.R. Effect of particle size on collision efficiency in the deposition of brownian particles with electrostatic energy barriers. *Langmuir* 1990;6:1153–1163.

75. O'Melia, C.R. Particle-particle interactions in aquatic systems. *Colloid Surf* 1989;39:255–271.

76. Lecoanet, H.F., Wiesner, M.R. Velocity effects on fullerene and oxide nanoparticle deposition in porous media. *Environ Sci Technol* 2004;38:4377–4382.

77. Xing, G., Zhang, J., Zhao, Y., Tang, J., Zhang, B., Gao, X., Yuan, H., Qu, L., Cao, W., Chai, Z., Ibrahim, K., Su, R. Influences of structural properties on stability of fullerenols. *J Phys Chem B* 2004;108:11473–11479.

78. Li, Y.H., Ding, J., Luan, Z.K., Di, Z.C., Zhu, Y.F., Xu, C.L., Wu, D.H., Wei, B.Q. Competitive adsorption of Pb^{2+}, Cu^{2+} and Cd^{2+} ions from aqueous solutions by multiwalled carbon nanotubes. *Carbon* 2003;41:2787–2792.

79. Lu, C., Chiu, H., Liu, C. Removal of zinc(II) from aqueous solution by purified carbon nanotubes: kinetics and equilibrium studies. *Ind Eng Chem Res* 2006;45:2850–2855.

80. Nguyen, T.H., Ball, W.P. Absorption and adsorption of hydrophobic organic contaminants to diesel and hexane soot. *Environ Sci Technol* 2006;40:2958–2964.

81. Cho, H.-H., Smith, B.A., Wnuk, J., Fairbrother, D.H., Ball, W.P. Effect of oxidation on the adsorption of naphthalene onto carbon nanotubes. *Environ Sci Technol* 2008;41:2899–2905.

82. Bond, T.C., Streets, D.G., Yarber, K.F., Nelson, S.M., Woo, J.H., Klimont, Z. A technology-based global inventory of black and organic carbon emissions from combustion. *J Geophys Res-Atmos* 2004;109:D14203.

83. Demirbas, E., Kobya, M., Senturk, E., Ozkan, T. Adsorption kinetics for the removal of chromium (VI) from aqueous solutions on the activated carbons prepared from agricultural wastes. *Water SA* 2004;30:533–539.

84. Zhang, X., Sreekumar, T.V., Liu, T., Kumar, S. Properties and structure of nitric acid oxidized single wall carbon nanotube films. *J Phys Chem B* 2004;108:16435–16440.

85. Shieh, Y.-T., Liu, G.-L., Wu, H.-H., Lee, C.-C. Effects of polarity and pH on the solubility of acid-treated carbon nanotubes in different media. *Carbon* 2007;43:1880–1890.

86. Peng, H., Alemany, L.B., Margrave, J.L., Khabasheku, V.N. Sidewall carboxylic acid functionalization of single-walled carbon nanotubes. *J Am Chem Soc* 2003;125:15174–15182.

87. Li, Y., Zhang, X., Luo, J., Huang, W., Cheng, J., Luo, Z., Li, T., Lui, F., Xu, G., Ke, X., Li, L., Geise, H.J. Purification of CVD synthesized single-wall carbon nanotubes by different acid oxidation treatments. *Nanotechnology* 2004;15:1645–1649.

88. Park, K., Hayashi, T., Tomiyasu, H., Endo, M., Dresselhaus, M.S. Progressive and invasive functionization of carbon nanotube sidewalls by dilute nitric acid under supercritical conditions. *J Mater Chem* 2005;15:407–411.

89. Barros, E.B., Filho, A.G.S., Lemos, V., Filho, J.M., Fagan, S.B., Herbst, M.H., Rosolen, J.M., Luengo, C.A., Huber, J.G. Charge transfer effects in acid treated single-wall carbon nanotubes. *Carbon* 2005;43:2495–2500.

90. Murphy, H., Papakonstantinou, P., Okpalugo, T.I.T. Raman study of multiwalled carbon nanotubes functionalized with oxygen groups. *J Vac Sci Technol B* 2006;24:715–720.

91. Kretzschmar, R., Borkovec, M., Grolimund, D., Elimelech, M. Mobile subsurface colloids and their role in contaminant transport. *Adv Agron* 1999;66:121–193.

92. Kretzschmar, R., Barmettler, K., Grolimund, D., Yan, Y.-D., Borkovec, M., Sticher, H. Experimental determination of colloid deposition rates and collision efficiencies in natural porous media. *Water Resour Res* 1997;33:1129–1137.

93. Nguyen, T.H., Cho, H.-H., Poster, D.L., Ball, W.P. Evidence for a pore-filling mechanism in the adsorption of aromatic hydrocarbons to a natural wood char. *Environ Sci Technol* 2007;41:1212–1217.

94. Akhter, M.S., Chughtai, A.R., Smith, D.M. Spectroscopic studies of oxidized soots. *Appl Spectrosc* 1991;45:653–665.

95. Smith, D.M., Chughtai, A.R. The surface structure and reactivity of black carbon. *Colloids Surf* 1995;105:47–77.

96. Langley, L.A., Fairbrother, D.H. Effect of wet chemical treatments on the distribution of surface oxides on carbonaceous materials. *Carbon* 2007;45:47–54.

97. Zhang, J., Zou, H., Qing, Q., Yang, Y., Li, Q., Liu, Z.-F., Guo, X., Du, Z. Effect of chemical oxidation on the structure of single-walled carbon nanotubes. *J Phys Chem B* 2003;107:3712–3718.

98. Jia, Y.F., Thomas, K.M. Adsorption of cadmium ions on oxygen surface sites in activated carbon. *Langmuir* 2000;16:1114–1122.

99. Arai, T., Aoki, D., Okabe, Y., Fujihira, M. Analysis of surface forces on oxides in aqueous solutions using AFM. *Thin Solid Films* 1996;273:322–326.

100. Boehm, H.P. Surface oxides on carbon and their analysis: a critical assessment. *Carbon* 2002;40:145–149.

101. Figueiredo, J.L., Pereira, M.F.R., Freitas, M.M.A., Orfao, J.J.M. Modification of the surface chemistry of activated carbons. *Carbon* 1999;37:1379–1389.

102. Vickerman, J.C. *Surface Analysis: The Principal Techniques*, West Sussex: John Wiley & Sons Ltd; 1997.

103. Langley, L.A., Villanueva, D., Fairbrother, D.H. Quantification of surface oxides on carbonaceous materials. *Chem Mater* 2006;18:169–178.

104. Chilkoti, A., Ratner, B.D., Briggs, D. Plasma-deposited polymeric films prepared from carbonyl-containing volatile precursors—Xps chemical derivatization and static sims surface characterization. *Chem Mater* 1991;3:51–61.

105. Chevallier, P., Castonguay, N., Turgeon, S., Dubrulle, N., Mantovani, D., McBreen, P.H., Wittmann, J.C., Laroche, G. Ammonia RF-plasma on PTFE surfaces: chemical characterization of the species created on the surface by vapor-phase chemical derivatization. *J Phys Chem B* 2001;105:12490–12497.

106. Povstugar, V.I., Mikhailova, S.S., Shakov, A.A. Chemical derivatization techniques in the determination of functional groups by X-ray photoelectron spectroscopy. *J Anal Chem* 2000;55:405–416.

107. Sutherland, I., Sheng, E., Brewis, D.M., Heath, R.J. Studies of vapor-phase chemical derivatization for XPS analysis using model polymers. *J Mater Chem* 1994;4:683–687.

108. Popat, R.P., Sutherland, I., Sheng, E.S. Vapor-phase chemical derivatization for the determination of surface functional-groups by X-ray photoelectron-spectroscopy. *J Mater Chem* 1995;5:713–717.

109. Attal, S., Thiruvengadathan, R., Regev, O. Determination of the concentration of single-walled carbon nanotubes in aqueous dispersions using UV-visible absorption spectroscopy. *Anal Chem* 2006;78:8098–8104.

110. Li, Z.F., Luo, G.H., Zhou, W.P., Wei, F., Xiang, R., Liu, Y.P. The quantitative characterization of the concentration and dispersion of multi-walled carbon nanotubes in suspension by spectrophotometry. *Nanotechnology* 2006;17:3692–3698.

111. Derjaguin, B.V., Landau, L. Theory of the stability of strongly charged lyophobic sols and of the adhesion of strongly charged particles in solutions of electrolytes. *Acta Physicochim* 1941;14:633–662.

112. Verwey, E.J., Overbeek, J.T.G. *Theory of the Stability of Lyophobic Colloids*, Elsevier; 1948.

113. Cheng, W.P., Huang, C., Pan, J.R. Adsorption behavior of iron-cyanide onto γ–Al_2O_3 interface: a coagulation approach. *J Colloid Interface Sci* 1999;213:204–207.

114. Elimelech, M., Gregory, J., Jia, X., Williams, R.A. *Particle Deposition and Aggregation: Measurement, Modelling and Simulation*, Oxford, UK: Butterworth-Heineman; 1995.

CHAPTER 8

Chemical and Photochemical Reactivity of Fullerenes in the Aqueous Phase

JOHN D. FORTNER, JAESANG LEE, JAE-HONG KIM, and JOSEPH B. HUGHES

School of Civil and Environmental Engineering, Georgia Institute of Technology, 200 Bobby, Dodd Way, Atlanta, GA 30332-0373, USA

8.1 INTRODUCTION

Carbon fullerenes represent a third allotrope of carbon different from both diamond and graphite in their physical and chemical properties (1). Among this class of materials, C_{60} has been the fullerene of choice in research and development as it is widely available in high purity and in relatively large quantities. As a result, C_{60} physical and chemical properties have been extremely well characterized (2–8). Molecularly, C_{60} consists of 60 carbon atoms arranged in a spherical cage structure, analogous to a soccer ball, with carbon vertices bonded via one double bond and two single bonds resulting in 20 hexagons and 12 pentagons. Similarly arranged, other carbon fullerenes have been widely identified and include C_{70}, C_{76}, C_{78}, C_{84}, and C_{90} among other larger structures (9). With increasing commercial interest in unique material properties, the manufacture and the use of fullerenes are expected to grow rapidly over the next decade. Projected uses of C_{60}, among other fullerenes, are expanding and currently include fuel cell development, diamond manufacturing, superconductivity devices, drug delivery agents, and high-temperature lubricants (10–19). Recent focus on industrial scale production reflects fullerenes' commercial potential (e.g., Frontier Carbon Corporation, a Mitsubishi Corporation subsidiary, planned to start producing multiton quantities of C_{60} in 2007 (13)).

While carbon fullerenes have been at the center of many recent investigations in nanoscale science and engineering, information available to accurately assess the effects of these materials on natural environments and human health is limited, which is currently the case for many engineered nanoscale materials. Concerns have increased due to indications that fullerenes can interact with living organisms with

Nanoscience and Nanotechnology   Edited by Vicki H. Grassian
Copyright © 2008 John Wiley & Sons, Inc.

deleterious effect(s) (20–29). Biological responses have been attributed to fullerene's capability to specifically bind and cleave DNA (30–32), inhibit enzyme activities (33–35), and quench radicals (36, 37). In particular, a growing number of studies have indicated direct evidences of fullerene impact on organisms *in vivo*; for example cellular damage was observed in juvenile bass brain tissue upon a low-dose exposure of a water-stable C_{60} aggregate suspension (38).

Paramount to fundamentally understanding biocompatibility is the concurrent development of the specific physical and chemical properties of fullerenes in the aqueous phase. For example, the photosensitizing capacity of fullerenes (39, 40) has been considered critical in recent years with regard to induced biological effects (24). Past studies have reported photoinduced DNA cleavage and lipid peroxidation activities by fullerene and fullerene derivatives (41–46). These studies also suggested reactive oxygen species (ROS) such as singlet oxygen and superoxide radical anion generated by fullerene as the primary cause for the observed oxidative damage. Interestingly, such photosensitizing properties are also critical in fullerene potential pharmaceutical applications (21).

Furthermore, understanding general fullerene reactivity in the aqueous phase is critical for the accurate assessment of environmental fate and transport. Fullerene transformation during natural or engineered (e.g., water and wastewater treatment) processes can alter molecular properties critical to behavior in aqueous systems including, but not limited to, size, hydrophobicity, and dipole moment, among others. For example, oxidative transformation during water treatment processes such as ozonation may occur due to the olefin-like reactivity of fullerenes (C_{60}) (7, 47). Little is known regarding the reaction of fullerene in the aqueous phase, primarily because neutral, molecular fullerenes are virtually insoluble in water (estimated C_{60} solubility $<10^{-9}$ mg/L) (48–50). However, as detailed below, fullerenes can become readily available in aqueous systems, thus new reaction routes are possible and must be considered.

In this chapter, a relevant literature review on fullerene physical and chemical properties, including aqueous aggregation phenomenon, is first presented. For this analysis, C_{60} was chosen as a model system based on current literature availability and scheduled industrial scale production. Background information in this literature review is meant to provide a framework from which the fate and reactivity of fullerenes in the environment can be evaluated further. Additionally, two recent studies regarding C_{60} photochemical reactivity and oxidative transformation in the aqueous phase are discussed in detail. Both of these areas are thought to be critical in evaluating the environmental impact of this material.

8.2 C_{60} PROPERTIES

8.2.1 Structure

The C_{60} molecule is a closed cage structure with 60 carbon atoms arranged in the most symmetrical arrangement possible. The diameter across (nucleus to nucleus of furthest carbon atoms apart) has been measured to be 0.71 ± 0.007 nm via NMR measurements (51). The electron cloud (primarily from π bonding) has been estimated to add an additional 0.335 nm (6) to the diameter of the molecule, giving C$_{60}$ an effective diameter of ca. 1.034 nm (6, 52).

Ring architecture is critical to the stability of C$_{60}$, as with all fullerenes. The arrangement of hexagon and pentagon rings in fullerenes was correctly hypothesized to be a function of Euler's theorem whereby each fullerene contains $2(10 + M)$ carbon atoms corresponding to exactly 12 pentagons and M hexagons (7). As a repeating hexagonal sheet will remain flat as observed for pure graphite (or it can even be "rolled" as observed for nanotubes' walls), pentagon ring inclusion in all fullerenes is obligatory for cage curvature (i.e., rounding) (1, 53). Similarly, both single-wall and multiwall nanotubes require pentagon inclusion in round-end formations (6, 54). Furthermore, the "more" isolated each pentagon is from each other the more cage stability is achieved as discussed elsewhere as the "isolated pentagon rule" (IPR) (53, 55). The IPR is based on the fact that adjacent pentagon rings will be thermo-dynamically unfavorable due to steric constraints, resulting in excess curvature and strain in structure as discussed by Kroto et al. and 8π-electron systems that would lead to resonance destabilization as discussed by Hirsch (7, 53). Out of about 15,000 possible isomer structures for C$_{60}$ in particular, only one possibility symmetrically incorporates 12 pentagons that are each "equally" isolated by 5 hexagon ring structures (20 total) minimizing the anisotropic contributions to the strain energy (7).

This stable, truncated icosahedral, 3D structure exhibits I$_h$ symmetry that is analogous to a soccer ball (1). While C—C bonding environments must meet the required carbon valence needs, one double bond and two single bonds for each carbon, the lowest energy Kekulé structure was calculated, keeping the IPR, such that all double bonds should to be located between hexagon vertices [6,6] that are tangential to each pentagon vertices [5,6] (7). These two types of bonds have been directly measured via neutron, electron, and X-ray diffraction along with ^{13}C-NMR and reported to be 0.144, 0.146, 0.147, and 0.145 for [5,6] bonds (pentagon ring, single bonds), respectively, and 0.139, 0.140, 0.136, and 0.137 nm for [6,6] bonds (hexagon ring, double bonds), respectively (7, 51, 56–58, Yannoni, 1991, #421). Differing averaged bond lengths between [5,6] and [6,6] are attributed the minimization strain energies by molecular symmetry and the π bonding environments (59). Interestingly, similar C—C bonds in larger fullerenes can be more variable as there are more C—C bonding environments to consider (i.e., higher degree of bond length alternation) (7). Basic C$_{60}$ chemical–physical properties, largely underpinned by the symmetrical 3D carbon ring structure and localized conjugated π system, including heat of formation, electron affinity, material density, and ionization potentials among others have been studied extensively and summarized in a number of review papers and books (6, 8, 52, 59–62) with selected properties and constants tabulated again here in Table 8.1.

8.2.2 Solid C$_{60}$

Physically, as a solid, pure C$_{60}$ is crystalline, with an FCC (face-centered cubic) unit cell ($a = 14.17$ Å) at room temperature and atmospheric pressure with a demonstrated density of 1.72 g/cm^3. In this crystal state, C$_{60}$ molecules rotate quite freely due to

TABLE 8.1 Selected Solid-State C_{60} Properties

Property	Value
Density	1.65 g/cm^3
Graphite density[a]	2.3 g/cm^3
Diamond density[a]	3.5 g/cm^3
Crystal structure (>255K)	FCC
Crystal structure (<255K)	SC
Nearest neighbor distance	10.04 Å (FCC)
Cage diameter	7.1 Å
Lattice constant	14.198 Å (FCC)
Electrical conductivity	Nonconductor (neutral state)
Index of refraction (RI)	2.2 at 630 nm λ

Constants summarized by Huffman (52). All values are at standard temperature and pressure unless otherwise noted.

[a]Allotrope comparison.

relatively weak molecular interactions (63). According to Prassides et al., as the temperature is lowered (ca. 261K), the crystal restructure undergoes a transition (rearrangement) to a simple cubic (SC) unit cell structure (6, 64). At atmospheric pressure, sublimation occurs around 1×10^3 K in inert atmosphere, which varies with pressure. In theory, at high temperature ($>1 \times 10^3$K) and elevated pressure ($>1 \times 10^4$ Torr), solid C_{60} can melt to a liquid state (6, 47) but has not been empirically verified. Molecular C_{60} begins to fragment at temperatures exceeding 4×10^3K and collapse at pressures greater than 1×10^8 Torr (6). It has been noted, though, that in the presence of oxygen, UV, or a catalyst such as TiO_2, thermal stabilities dramatically lowered as oxidative degradation occurs (47). Furthermore, a number of cocrystallites and clathrate structures with other molecules (e.g., benzene, CCl_4, etc.), usually as solvents, have been widely observed and described (6, 65).

8.2.3 Solubility

C_{60} solubility has been widely studied, with an estimated 150 known values for various solvents (65). However, as Korobov and Smith explain (65), "There is still no good theory to explain or predict absolute values for fullerene solubility and changes in solubility when changing the solvent or the fullerene itself." Nevertheless, careful empirical observations and theoretical discussions regarding specific trends exist that are of value. Ruoff et al. (50) attempted to correlate C_{60} solubility to the polarizability, polarity, molecular size, and cohesive energy density of the solvent. In this study, no one distinctive parameter universally explained C_{60} solubility; however, positive correlations were observed with similar solvent parameters to that of C_{60} such as molecular size (50). Multivariate analyses employed by Murray et al. (66) using 22 solvents achieved a linear coefficient of 0.95 (r^2), indicating that solvent surface area was critical. Studies by others indicate similar positive correlations with solvent size (expressed as molar volume) in addition to positive correlation to solvents that are a good Lewis acid and Lewis base (both) but with minimal polarity or charge (65). In

TABLE 8.2   Solubility of C$_{60}$ in Various Solvents

Solvent	[C$_{60}$](mg/mL)
Benzene	1.7
Toluene	2.8
Cyclohexane	0.036
TCE	1.2
Chloroform	0.16
Ethanol	0.001
Acetone	0.001
Pyridine	0.89
Carbon disulfide	7.9
n-decane	0.071

Values taken from Ref. 80.

addition, organic solvent electron donor (ED) capacity (Lewis base) was further identified by Talukdar et al. as being positively correlated with C$_{60}$ solubility, along with more evidence supporting that solvent size and polarizability are critical (65). In addition, C$_{60}$ solubility displays an abnormal temperature dependence in a number of solvents (67), actually decreasing with increased temperature. While the solubility of C$_{60}$ in water has not been measured directly, it has been estimated at less than 10^{-9} mg/L by Heymann (48). As described by Kadish and Ruoff (65), a list of selected solubilities at 298 K are tabulated in Table 8.2.

8.2.4 Spectroscopic Properties

Unique spectral properties were critical in the initial fullerene identification and remain so for current analytical analyses. C$_{60}$ has been well characterized with a range of spectral analyses including mass spectroscopy (MS), UV–Vis, FTIR, Raman, and NMR (7, 9, 68–73). Mass spectroscopy has been employed in fullerene identification via molecular weight(s) since the conjecture of C$_{60}$ by Kroto et al. in 1985 (1). With a molecular weight of 720 (*m/z* 720 @ monoionized) and being fairly easily ionized due to HOMO–LUMO molecular orbital situation, MS identification of C$_{60}$ has been reported throughout the literature using both positive and negative ion modes with laser desorption/ionization (LDI), with and without matrix assistance, FAB, and EI (7, 68, 69).

Based on inherent electronic and vibrational structures, IR and UV–Vis spectra confirmed the structure of molecule by matching values calculated previously for an icosahedral 60 carbon cage structure by others (9, 62, 71, 72, 74). C$_{60}$ vibrational structure (IR and Raman spectroscopy) was predicted to have four IR-active frequencies (T$_{1u}$ modes, dipole) based on four spring constants (C–C (pentagon bond), C=C (hexagon bond), C–C=C (angle), and C–C–C (angle)) and derived eigenvalues along with 10 theoretically active Raman frequencies (H$_g$ and A$_g$ modes) despite being 60 carbons (360 total electrons and a calculated 174 active modes) as described by Weeks and Harter among others (62, 74). Four such IR-active absorption bands were empirically observed to be at 1429, 1183, 577, and $528\,\text{cm}^{-1}$ indicating a high molecular symmetry and confirming a free, truncated icosahedral molecule (9, 72, 74). In comparison, a 60 carbon graphite isomer (D_{6h} symmetry) has some 20 IR-active frequencies (72). Following IR reports, 10 empirical Raman shifts were later observed, matching vibrational predictions (albeit slightly shifted) (74) further confirming the electronic structure (75).

A C_{60} UV–Vis absorption spectra was suggested as early as 1987 and then confirmed in 1990 (9, 71). Larsson et al. calculated oscillator strengths, which correspond to the probability of orbital transitions, of the allowed transitions as a function of eV (which can be converted to wavelength). These predictions match absorbance spectra (overlaid) by Kroto et al. in the UV–Vis (200–700 nm) range (C_{60} as a thin film). In addition, Kroto et al. observed broad absorption at 450 nm, which is also discussed by Larsson et al. and Dresselhaus et al. as symmetry-forbidden transitions and observed by others normally for solid-state C_{60} films, indicating probable solid-state C_{60}–C_{60} interactions. Dissolved C_{60} UV–Vis spectra have been published extensively and exhibit slight red- or blue-shifts depending on the solvent (50). Specifically, dissolved in hexanes (low background UV absorbance), C_{60} exhibits strong absorbance in the UV at 211, 256, and 328 nm, a lesser absorbance peak at 410, and low visible absorbance peaks centered 492, 540, 568, 591, 598, and 621 nm (69).

^{13}C NMR C_{60} spectra were reported independently by three groups almost simultaneously in 1990 (68, 69, 73). All three reports, as hypothesized, observed C_{60} as a single-line response, ranging from 143.3 to 143.68 ppm shift from TMS reference, indicating an icosahedral symmetry (i.e., I_h symmetry), and further confirming the exact equivalency of all carbon atoms. Such as downfield shift from the benzene or other PAH analogues was further hypothesized to be a function of strain-induced hybridization changes similar to indane (143.9 ppm) and benzocyclobutene (146.3 ppm) (73).

8.2.5 Aromaticity

Kroto suggested in 1985 that C_{60} could be, "the first example of a spherical aromatic molecule" (1). Since then, the question whether fullerenes are truly aromatic has been the source of many studies and is still debated (59). The problem lies in the fact that fullerenes in many ways are difficult to compare with planar aromatics by which aromaticity has been defined. Moreover, the actual definition of aromaticity has changed over time and is still even considered controversial in some respects (76). Buhl and Hirsch (59) recently published a review article evaluating fullerene aromaticity and discuss some of the problems with defining and/or characterizing fullerene aromaticity:

> Compared to benzene, being the archetype of a two-dimensional aromatic molecule, the discussion of aromaticity of fullerenes must take into account the strain provided by the pyramidalization of the C atoms. The rich exohedral chemistry of fullerenes, which is basically an addition to the conjugated π system, is to a large extent driven by the reduction of strain. Analysis of the reactivity and region chemistry of addition reactions

reveals behavior reminiscent of electron-deficient olefins. However, especially the magnetic properties of fullerenes clearly reflect delocalized character of the conjugated π system, which, depending on the number of π electrons, can cause the occurrence of diamagnetic or paramagnetic ring currents within the loops of the hexagons and pentagons. Neutral C$_{60}$ for example, containing diatropic hexagons and paratropic pentagons was labeled by others 'ambiguously aromatic' (59).

Following the Buhl and Hirsch text, fullerene aromaticity is discussed in classic aromatic criteria, which include structure (including count numbers), energy, reactivity, and magnetism (59). Briefly, C$_{60}$ properties for each criterion are discussed, but should only be interpreted as a brief summary of extensive studies by others on the subject.

8.2.5.1 *Structure*

Ring structure, particularly with regard to bond lengths ([5,6] pentagon bonds and [6,6] hexagon bonds), and induced strain from the pyramidalization are critical to fullerene aromaticity properties. As discussed before, neutral C$_{60}$ [5,6] and [6,6] bond lengths differ, which is attributed to the filling of h$_u$, t$_{1u}$, and the t$_{1g}$ orbitals within the $I=5$ shell (π) (59). When C$_{60}$ is neutral, the $I=5$ shell is incompletely filled, as it requires 22 π electrons, 72 total ($I=4$ shell requires 18 π electrons, 50 total). This fact actually causes higher incidence of electron *localization*, giving rise to single bond characteristics that are physically longer along the [5,6] and double bond characteristics that are shorter and more reactive along the [6,6] (59). Intuitively, charged fullerenes such as the theoretical C$_{60}^{10+}$ and even C$_{60}^{n-}$ ($n=2$–12) are considered "more" aromatic in this regard as the $I=5$ shell is varied. C$_{60}^{10+}$ [5,6] and [6,6] C−C bonds have been calculated to relax based on the loss of electrons from the $I=5$ shell electrons giving rise only to a completely filled $I=4$ shell, which corresponds to the lower bond length difference, thus increasing delocalization. Similarly, the addition of electrons to the $I=5$ shell has also been calculated to increase aromaticity as bond lengths would also relax based on increased delocalization as the filled shell (59). However, it should be noted that electron addition above six is unlikely as the $I=6$ nodal plane takes on electrons after C$_{60}^{6-}$, instead of complete $I=5$ filling (59). These bond length differences in neutral C$_{60}$ contribute to the weakly antiaromatic pentagons and weakly aromatic hexagons based on the HOMA indexing (harmonic oscillator model of aromaticity).

8.2.5.2 *Energetics*

Aromaticity can also be investigated from energetics. A number of methods have been designed to measure the molecular orbital (MO) resonance energy (RE) or the molecular stabilization of the cyclic bonding π delocalization. As Buhl and Hirsch point out, however, such measurements have been classically zeroed to or designed for well-characterized planar PAH molecules such as benzene. While work has been done to normalize the caged structure for such measurements, including strain energies from pyramidalization (i.e., bond angle contributions) (77), disparities exist when comparing C$_{60}$ characteristics with the planar benzene structure, indicating that while C$_{60}$ is aromatic, just how aromatic (compared to benzene) depends on the methodology of choice (59). For larger fullerenes, such MO-RE calculations begin to converge with aromatic values for graphite (53) as bond angles relax and the ratio of hexagons to pentagons increases. Other, more straightforward, estimates of aromaticity stabilization contributions measure the bond separation energy or heat of hydrogenation. Through the use of isodesmic equations, which maintain the number of formal single and double bonds between the carbon atoms, the energy of separating the bonds can be predicted. Such analysis was described by Fowler et al. and demonstrates C_{60} to be generally less aromatic when compared to benzene and based on the energy associated with the separation (2245 kJ/mol for C_{60}) (78). Similarly, the heat of hydrogenation for C_{60}, which can be taken as a measure of energetic stability (the higher the heat the less stable), has been theoretically calculated to be about double (normalized per C) that of benzene (-206 kJ/mol), indicating that C_{60} is to a "lesser" degree aromatic (78, 79).

8.2.5.3 *Reactivity*
Generally speaking, classic aromatic reactivity such as substitution reactions cannot be considered as fullerenes lack boundary hydrogen, thus aromatic substitution reactions are not possible, leaving only addition and redox reactions to consider (again making traditional PAH comparisons difficult). In addition, caged molecules differ from traditional planar aromatic in that bond strain can alter reactivity from what is expected (6, 7, 47). Traditional indexes of molecular reactivity such as chemical hardness, defined as η where $\eta = (HOMO - LUMO)/2$, are based on molecular orbital energy gaps. A larger η equates, normally, to less reactive molecule. Compared to benzene, C_{60} has an approximately three times lower hardness index, indicating a higher reactivity than benzene. While there are a number of reactions to consider, Buhl and Hirsch summarized reactivity as it relates to aromaticity (59):

> The reversibility of several addition reactions could be an indication of the propensity of the retention of the structural type which is considered to be a reactivity criterion of aromaticity. The regioselectivity of two-step additions, such as the addition of a nucleophile followed by the trapping of the fullerenide intermediate with an electrophile, on the other hand, shows that the charge is not delocalized over the whole fullerenide. This speaks against a pronounced aromatic character. Also, other reactions such as the additions of transition-metal fragments are reminiscent of olefins rather than of aromatics. In many cases an important driving force for the regioselectivity of multiple addition reactions is the formations of the substructures that are more aromatic than the substructures of the parent fullerene.

8.2.5.4 *Magnetic Properties*
Electron delocalization in aromatic compounds is the source of characteristic magnetic properties and can be observed via NMR shifts and diamagnetic susceptibilities. Fullerenes were first predicted by Kroto et al. to have unusual magnetic properties in 1985 (1). Since then, numerous studies have been conducted evaluating fullerene magnetic properties, many of which are beyond the scope of this review. However, as they relate to aromaticity, relevant fullerene magnetic properties must be acknowledged. Ring current theories, which are based on π electrons freely and separately along a σ-bonded ring framework, allow for the computation of magnetic susceptibilities, which are difficult to observe, and NMR chemical shifts.

Empirically, ring currents can be measured indirectly though proton shielding (which only applies to fullerenes upon hydrogenation). When a perpendicular magnetic field is applied to an aromatic ring, delocalized π electrons are induced to circulate around the ring producing a small local magnetic field (76). According to McMurray (79), this induced field opposes the applied field in the middle of the ring, but actually reinforces the applied field outside of the ring, thus effectively deshielding aromatic protons (on the ring exterior), which can be observed via ^{1}H NMR. For example, when a strong magnetic field is applied to [18] annulene ($C_{18}H_{18}$), a cyclic conjugated polyene with 18 π electrons, the 6 interior protons are strongly shielded by the ring current (-3.0 δ relative to TMS), whereas the 12 exterior (outside) protons are deshielded and observed in the typical aromatic region (9.3 δ TMS) (79). Benzene as a model PAH has demonstrated similar proton deshielding (7.2 δ TMS) consistent with a near-free π circular electron current (59, 80).

However, for underivatized fullerenes, deshielding measurements are not possible to observe as there are no boundary bonds, thus, other estimates are necessary. Proper use of derivatized fullerenes appears to be quite difficult as it is hard to know the exact architecture of the product. Generally though, gainful employment of appropriate covalent adducts has shed insight of the shielding capacity via an applied magnetic field (59). The application of quantum-chemical methods such as Hückel-MO-based London method and Hartree–Fock *ab initio* calculations for current densities) can calculate localized (individual ring) aromatic, also termed diatropic, π electron ring currents and antiaromatic, also termed as paratropic, ring currents (59). Conjugated antiaromatic molecules (e.g., cyclobutadiene) increase in molecular energy when π electrons are delocalized in contrast to aromatic conjugated molecules that are lower in molecular energy with increased delocalization (76, 79).

Taken together over the whole molecule, ring current calculations indicate that C$_{60}$ is slightly aromatic with regard to magnetic susceptibility (81). However, individual ring current calculations indicate that the hexagon rings are diatropic, with similar values to benzene, and pentagons are paratropic, which, when taken together, account for a small *net* aromatic index (magnetic susceptibility). C$_{70}$, however, with an equatorial band of additional paratropic hexagons (same 12 pentagons as C$_{60}$) is calculated to be more susceptible to magnetic fields and thus more aromatic when considered as a whole (*net*) (77, 81). Similarly, "giant" fullerenes, such as C$_{240}$, C$_{540}$, and higher with increasing number of hexagons and less bond angle strain have been calculated to have π electron ring current susceptibilities that begin to approach graphite (i.e., higher than C$_{60}$ and C$_{70}$) (59).

8.2.5.5   *Count Rules* For simple, planar annulene aromatics, the Hückel count rule of aromaticity that simply states that $4n + 2$ π electrons are necessary, where $n =$ succeeding energy levels (79). For such planar systems, $4n + 2$ π electrons represent a closed-shell system. However, this rule is general, and not applicable for more complex polycyclic systems that contain benzenoid rings joined by four or five member rings, as demonstrated for biphenylene (which still demonstrates aromatic properties) (76). Similarly, fullerene aromaticity cannot be considered by the $4n + 2$ Hückel rule. A spherical closed-shell situation is realized if the fullerene contains

$2(N+1)^2$ π electrons, similar to noble gas configurations (59). For C_{60} to reach a $2(N+1)^2$ π electron configuration, it must exist as C_{60}^{10+} ($I = 4$, filled) or C_{60}^{12-} ($I = 5$, filled), the later of which would not likely occur as outer shell orbitals (t_{1g}, $I = 6$) of C_{60} are filled before inner shells (t_{2u}, $I = 5$) (59). Theoretical calculations of ^{3}He (endohedral) chemical shifts ($-81.4\,\delta$) in the center of C_{60}^{10+} support this, indicating extreme diamagnetic shielding (59).

In conclusion, the consideration of fullerene aromaticity is a complex subject. As Buhl and Hirsch summarize, "When π MOs on the surface of a sphere are grouped into shells according to their nodal properties, a simple model predicts that maximum aromaticity is reached when the individual shells are filled (i.e., $2(N+1)^2$ π electrons). . ... Such complete shell filling occurs less often than in cyclic annulenes, with many intermediate situations exhibiting both aromatic and antiaromatic properties (59)." This is the case for neutral C_{60}, as it demonstrates both aromatic properties including calculated localized ring current, diamagentic shielding (as demonstrated by adduct additions), and increased molecular stabilization; and also antiaromatic properties including reactivity behaviors, similar to electron-deficient olefins, bond length alterations and paratropic ring current regions (pentagons). Based on these criteria, the term "ambiguously aromatic" or as Fowler et al. states, "a more or less aromatic system" seems to describe neutral C_{60} well (81).

8.2.6 Reactivity

C_{60} reactivity is a function of electronic (e.g., HOMO–LUMO energy gaps, incomplete $I = 5$ nodal shell, and paratropic/diatropic ring currents) and physical structure (e.g., bond strain, bond length alterations, etc.), which are irreducibly intertwined. Briefly, electrochemistry (redox), covalent additions, and photophysical reactivity among others are discussed below. Each area of fullerene reactivity has been studied and discussed extensively in the literature. For the sake of brevity, discussion of each is limited to fundamental, representative studies.

8.2.6.1 Electrochemistry Based on degenerate low-lying unoccupied molecular orbitals (LUMO t_{1u} symmetry and LUMO + 1 t_{1g} symmetry), C_{60} is predicted to be an electronegative molecule capable of accepting up to six electrons upon reduction. This was shown to be true via cyclic voltammetry, producing the anionic forms (fullerides) consisting of one to six electron additions that fill the LUMO t_{1u} (6, 70, 82–85). C_{60} potentials ($E_{1/2}$) in a mixed solvent system of acetonitrile:toluene (1 : 5) under vacuum at $-10°C$ with TBAPF$_6$ (tetra-n-butylammoniumhexafluorophosphate) as a supporting electrolyte were measured by Echegoyen et al. to be -0.98, -1.37, -1.87, -2.35, -2.85, and -3.26 V versus Fc/Fc$^+$ (Ferricinium/Ferricinium ion) (85). While mono- through tetraanions have been shown to be stable on the time frame of hours to days depending on the solvent temperature and supporting electrolyte; penta- and hexaanions quickly reverse back to lower anions (85). Conversely, it was predicted to be difficult to oxidize C_{60} giving rise to a charged cation species. Voltammetry studies showed a one electron oxidation in TCE (under vacuum) occurs at $+1.26$ eV versus Fc/Fc$^+$ (85). The difference between the first reduction and first oxidation, which is a measure of the

HOMO–LUMO energy gap, can be calculated to be 2.32 eV for C$_{60}$ under these described conditions (65). Similar calculations for C$_{70}$ indicated a slightly smaller HOMO–LUMO gap of 2.22 V (65, 85). Additionally, it has been demonstrated that the choice of solvent and supporting electrolyte can have significant effect on the potential at which fullerene ions are generated (7, 70, 85). Furthermore, based on electron symmetry and orbital levels filled, C$_{60}^{n-}$ ($n = 1$–4) anionic forms of C$_{60}$ take on unique spectral properties. For example, adding one or more electron to the degenerated LUMO molecular orbital results in a new IR transitions, widening and redshifting of UV absorbance, change in visible color (C$_{60}^{-}$ red purple; C$_{60}^{2-}$ red orange; C$_{60}^{3-}$ dark red brown), ^{13}C NMR downfield shifts (7), and unique ESR (electron spin resonance) spectra. Such reductions can also be accomplished via traditional chemical reaction, for example, with Rb, Li (Birch reaction), and other alkali earth metals in liquid ammonia, potassium napthalenide, naphthalene, and mercury among other donors (6, 7). Furthermore, upon ionization (C$_{60}^{n-}$), it has been well established that fullerides readily react with a variety of electrophiles to form covalently bonded adducts (6–8).

8.2.6.2 Nucleophilic Additions

Based on electronic structure, neutral C$_{60}$ undergoes a number of nucleophilic additions. Such an addition gives rise to an intermediate anion species that is then subsequently stabilized with an electrophile addition, such as H^{+} (7). Such addition(s) gives rise to many possible isomers. However, based on electron density calculations (Muliken charges based on AM1 calculations) after a nucleophile addition (t-BuOH forming the intermediate t-BuC$_{60}^{-}$ anion), the carbon (across the double bond [6,6]) directly across the once double bond becomes the most negative (increase e^{-} density) and thus most probable electrophilic addition position (7). Simple examples of nucleophile additions include early observations of reactions with typical nucleophiles such as primary amines, as RNH$_2$, (where R is an aryl or alkyl group); alkyls (simple to complex); cyanide; organolithium compounds; and Grignard reagents among others (6, 7, 61, 86). Number and cage placement of such additions depend on the reaction conditions and type of nucleophile. In addition, some nucleohphiles with leaving groups (e.g., Br) can be electrophilic enough to undergo intramolecular substitution at the electron-dense carbon (along the fullerene [6,6] C–C) and form a cycloadduct. Additionally, other cycloadditions can be accomplished via Diels–Alder-type reactions with appropriate dienes (e.g., cyclo-pentadiene) (6–8, 79). Such cycloadditions can happen via a number of reactants and through a number of shared pathways (4 + 2, 3 + 2, 2 + 2, and 2 + 1 types in reference to shared π electrons) that are beyond the scope of this review.

8.2.6.3 Hydrogenation

Fullerene reduction via hydrogen addition (hydro-fullerene) can be accomplished via a number of standard hydrogenation reactions. Degrees and placement of proton additions vary according to reactant types, ratios, and solvents (6–8, 61). Early reports of hydrogen addition included C$_{60}$H$_{18}$ and C$_{60}$H$_{36}$ via Birch Reduction (82) along the [6,6] C=C double bond (termed by some as a 1,2 addition). Reducing metal treatment with Zn/Cu, Mg, Ti, Al, or Zn with a proton source produces a number of lower hydrofullerenes including C$_{60}$H$_2$, C$_{60}$H$_4$, and C$_{60}$H$_6$ (7). Additionally, C$_{60}$ can be hydrogenated with molecular hydrogen in the presence of a solid reductant (7). Such reactions proceed with an elevated H_2 pressure under heat, pressure, and a reducing environment such as supported (alumina or activated carbon) Ru, Rh, Ir, Pd, Pt, Co, and Ni giving rise to $C_{60}H_{18}$ (Ru, Rh, and Ir) and $C_{60}H_{36}$ (Pd, Pt, Co, and Ni) isomers (87). Other organic reactants such as an NADH analogue (BNAH under photoexcitation) can be used to accomplish lower numbered hydrofullerenes ($C_{60}H_2$) under mild conditions (88). It has also been shown that hydrofullerenes can be oxidized back to the parent fullerene via treatment with DDQ (2,3-dichloro-5,6-dicyanobenzoquinone) (7).

8.2.6.4 *Halogenation*   Exothermic halogenation reactions readily occur with fluorine, chlorine, and bromine giving rise to final products that are limited by the size and affinity of the halogen. Fluorination $(C-F)$, the most exothermic of the halogenation reactions, can be accomplished via fluorine gas, noble gas fluorides, halogen fluorides, and metal fluorides among others (7, 8, 61, 89). While a variety of fluorinated fullerenes have been reported, including $C_{60}F_{48}$ as observed by a number of groups, however, higher derivatives including $C_{60}F_{60}$ have been reported (89). Direct chlorination occurs in a similar manner via chlorine gas under elevated temperatures or with liquid chlorine at lowered temperatures $(-35°C)$ resulting in $C_{60}Cl_n$ ($n = 6–26$ depending on reaction conditions) (7, 90). Bromination occurs less readily, however, it can be accomplished directly with liquid bromine giving rise to $C_{60}Br_{24}$ or in benzene, giving lower derivatives such as $C_{60}Br_6$ (7, 91, 92). Halogen addition patterns vary by reaction conditions and halogen affinity and size as described by Taylor et al. (61).

8.2.6.5 *Electrophilic Additions and Oxidation*   While cage reduction reactions consisting of nucleophilic additions are favored for neutral fullerene species, oxidation reactions (electron withdrawing) can occur readily with appropriate electrophiles (strong). Simple epoxide adducts ($C_{60}O_n$, where $n = 1–12$) can be synthesized via a range of oxidizing agents such as oxygen (under UV), hydrogen peroxide, *m*-chloroperoxybenzoic acid (MCPBA), dimethyldioxirane, and methyl(trifluoromethyl) dioxirane, among others, in various solvents (47). Additionally, direct facile oxidation can be accomplished via osmylation (OsO_4) and ozonation (O_3) whereby an electron-withdrawing cycloadduct is formed, which for ozone (ozonide) quickly degrades (thermolysis) into a fullerene epoxide in organic solvents (93). Hydroxylated fullerenes (fullerols) can be synthesized similarly beginning with an electron-withdrawing agent such as a nitronium ion in the presence of H_2SO_4 and KNO_3 under elevated temperatures $(95°C)$ or with nitronium tetrafluoroborate in the presence of alkyl or aryl carboxylic acid in methylene chloride under N_2 atmosphere and subsequently hydrolyzed in basic water (via NaOH) (94–96). Similarly, polysulfonated fullerene intermediates, which can be synthesized via neat fuming sulfuric acid at $55°C$ under N_2, are hydrolyzed in aqueous NaOH resulting in $C_{60}(OH)_x$ (fullerols) (97). Other methods have been demonstrated, substituting halogen additions with hydroxyl groups and even a simple two-phase (C_{60} dissolved in toluene–water) interface whereby an aqueous pH was elevated and a phase transfer agent (TBAH) employed (98, 99). Depending on preparation conditions, fullerols can be composed of 8–24 hydroxyl groups, can also contain carbonyl groups, are generally hydrophilic, and readily dissolve in water (47).

8.2.6.6 *Photochemical Reactivity* C_{60} has been reported to exhibit chemical reactivity toward various radicals, producing stable multiple radical adducts species (100, 101). Especially, the photoinduced excitation of C_{60} can facilitate electron transfer from electron donors to acceptors and mediate energy transfer to oxygen with high quantum yield, leading to the formation of singlet oxygen. These photochemical properties (39, 40) have given C_{60} priority as possible applications in photodynamic therapy (32, 102), oxidative synthetic reaction (103), and electron-shuttling action in dye-sensitized solar cell (60). Singlet oxygen (104) and superoxide radical (27, 31, 32) generated in C_{60} suspension can trigger DNA damage and cellular structural destruction, (27, 31, 32, 104–106). While pristine C_{60} in hydrocarbon solvent has been known as an efficient singlet oxygen precursor under the UV irradiation (39), there have been some controversial reports about photoactivity of C_{60} suspended in aqueous phase; Beeby et al. (107) demonstrated photochemical generation of singlet oxygen was efficiently performed by C_{60} stabilized into aqueous solution of Triton X100 (nonionic surfactant, TX) applied above critical micelle concentration (CMC). On the other hand, Yamakoshi et al. (31, 32) suggested that based on ESR analysis, C_{60} solvated into water with PVP (polyvinylpyrrolidone)- generated superoxide and ·OH radical in the presence of electron donor, with no detectable singlet oxygen produced.

8.2.6.7 *Other Chemistries* Fullerene chemistry is a broad topic. In addition to the aforementioned chemistries, a number of articles (including review articles) and books have been published demonstrating similar reaction pathways albeit with higher fullerenes (i.e., C_{70}) and other more complex chemistries, including, but not limited to, fullerene polymer synthesis, derivatized fullerene parent materials, endohedral fullerene parent materials, and radical chemistries (101), among others. Again, thorough fullerene chemistry reports by fullerene chemists including Buhl and Hirsch (7), Taylor (47), Dresselhaus et al. (108), and Kadish and Ruoff (65) are widely available and provide an excellent starting point for many cases.

8.3 FULLERENES IN WATER

Incidentally, the behavior of C_{60} in the aqueous phase can differ from what is expected as C_{60} and other pristine, neutral fullerenes are virtually nonwettable (estimated solubility of $C_{60} < 10^{-9}$ mg/L) (49, 50, 109). The unfavorable free energy of remaining in water should drive hydrophobic C_{60} to adsorb onto/or within (absorption) matrices, which corresponds to large octanol-water and organic carbon–water partition coefficient estimates (Table 8.3). Based on classic estimation parameters, the aqueous behavior is expected to be similar to other large, nonpolar hydrophobic molecules (e.g., polyaromatic hydrocarbons), which tends to associate strongly with organic carbon (OC) and in natural systems. However, apparent solubility or availability of C_{60} in water can be significantly altered by the following three pathways. First, C_{60} can be chemically modified to include hydrophilic functionalities such as hydroxyl groups that have been shown to substantially increase its aqueous solubility (94, 110, 111). Numerous other water-soluble fullerene derivatives (e.g., fullerols)

TABLE 8.3 Selected C_{60} Properties and Partitioning Estimations

Property	Value
Molecular weight (g/mol)	720 g/mol
Aqueous solubility (g/L)	$<10^{-12}$ g/L
Vapor pressure (atm)	6.6×10^{-9} atm

Partitioning Estimations	Value
Henry's law constant (atm m^3/L)	Estimated: 10^{-11}, based on structural unit contributions
Octanol–water partition coefficient (K_{ow})	Estimated: $10^{14.7}$–$10^{16.7}$, based on $\log K_{ow} = -\log C^{sat}_w(1, L) - \log \gamma_0 - \log V_0$
Organic carbon–water partition coefficient (K_{oc})	Estimated: $10^{14.2}$–$10^{16.2}$ based on $\log K_{oc} = 1.01 \times \log K_{ow} - 0.72$ (suggested for aromatic hydrocarbons)

have been described and include organic acids (112, 113) and electrophilic additions (96) as discussed above. Second, C_{60} can be rendered available in water by surfactants, polymers (i.e., PVP), and even natural macromolecules including γ-cyclodextrin, sucrose, and dextran (40 kDa), which can effectively shield the hydrophobic surface of the C_{60} molecules from water (114–116). Third, C_{60} in polar solvents, including water, can form stable aggregate suspensions at the micro- to nanoscales (26, 117–120). Compared to hydrophobic sorption alone, C_{60} rendered "available" in water by any of these methods will be subject to dispersive processes, increasing the potential volume of the media exposed.

8.3.1 Water-Stable C_{60} Aggregates

Recent reports have demonstrated aggregation phenomenon of C_{60} in water (case 3 above) occurring in relatively simple systems such as solid C_{60} powder (>99.9 purity; MER Corp., Tucson, AZ, USA) was mixed in water at high sheer rates for an extended period of time (days to weeks) (120). C_{60} aggregate suspensions have been reported via a number of methods and have demonstrated dimensions at, or near, the nanoscale diameters, ca. 5–500 nm, with a net negative surface charge thus allowing for stable ppm concentrations (26, 117–119, 121, 122). A typical suspension prepared by this lab is shown in Fig. 8.1 (prepared by the modified Deguchi method as described by Fortner et al.) (119, 123). These aggregates allow for concentrations up to ca. 150 mg/L, which is ~11 orders of magnitude more than the estimated molecular solubility (48, 50, 124). Though this material has been the subject of some study over the past decade, questions concerning the formation, composition, and stability remain unresolved/ambiguous (117–119, 121, 125). Water-stable C_{60} aggregates' properties are summarized here, which are being being organized into formation pathways and properties along with a brief summary of the current surface charge (source) hypotheses.

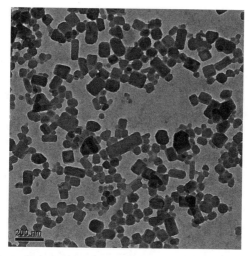

FIGURE 8.1   Transmission electron micrograph of nano-C_{60} (Deguchi, THF method).

8.3.2   Formation and Properties

First described by Scrivens et al. (26) in 1994, water-stable C_{60} aggregates were observed via a solvent-exchange protocol beginning with toluene containing dissolved C_{60}, which was diluted into THF, then diluted further into acetone, and finally into water whereby a fine mustard- colored suspension was formed. UV–Vis spectra of the suspension showed characteristic C_{60} peaks at 227, 280, and 360 nm (redshifted, broadened) and an average size of approximately 300 nm (SEM). In addition, radio-labeled ($^{12}C : ^{14}C = 200 : 1$) water-stable C_{60} aggregates were synthesized with this method for *in vitro* uptake studies with human keratinocytes (26). To date, this is the only literature report of successfully synthesizing ^{14}C-labeled C_{60}. A year later, in 1995, Andrievsky et al. (117) observed a similar aqueous suspension after sonicating a C_{60}-laden toluene suspension overlaying water. A similar broad UV–Vis spectrum characteristic of C_{60} was observed, with aggregates reported below 200 nm (117). Starting with fullerene anions, Wei et al. (122) reports a similar aqueous suspension (referred to as a sol in their work) after exposing C_{60}^{-} and C_{60}^{2-} to undegassed water (oxygen present). To begin, C_{60} was reduced to the fulleride in THF via Al–Ni alloy, tin, zinc powder, or sodium hydrosulfite in the presence of NaOH. Fulleride presence was confirmed through NIR, cyclic voltammetry, electron paramagnetic resonance (EPR), and NMR (which shifts dramatically downfield to 188.2 ppm). Fulleride/THF solutions were then added dropwise into stirring water at ambient temperatures, quickly forming a dark red-brown aqueous suspension at relatively high concentrations (1 mg/mL) (122). A change in color was reported by varying concentration: 0.33 mg/mL, brown red; 0.11 mg/mL, pale red; and 0.04 mg/mL, pale yellow. Suspensions demonstrated characteristic IR spectrum (KBr pellet) of underivatized, neutral C_{60} (1428, 1183, 573, and 527 cm^{-1}, as reported). Interestingly when degassed water was added to THF fulleride solution, voltammetry results suggest that water alone cannot oxidize the fulleride, nor was it susceptible to proton donation from water, thus leading the authors to conclude that a fullerene anion (C_{60}^-) is stable in oxygen-free, pure water (122). Furthermore, this method reports much smaller aggregate size(s) (average aggregate size to be ca. 10 nm) via transmission electron microscopy (TEM) analysis.

Other groups have since reported similar aggregate suspensions, prepared through solvent transfers, including Deguchi et al. in 2001, who reported the simple method of saturating a THF solution with C_{60} and adding it to water (1:1 volume) whereby a yellow suspension was formed (dilute) (126). THF was then removed via rotary distillation, leaving an aqueous C_{60} suspension (ca. 3–5 mg/L C_{60}) that could then be concentrated. As described, this method produced aggregates on the order of 60 nm in diameter, with similar spectral properties of other aqueous C_{60} aggregate systems. Furthermore, C_{70} clusters were produced in a similar manner (THF/water) (119). Similarly, aggregates have been produced in other polar solvents according to Alargova et al. (118), including acetone, acetonitrile, and ethanol via similar pathways (i.e., starting with a dissolved C_{60} solution in toluene or benzene, which is added sequentially to more polar solvents). This study was also the first to observe aggregate size as a variable that was a function of initial C_{60} concentration, leading to a conceptual model for C_{60} aggregation in polar systems (118). By lowering initial C_{60} concentrations, smaller aggregates were formed (118). Perhaps most interesting, though, is the observation by Cheng et al. (120) that by simply adding C_{60} to DI water at ambient temperatures and pressure, under a high sheer rate (rapidly stirred), an orange-yellow suspension of water-stable C_{60} aggregates can be found. While producing a broader size range of aggregates (25–300 + nm, polydispersed) than other methods and taking longer to aggregate (weeks), these suspensions demonstrated similar chemical and physical properties as observed by others (120, 126).

8.3.3 Surface Chemistry

The surface chemistry of these aggregate species is of interest as it is this interface where hydrophobic fullerenes are in some way rendered hydrophilic. In fact, these aggregates as noted by a number of studies, do not readily appreciate back into a polar solvent (0.8–3.6%) (26, 123, 127). It has been confirmed that the surfaces of these materials are negatively charged by performing electrophoretic mobility studies (ζ potential ranging from ca. −9 to −40 mV) (119, 123, 128). A number of possible explanations have been proposed for the surface charge source: It may be that upon contact with water, pristine fullerenes undergo a chemical reaction to create a small population of partially oxidized, and hence more polar, amphiphilic fullerenes that are able to stabilize the hydrophobic underivatized core (26). C_{60} is a relatively reactive species; its degradation by light and oxygen has been noted (3, 129) In particular, one inefficient method to form partially hydroxylated fullerenes relies on the introduction of THF/C_{60} solutions to water at high pH values (>12) (130). Such a reaction may also proceed, albeit at lower yields, when THF/C_{60} solutions are introduced to neutral water. Alternatively, C_{60} is an excellent electron acceptor. Both Andrievsky et al. and Deguchi et al., along with our own findings, suggest that water itself (or another polar solvent) may form a donor–acceptor complex with C_{60} leading to a weakly charged colloid (117, 119, 125). Other mechanisms proposed include hydroxyl ion adsorption to the surface of the aggregate, which seems improbable as Deguchi et al. point out, citing the fact that similar aggregates are formed in polar solvents without hydroxyl ions (e.g., ethanol, acetone, and ACN) (118, 119). Moreover, Avdeev et al. along with Andrievsky et al. discuss possible molecular C_{60} hydration, which they termed as $[C_{60}@(H_2O)_n]_m$, suggesting each C_{60} molecule being hydrated by 20–24 organized water molecules, as it may relate aggregate stabilities (125, 132). A number of identifying names have been assigned to the water-stable C_{60} suspensions based on preparation methods and author's discrepancies. Such names include nC_{60}, aqu/nC_{60}, SON/nC_{60}, THF/nC_{60}, TTA/nC_{60} (127, 133); C_{60} dispersions (118, 119); C_{60} hydrosol, C_{60} FWS (117, 121, 125); nano-C_{60} (27, 123); and nC_{60} (134).

Based on a net negative surface charge, as measured via zeta potential (ζ) of the C_{60} aggregate, studies have been conducted with variable ionic strengths and types (divalent and monovalent) to better understand the surface chemistry (119, 123, 135). In particular, a recent study by Brant et al. (135) is almost completely focused on this topic. Comparing C_{60} aggregate solutions formed through a THF intermediate (Deguchi method) and by simply adding C_{60} to water and stirring over time (Cheng method), a range of ionic strengths and types was investigated. Results indicated that as ionic strength was increased from 0 to $0.1I$, ζ potential values increased (less negative) until aggregates settled out of solutions with ionic strength at least $0.1I$. Furthermore, it was interestingly noted that with divalent cations such as calcium, and within a pH range of 5–8, aggregates could actually take on a net positive effective surface charge. Furthermore, as pH was adjusted for suspensions with monovalent ions (NaCl, 0.001–$0.1I$), the ζ potential values generally decreased (less negative) for the aggregates consistent with classic electrostatic theory. Our lab observed similar results in terms of aggregate stability range as a function of ionic strength (THF prepared) noting macroscale C_{60} aggregate settling above $0.5I$ (monovalent NaCl, pH 5) (123).

The early interest in aggregate forms of fullerenes, particularly C_{60}, was motivated by their applications in technologies, particularly those requiring fullerenes in water (117, 118, 136). Aqueous stable C_{60} suspensions have shown little promise in this regard, as typical concentrations (<100 ppm) are far below the 10,000–100,000 ppm levels that are achievable when the carbon cage is intentionally altered to include polar functionalities (97, 98, 110, 137–139). Still, such aggregate generation through unintentional exposure of fullerenes to water is possible, and the amounts generated may be significant for ecological effects (≤100 ppm). Other lipophilic organic molecules, which might be analogous to C_{60}, have significant ecological impact in aqueous systems at concentrations of 1–10 ppm (140). Once available in water, these materials may be transported and react readily with both abiotic and biological processes/receptors. For instance, a recent study by Lecoanet et al. (141) on the mobility of suspended nanomaterials in porous media showed that both functionalized and C_{60} aggregates were capable of migrating through a well-defined porous medium. Such water-available fullerenes have been reported to cause varying biological responses upon exposure (104, 142–145). Specifically, previous reports by this lab and others have demonstrated that fullerenes, as suspended C_{60} in aggregate form, can elicit a biological response at relatively low concentrations (<1 ppm) (27, 38, 133). Considering these facts, fullerenes stable in water, including those in nanoscale aggregate forms, should be well understood in terms of both chemical and physical properties for appropriate consideration during risk and life cycle assessments of the material.

8.4 PHOTOCHEMICAL REACTIVITY OF FULLERENE IN THE AQUEOUS PHASE

8.4.1 Introduction

While C_{60} does not fluoresce or phosphoresce in ambient environment, C_{60} in organic solvent in its ground singlet state ($^1C_{60}$) was found to be readily excited to singlet state $^1C_{60}^{\bullet}$ very efficiently with quantum yield of nearly 1.0 (100%) upon visible light irradiation (39). The produced $^1C_{60}^{\bullet}$ can either convert back to the ground state by fluorescence emission and internal conversion process or transform into triplet state ($^3C_{60}^{\bullet}$) through intersystem crossing (ISC) (39). The resulting triplet state $^3C_{60}^{\bullet}$ is subject to three quenching pathways: (1) triplet quenching pathways including energy transfer to ground-state triplet oxygen ($^3O_2$) resulting in photochemical generation of singlet oxygen ($^1O_2$); (2) self-quenching mechanism through interaction between triplet and ground state C_{60}; and (3) triplet–triplet annihilation between triplet state C_{60} and another adjacent triplet state C_{60}. In the presence of oxygen, energy transfer to oxygen is dominant with very high yield (39). Essentially, the same process has been observed with C_{70} (39). This scheme is illustrated below:

$$^1C_{60} \xrightarrow{h\nu} {}^1C_{60}^{\bullet} \xrightarrow[\text{crossing}]{\text{Intersystem}} {}^3C_{60}^{\bullet} \xrightarrow{\overset{^3O_2 \quad {}^1O_2}{\diagdown\diagup}} {}^1C_{60}$$

Photoexcited C_{60} has been also reported as an excellent electron mediator, since excited-state triplet C_{60} is a better electron acceptor than ground-state singlet C_{60}. During electron mediation, C_{60} accepts an electron efficiently from electron donors such as amines (40, 146), alcohols (19), and photoexcited metal oxides (147–149), and transfers it to easily reducible chemical species, especially oxygen, to produce superoxide radical anion ($O_2^{\bullet-}$) (32), even though such a reaction is thermodynamically unfavorable ($E^0(C_{60}/C_{60}^{\bullet-}) = -0.2V_{NHE}$ (148) and $E^0(O_2/O_2^{\bullet-}) = 0.33V_{NHE}$). The scheme for the generation of $O_2^{\bullet-}$ by C_{60} is illustrated below:

$$^1C_{60} \xrightarrow{h\nu} {}^1C_{60}^{\bullet} \xrightarrow[\text{crossing}]{\text{Intersystem}} {}^3C_{60}^{\bullet} \xrightarrow[\underset{ED \quad ED^{\bullet+}}{}]{} {}^3C_{60}^{\bullet\bullet} \xrightarrow{\overset{O_2 \quad O_2^{\bullet\bullet}}{\diagdown\diagup}} {}^1C_{60}$$

These unique photochemical properties have expanded C_{60} applications into photodynamic therapy (32, 102), oxidative synthetic reactions (150), and electron shuttling action in dye-sensitized solar cell (60). Such photochemical properties may also confound ecological risk assessment, as $^1O_2$ (104) and $O_2^{\bullet-}$ (27, 31, 32, 151) produced by C_{60} can be responsible for cytotoxicity (27, 152) as a result of DNA (31, 32, 151) and/or cellular structure (e.g., lipid peroxidation) damage (104).

On the basis of above established photochemistry in organic phase, some researchers have suspected that C_{60} might be responsible for oxidative stress and lipid peroxidation in cell culture through the same ROS production mechanism (27). Consistently, the cytotoxicity of various forms of C_{60} in the aqueous phase (e.g., chemically derivatized C_{60} or C_{60} encapsulated by polymer) has been reported to be greatly enhanced by photoirradiation (152). In particular, Yamakoshi et al. demonstrated photoinduced DNA cleavage in the presence of NADH (common reductant *in vivo*) by C_{60} associated with polyvinylpyrrolidone by the formation of superoxide and OH radical, the presence of which was confirmed by EPR spectra (31, 32). Since the bandgap energy of pure C_{60} is relatively low (2.3 eV) and quantum yield for the conversion into triplet state is close to unit (39), it is possible that C_{60} in the biosystem might be easily excited to produce ROS under sunlight illumination (<540 nm). Such a route might be of particular concern for human dermal exposure of fullerenes when the light can easily penetrate into the exposure site. Also, ROS production mechanism could be responsible for the interaction of C_{60} with (micro)organisms present in relevant waters.

While the aforementioned photoinduced ROS production by C_{60} in organic solvent is relatively well known (39, 153), there have been somewhat conflicting reports regarding photoactivity in the aqueous phase. For example, Beeby et al. (107) demonstrated aqueous photochemical generation of $^1O_2$ by C_{60} stabilized by a nonionic surfactant (Triton X100), applied above critical micelle concentration. In contrast, Yamakoshi et al. (31, 32) reported that $^1O_2$ was not produced when C_{60} was stabilized with a polymer (PVP) in water, while $O_2^{\cdot-}$ and $OH^\cdot$ were generated in the presence of electron donors such as EDTA and NADH. Since C_{60} under these conditions is not functionally derivatized, properties (e.g., chemistries, sizes, etc.) of the stabilizing molecules might play an important role in determining photoactivity of C_{60} in the aqueous phase. In natural aqueous matrices, the effect of natural organic matter (NOM) is of primary interest during C_{60} aggregation and photochemistry processes, but little is known to date. Moreover, few reports are currently available regarding the photochemical reactivity of C_{60} stabilized in water as colloidal aggregate suspensions. This section presents the recent study by our research group that investigated photochemical generation of ROS by various preparations of C_{60} in the aqueous phase (131).

8.4.2 Experimental

8.4.2.1 Preparation of Different C_{60} Samples
Aqueous stable C_{60} aggregates were prepared according to Fortner et al. (123). C_{60} sample prepared according to this specific method is herein referred to as nC_{60}. The physical and chemical characteristics of nC_{60} have been well reported in the literature (118, 123, 154). The second preparation method involved application of ultrasound to a heterogeneous mixture of toluene containing C_{60} and ultrapure water in a sealed bottle (117). Aqueous suspensions of C_{60} prepared according to this specific method is herein referred to as

son/C_{60}. C_{60} was also stabilized in water using polymer (PVP) and surfactants (TX and Brij 78 (Brij)) following the methods by Yamakoshi et al. (155) and Beeby et al. (107), respectively. These C_{60} samples are referred to as C_{60}/PVP, C_{60}/TX, and C_{60}/Brij. C_{60} associated with model NOM was prepared by applying an ultrasound (50/60 Hz, 125 W) on the mixture of toluene containing C_{60} and Milli-Q water containing SRNOM for 24 h.

8.4.2.2 Photochemical Experiments Photochemical experiments were carried out using a magnetically stirred 60-mL cylindrical quartz reactor surrounded by six 4-W black light bulbs (BLBs, Philips TL4W) at ambient temperature (22°C). Emission wavelength region of 350–400 nm was confirmed by a Spectropro-500 spectrophotometer (Acton Research Co., USA). The incident light intensity in this active wavelength region was measured at 3.33×10^{-4} Einstein (min L) by ferrioxalate actinometry (156). Reaction solutions contained 5 mg/L C_{60} and indicator chemicals discussed below and were buffered at pH 7 using phosphate (10 mM for $^1O_2$ and 50 mM for $O_2^{\cdot-}$ experiments). As the photochemical reaction proceeded, sample aliquots of 1 mL were withdrawn from the reactor using a syringe, filtered through a 0.45-μm PTFE filter (Millipore), and injected into a 2-mL amber glass vial for further analyses. All experiments were run in duplicate. Concentrations of $^1O_2$ were measured using furfuryl alcohol (FFA) as an indicator [k(FFA+$^1O_2$) = 1.2×10^8 M^{-1} s^{-1} (157)]. The capacity of each C_{60} sample to mediate electron from electron donor (triethylamine) to oxygen was evaluated using NBT (nitro blue tetrazolium) method, commonly applied for monitoring superoxide radical [k(NBT^{2+} + $O_2^{\cdot-}$) = 5.88×10^4 M^{-1} s^{-1}].

8.4.3 Comparing Photochemical Production of $^1O_2$ by C_{60} in Organic Solvent and Water-stable C_{60} Aggregates

Experimental results shown in Fig. 8.2 confirmed that $^1O_2$ was effectively produced by pristine C_{60} in toluene. With all six lamps on (typically, $I = 3.33 \times 10^{-4}$ Einstein (min L)), over 90% of FFA was degraded within 15 min. The presence of excess 2,5-DMF as $^1O_2$ scavenger [k(2,4-DMF + $^1O_2$) = 6.8×10^8 M^{-1} s^{-1} (157)] drastically retarded the degradation rate of FFA, indicating that $^1O_2$ was a dominant oxidizing species. Because the reactions were performed in toluene, it was possible to exclude the contribution of $^\cdot$OH to the oxidative degradation of FFA [k(toluene + $^\cdot$OH) = 3.0×10^9 M^{-1} s^{-1} (158)]. When the experiments were performed using one lamp ($I = 5.3 \times 10^{-5}$ Einstein (min L)), FFA degradation rate was decreased. Correspondingly, as C_{60} dose was increased, faster FFA oxidation rates were observed (Fig. 8.2).

Contrary to pristine C_{60} in organic solvent, $^1O_2$ was not generated by nC_{60} and son/C_{60} in water under the same UV irradiation condition as shown in Fig. 8.3. UV–Vis absorption spectra of nC_{60} and son/C_{60} showed appearance of a new broad band UV absorption in the wavelength region from 400 to 500 nm compared to C_{60} in toluene (results not shown). This broadband absorption is known to occur in C_{60} in solid-state film and aggregate due to symmetry-forbidden transitions, hence indicating solid-state C_{60}–C_{60} interactions (71, 123, 125, 154). Accordingly, average population

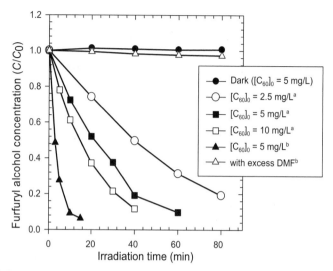

FIGURE 8.2   Photochemical degradation of FFA in pristine C_{60}-toluene ([FFA]$_0$ = 0.85 mM; [2,4-DMF]$_0$ = 0.1 M; [a]I = 5.3 × 10^{-5} Einstein/(min L); [b]I = 3.33 × 10^{-4} Einstein/(min L).

diameters, taken by dynamic light scattering (DLS), were 100.1 and 86.2 nm for nC_{60} and son/nC_{60}, respectively. Under the same conditions, derivatized C_{60} (fullerol, polyhydroxyfullerene) induced photochemical degradation of FFA, albeit the kinetics was much slower compared to C_{60} in organic solvent, consistent with the earlier findings (159). This result was unexpected since (1) nC_{60} and son/C_{60} absorbed UV

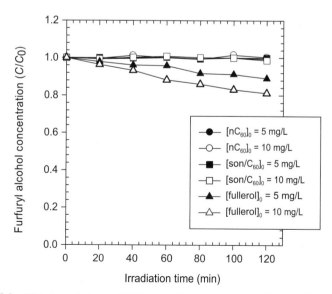

FIGURE 8.3   FFA degradation in UV-irradiated suspensions of nC_{60}, son/C_{60}, and fullerol ([FFA]$_0$ = 0.85 mM; [phosphate]$_0$ = 10 mM; pH$_i$ = 7).

light in the active wavelength region much more efficiently than fullerol and (2) nC_{60} would not be chemically modified according to recent ^{13}C NMR analysis (123). On the other hand, considering the fact that morphological modification is most conspicuous difference between nC_{60} or son/C_{60} and molecular C_{60}, it suggests that colloidal aggregation of C_{60} in the aqueous phase may have a strong influence on its photochemical properties.

8.4.4   Photochemical Production of $^1O_2$ by C_{60} Associated with Polymer and Surfactant in Aqueous Phase

C_{60}/PVP at weight ratio of C_{60} to PVP at 0.008 and 0.004 did not produce any measurable amount of $^1O_2$ under UV irradiation in water, consistent with the finding by Yamakoshi et al. (32) in which the same C_{60}/PVP ratio was examined (Fig. 8.4a). However, when the ratio was decreased to 0.0008 (i.e., a higher amount of PVP), at equivalent C_{60} concentrations, $^1O_2$ production occurred at a rate corresponding to ca. 5% of that in toluene (estimated according to References 157 and 160). Production of $^1O_2$ by C_{60} in this case was confirmed as (1) FFA degraded at ca. twice faster rate in

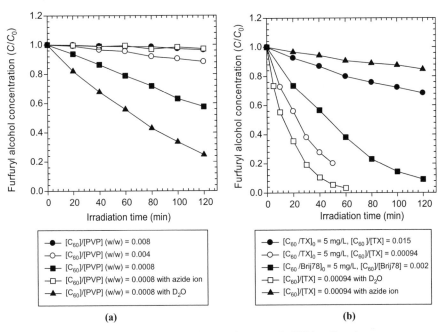

(a)   (b)

FIGURE 8.4 Comparison of the FFA degradation rates in UV-irradiated aqueous suspension of (a) C_{60}/PVP, prepared at different weight ratios ($[C_{60}]/[PVP])(w/w)$ of 0.008, 0.004, and 0.0008 and (b) C_{60}/TX at two different weight ratios ($([C_{60}]/[TX])(w/w) = 0.015$ and 0.00094) and $C_{60}/Brij78$ ($([C_{60}]/[Brij78])(w/w) = 0.002$) ($[C_{60}/PVP]_0 = [C_{60}/TX]_0 = [C_{60}/Brij]_0 = 5\,mg/L$; $[FFA]_0 = 0.85\,mM$; $[D_2O] = 70\%(v/v)$; $[phosphate]_0 = 10\,mM$; $pH_i = 7$). $[azide\ ion]_0 = 50\,mM$ for (a) and 5 mM for (b).

70% (v/v) D_2O, which is a less effective $^1O_2$ quenching solvent than H_2O [$k_d(H_2O)=$ $2.4 \times 10^5 \, s^{-1}$; $k_d(D_2O) = 1.6 \times 10^4 \, s^{-1}$ (157, 160)]; (2) the presence of excess azide ion as $^1O_2$ scavenger completely suppressed FFA decay; and (3) PVP solution in the absence of C_{60} did not produce $^1O_2$ under identical conditions (data not shown). It might indicate that high concentration of PVP as an encapsulant could control the clustering phenomenon of C_{60} in water, which possibly made an influence on its photoactivity.

Effect of dispersion status of C_{60} in the aqueous phase on its photochemical reactivity was further examined using C_{60}/TX prepared at two different weight ratios. It has been reported that C_{60} tends to disperse as aggregates at lower TX concentration, while it is molecularly solubilized within micelles at higher TX concentration above CMC (107). Consistently, C_{60}/TX at [C_{60}]/[TX](w/w) of 0.00094 (i.e., higher TX concentration) exhibited a sharp absorption centered at 330 nm, similar to C_{60} in organic solvent (123) and no broadband absorption in 400–500 nm. In contrast, C_{60}/TX at [C_{60}]/[TX](w/w) of 0.015 (i.e., higher ratio, lower TX concentration with the same C_{60} concentration) showed a slightly redshifted, blunt specific peak at ca. 340 nm and a broadband absorption over the wavelength region between 400 and 500 nm, suggesting the presence of aggregate forms of C_{60} (123, 125, 154). DLS analysis supported C_{60}/TX spectral observations, as higher TX concentration suspensions were monodispersed with an average diameter below 5 nm that approaches the instrument detection limit. Lower TX concentrations gave rise to a polydispersed C_{60}/TX population with both small (<10 nm) fractions along with larger (>50 nm) fractions presumably due to partial C_{60} aggregation. Under UV irradiation, C_{60}/TX at lower ratio (more molecularly dispersed) produced $^1O_2$ much more efficiently than that at higher ratio (more aggregated) (Fig. 8.4b). Production of $^1O_2$ by C_{60}/TX was further confirmed as (1) addition of D_2O increased FFA degradation rate, (2) azide ion inhibited FFA degradation, and (3) no FFA degradation was observed by TX alone under UV irradiation (data not shown).

A few other types of surfactants were also examined using the preparation method suggested by Beeby et al. (107). A cationic surfactant, cetyltrimethylammonium bromide (CTAB), stabilized C_{60} at a relatively high concentration (i.e., 2 g CTAB per 5 mg C_{60}), but the resulting suspension precipitated overnight perhaps due to Coulombic interaction between cationic CTAB and negatively charged C_{60} aggregates (135). Sodium dodecyl sulfate (SDS) is an anionic surfactant that is frequently used to disperse carbon nanotubes in the aqueous phase (161). SDS applied directly into water (SDS is insoluble in organic solvent and the method by Beeby et al. (107) was not applicable) at concentration 5 g/L (above CMC) did not stabilize C_{60}. While nonionic Tween did not stabilize C_{60} in the aqueous phase, another nonionic surfactant, Brij 78, efficiently dispersed C_{60} in water, producing a yellowish suspension. Accordingly, C_{60}/Brij did not show a broadband in the wavelength region from 400 to 500 nm and effectively produced $^1O_2$ under UV irradiation.

Similar to surfactants that promote hydrophobic environment within the micelle structure, NOM may be expected to facilitate C_{60} stabilization or change its dispersion status in the aqueous phase. But UV–Vis spectra of 20 mg/L C_{60} associated with 40,

100, and 200 mg/L of SRNOM showed a presence of broadband absorption over 400–500 nm, although blueshifted specific peak at 350 nm, indicative of concentration of aqueous C_{60} colloid, somewhat increased (data not shown). Accordingly, the average population diameters of 103.8 and 104.7 nm were measured for 100 and 200 mg/L NOM, respectively, by DLS, indicating C_{60} aggregation occurred. Eventually, no inhibition of aggregation of C_{60} by NOM resulted in the negligible production of $^1O_2$ by C_{60}/NOM (data not provided).

Through these studies, it is demonstrated that $^1O_2$ was produced efficiently by molecularly dispersed C_{60} in toluene and less efficiently by C_{60} dispersed and stabilized in the aqueous phase through association with TX and Brij. In contrast, no measurable amount of $^1O_2$ was produced when C_{60} was present as aggregate (nC_{60} and son/C_{60}). When C_{60} was present with PVP, $^1O_2$ was produced only when PVP concentration was high, minimizing C_{60} aggregation. These observations suggest that inhibition of $^1O_2$ production by C_{60} under UV irradiation may result from C_{60} aggregation and the degree of C_{60} aggregation is influenced by the type of solvent and dispersion method (162). Specifically, the clustering of C_{60} derivatives in aqueous media was reported to remarkably accelerate self-quenching mechanism(s) and triplet–triplet annihilation, leading to completely deplete $^3C_{60}^\bullet$ available for production of $^1O_2$, which, in our studies with nC_{60}, has not been observed either using nanosecond transient spectroscopy.

8.4.5 Photochemical Production of $O_2^{\bullet-}$ by nC_{60} and C_{60} Associated with PVP and TX 100 in Aqueous Phase

Experiments for probing $O_2^{\bullet-}$ generation by C_{60} with TEA as ED and NBT^{2+} as $O_2^{\bullet-}$ indicator were conducted. Changes in the absorption spectra as a function of time indicated production of purple-colored monoformazan (MF^+) with $\lambda_{max} = 530$ nm (i.e., product of NBT^{2+} reduction by $O_2^{\bullet-}$) and consequently, $O_2^{\bullet-}$. Similar to $^1O_2$ results, nC_{60} did not produce any measurable amount of $O_2^{\bullet-}$, whereas also consistent with the observation made for production of $^1O_2$, C_{60}/TX $(([C_{60}]/[TX])_0 = 0.00094)$ generated $O_2^{\bullet-}$ at an estimated concentration of ca. 50 μM after 60 min of UV irradiation. However, C_{60}/PVP $(([C_{60}]/[PVP])_0 = 0.0008)$ could not produce $O_2^{\bullet-}$, even though $^3C_{60}^\bullet$ should be available for further reaction (i.e., accept electron) as evidenced by $^1O_2$ production.

8.4.6 Environmental Significance

While our results suggest that C_{60} as a water-stable aggregate might not trigger cytotoxicity through ROS production, as observed for nC_{60}, it may very well be possible, however, for C_{60} to produce ROS under specific environments, for example, within lipid bilayers, a condition similar to surfactant micelles. Therefore, these findings suggest that the modes of exposure could be a determining factor for observed toxicity, which have not been clearly studied to date. These results further highlight the need for in-depth mechanistic study on nC_{60}'s chemical reactivity at the biological interface in aqueous systems.

8.5   REACTION OF WATER-STABLE C$_{60}$ AGGREGATES WITH OZONE

8.5.1   Introduction

As mentioned above, C$_{60}$ dissolved in organic solvents is known to be readily oxidized by ozone. Reaction kinetics and product identities have been described under varying reaction conditions (93, 163–165). Unlike other olefins though, steric constraints of the cage structure do not allow a typical Criegee's mechanism to occur (166). Instead, it was suggested that O$_3$ attack under typical conditions occurs primarily along the [6,6] C–C double bond (part of the hexagon ring structure) resulting in a short-lived ozonide (C$_{60}$O$_3$) that dissociates via either thermolysis (164) or photolysis (93) into epoxide ([6,6] closed epoxide) or ether ([5,6] open oxidoannulene), respectively. Multiple oxygen adducts (C$_{60}$O$_x$) have also been observed through such mechanisms as a function of O$_3$ availability and reaction time. The molar ratio of oxygen added per molecule of C$_{60}$ has been reported as high as 21 (163, 167), but are usually reported at lower ratios of 1–10 (163, 165, 168). Specific molecular characterization of highly ozonated fullerenes is complex, as the number of possible congeners and isomers are exceedingly large. Nevertheless, a range of functional groups such as epoxide, ether, carbonyl, and hydroxyl have been previously reported (93, 163, 169, 170).

Despite the available literature regarding fullerene ozonation in organic phases, little is known about a direct fullerene reaction with ozone in the aqueous phase due to solubility limitations. In this section, a recent study by Fortner et al. on the reaction of nanoscale, aqueous stable C$_{60}$ aggregates (nC$_{60}$) with ozone is presented. Briefly, this study indicates that nC$_{60}$ is reactive with dissolved ozone resulting disaggregation concurrent with the formation of highly derivatized (C:O $\cong$ 2:1), water-soluble fullerene oxide(s) with symmetrical, repeating hydroxyl and hemiketal functionalities (134).

8.5.2   Experimental

8.5.2.1   Semibatch   Ozonation   Experiments   Semibatch experiments, defined as an open system maintaining the ozone concentration constant in the liquid phase by constant bubbling of ozone gas, were performed at pH 5.4, 6.8, and 8.9 and at 19.5 $\pm$ 0.5°C. Ozone gas was generated by a Wedeco Model GSO 10 ozone generator (Herford, Germany) from pure oxygen. Prior to contact with the reaction medium, the ozone gas was washed through a solution containing 10 mM phosphate buffer solution at pH 6. Concentrations of dissolved ozone was measured by the indigo colorimetric method described by Chiou et al. (171). Selected experiments were performed using excess *tert*-butanol (*t*-BuOH) as the hydroxyl radical scavenger ($k[t$-BuOH + $^{\bullet}$OH]$= 5 \times 10^8 \, \text{M}^{-1} \, \text{s}^{-1}$) (172, 173). Typical semibatch reactions were initiated by adding 10–20 mL of nC$_{60}$ concentrate solution to the ozone solution resulting in 5–10 mg/L suspensions. Sample aliquots of 2 mL were taken at appropriate time intervals, with a total sample volume less than 10% of the reaction volume. For consistency, reaction conditions (reaction time and ozone concentration) are normalized as *CT* ozone dissolved (mg) per liter multiplied by time in minutes). Additional experiments were performed in a batch mode using a custom built, multichannel stopped-flow reactor at $20.0 \pm 0.1°C$ (174). Stock solutions of dissolved ozone concentration ranging from 2.0 to 15 mg/L were prepared by bubbling ozone gas through ultrapure water at pH 5. For both semibatch and batch experiments, ozone in the sample was quenched using excess sodium thiosulfate (4:1 $Na_2S_2O_3:O_3$) (172) or stripped with N_2 (UHP) (for ^{13}C NMR, MS, XPS, IR and total organic carbon (TOC) analyses to avoid potential background interferences). All chemicals used throughout were reagent grade or higher unless otherwise noted.

8.5.2.2 Product Characterization

Size and shape of nC_{60} aggregates were analyzed by two primary methods as previously demonstrated (123): dynamic light scattering (ZetaPALS, Brookhaven Instruments Corporation, and a Zetasizer Nano ZS, Malvern Instruments Ltd.) and transmission electron microscopy. TEM images were prepared by evaporating 40 µL of concentrated suspension on a 400 mesh carbon-coated copper grid and imaged with a JEOL FasTEM 2010 at 100 kV calibrated to an aluminum standard. Spectral analyses were performed on product samples prepared from 100 to 180 mg/L C_{60} (as nC_{60}) in contact with dissolved ozone (3–6 mg/L) for 30–50 min resulting in a complete reaction. UV–Vis absorption spectra were taken within the range of 190–800 nm at 0.5 nm intervals using a Varian Cary 50 UV-Vis Spectrophotometer and corrected for the appropriate background. ^{13}C-NMR spectrum of nC_{60} suspension prepared with ^{13}C enriched C_{60} (25%) in D_2O before and after ozonation was obtained using a Varian Inova 600 MHz NMR equipped with a carbon-enhanced cold probe. XPS analysis was performed using a PHI Quantera SXM™ (scanning X-ray microprobe) (ULVAC-PHI) with an Al mono, 24.8 W X-ray source, and 100.0 µm X-ray spot size at 45.0 degrees (26.00 eV for 1 h). Samples were prepared by first sputter coating a clean silicon substrate with Au for 2–10 min at 100 mA and evaporating ca. 100 µL of concentrated liquid samples on the substrate overnight at room temperature in dust-free atmosphere. Data were analyzed with PeakFit® to estimate peak position and relative peak areas. ATR-FTIR spectroscopy was taken with a Thermo Nicolet Nexus 870 FTIR and a Pike HATR equipped with germanium (Ge) trough. TOC was measured using a Shimadzu TOC-5050A Total Carbon Analyzer (Shimadzu Scientific Instruments, Inc., Columbia, MD, USA) performing catalytic combustion at 680°C and equipped with infrared detector. MS analyses (both nC_{60} and reaction products) via a laser desorption/ionization setup was performed using a tandem time-of-flight (TOF/TOF) mass spectrometer equipped with a 200 Hz laser (Applied Biosystems 4700 Proteomics Analyzer) under a positive ion mode. For matrix assisted (MALDI) analyses, samples were temporarily dried and dissolved into an organic matrix (cyano-4-hydroxycinnamic acid (CHCA)) for increased sensitivity toward less polar products.

8.5.3 Reaction Kinetics

Semibatch experimental results shown in Fig. 8.5 indicate that nC_{60} is susceptible to reaction with ozone. The loss of characteristic $^1T_{1u}$–^1A_g transition peaks at 450, 340, and 260 nm (175) suggests molecular alteration upon reaction (Fig. 8.5a). In parallel with UV–Vis absorbance change, the solution gradually lost its characteristic

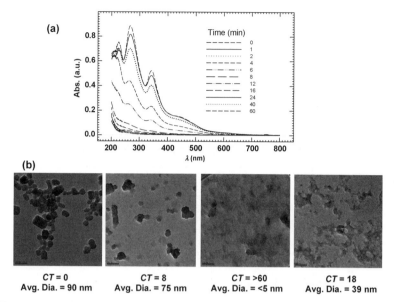

FIGURE 8.5 Ozonation of nC$_{60}$ (a) nC$_{60}$ (5mg/L) spectra as a function of reaction time at pH 6.8 during semibatch ozonation ([O$_3$] = 4.3 mg/L). (b) Changes in aggregate size and shape during semibatch ozonation. nC$_{60}$ (C_0, CT = 0) 100 mg/L. TEM scale bar = 100 nm.

yellow-orange color and became clear without accumulation of surface residues or precipitates. Also noticeable through DLS and TEM analyses as reaction proceeded was the decrease in aggregate sizes (Fig. 8.5b), which suggested that the products were no longer aggregates but probably molecularly dispersed (below detection limit of DLS, <5 nm) in water. Morphological changes during the reaction, such as loss of relatively sharp facets in aggregates of parent compound, were also observed (e.g., Fig. 8.1b, CT = 8 mg min/L).

8.5.4 Product Characterization

The ^{13}C NMR spectrum of ^{13}C-labeled parent compound showed a single nC$_{60}$ peak at 146 ppm, characteristics of underivatized material with I$_h$ symmetry (Fig. 8.6a) (68, 69, 73, 123). After the reaction with ozone, the peak at 146 ppm was no longer observed. In contrast, four new peaks were observed at 176, 168, 128, and 95 ppm (Fig. 8.6b), indicating high level of functional derivatization and loss of I$_h$ symmetry (7, 47). Product peak shifts indicate the presence of different oxygen moieties in the products. These peaks might be assigned as carbonyl carbon at 176 ppm, vinyl ether ($-$C$=$$\underline{C}$$-O-$) carbon at 168 ppm, and C$-$C cage carbon at 128 ppm (as C$=$C), which is shifted downfield as a result of decreased molecular strain, and C$-$O (or $-$O$-$C$-$O$-$) carbon at 95 ppm according to previous reports (80, 97). The relatively small number of peaks present at different chemical shifts suggests a high level of molecular symmetry with repeating functional arrangements in these products.

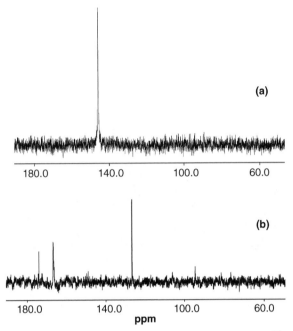

FIGURE 8.6 ^{13}C NMR spectra of (a) parent nC$_{60}$ (150 mg/L, 25% ^{13}C C$_{60}$) in water; (b) ozonation product (180 mg/L) in water.

The ATR-FTIR spectrum of final products provides further information on the identity of oxygen moieties present in the products (Fig. 8.7). A broad OH absorption at $3400 \, \text{cm}^{-1}$, C–OH in-plane bending at $1360 \, \text{cm}^{-1}$ and C–O stretching at $1315 \, \text{cm}^{-1}$ collectively indicate the existence of hydroxyl functionalities. A sharp and strong absorption at $1630 \, \text{cm}^{-1}$ along with weaker shoulder at $1760–1770 \, \text{cm}^{-1}$

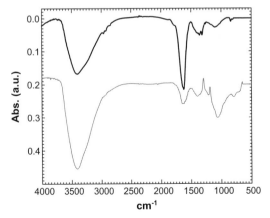

FIGURE 8.7 Top: ATR-FTIR product spectrum of nC$_{60}$ (180 mg/L) at pH 3.4 (unadjusted) after ozonation. Bottom: ATR-FTIR spectrum of C$_{60}$(OH)$_{22}$ (MER Corp.) (200 mg/L).

suggested other forms of oxygen moieties, as similar peaks have been observed for other fullerene oxides including fullerenes ozonated in organic solvents and hemiketal fullerols. In particular, these peaks have been attributed to carbonyl moieties such as carboxyl groups and ketones (168, 169) or as part of a hemiketal moiety (R—O—C—OH) (97, 169, 176). Ether functionality could be responsible for the broad and weak band centered at $1110 \, \text{cm}^{-1}$ (80). In any case, presently there are few absolute reports of IR functional identification of fullerene oxide groups, as the types of possible derivatives vary widely depending on oxidation conditions (95, 97, 111, 169). For example, an IR spectrum for a commercial fullerene oxide from MER Corp. prepared through the substitution of brominated C$_{60}$ at pH 3 (C$_{60}$(O)$_x$(OH)$_y$, where $x + y = \sim 22$) (98) is provided in Fig. 8.7 (bottom) for comparison.

The XPS absorption spectrum of the product shown in Fig. 8.8 is consistent with the above observations. In addition to C(1s) absorption with binding energy at 285.5 eV by underivatized carbon, two additional peaks appeared at higher energy levels. As C$_{60}$ was the only carbon source in the suspension, additional peaks represented carbon at different oxidation states, that is, monooxygenated carbon (C—O) at 287.45 eV and dioxygenated carbon (O—C—O, C=O) at 298.73 eV (111). Dioxygenated carbon has been observed in fullerene oxides particularly in hemiketal forms by Chiang et al. (111). Spectrum deconvolution by curve fitting indicates that the ratio between nonoxygenated and oxygenated carbons was ca. 31:29, suggesting an average derivative structure as C$_{60}$(O)$_x$(OH)$_y$, where $x + y = \sim 29$. Among 29 oxygenated carbons, approximately 11 were monooxygenated, while the remaining 18 were dioxygenated.

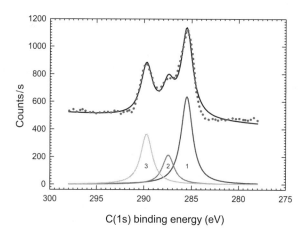

Peak	Position	Area	% C(1s)	Carbon-ID
1	285.50	1457.1	52	Underivatized C
2	287.45	489.75	18	Monooxidized C
3	289.73	836.67	30	Dioxidized C

FIGURE 8.8   C(1s) XPS spectrum and curve-fitting analysis of the ozonation product of nC$_{60}$. Top: Points represent spectral data. Bottom: Curve-fitting deconvolution expressed in relative intensity. Table: Identification and integrated peak areas (relative intensities).

8.5.5 Environmental Significance

These results collectively indicate that nC_{60} readily reacts with ozone in the aqueous phase over a range of pH values to form highly oxidized (average ~29 oxygen additions) and molecularly soluble products. MS (not shown), NMR, ATR-FTIR, XPS, and UV (not shown) analyses suggested that these products maintain C_{60} cage structure and the oxygen-based functionalities in the products are most likely in repeating hemiketal and hydroxyl forms, while the presence of other minor functional groups should not be excluded. Based on similar reaction mechanisms proposed for organic phase reactions, the aqueous phase reaction is likely to proceed through initial formation of a [6,6] primary ozonide (164), subsequent dissociation into a [6,6]-closed epoxide (concurrent with loss of O_2) (164), and further hydrolysis to form hemiketal arrangements (97).

The work presented herein demonstrates the feasibility of chemically functionalizing C_{60} in the aqueous phase, through the formation of water-stable, nanoscale, aggregate intermediates, which overcome solubility limitations of the parent molecule. Accordingly, a wide spectrum of other fullerene chemistries (e.g., electrophilic and nucleophilic additions, electron transfer, etc.) may also be possible via similar aqueous phase route(s), expanding the applicability of C_{60} and its polar derivatives. Conversely, aqueous availability and reactivity of C_{60} take on additional significance as inadvertent reactions may also result in soluble products. For example, an ozone disinfection process with *CT* of 7.8 mg min/L for 2-log removal of *Cryptosporidium parvum* oocysts at 20°C according to the Long Term 2 Surface Water Treatment Rule by USEPA (50) might result in approximately 35–45% of C_{60} in aggregate forms to undergo chemical derivatization at pH 8.9. These soluble products will behave significantly differently from the parent aggregate in aqueous environments as they are molecularly dispersed (141). Consequences of such reactivity, which is certainly not limited to ozonation only, should be considered appropriately in relevant environmental fate and transport studies.

8.6 CONCLUSIONS

Understanding aqueous fullerene chemistry is critical for accurate life cycle and risk assessments. While there is a substantial body of literature regarding the behavior of fullerenes dissolved in an organic phase, photochemical and chemical reactivities of the aqueous stable forms, including those associated with foreign molecules (e.g., surfactants, polymers, and natural organic matter) and those present as stable aggregates, remain underdeveloped at this time. Accordingly, as this area continues to mature, appropriate studies will provide valuable insight into both potential biological interactions (e.g., mechanism(s) of activity, toxicity, and accumulation, etc.) and material fate and transport in both natural and engineered aqueous environments.

It should be noted that at current production levels, longer term global exposure to fullerenes does not pose an immediate general threat, while near-term concerns are likely to be confined to manufacturing workers (152). However, as recommended by industry and academic leaders, a prudent proactive approach to minimize negative implications of engineered nanomaterials (including fullerenes) should be developed concurrently with material advances to ensure a safe and sustainable industry. This further emphasizes the importance of ongoing work in related fields.

REFERENCES

1. Kroto, H.W., Heath, J.R., Obrien, S.C., Curl, R.F., Smalley, R.E. *Nature* 1985;318: 162–163.
2. Hebard, A.F. *Annu Rev Mater Sci* 1993;23:159–191.
3. Diederich, F., Thilgen, C. *Science* 1996;271:317–323.
4. Bagrii, E.I., Karaulova, E.N. *Petrol Chem* 2001;41:295–313.
5. Karaulove, E.N., Bagrii, E.I. *Russ Chem Rev* 1999;68:889–907.
6. Dresselhaus, M.S., Dresselhaus, G., Eklund, P.C. *Science of Fullerenes and Carbon Nanotubes*. San Diego: Academic Press; 1996.
7. Hirsch, A., Brettreich, M. *Fullerenes: Chemistry and Reactions*. Weinheim: Wiley-VCH Verlag GmbH & Co; 2005.
8. Taylor, R., Walton, R.M. *Nature* 1993;363:685–693.
9. Kratschmer, W., Lamb, L.D., Fostiropoulos, K., Huffman, D.R. *Nature* 1990;347: 354–358.
10. Regueiro, M.N., Monceau, P., Hodeau, J.L. *Nature* 1992;355:237–239.
11. Bocquillon, G., Bogicevic, C., Fabre, C., Rassat, A. *J Phys Chem* 1993;97:12924–12927.
12. Mort, J., Ziolo, R., Machonkin, M., Huffman, D.R., Ferguson, M.I. *Chem Phys Lett* 1991;186:284–286.
13. Tremblay, J. *Chem Eng News* 2002;80:16–17.
14. Wharton, T., Kini, V.U., Mortis, R.A., Wilson, L.J. *Tetrahedron Lett* 2001;42: 159–5162.
15. Haddon, R.C., Hebard, A.F., Rosseinsky, M.J., Murphy, D.W., Duclos, S.J., Lyons, K.B., Miller, B., Rosamilia, J.M., Fleming, R.M., Kortan, A.R., Glarum, S.H., Makhija, A.V., Muller, A.J., Eick, R.H., Zahurak, S.M., Tycko, R., Dabbagh, G., Thiel, F.A. *Nature* 1991;350:320–322.
16. Iqbal, Z., Baughman, R.H., Ramakrishna, B.L., Khare, S., Murthy, N.S., Bornemann, H.J., Morris, D.E. *Science* 1991;254:826–829.
17. Kelty, S.P., Chen, C.C., Lieber, C.M. *Nature* 1991;352:223–225.
18. Gupta, B.K., Bhushan, B., Capp, C., Coe, J.V. *J Mater Res* 1994;9:2823.
19. Hwang, K.C., Mauzerall, D. *J Am Chem Soc* 1992;114:9705–9706.
20. Chen, B.X., Wilson, S.R., Das, M., Coughlin, D.J., Erlanger, B.F. *Proc Nat Acad Sci*, USA 1998;95:10809–10813.
21. Da Ros, T., Prato, M. *J Chem Soc Chem Commun* 1999;663–669.
22. Bullard-Dillard, R., Creek, K.E., Scrivens, W.A., Tour, J.M. *Bioorg Chem* 1996;24: 376–385.
23. Foley, S., Crowley, C., Smaihi, M., Bonfils, C., Erlanger, B.F., Seta, P., Larroque, C. *Biochem Biophys Res Commun* 2002;294:116–119.

24. Irie, K., Nakamura, Y., Ohigashi, H., Tokuyama, H., Yamago, S., Nakamura, E. *Biosci Biotechnol Biochem* 1996;60:1359–1361.

25. Mashino, T., Okuda, K., Hirota, T., Hirobe, M., Nagano, T., Mochizuki, M. *Bioorg Med Chem Lett* 1999;9:2959–2962.

26. Scrivens, W.A., Tour, J.M., Creek, K.E., Pirisi, L. *J Am Chem Soc* 1994;116: 4517–4518.

27. Sayes, C.M., Fortner, J.D., Guo, W., Lyon, D., Boyd, A.M., Ausman, K.D., Tao, Y.J., Sitharaman, B., Wilson, L.J., Hughes, J.B., West, J.L., Colvin, V.L. *Nanoletters* 2004;4:1881–1887.

28. Cui, D.X., Tian, F.R., Ozkan, C.S., Wang, M., Gao, H.J. *Toxicol Lett* 2005;155:73–85.

29. Jia, G., Wang, H.F., Yan, L., Wang, X., Pei, R.J., Yan, T., Zhao, Y.L., Guo, X.B. *Environ Sci Technol* 2005;39:1378–1383.

30. Zhao, X.C., Striolo, A., Cummings, P.T. *Biophys J* 2005;89:3856–3862.

31. Yamakoshi, Y., Sueyoshi, S., Fukuhara, K., Miyata, N. *J Am Chem Soc* 1998; 12363–12364.

32. Yamakoshi, Y., Umezawa, N., Ryu, A., Arakane, K., Miyata, N., Goda, Y., Masumizu, T., Nagano, T. *J Am Chem Soc* 2003;125:12803–12809.

33. Friedman, S.H., Decamp, D.L., Sijbesma, R.P., Srdanov, G., Wudl, F., Kenyon, G.L. *J Am Chem Soc* 1993;115:6506–6509.

34. Toniolo, C., Bianco, A., Maggini, M., Scorrano, G., Prato, M., Marastoni, M., Tomatis, R., Spisani, S., Palu, G., Blair, E.D. *J Med Chem* 1994;37:4558–4562.

35. Iwata, N., Mukai, T., Yamakoshi, Y.N., Hara, S., Yanase, T., Shoji, M., Endo, T., Miyata, N. *Fullerene Sci Technol* 1998;6:213–226.

36. Chiang, L.Y., Lu, F.J., Lin, J.T. *J Chem Soc Chem Commun* 1995;1283–1284.

37. Lai, Y.L., Chiou, W.Y., Chiang, L.Y. *Fullerene Sci Technol* 1997;5:1337–1345.

38. Oberdörster, E. *Environ Health Perspect* 2004;112:1058–1062.

39. Arbogast, J.W., Darmanyan, A.P., Foote, C.S., Rubin, Y., Diederich, F.N., Alvarez, M.M., Anz, S.J., Whetten, R.L. *J Phys Chem* 1991;95:11–12.

40. Arbogast, J.W., Foote, C.S., Kao, M. *J Am Chem Soc* 1992;114:2277–2279.

41. Tokuyama, H., Yamago, S., Nakamura, E. *J Am Chem Soc* 1993;115:7918–7919.

42. Boutorine, A.S., Tokuyama, H., Takasugi, M., Isobe, H., Nakamura, E., Helene, C. *Angew Chem Int Ed* 1995;33:2462–2465.

43. Nakamura, E., Tokuyama, H., Yamago, S., Shiraki, T., Sugiura, Y. *Bull Chem Soc Jpn* 1996;69:2143–2151.

44. An, Y.Z., Chen, C.H.B., Anderson, J.L., Sigman, D.S., Foote, C.S., Rubin, Y. *Tetrahedron* 1996;52:5179–5189.

45. Yamakoshi, Y.N., Yagami, T., Sueyoshi, S., Miyata, N. *J Org Chem* 1996;61:7236–7237.

46. Sera, N., Tokiwa, H., Miyata, N. *Carcinogenesis* 1996;17:2163–2169.

47. Taylor, R., editor. *The Chemistry of Fullerenes*, Vol. 4. Singapore: World Scientific Publishing Co. Pte. Ltd; 1995.

48. Heymann, D. *Fullerene Sci Technol* 1996;4:509–515.

49. Heymann, D. *Carbon* 1996;34:627–631.

50. Ruoff, R.S., Tse, D.S., Malhotra, R., Lorents, D.C. *J Phys Chem* 1993;97:3379–3383.

51. Yannoni, C.S., Bernier, P.P., Bethune, D.S., Meijer, G., Salem, J.R. *J Am Chem Soc* 1991;113:3190–3192.

52. Huffman, D.R. *Phys Today* 1991;44:22–29.

53. Kroto, H.W. *Nature* 1987;329:529.

54. Smalley, R.E. Discovering the fullerenes. In: Grenthe, I., editor. *Nobel Lectures, Chemistry 1996–2000.* Singapore: World Scientific Publishing Co.; 1996.

55. Schmalz, T.G., Seitz, W.A., Klein, D.J., Hite, G.E. *Chem Phys Lett* 1986;130:203.

56. Liu, S.Z., Lu, Y.J., Kappes, M.M., Ibers, J.A. *Science* 1991;254:408–410.

57. Hedberg, K., Hedberg, L., Bethune, D.S., Brown, C.A., Dorn, H.C., Johnson, R.D., Devries, M. *Science* 1991;254:410–412.

58. David, W.I.F., Ibberson, R.M., Matthewman, J.C., Prassides, K., Dennis, T.J.S., Hare, J.P., Kroto, H.W., Taylor, R., Walton, D.R.M. *Nature* 1991;353:147–149.

59. Buhl, M., Hirsch, A. *Chem Rev* 2001;101:1153.

60. Kamat, P.V., Haria, M., Hotchandani, S. *J. Phys. Chem. B* 2004;108:5166–5170.

61. Taylor, R. *Philos Trans R Soc Lond A Math Phys Eng Sci* 1993;343:87–101.

62. Koruga, D., Hameroff, S., Withers, J., Loufty, R., Sundareshan, M. *Fullerene C$_{60}$: History, Physics, Nanobiology, Nanotechnology.* Amsterdam: Elsevier Science Publishers; 1993.

63. Yannoni, C.S., Johnson, R.D., Meijer, G., Bethune, D.S., Salem, J.R. *J Phys Chem* 1991;95:9–10.

64. Prassides, K., Kroto, H.W., Taylor, R., Walton, D.R.M., David, W.I.F., Tomkinson, J., Haddon, R.C., Rosseinsky, M.J., Murphy, D.W. *Carbon* 1992;30:1277–1286.

65. Kadish, K.M., Ruoff, R.S., editors. *Fullerenes: Chemistry, Physics, and Technology.* New York: Wiley-Interscience; 2000.

66. Murray, J.S., Gargarin, S.G., Polititzer, P. *J Phys Chem* 1995;99:12081–12083.

67. Ruoff, R.S., Malhotra, R., Huestis, D.L., Tse, D.S., Lorents, D.C. *Nature* 1993;362:140–141.

68. Taylor, R., Hare, J.P., Abdul-Sada, A.K., Kroto, H.W. *J Chem Soc Chem Commun* 1990; 1423–1425.

69. Ajie, H., Alvarez, M.M., Anz, S.J., Beck, R.D., Diederich, F., Fostiropoulos, K., Huffman, D.R., Kratschmer, W., Rubin, Y., Schriver, K.E., Sensharma, D., Whetten, R.L. *J Phys Chem* 1990;94:8630–8633.

70. Cox, D.M., Behal, S., Disko, M., Gorun, S.M., Greaney, M., Hsu, C.S., Kollin, E.B., Millar, J., Robbins, J., Robbins, W., Sherwood, R.D., Tindall, P. *J Am Chem Soc* 1991;113:2940–2944.

71. Larsson, S., Volosov, A., Rosen, A. *Chem Phys Lett* 1987;137:501–504.

72. Stanton, R.E., Newton, M.D. *J Phys Chem A* 1988;92:2141–2145.

73. Johnson, R.D., Meijer, G., Bethune, D.S. *J Am Chem Soc* 1990;112:8983–8984.

74. Weeks, D.E., Harter, W.G. *Chem Phys Lett* 1988;144:372.

75. Bethune, D.S., Meijer, G., Tang, W.C., Rosen, H.J., Golden, W.G., Seki, H., Brown, C.A., Devries, M.S. *Chem Phys Lett* 1991;179:181–186.

76. Schleyer, P.v.R. *Chem Rev* 2001;101:1115–1118.

77. Haddon, R.C. *Science* 1993;261:1545–1550.

78. Fowler, P.W., Collins, D.J., Austin, S.J. *J Chem Soc, Perkins Trans 2* 1993;275–277.

79. McMurray, J. *Organic Chemistry*, 4th ed. Pacific Grove, CA: Brooks/Cole Publishing Company; 1996.

80. Silverstein, R.M., Bassler, G.C., Morrill, T.C. *Spectrometric Identification of Organic Compound*, vol. 4. New York: John Wiley & Sons, Inc.; 1981.

81. Fowler, P.W., Lazzeretti, P., Malagoli, M., Zanasi, R. *Chem Phys Lett* 1991;179:174–180.

82. Haufler, R.E., Conceicao, J., Chibante, L.P.F., Chai, Y., Byrne, N.E., Flanagan, S., Haley, M.M., Obrien, S.C., Pan, C., Xiao, Z., Billups, W.E., Ciufolini, M.A., Hauge, R.H., Margrave, J.L., Wilson, L.J., Curl, R.F., Smalley, R.E. *J Phys Chem* 1990;94:8634–8636.

83. Dubois, D., Kadish, K.M., Flanagan, S., Haufler, R.E., Chibante, L.P.F., Wilson, L.J. *J Am Chem Soc* 1991;113:4364–4366.

84. Allemand, P.-M., Koch, A., Wudl, F., Rubin, Y., Diederich, F., Alavarez, M.M., Anz, S. J., Whetten, R.L. *J Am Chem Soc* 1991;113:1050–1051.

85. Echegoyen, L., Echegoyen, L.E. *Acc Chem Res* 1998;31:593–601.

86. Klos, H., Rystau, I., Schutz, W., Gotschy, B., Skiebe, A., Hirsch, A. *Chem Phys Lett* 1994;224:333–337.

87. Song, H., Lee, C.H., Lee, K., Park, J.T. *Organometallics* 2002;21:2514–2520.

88. Fukuzumi, S., Suenobu, T., Patz, M., Hirasaka, T., Itoh, S., Fujitsuka, M., Ito, O. *J Am Chem Soc* 1998;120:8060–8068.

89. Holloway, J.H., Hope, E.G., Taylor, R., Langley, G.J., Avent, A.G., Dennis, T.J., Hare, J.P., Kroto, H.W., Walton, D.R.M. *J Chem Soc Chem Commun* 1991; 966–969.

90. Olah, G.A., Bucsi, I., Lambert, C., Aniszfeld, R., Trivedi, N.J., Sensharma, D.K., Prakash, G.K.S. *J Am Chem Soc* 1991;113:9387–9388.

91. Tebbe, F.N., Harlow, R.L., Chase, D.B., Thorn, D.L., Campbell, G.C.J., Calabrese, J.C., Herron, N., Young, R.J., Jr: Wasserman, E. *Science* 1991;256:822–825.

92. Birkett, P.R., Hitchcock, P.B., Kroto, H.W., Taylor, R., Walton, D.R.M. *Nature* 1992;357, 479–482.

93. Weisman, R.B., Heymann, D., Bachilo, S.M. *J Am Chem Soc* 2001;123:9720–9721.

94. Chiang, L.Y., Bhonsle, J.B., Wang, L., Shu, S.F., Chang, T.M., Hwu, J.R. *Tetrahedron* 1996;52:4963–4972.

95. Chiang, L.Y., Swirczewski, J.W., Hsu, C.S., Chowdhurry, S.K., Cameron, S., Creegan, K. *J Chem Soc Chem Commun* 1992;24:1791–1793.

96. Chiang, L.Y., Upasani, R.B., Swirezewski, J.W. *J Am Chem Soc* 1992;114:10154–10157.

97. Chiang, L.Y., Upasani, R.B., Swirezewski, J.W., Soled, S. *J Am Chem Soc* 1993;115:5453–5457.

98. Schneider, N.S., Darwish, A.D., Kroto, H.W., Taylor, R., Walton, D.R.M. *J Chem Soc Chem Commun* 1994;4:463–464.

99. Husebo, L., Sitharaman, B., Furukawa, K., Kato, T., Wilson, L. *J Am Chem Soc* 2004;126:12055–12062.

100. Mcewen, C.N., Mckay, R.G., Larsen, B.S. *J Am Chem Soc* 1992;114:4412–4414.

101. Krusic, P.J., Wasserman, E., Keizer, P.N., Morton, J.R., Preston, K.F. *Science* 1991;254:1183–1185.

102. Jensen, A.W., Wilson, S.R., Schuster, D.I. *Bioorg Med Chem* 1996;4:767–779.

103. Jensen, A.W., Daniels, C. *J Org Chem* 2003;68:207–210.

104. Kamat, J.P., Devasagayam, T.P.A., Priyadarsini, K.I., Mohan, H. *Toxicology* 2000;155: 55–61.

105. Sayes, C.M., Gobin, A.M., Ausman, K.D., Mendez, J., West, J.L., Colvin, V.L. *Biomaterials* 2005;26:7587–7595.

106. Colvin, V.L. *Nat Biotechnol* 2004;22:760–760.

107. Beeby, A., Eastoe, J., Heenan, R.K. *J Chem Soc Chem Commun* 1994;173–175.

108. Dresselhaus, M.S., Dresselhaus, G., Eklund, P.C. *Science of Fullerenes*. San Diego: Academic Press;1996.

109. Heymann, D. *Fullerene Sci Technol* 1996;4:509–515.

110. Arrais, A., Diana, E. *Fullerenes Nanotubes &Carbon Nanostruct* 2003;11:35–46.

111. Chiang, L.Y., Wang, L.-Y., Swirezewki, J.W., Soled, S., Cameron, S.J. *Org Chem* 1994;59:3960–3968.

112. Lamparth, I., Hirsch, A. *J Chem Soc Chem Commun* 1994, 1727.

113. Dugan, L.L., Turetsky, D.M., Du, C., Lobner, D., Wheeler, M., Almli, C.R., Shen, C.K., Luh, T.Y., Choi, D.W., Lin, T.S. *Proc Nat Acad Sci, USA* 1997;94:9434–9439.

114. Murthy, C.N., Choi, S.J., Geckeler, K.E. *J Nanosci Nanotechnol* 2002;2:129–132.

115. Ungurenasu, C., Airinei, A. *J Med Chem* 2000;43:3186–3188.

116. Litvinova, L.S., Ivanaov, V.G., Mokeev, M.V., Zgonmik, V.N. *Mendeleev Commun* 2001; 1–2.

117. Andrievsky, G.V., Kosevich, M.V., Vovk, O.M., Shelkovsky, V.S., Vashchenko, L.A. *J Chem Soc Chem Commun* 1995; 1281–1282.

118. Alargova, R.G., Deguchi, S., Tsujii, K. *J Am Chem Soc* 2001;123:10460–10467.

119. Deguchi, S., Alargova, R.G., Tsujii, K. *Langmuir* 2001;17:6013–6017.

120. Cheng, X.K., Kan, A.T., Tomson, M.B. *J Chem Eng Data* 2004;49:675–683.

121. Andrievsky, G.V., Klochkov, V.K., Karyakina, E.L., McHedlov-Petrossyan, N.O. *Chem Phys Lett* 1999;300:392–396.

122. Wei, X., Wu, M., Qi, L., Xu, Z. *J Chem Soc Perkins Trans 2* 1997;2:1389–1393.

123. Fortner, J.D., Lyon, D.Y., Sayes, C.M., Boyd, A.M., Falkner, J.C., Hotze, E.M., Alemany, L.B., Tao, Y.J., Guo, W., Ausman, K.D., Colvin, V.L., Hughes, J.B. *Environ Sci Technol* 2005;39:4307–4316.

124. Heymann, D. *Carbon* 1996;34:627–631.

125. Andrievsky, G.V., Klochkov, V.K., Bordyuh, A.B., Dovbeshko, G.I. *Chem Phys Lett* 2002;364:8–17.

126. Lyon, D.Y., Adams, L.K., Falkner, J.C., Alvarez, P.J.J. *Environ Sci Technol* 2006;40:4360–4366.

127. Brant, J.A., Labille, J., Bottero, J.-Y., Wiesner, M.R. *Langmuir* 2006;22:3878–3885.

128. Mchedlov-Petrossyan, N.O., Klochkov, V.K., Andrievsky, G.V. *J Chem Soc Faraday Trans* 1997;93:4343–4346.

129. Taylor, R., Parsons, J.P., Avent, A.G., Rannard, S.P., Dennis, T.J., Hare, J.P., Kroto, H.W., Walton, D.R.M. *Nature* 1991;351:277–277.

130. Li, J., Takeuchi, A., Ozawa, M., Li, X., Saigo, K., Kitazawa, K. *J Am Chem Commun* 1993;1784–1786.

131. Lee, J., Fortner, J.D., Hughes, J.B., Kim, J.H. *Environ Sci Technol* 2007;41:2529–2535.

132. Avdeev, M.V., Khokhryakov, A.A., Tropin, T.V., Andrievsky, G.V., Klochkov, V.K., Derevyanchenko, L.I., Rosta, L., Garamus, V.M., Priezzhev, V.B., Korobov, M.V., Aksenov, V.L. *Langmuir* 2004;20:4363–4368.

133. Lyon, D.Y., Fortner, J.D., Sayes, C.M., Colvin, V.L., Hughes, J.B. *Environ Toxicol Chem* 2005;24:2757–2762.

134. Fortner, J.D., Kim, D.I., Boyd, A.M., Falkner, J.C., Moran, S., Colvin, V.L., Hughes, J.B., Kim, J.H. *Environ Sci Technol* 2007;41:7497–7502.

135. Brant, J., Lecoanet, H., Hotze, M., Wiesner, M. *Environ Sci Technol* 2005;39:6343–6351.

136. McHedlov-Petrossyan, N.O., Klochkov, V.K., Andrievsky, G.V. *J Chem Soc Faraday Trans* 1997;93:4343–4346.

137. Li, T.B., Huang, K.X., Li, X.H., Jiang, H.Y., Li, J., Yan, X.Z., Zhao, S.K. *Chem J Chin Univ Chin* 1998;19:858–860.

138. Cusan, C., Da Ros, T., Spalluto, G., Foley, S., Janto, J.M., Seta, P., Larroque, C., Tomasini, M.C., Antonelli, T., Ferraro, L., Prato, M. *Eur J Org Chem* 2002; 2928–2934.

139. Da Ros, T., Prato, M. *J Org Chem* 1996;61:9070–9072.

140. Chung, N., Alexander, M. *Environ Sci Technol* 1999;33:3605–3608.

141. Lecoanet, H.F., Bottero, J.Y., Wiesner, M.R. *Environ Sci Technol* 2004;38:5164–5169.

142. Yamago, S., Tokuyama, H., Nakamura, E., Kikuchi, K., Kananishi, S., Sueki, K., Nakahara, H., Enomoto, S., Ambe, F. *Chem Biol* 1995;2:385–389.

143. Yang, A., Cardona, D.L., Barile, F.A. *Cell Biol Toxicol* 2002;18:97–108.

144. Yang, X.L., Fan, C.H., Zhu, H.S. *Toxicol In Vitro* 2002;16:41–46.

145. Takenaka, S., Yamashita, K., Takagi, M., Hatta, T., Tsuge, O. *Chem Lett* 1999; 321–322.

146. Sension, R.J., Szarka, A.Z., Smith, G.R., Hochstrasser, R.M. *Chem Phys Lett* 1991;185:179–183.

147. Kamat, P.V. *J Am Chem Soc* 1991;113:9705–9707.

148. Kamat, P.V., Bedja, I., Hotchandani, S. *J Phys Chem* 1994;98:9137–9142.

149. Stasko, A., Brezova, V., Biskupic, S., Dinse, K.P., Schweitzer, P., Baumgarten, M. *J Phys Chem* 1995;99:8782–8789.

150. Jensen, A.W., Daniels, C. *J Org Chem* 2003;68:207–210.

151. Nakanishi, I., Fukuzumi, S., Konishi, T., Ohkubo, K., Fujitsuka, M., Ito, O., Miyata, N. *J Phys Chem B* 2002;106:2372–2380.

152. Colvin, V.L. *Nat Biotechnol* 2003;21:1166–1170.

153. Orfanopoulos, M., Kambourakis, S. *Tetrahedron Lett* 1995;36:435–438.

154. Deguchi, S., Alargova, R.G., Tsujii, K. *Langmuir* 2001;17:6013–6017.

155. Yamakoshi, Y.N., Yagami, T., Fukuhara, K., Sueyoshi, S., Miyata, N. *J Chem Soc Chem Commun* 1994; 517–518.

156. Hatchard, C.G., Parker, C.A. *Proc R Soc A* 1956;235:518–536.

157. Haag, W.R., Hoigne, J. *Environ Sci Technol* 1986;20:341–348.

158. Buxton, G.V., Greenstock, C.L., Helman, W.P., Ross, A.B. *J Phys Chem Ref Data* 1988;17:513–886.

159. Pickering, K.D., Wiesner, M.R. *Environ Sci Technol* 2005;39:1359–1365.

160. Studer, S.L., Brewer, W.E., Martinez, M.L., Chou, P.T. *J Am Chem Soc* 1989;111:7643–7644.

161. Islam, M.F., Rojas, E., Bergey, D.M., Johnson, A.T., Yodh, A.G. *Nano Lett* 2003;3:269–273.

162. Bensasson, R.V., Bienvenue, E., Dellinger, M., Leach, S., Seta, P. *J Phys Chem* 1994;98: 3492–3500.

163. Malhotra, R., Kumar, S., Satyam, A. *J Chem Soc Chem Commun* 1994;1339–1340.

164. Heymann, D., Bachilo, S.M., Weisman, R.B., Cataldo, F., Fokkens, R.H., Nibbering, N.M.M., Vis, R.D., Chibante, L.P.F. *J Am Chem Soc* 2000;122:11473–11479.

165. Heymann, D. *Fullerene Nanotubes Carbon Nanostruct* 2004;12:715–729.

166. Shang, Z., Pan, Y., Cai, Z., Zhao, X., Tang, A. *J Phys Chem A* 2000;104:1915–1919.

167. Anachkov, M.P., Cataldo, F., Rakovsky, S. *Fullerenes Nanotubes Carbon Nanostruct* 2004;12:745–752.

168. Cataldo, F., Heymann, D. *Polym Degrad Stab* 2000;70:237–243.

169. Cataldo, F. *Carbon* 2002;40:1457–1467.

170. Cataldo, F. *Fullerenes Nanotubes Carbon Nanosctruct* 2003;11:1–13.

171. Chiou, C.F., Marinas, B.J., Adams, J.Q. *Ozone Sci Eng* 1995;17:329–344.

172. Buxton, G.V., Greenstock, C.L., Helman, W.P., Boss, A.B. *J Phys Chem Ref Data* 1988;17:2560–2564.

173. Staehelin, J., Hoigne, J. *Environ Sci Technol* 1985;19:1206–1213.

174. Kim, D.I., Fortner, J.D., Kim, J.H. *Ozone Sci & Eng* 2007;29:121–129.

175. Gasyna, Z., Schatz, P.N., Hare, J.P., Dennis, T.J., Kroto, H.W., Taylor, R., Walton, D.R.M. *Chem Phys Lett* 1991;183:283–291.

176. Xing, G., Zhang, J., Zhao, Y., Tang, J., Zhang, B., Gao, X., Yuan, H., Qu, L., Cao, W., Chai, Z., Ibrahim, K., Su, R. *J Phys Chem B* 2004;108:11473–11479.

Bacterial Interactions with CdSe Quantum Dots and Environmental Implications

JAY L. NADEAU[1], JOHN H. PRIESTER[2], GALEN D. STUCKY[3], and PATRICIA A. HOLDEN[2]

[1]Department of Biomedical Engineering, McGill University, Montreal, Canada,University of California, Santa Barbara, CA 93106, USA
[2]Donald Bren School of Environmental Science and Management, University of California, Santa Barbara, CA 93106, USA
[3]Department of Chemistry and Biochemistry, University of California, Santa Barbara, CA 93106, USA

9.1 INTRODUCTION

9.1.1 Nanoparticles

Nanoparticles are particles whose largest dimension is less than or equal to 100 nm (113). Nanoparticles may be either natural, incidental, or engineered (153), and either amorphous, crystalline (9), polymeric (152), or composites (131). Their chemistries may be predominantly nonmetal (e.g., carbon), metallic (e.g., Au, Ag), semiconductor (e.g., CdSe), or a combination (153). Their shapes include spheres, tubes, rods, horns, and platelets. Their physical properties are related to their size and chemical composition. Their surface chemistries, including surface defects and impurities, contribute to their reactivity (9).

Natural nanoparticles arise abiotically from earth processes contributing to rock and soil formation, including volcanism (60, 106), dissolution and precipitation (9), abrasion, and erosion (5). As summarized by Gilbert and Banfield (57), nature's abiotically derived natural nanoparticles include weathered and unweathered silicates (e.g., clays such as smectite), iron oxides such as ferrihydrite, hematite, and magnetite (127), uranite (154), calcium and magnesium carbonate (2), and carbon black (19).

Nanoscience and Nanotechnology   Edited by Vicki H. Grassian

Crystal nucleation, growth, and mineral formation, well-studied topics in geochemistry, lead to nanoparticulates when continued crystal growth is limited by temperature or ion content (9). Nanosized silica occurs widely in organisms, including nanosized silica in rice husks (28) and nanoparticulate silica in diatoms (75, 120). Biogenic nanoparticles also arise from bacteria, for example, magnetite (10, 53), polyphosphate (38), sulfurous particles (69, 135), elemental selenium (114), CdS crystals (144, 160, 164), and manganese oxides (146, 147). Some of these particles are formed through "biomineralization" externally to cells, typically as a result of indirect effects of cells on precipitation, crystal growth, and solubility (135, 160). However, other particles are formed directly by bacterial cells through a series of enzymatically catalyzed steps (47). Viruses are biological nanoparticles used as templates for inorganic nanoparticle synthesis (101). Biofilm extracellular matrices also contain membrane vesicles that range from approximately 10 nm to 350 nm in size (134).

"Incidental" or "accidental" describes unintentional human-derived nanoparticles (153) such as nanoparticulate products of combustion, for example, the smallest particles in motor vehicle exhaust (31) or in smoke from coal-fired power plants (56, 154). Incidental nanoparticles are anthropogenic but have no commercial application and can be serious air pollutants (5). Carbon soot and organic particulate carbon are released in large quantities into the atmosphere from transportation and furnaces (e.g., Reference 88).

In contrast, "engineered" or "manufactured" nanoparticles are concertedly produced for commercial purposes and their composition, size, and surface characteristics are designed for specific applications (153). They are, thus, more homogeneous than the ultrafine particle mixtures that arise "by accident" (113) and thus may have more specific interactions with biological systems that directly arise from easily identifiable characteristics such as their common size, shape, or surface chemistry.

Given so many sources, it would seem that nanoparticles are everywhere (9), arising geologically, biologically, and anthropogenically. Along with such diverse origins, nanoparticles are extremely diverse in bulk and surface chemical composition, sizes and morphologies, and crystal and electronic structures. Because their properties are highly related to such characteristics, functions of nanoparticles in the environment may vary extensively, possibly to the point of precluding generalizations based on size and thus making irrelevant a general discourse on "nanoparticles." However, at this early stage of research and understanding, broadly considering findings across a nanoparticle spectrum in addition to focusing on one nanoparticle type is needed for discovering paradigms in nanoparticle interactions with biological and environmental systems. The focus of this chapter is on various CdSe quantum dots, but it also draws from literature pertaining to other solid nanoparticles, as appropriate.

9.1.2 CdSe Quantum Dots

Quantum dots (QDs) are photoluminescent semiconductor nanocrystals that promise an extraordinary advance in computing (161), photovoltaic materials (81), and biological labeling at the nanoscale (27, 86). The most common QD materials for biological use are made from cadmium and selenium, in the form of CdSe wurtzite

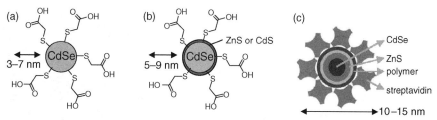

FIGURE 9.1 Chemical composition of the most commonly used QDs in biological applications. (a) CdSe QDs (referred to as bare or core QDs in this chapter). Solubilization is accomplished by self-assembly of an alkanethiol (such as mercaptoacetic acid or MAA, shown) whose $-SH$ bonds directly to the semiconductor, leaving the carboxylate group free to interact with aqueous solution. (b) "Core–shell" QDs are overlaid with a 1–2-nm-thick layer of ZnS or CdS. Solubilization is identical to the case of the bare crystals. In both cases, the free carboxylates can be covalently bonded to proteins or other organic molecules of choice. (c) Commercially available QDs (Invitrogen) consist of the CdSe/ZnS "core–shell" further coated with polymers and the protein streptavidin. Overall nanocrystal diameter is two- to three-fold that of the "core–shell" QDs.

nanocrystals (bare or core QDs). These nanocrystals are often overlaid with another semiconductor material, such as ZnS or CdS, to increase the intensity of the light emission; these are referred to as "core–shell" QDs and the outer shell may be no more than a single atomic monolayer. QDs for biological applications are nearly always rendered water soluble by a single-chain, functionalized thiol of varying length (mercaptoacetic acid, mercaptopriopionic acid, mercaptoundecanoic acid) or by a dithiol such as dihydrolipoic acid (Fig. 9.1a and b). This surrounding solubilizing layer is referred to as a "cap." Additionally, commercially available quantum dots are further coated with proprietary polymer and protein layers to render them biologically compatible and water soluble (http://probes.invitrogen.com/products/qdot/; http://www.evidenttech.com/qdot-definition/quantum-dot-introduction.php) (Fig. 9.1c). CdSe QDs may also be synthesized with citrate as the stabilizing ligand (130). However, citrate-stabilized CdSe QDs are weakly fluorescent when compared to QDs stabilized using thiol chemistry.

Absorption spectra of QDs are broad and their emission spectra are narrow, allowing for multiwavelength labeling that is easily distinguished (Fig. 9.2), and is controllable by varying the size of the particle (26, 54, 116). Quantum dot composition can be varied (123, 124, 169, 171) to enable targeted binding to a wide variety of biological molecules such as peptides, DNA, and antibiotics while preserving fluorescence (26, 55, 59, 117, 125, 151). The solubilizing thiol cap of QDs can be readily bound to organic molecules (103) (27) (23) resulting in a spectrum of biologically compatible labels that are all excited with a single wavelength. This feature has been used to create DNA microarrays (62), to label fixed and permeabilized mammalian cells (8, 27, 51, 61, 90, 91, 94, 116, 167), and to observe living slime molds (76) and bacteria (86). Consistently, bare (172) and core–shell (with a ZnS coating) (61, 85, 86) CdSe quantum dots have been conjugated with biomolecules for bacterial and eukaryotic cellular labeling and subsequent imaging.

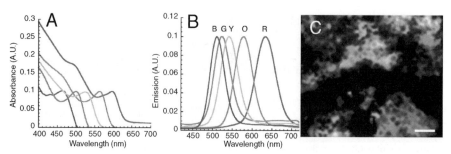

FIGURE 9.2 CdSe QD photoluminescence in physiological saline at 25–37°C. (a) Visible absorbance spectra from five independent preparations of CdSe QDs. (b) Emission spectra of the samples in panel (a). The peaks almost span the visible range and are labeled B(lue) G(reen) Y(ellow) O(range) R(ed), although the blue, green, and yellow appear similar to the eye under a long-pass filter. Excitation, 400 nm. (c) Epifluorescence image of triple labeling of *S. aureus* cultures with red, orange, and green QDs (scale bar = 5 μm). (All images adapted from (86) and (107)).

Because biological QDs are synthesized and stored in liquids, inhalation in the workplace is of minor concern; greater risks may result from ingestion (either inadvertently or for medical procedures), and also contamination of water and soil with possible consequences to humans, microorganisms, and ecosystems. Quantum dots are variable in their composition, which may preclude sweeping generalizations regarding their toxicity (46, 63, 84), but CdSe nanoparticle cytotoxicity is not typically observed over the short timescales (hours) used for biological labeling and imaging (as reviewed by Hardman (63)). Obvious environmental risks of CdSe quantum dots are associated with toxic concentrations of the constituent metals Cd and Zn, and of the metalloid Se (63). Cadmium has no known nutritional value to any living organisms, with the possible exception of a species of diatom (92), and is considered a serious pollutant (79). The primary health effect of cadmium exposure to humans is that it accumulates in the kidneys and causes renal tubular damage (78). Selenium and zinc, although essential trace elements, can be toxic to birds, fish, and humans (70, 159). However, possible environmental risks of QDs arise not only from the heavy metals and metalloids from which they are made, but also from specific properties that are engendered by the nanometer scale of the particles. In particular, two properties of QDs are not possessed by bulk semiconductors: (1) their small size permits them to intercalate into or to cross cell membranes; (2) they are very strong oxidizing and reducing agents. Both of these issues are considered in this chapter.

9.1.3 Bacteria

All nanoparticles, whether natural or anthropogenic, accidental or engineered, biotic or abiotic, occur, or will eventually occur, in soil, sediments, and water. In these environmental compartments, nanoparticles will most certainly encounter bacteria, the most abundant life forms on Earth (165). Bacteria catalyze all major reactions in nutrient cycling, including biodegradation, and are the biogeochemical basis for the biosphere as we know it (132). Among the many bacterially-catalyzed reactions are those that involve metals, for example, where metals are either electron donors or acceptors (109, 110). To the degree that bacteria exchange electrons with or otherwise alter nanoparticulate metals, the consequences of nanoparticles for the environment will be shaped by bacterial–nanoparticle interactions. Bacteria, as they grow fast and are relatively simple structures, are useful model systems for studying cellular interactions with nanoparticles. Interestingly, bacteria are also known to, in some environments and by some species, naturally assemble nanoparticulates (as above) suggesting that they are highly adapted to such materials and in fact may benefit from them. Thus, mechanisms of nanoparticle synthesis, tolerance, and metabolism in bacteria may have evolved to the extent that such mechanisms are widespread and even beneficial in microbial ecology. Such mechanisms may also be at work when bacteria encounter novel, engineered nanoparticles released into nature. Studying bacterial interactions with engineered nanoparticles is necessary to predict nanoparticle fate in the environment and how bacteria will potentially facilitate bioremediation of nanoparticle-based pollution.

This chapter is about engineered metal nanoparticle interactions with bacteria in aqueous and soil environments, specifically considering

- Abiotic fates of CdSe QDs in the environment.
- Bacterial growth habits in soil and water as conditions for CdSe QD encounters in the environment.
- Interactions between bacterial cells and biofilms with QDs including binding, uptake of nanoparticles or metals, intracellular processes, expulsion, and toxicity.
- Microbial ecological implications.
- Implications of bacterial-QD interactions to the larger scale environmental fate and transport of QDs.
- Prospective research questions.

9.2 EFFECTS OF ABIOTIC FACTORS ON QD FLUORESCENCE AND STABILITY

The thiol ligands used to solubilize QDs quench the nanoparticle fluorescence by acting as electron donors to a degree that varies according to the nature of the ligand and its solution environment. Thus, in typical biological solutions, QDs are highly sensitive to pH, being essentially nonfluorescent at or below pH 5 where protonation of the sulfur atoms becomes important. Quenching due to low-pH conditions is irreversible, whereas the moderate quenching seen as the pH is raised above 8 can be reversed upon restoring to neutral conditions (Fig. 9.3a).

QDs stabilized with other ligands, for example, citrate (130), are also expected to be strongly affected by pH. The citrate ion has three carboxylate groups ($pK_a 1 = 3.13$, $pK_a 2 = 4.76$, and $pK_a 3 = 6.14$). The greater ionic charge of the citrate anion above a pH of approximately 7.7 plays a major role in both its ability to act as a capping agent

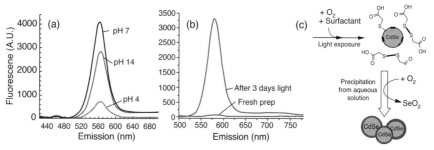

FIGURE 9.3 Quenching of bare CdSe QDs under specific environmental conditions. (a) Effects of pH. Even pure NaOH solutions (pH 14) cause only a moderate decrease in QD fluorescence (30–40%), while acid solutions (pH 4) lead to a near-total quenching accompanied by blueshift. This quenching is accompanied by precipitation of insoluble material. (b) Exposure to room light in aqueous solutions under room air causes surface oxidation of the QDs and large increases in fluorescence with loss of solubility. Shown, bare CdSe QDs freshly solubilized into aqueous solution (fresh prep.) and after 3 days under room light. (J. Nadeau, unpublished data). (c) Schematic of probable changes due to light and oxygen. QDs as in Fig. 9.1a first undergo "cap decay," where formation of dithiols causes the capping molecules to detach from the surface. The thick lines indicate the formation of cadmium oxide, which increases fluorescence. As photooxidation progresses, the caps detach completely, leaving insoluble but very bright cadmium oxide-capped particles. Different degrees of core decay lead to a broadening in the particle size distribution and thus to a broad emission spectrum as seen in panel (b).

and as a chelating agent that might remove $CdOH^+$ ions from the nanoparticle. These potentially competing effects have not yet been carefully studied.

In addition to pH effects, the thiol–semiconductor bond is unstable in the presence of molecular oxygen, so QDs show photochemical instability as a result of oxidation first of the surrounding thiol ligand cap, then eventually of the nanocrystal shell and/or core. As the size of the particle decreases, it may escape from its solubilizing cap and precipitate from the solution over a time course of hours to days (Fig. 9.3b and c). At the same time, component ions from the nanocrystal shell and/or core are released into the solution. Inductively coupled plasma optical emission spectroscopy (ICP/OES) measurements of ionic concentrations in solution show cytotoxic levels of free Cd^{2+} ions in solutions of bare–core CdSe exposed to UV irradiation, from 10 ppm in 1 h of irradiation to greater than 80 ppm in 8 h (46). Core–shell QDs do not release detectable Cd^{2+} in such time periods, but liberate comparable amounts of Zn^{2+} (36).

The redox potential of the solution, or the presence of electron- or hole- scavenging molecules, also has a significant effect on QD fluorescence and probably breakdown. Semiconductors, because of their band structure, may donate electrons or holes to adjacent molecules, that is, they may behave as reducing or oxidizing agents. As particle sizes become smaller, the oxidizing and reducing potentials of the semiconductor materials increase because of the size-confined broadening of the bandgap (Fig. 9.4a). It has been shown that photogenerated electrons from CdSe are able to cross a water–organic solvent interface and interact with electron acceptors, whereas photogenerated holes are not (142). Transfer of either electrons or holes to a

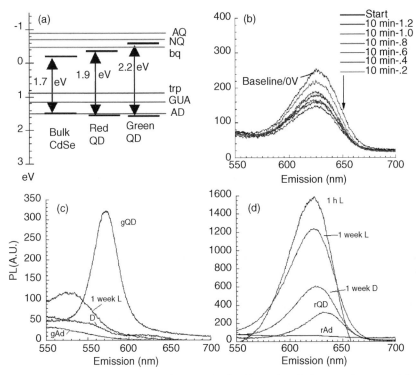

FIGURE 9.4 (a) Energy levels of QD samples in this study relative to redox potentials of some commonly used electron donors and acceptors. *Upper lines*: reduction potentials of naphthoquinone (NQ), benzoquinone (bq), and anthraquinone (AQ). In a previous study (24), the first two compounds were able to quench QD fluorescence, while the last was not (QDs used were slightly blueshifted from our green sample, conduction band edge $= -0.66$ eV). *Lower lines*: oxidation potentials of purines and amino acids. GUA, guanine (pH 12); AD, adenine (pH 7); trp, tryptophan (pH 6; oxidation potential decreases as pH rises). The short lines show band edge positions for the QD samples used in this study, calculated from absorbance spectra. The valence band edge of the red QDs is very close to the redox potential of adenine. (b) Bare CdSe are quenched partially by exposure to negative potentials (reduction), but this quenching is reversible after the reducing environment is removed (shown: "start," CdSe QDs in saline; then 10 min of exposure to increasingly negative potentials, from -0.2 to -1.2 V. After the potential was removed, fluorescence returned to baseline and overlapped the original spectrum after 1 h (baseline/0V). (c) Green-emitting QDs (gQD) are completely quenched by conjugation to adenine (gAd). The quenching remains complete after 1 week in the dark (d), and only slightly returns after 1 week of room light (1 week L). (D), Red-emitting QDs (rQD) show quenching immediately after conjugation to adenine (rAd), but strong, slightly blueshifted fluorescence has returned after 1 h at RT in room light (1 h L). In identical samples prepared at the same time, 1 week in room light appears similar to the 1 h case (1 week L). Significant fluorescence is seen even if the samples are kept in the dark (1 week D); the peak is identical to that of the light-exposed samples, and is more than twofold as strong as the original fluorescence of the sample.

surrounding molecule will quench QD fluorescence because it prevents radiative recombination of the electron–hole pair. However, exposure of QDs to reducing agents or highly reducing potentials has been shown to lead to reversible quenching, whereas the quenching seen under highly oxidizing conditions is irreversible (Fig. 9.4b). An investigation of particle breakdown under these circumstances has not yet been done, although it would be of great importance as redox agents are often found as environmental pollutants. An example of an oxidant that has been shown to accept electrons from QDs is methyl viologen, or paraquat (24).

The redox activity of semiconductor nanocrystals has a wide variety of applications and implications. The smallest particles of TiO_2, for example, are capable of oxidizing hydrogen; hence their preliminary applications in photovoltaic cells (111). More commonly used QD materials, such as CdSe, are weaker oxidizers than this, but are still capable of oxidizing important biomolecules such as DNA purine bases (Fig. 9.4c and d) (166).

All of the data we have presented were taken with thiol-solubilized nanoparticles. Many alternative methods of solubilizing QDs have been sought because of the unstable nature of the nanoparticle–thiol bond. Addressing each of these methods would be outside the scope of this chapter; they are systematically reviewed in (136). Most of these are time consuming or difficult to perform, and none has succeeded in replacing the simple thiol solubilization methods for the time being. One possible exception is the citrate ligand chemistry previously mentioned that, although even simpler than thiol chemistry, is based on weak electrostatic interactions between the citrate molecule and the QDs. Briefly, all QDs in aqueous solution were sensitive to photooxidation, while there was greater variability in sensitivity to chemical oxidation. Low pH will dissolve nearly all QDs except those coated by polyethylenimine, which has been shown to act as a "proton sponge," preventing the acid from etching the semiconductor particle.

In summary, dissolution of CdSe nanoparticles may be slow in the aqueous, abiotic environment, given the very low solubility product for bulk CdSe at 298°K in distilled water (6.3E-36) (37), but dissolution becomes much more important as particle sizes are reduced. Moreover, dissolution rates of nanomaterials will vary with environmental conditions and are expected to be enhanced by biological ligands, especially chemical oxidants (16).

9.3   BACTERIAL MICROENVIRONMENTS AND PHYSICAL ASSOCIATIONS WITH NANOPARTICLES

At the pore scale in water and soil, bacteria occur in solution and at interfaces (air/water or neuston, solid/water) (Fig. 9.5). Most bacteria in nature and in disease exist as biofilms (39), that is, cells attached to surfaces where they multiply and encapsulate themselves in extracellular polymeric substances (EPS) of their own making (30, 39) that are necessary for biofilm formation (43). The biofilm matrix is a mixture of cells, both live and dead, polysaccharides, nucleic acids (139, 163), proteins (122, 140), and inorganic material. Most of the biofilm is polyanionic because of the presence

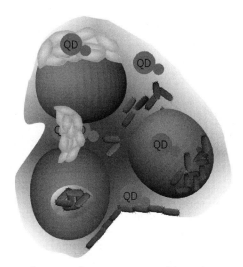

FIGURE 9.5 Conceptual pore scale arrangements of bacteria (dark grey rods), biofilms (white multicellular structures), and QDs (spheres labeled QD with smaller conjugate circles) with organomineral complexes (large ovoids) in water-filled pores (grey background) or adsorbed pools (silvery splotch on lower left ovoid). Bacterial physiology varies substantially depending on the growth habit (i.e., attached, biofilm, or free living), thus altering local interactions with QDs. Also, microbes respond readily to shifts in organic and inorganic nutrients, energy sources, terminal electron acceptors, and environmental factors (e.g., pH, temperature, water availability, and oxidation–reduction potential). Thus biotic fates and effects of QDs will vary predictably, for example seasonally, in open environmental systems.

of carboxylate or phosphate groups (12), readily binds metals (50), and may form a protective shield for the enclosed bacteria, imparting increased heavy metal resistance (148). Under fluid flowing conditions, the structure of biofilms can be very complex, including towers and fluid channels (40, 44). Also, biofilms in the environment (e.g., on soil, in aquifers, in pipes) consist of many species of bacteria and eukaryotes that form an aggregated community.

Unsaturated biofilms grow in transiently wet environments like soil (71, 73) or on stone surfaces (83) and can be studied in mono- or cocultures in the lab using membranes (0.2 μm pore size) overlying solid media (138), by cultivating in porous media such as sand (42, 74), and in reactors that mimic solid media but with a nearly infinite, well-mixed reservoir of nutrients (71, 73). The culture format must be carefully selected to mimic the environmental conditions, including nutrient and water activity conditions, of the system being simulated (71). While soil bacteria are more likely in microcolonies (49) and not in spatially extensive biofilms, cultivation and study of unsaturated biofilms under laboratory conditions yields representative and conveniently sufficient biomass quantities for analysis.

Different from the highly channelized morphologies observed in laboratory flow-cell culture systems (141, 157), unsaturated biofilms that lack the selective pressure of fluid shear are flat (Fig. 9.5), dense, and unchannelized (Fig. 9.6) with similar

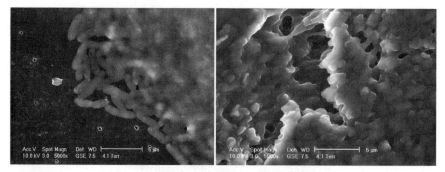

FIGURE 9.6 Environmental scanning electron micrographs of *Pseudomonas aeruginosa* PG201 cultivated as unsaturated biofilm on membranes overlying solid media, where glucose was the carbon source. Left: edge of biofilm with pioneer cells to the right of the dense mat (2500X). Right: Stretched EPS binding cells together (5000X), where drying led to the development of large crevices. Biofilms were treated with glutaraldehyde, ruthenium red, lysine, and osmium tetroxide to enhance resolution of cells and polysaccharides as per Reference 121.

architecture to "colony biofilms" described previously (173). Staining with a heavy metal such as ruthenium enhances the appearance of biofilm EPS for SEM visualization (121), presumably because there are strong physicochemical interactions between the metals and EPS polymers. Unsaturated biofilm architecture likely creates different responses, versus saturated biofilms, to QDs because of diffusional limitations resulting from the relatively higher biofilm density. The biofilms in Fig. 9.7 impeded diffusion of benzene, that is, conferring 100 times more resistance to mass transfer than water or any saturated biofilm (73). The biofilm–air interface contributed little to this extreme mass transfer resistance (163). Rather, the cells and composite EPS are the main barriers to rapid diffusion. Resistance to biofilm penetration is expected with solid QDs, but some penetration should occur based on observations of natural nanoscale minerals that were distributed throughout stone monument biofilm EPS

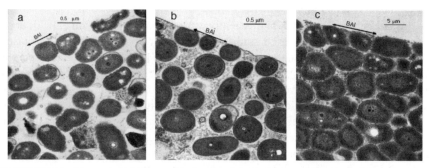

FIGURE 9.7 Transmission electron micrographs of unsaturated *P. putida* mt-2 biofilms cultivated on membranes with PEG-8000-controlled water activity opposite the membrane (71, 73). BAI = biofilm–air interface. (a) 0.0 MPa (100% relative humidity). (b) −0.25 MPa (moderately dry). (c) −1.5 MPa (very dry, the lower limit of physiological activity for many bacteria).

matrices and on cells (83). Given that unsaturated biofilm EPS contains chromosomal DNA (139), QDs and their dissolution products may strongly associate with these or other EPS macromolecules as was observed for extracellular DNA (eDNA) and chromium ions (122).

9.4   BIOPHYSIOCHEMICAL INTERACTIONS BETWEEN BACTERIA AND QUANTUM DOTS

Cellular bacterial interactions with conjugated, bare or core–shell, CdSe QDs include effects from extracellular decomposition and internalization of constituent metals and metalloids, possible production of reactive oxygen species (ROS) that disrupt the cell envelope and allow for intracellularization of nanoparticles, receptor-mediated binding to the cell surface, nonspecific binding, shedding, and expulsion (Fig. 9.8). Evidence for these various processes and effects are discussed in this section.

9.4.1   QD Labeling, Uptake, Breakdown, and Toxicity in Bacteria

The laboratory of Nadeau was the first to show that when living bacteria are exposed to QDs, internalization and/or degradation of the nanocrystals may occur even more rapidly than in ordinary solutions, although the final end points were not determined (85, 86). When QDs conjugated to a biomolecule are incubated with microorganisms

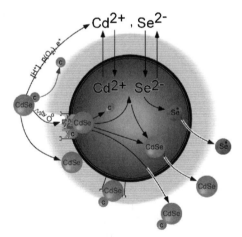

FIGURE 9.8   Conceptual diagram of CdSe quantum dot (QD, spheres with small sphere conjugated attached) interactions with a bacterial cell. Bioconjugated, bare or core–shell, CdSe QDs are subject to abiotic decomposition processes in the vicinity of cells that can lead to heavy metal and metalloid release; released metals and metalloids may be intracellularized, causing toxicity and possibly reassembling inside cells where they are retained or expelled. Extracellular labeling by whole QDs can be nonspecific. Receptor-mediated labeling may lead to whole nanoparticle uptake, especially through damaged membranes that are otherwise insufficiently porous to allow diffusive passage. Once inside cells, whole QDs may be expelled or metabolized to various extents including conjugate loss and complete breakdown.

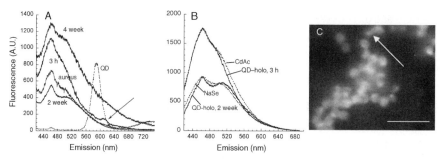

FIGURE 9.9 QDs are processed and broken down by *S. aureus*. (a) Emission spectra of *S. aureus* alone (aureus); incubated for 3 h with freshly prepared (unoxidized) QD-holotransferrin conjugates (3 h); incubated for 3 h with unoxidized conjugates and allowed to sit at 4°C under room light for 2 week or 4 week. There is a large feature appearing over time at ~520 nm. A small peak at ~615 nm is seen right after the incubation (arrow), but disappears with time. Incubation with oxidized holo-QDs leads to a signal much like the 2 week spectrum (not shown). The spectrum at 4 week suggests that the cells are beginning to break up, as it is similar to the spectrum seen from cells killed by heat shock or freeze-thaw (not shown). (b) Comparison of holo-QD incubated *S. aureus* spectra with spectra of *S. aureus* incubated with heavy metal salts (1 mM for 3 h). NaSe, sodium selenide; CdAc, cadmium acetate. (c) Epifluorescence image of *Staphylococcus aureus* incubated with red QD–holotransferrin for 3 h at 37°C, then stored for 2 week in the dark at 4°C. The labeling permeates the entire bacterial cell, often obscuring autofluorescence. QDs outside the cell retain their original color (as indicated by an arrow), whereas those inside the bacteria are significantly blueshifted. Similar labeling is seen when the QD–transferrin conjugate is allowed to photooxidize before incubation. Labeling is always a diffuse, broad orange in color and is unrelated to the original spectrum of the QDs (data adapted from (86)).

bearing specific receptors to that biomolecule, fluorescent labeling is often seen internal to the cells. Examples are human pathogens (such as *Staphylococcus aureus*) incubated with human transferrin–QD conjugates (Fig. 9.2c and Fig. 9.9), most cocci incubated with QD–tryptophan, wild-type *Bacillus subtilis* incubated with QD–adenine conjugates, and adenine import mutants of *B. subtilis* incubated with QD–adenosine monophosphate (AMP) conjugates (Fig. 9.10). This internalization was not seen with unconjugated QDs nonspecifically bound to organisms (86).

Tracking of the fluorescence with time, light exposure, and other variables provides valuable insight into the fate of the particles in the cells. In many cases, the fluorescence seen is not due to intact QDs, and peak shifts and broadening can be used to infer QD breakdown. The rate and type of breakdown differ markedly among species and conjugates. For instance, most of the fluorescent signal of *S. aureus* incubated with QD–transferrin is not a narrow peak corresponding to the original QDs, but rather a broad peak significantly shifted from the peak of either the original or oxidized QDs (Fig. 9.9a). The spectrum of this peak does not vary with the emission spectrum of the QDs (i.e., green QDs are redshifted and red QDs are blueshifted). In fact, the spectrum seen is similar to what is seen when *S. aureus* is incubated with Cd and Se salts instead of nanocrystals. Immediately after QD–holotransferrin incubation, the peak seen corresponds to Cd with Se increasing over time (Fig. 9.9).

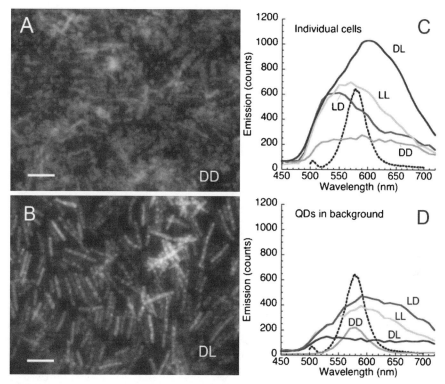

FIGURE 9.10 Effects of light exposure and pH on uptake of QD-AMP by *Bacillus subtilis* adenosine phosphoribosyltransferase (aprt) single mutant. Scale bar = 5 mm. A, When QD conjugates prepared in the dark and incubated with bacteria in the dark, virtually no QD material is taken up into the cells, as evidenced by lack of cellular labeling and clumps of QDs in the background. B, When the same QD conjugate as in A is incubated with the same culture under room light, the material is taken up into the cells. C, Spectra taken from individual cells after incubation with QD-apt. DD and DL = prepared in dark, incubated in dark and light respectively; LD and LL = conjugates prepared in light, incubated in dark and light respectively. Note that in all cases, the spectra of the internalized material is broadened from that of the original QDs (shown by a dashed line). D, Spectra of QDs in the background with each labeling technique. Note that the DD case retains the spectrum of the original QDs; also note the relative dimness compared with the internalized material in C. E, (no change).

The rate and nature of the QD degradation may depend on bacterial strain and metal-dependent enzymes; it did not occur with killed bacteria (86). With *Pseudomonas aeruginosa* grown in the presence of citrate stabilized, but otherwise bare, CdSe QDs, nanocrystals were decomposed (Fig. 9.11) and the metals internalized. The breakdown process was cell associated because it did not occur when dialysis tubing separated the cells from the QDs. However, such breakdown could be either "indirect" or "direct" (21), with the former referring to large biological ligands that

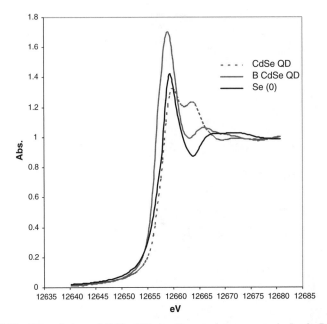

FIGURE 9.11   Degradation of CdSe QDs by *P. aeruginosa* grown in Luria Bertani broth (30°C, aerobic) with citrate-stabilized CdSe QDs as evidenced by XANES of cells showing cellular Se spectrum (B CdSe QD) more closely resembling that of the Se^0 standard (Se(0)) than the CdSe QD standard to the right. (J. Priester et al., unpublished data).

could chelate and remove surface Cd ions on QDs extracellularly or on the membrane and the latter referring to enzymatic processes mediated on the cell envelope or in the cell. Indirect extraction of metal ions from solids is facilitated by microbially produced high-affinity ligands such as the siderophores that bind iron (68) and molybdophores that bind molybdenum (95). Wherever breakdown occurred relative to *P. aeruginosa*, the end point was clear in that cellular Se was elemental (Se^0) by X-ray absorption near-edge spectroscopy (XANES, Fig. 9.11). Also, cells experienced toxicity at magnitudes similar to that caused by similar concentrations of Cd ions (Fig. 9.12). Both Se and Cd ions were accumulated in cells in the QD treatment and appeared to be localized by scanning transmission electron microscopy coupled with energy dispersive spectroscopy (STEM/EDS); however, the selenium oxidation state definitively traced the breakdown of the quantum dots and thus similarly to before (86) suggested that bare quantum dots may not enter cells intact. The one caveat in this interpretation is that cell membranes, if compromised sufficiently, could allow passive diffusion of QDs into cells. If this occurred, one would conclude that QDs were broken down within cells.

For bare quantum dots, toxicity was not reported over short time scales: this may be because the nanoparticles or their constituent metals are not internalized (86). Yet, over long timescales associated with environmental processes, caps or conjugates may be lost (99) and this can lead to extracellular Cd ion release and subsequent

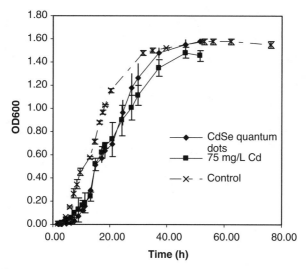

FIGURE 9.12 Growth curve (optical density @ 600 nm = OD600 versus time) for *P. aeruginosa* cultivated aerobically in LB broth (30°C) with no metals (control), 75 mg/L Cd(II), or bare citrate-stabilized CdSe QDs (75 mg/L as Cd). (J. Priester et al., unpublished data).

mammalian, and perhaps bacterial, cell toxicity (as reviewed in Reference 63). Over longer timescales in the lab, toxicity to *P. aeruginosa* by CdSe QDs is quantitatively similar to that imposed by equivalent concentrations of Cd ions (Fig. 9.12). Also, the appearance of *Escherichia coli* cells after several hours of QD exposure suggests expulsion of Cd from cells in vesicular occlusions (Fig. 9.13). Resistance to Cd ion toxicity in bacteria is well studied and involves many processes including efflux pumps (111). Whether or not observed expulsion (Fig. 9.13) is demonstrative of established Cd ion resistance processes is unknown.

Toxicity of similarly, positively charged CdTe quantum dots was size-dependent to mammalian cells and cellular responses included chromatin condensation and membrane blebbing (98). Subcellular uptake depended on particle size (98). In mammalian cell culture CdSe quantum dot toxicity, as observed both by cell viability and by gene expression through high density array analysis, was greatly reduced with particle modification with PEG and silica coating (170). However, the relationships between toxicity and silica nanoparticle characteristics also vary significantly with cell line such that rapidly growing cancer cells were more resistant to toxicity than slow growing fibroblast cells (29). This may imply that vastly different effects could result with different groups of microbes.

Analogously to findings with CdSe QDs, Thill et al. (150) found that 7 nm CeO_2 nanoparticles adsorbed to the outer membrane of washed, stationary-phase *E. coli* grown in rich media [Luria–Bertani(LB)]. Sorption was more than a monolayer, and the majority of the nanoparticles were reduced to Ce(III) within 3 h of contact between cells and nanoparticles. Contact resulted in cell death, but toxicity was not observed when cells were resuspended in rich media during nanoparticle contact experiments.

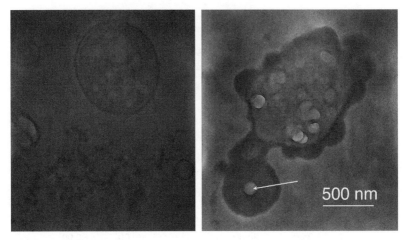

FIGURE 9.13 Blebbing of Cd from adenine-auxotrophic *E. coli*. Left: TEM image of a thin section of an adenine auxotrophic mutant of *E. coli* (ATCC 23804) immediately upon addition of adenine-conjugated QDs. The individual QDs are too small to resolve. Right: After 3 h on incubation with the conjugated QDs, the *E. coli* cells have formed inclusion bodies many times the size of the QDs. Some of these bodies are being expelled from the cells (arrow). (J. Nadeau, unpublished data).

Also, when cells were suspended in rich media, the reduction of Ce(IV) to Ce(III) was more extensive. The authors reason that Ce(IV) reduction must be happening in this latter case away from the cell, but still with cell mediation since LB alone did not reduce Ce(IV). An alternative explanation is that the energy provided in LB was sufficient to stimulate growth and either cells were growing, and thus not obviously succumbing to cytotoxicity, or energy depletion by CeO_2 reduction at the membrane was offset by bacterial energy generation in rich media. In other words, cells might have overcome the stress of CeO_2 toxicity by being provided LB, even though reduction was occurring. This would potentially explain the observation that reduction of CeO_2 in cells with LB was highest at low, and decreased with increasing, CeO_2 concentrations (150). However, the conclusion of the authors that direct contact of nanoparticles and bacteria is a prerequisite to cytotoxicity is not necessarily supported. They observed that bacteria with sorbed nanoparticles in the absence of LB showed toxicity in a dose-dependent fashion, but they did not test if unsorbed nanoparticles, perhaps kept distant by a dialysis membrane, conferred toxicity. Further, with LB present, they did not observe toxicity and yet nanoparticles were sorbed. However, toxicity might not have been measured if bacteria were growing and dying at the same rate. Even if it were measured, that is, if bacteria were not growing, then the interaction of sorption and toxicity would appear to depend on the growth state and nutrition conditions of the bacteria; that is, whether nutrients are available to overcome toxicity even to a nongrowing population. One important conclusion by the authors of this study is that the physicochemical behaviors of nanoparticles in various media must be understood before, or at least at the same time as, biological tests if the latter are to be rationally interpreted (150).

9.4.2   Electron Transfer from Bacteria to Nanoparticles

Many microbes use metals as terminal electron acceptors, thereby reducing them and solubilizing solid surfaces including metal oxides (109), for example, *Shewanella* sp. decomposing magnetite nanocrystals (93). Electron transfer from bacteria to external electron acceptors is the basis for the microbial fuel cell concept (13, 32, 96, 97, 128, 149). That microbial communities, that is, mixtures of varying populations, can create fuel cells is informative to the general idea of electron transfer to nanoparticulates in the environment, suggesting that this could occur broadly with different taxa (97).

The decrease in metabolic activity of mammalian cells with photoactivation of quantum dots is suggestive of electron transfer from cellular energy-generating systems to the nanoparticles (143). This observation was exploited by Clarke et al. who modified CdSe quantum dots with a dopamine bioconjugate that shuttles electrons to the quantum dot from cells (36). Phototoxicity is observed but is controlled with the administration of antioxidants (36). The importance of this work is mainly that fluorescent biosensors are shown to be highly redox sensitive within cells, thus fluorescing maximally where cellular subcompartments are most oxidizing and thus electrons are flowing away from the quantum dot (36). However, this work was also applied to investigating the possibility of electron transfer between membranes of actively respiring bacteria and nanoparticles, as has been shown recently for TiO_2 and *E. coli* (108). Extensive outer labeling can be observed (Fig. 9.14a and b) of *E. coli* with dopamine-conjugated core–shell (ZnS) CdSe QDs. QDs also appear clustered away

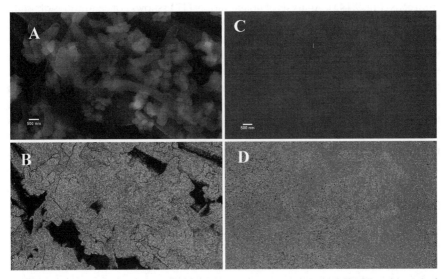

FIGURE 9.14   Association of dopamine-conjugated core–shell (ZnS) CdSe QDs with washed, exponential phase *Escherichia coli* cells as visualized with ESEM, showing (a) clusters of QDs (bright regions) around cells, (b) enhanced image variety (121) at the cell and QD cluster perimeters, (c) the relative absence of electron-dense regions in control (no QD) cells, and (d) lack of image variety (121) with control cells. (Priester et al. unpublished data).

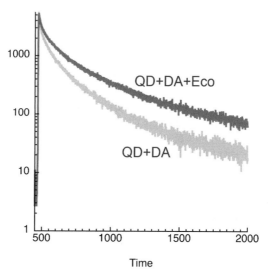

Time

FIGURE 9.15 Fluorescence decay (lifetime fluorescence) of solutions containing dopamine-conjugated green CdSe-ZnS core–shell quantum dots incubated (30 min) either with washed, exponential phase *Escherichia coli* cells (QD + DA + Eco) or without cells (QD + DA). Excitation was with a 100 fs laser pulse (5–13 mW, 400 nm) and the buffer is phosphate (Sigma) (J. Nadeau, S. Clarke, J. Priester, P. Holden, et al. unpublished data).

from cells (Fig. 9.14a). Analysis of image variety (121) shows concentrations of bright pixels, indicative of intense metal labeling, on the perimeters of cells (Fig. 9.14b), suggesting close physical association between cells and QDs. Lifetime fluorescence measurements (Fig. 9.15) using time-resolved single photon counting (TRSPC, i.e., laser spectcroscopy), similarly to Nadtochenko et al. (108), are suggestive of fluorescence enhancement for QDs associated with cells. This was not observed to a similar extent with either NaN_3-treated cells or with *P. aeruginosa*, nor with live, unstained cells (data not shown). While this preliminary data are suggestive of electron transfer between cells and dopamine-conjugated QDs, additional experimentation is needed to rule out confounding effects of other, perhaps media-associated, constituents.

9.4.3 Nanocrystal Formation in Cells

Bacteria are known to "biomineralize," that is, biotically form (as reviewed in Reference 162) a variety of iron and manganese minerals (52, 100) and many other minerals (9). These include intracellular inclusions called "magnetosomes" consisting of vesicular nanoparticulate magnetic magnetite (Fe_3O_4) or greigite (Fe_3S_4) whose function is to orient bacteria within a magnetic field (10). Variations in the mineral structure, magnetism, and intra- versus extracellular (i.e., on the envelope) formation location appear depending on the strain involved and growth conditions (e.g. 118). Other types of metals are certainly intracellularly packaged, for example, *Ralstonia* bacteria isolated from gold mines accumulate nanoparticulate gold in and on cell surfaces when grown with gold ions in solution (129).

When coexposed to Se and Cd ions, cells may take up Se as a nutrient and sequester or extrude Cd; often the toxic Cd is enclosed into "nanocrystals" of the cells' own making. Similarly, bacteria exposed to Zn may create ZnS nanocrystals (6, 7). These properties have been exploited to create semiconductor nanocrystals from microbial cultures (87), but the environmental impact on soil or water bacteria that can be either planktonic or biofilm, or the impact on bacterial–eukaryotic symbioses, has not been addressed.

Possibly because of decomposition, otherwise bare citrate-stabilized CdSe QDs appear to be as toxic to *P. aeruginosa* as Cd ions. However, coadministration of selenite and cadmium acetate to *P. aeruginosa* results in enhanced toxicity and increased cellular cadmium. Cd- and Se-rich regions (by STEM-EDS) in the cells implied there are interactions between the two metals. Also, XANES spectra were consistent with CdSe standards, and XRD patterns suggested that at least some wurtzite was present (unpublished data, Holden). While toxicity was not reported, the metal accumulation result is similar to that of Wang et al. (160), who isolated a marine *Pseudomonas* sp. that aerobically reduced sulfate in the presence of Cd(II) and accumulated CdS on the cell envelope in liquid culture. Coadministration of cadmium ions and selenite to mice resulted in reduced toxicity and lowered accumulation of Cd ions in the liver, presumably due to the interactions of the two elements (77). In sulfate-reducing bacterial biofilms, however, CdS never reaches cells and instead accumulates on the surface of the biofilm (164). Taken together, unpublished and published reports suggest that while bacteria can take up and break down QDs, they can also reform them. This point is of interest to biotechnologists in material science, but it is also an interesting context for considering fates of QDs in the environment and exactly what is "engineered" versus "natural" when it comes to nanoparticles.

9.5 MICROBIAL ECOLOGICAL IMPLICATIONS

Interactions between microbes and their environment may occur as a result of CdSe QD release. Essentially no specific research has been performed on this exact issue. However, implications to microbial communities of Cd and Se ions released by QD decomposition can be discussed generally. Also, and perhaps a new area of research obviated by concerns for engineered nanoparticles in the environment, there are many physicochemical relationships between nanoparticles, solution-phase nutrients, and water that could have interesting implications to microbes in nature. To the degree that similar interactions occur with natural, either biotic or geological, nanoparticles, such implications can be extrapolated to understanding roles of nanoparticles in natural ecosystems.

If transformation results in breakdown of metal nanoparticles and release of dissolved metals, toxicity to soil microbial communities could occur. Heavy metals interfere with C and N mineralization in even organic-amended soils (82), also causing microbial community shifts (82) that could have longer term effects on soil ecosystem function. Ecosystem effects could also be manifested by released metals interfering with extracellular enzymes, as demonstrated for a variety of trace elements affecting, for example, soil arylamidase (1) and cellulase (45) activities. The sorption of metals to bacterial cell surfaces will, as long as the bacteria are mobile, result in the movement of metals (168). If bacteria are strained or metal distribution does not favor bacterial metal binding, for example, due to pH changes or other factors influencing metal speciation, bacterially enhanced metal mobility may not be observed (168).

Indirect effects could stem not only from the high specific surface of nanoparticles but also from the nature of surfaces and surface defects. If crystalline, the atomic exposure on the nanoparticle surface is defined by the crystal structure, that is, the crystallographic planes that determine the morphology. The surface chemistry of atoms on the crystallographic planes or at the intersections of these planes can vary greatly. It should also be noted that the morphology of a given structure and composition can also vary as a function of the electrolyte, and that the nanoparticle composition can acquire the composition of the media in which the nanoparticles are formed. In addition, metal oxide surfaces can be defective and, consequently, highly reactive (21). For example, defects are the sites of reactions involving the chemisorption and dissociation of water (21). Atmospheric nanoparticles, depending on the exact chemistry, are very hydroscopic (41). Thin films of water, as occur even with reasonably high water contents in porous media dominated by fine-grained materials, are more ordered than thicker films (115), owing to surface interactions with water molecules (58). Bacteria and other microbes are highly sensitive to shifts in local water activity (20, 64, 119), even to the point of investing energy in making intracellular solutes to reduce osmotic water loss in dry environments (67). Water activity affects the rate of bacterial growth (133) and utilization of substrates (72). Enzyme–substrate interactions appear constrained when water is highly ordered (126). Implications of this to nanoparticle interactions in the environment could extrapolate to extracellular enzyme activity within water films around nanoparticles. In addition, direct nanoparticle –protein binding may also occur (105) that could limit nanoparticle dispersion outside cells (105), and also reduce the catalytic activity of extracellular enzymes (4, 102). Lastly, some manufactured nanoparticles are intended for water purification, that is, nutrient P removal (15) indicating that particles in the environment can have an effect on nutrient availability to microbes.

9.6 ENVIRONMENTAL IMPLICATIONS

Bacterial interactions with QDs could have implications for either QD or heavy metal transport over long distances in the environment. As bacteria can decompose quantum dots and thereby release Cd ions, they are essentially toxicologically activating (3) these solids. However, they also have the ability to sequester metal ions that can reduce their bioavailability to higher biota (Fig. 9.16).

In model laboratory pure culture systems, dissolved metals bind to bacterial cells (35) and to extracellular biomacromolecules including DNA (122) with the extent depending on the metal, bacterial strain, and growth conditions. Competition between

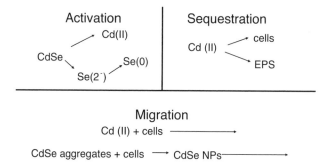

FIGURE 9.16 Three potential environmental implications of research to date regarding bacterial interactions with CdSe QDs. Activation (upper left) refers to the increase in toxicity of a substance, in this case resulting from Cd(II) ion release during bacterial decomposition of CdSe QDs. Sequestration (upper right) occurs when released Cd(II) binds with cells or to EPS associated with bacterial biofilms. Sequestration could limit migration. Migration (lower panel) could also be enhanced if Cd(II) ions or CdSe QDs are transported with cells when the metals or particles would otherwise stick to nonmoving inorganic surfaces in the environment. Lastly, migration of primary particles could be enhanced if aggregated CdSe QDs (e.g., Fig. 14a) distribute onto cells as primary particles (Fig. 14b) wherein cellular, and thus QD, transport could occur.

metals for specific binding sites on cell walls has been observed, as well as strong influences of pH on binding (35). In natural environments, the amount of dissolved metals sorbed to bacterial cell walls may be similar to that bound to organic matter (35). But the influence of bacterial species and growth phase, and even environmental factors such as ionic strength, may be less important to the total amount of metal sorption to bacterial cell surfaces than the total cells present in a system (17). A consequence of metal binding might be sequestration within biofilms if fluid flow is not occurring (122). On the contrary, metal binding to cells that are initially attached to surfaces, then subsequently mobilized and advected through porous media, could facilitate metal translocation in, for example, subsurface porous media (18).

Soil bacteria (48, 89, 155), microbial mats (11), fungi, and algae (158) sorb Cd ions and metalloids like Se ions (11). Coadministration of Cd(II) and sulfate resulted in cell-associated, nearly complete, removal of Cd(II) from solution by a marine *Pseudomonas* sp. (160). Preferential sorption onto microbial, as opposed to geological, surfaces can enhance metal migration in the saturated subsurface (22, 80). Besides intact cells, bacterial extracellular polymers (33, 80) including proteins (89), polysaccharides (14), and glycolipids (145) are also implicated in preferential metal binding, reduced microbial toxicity (14), and enhanced metal mobility (33, 66, 80). Binding constants for Cd ions to extracellular polymers are certainly higher than for sand (33) and soil (66) and, once desorbed from solid surfaces by extracellular ligands, Cd ions may be quite stably associated with, for example, bacterial glycolipid vesicles (25). High affinities between bacterial exopolymers (104), cells (156), and heavy metals could be applied toward environmental remediation. In biofilms, however, cells may compete with extracellular polymers for Cd(II) binding (137).

TABLE 9.1 Information Typically Used when Designing or Executing Conventional Pollutant Remediation and the Relative Availability of Such Information for Engineered Nanoparticles as Pollutants

Useful Information	Availability for Conventional Pollutants	Relative Availability for Engineered Nanoparticles
Structure and composition including primary particle and conjugates	Well-defined measurement approaches; databases available	Many are proprietary Objectives ill-defined Impurities. may control characteristics
Physiochemical properties (aqueous solubility, vapor pressure, K_{ow}, charge and oxidation state, radioactivity)	Compendia available Information relevant to environmental conditions	Mostly unknown for environmental purposes Other characteristics (surface area, charge, chemistry) may be most relevant
Mass transfer characteristics (diffusivity in water, air; sorptivity)	Measurements and correlations available Well-defined principles	Mostly unknown for environmental purposes
Environmental processes (speciation/complexation, reactivity, toxicity, biodegradability)	Characterized for many conditions Enzymatic pathways often established	Unknown for the most part

A constraint on subsurface delivery of metal nanoparticles for groundwater cleanup, such as solvent dechlorination, has been nanoparticle aggregation and subsequent loss of reactivity. Dispersion and reactivity can be maintained by coating nanoparticles with water-soluble organics (65). However, as in the case of hematite nanoparticles and alginate, the organic polymeric coating may enhance aggregation if solution ionic chemistry and strength favor the formation of nanoparticle-entrapped polymeric structure formation (34). In this situation, adhesion to bacteria, perhaps instead of adding a coating, might be advantageous for maintaining nanoparticle dispersion and facilitating nanoparticle delivery.

More generally, compared to conventional pollutants such as chlorinated solvents, hydrocarbons, and dissolved metals, little is known about the overall characteristics and processes governing engineered nanoparticles in the context of environmental cleanup (Table 9.1). Efforts to understand the role of bacteria in affecting nanoparticle pollution should be complemented by filling in data gaps identified in Table 9.1.

9.7 RESEARCH NEEDS

Research into QD interactions with bacteria is ongoing and there is much to be learned. Many questions that seem specific to QD-bacterial interactions easily apply to other nanoparticles and their interactions with bacteria. Questions that remain unanswered include

- *Size Effects:* What is the relationship between QD size and effects on cells? Are relationships additive across size ranges? Can the effects be predicted for mixtures of sizes and shapes by knowing the effects of individual particles? For both prokaryotes and eukaryotes, is there a threshold nanoparticle size beyond which the nanoparticles are not taken up into cells and cellular compartments? How does this vary with nanoparticle material? Can nanoparticles be considered as "safe" as their bulk counterparts when they are sufficiently large? Is there a threshold number of conjugates below which nanoparticles cannot enter a cell by receptor-mediated mechanisms? If so, how does this number vary according to nanoparticle size and/or material? Additionally, do ROS directly determine cellular uptake, irrespective of size, and how does this vary with conjugation?

- *Nanoparticle and Conjugate Stability:* How do bacteria influence nanoparticle and conjugate stability *outside* of the cell? In bacteria, how does extracellular nanoparticle stability affect intracellularization? Once inside cells, what processes affect nanoparticle stability *within* the cell?

- *QD Decomposition:* What is the nature of the QDs and QD material found inside bacteria that have taken up particles? Do bacteria break down or otherwise substantially alter nanoparticles? What are the breakdown products, and are they released from cells before or after cell death? What are the possible mechanisms of nanoparticle breakdown by bacteria? For example, do bacteria accelerate abiotic dissolution by either intracellularly sequestering breakdown products or by releasing chelating organic molecules that favor dissolution? Is bacterial decomposition of nanoparticles relatively fast and complete? What are the relative rates and extents of abiotic versus bacterially facilitated nanoparticle decomposition?

- *Decomposition Outcomes:* What are the consequences of nanoparticle breakdown and other interactions preceding breakdown by bacteria? For example, what are the products of bacterially mediated nanoparticle decomposition? Are the products more or less toxic than the particles themselves?

- *Cellular Responses:* How are nanoparticle uptake and stability affected by membrane potential in either prokaryotic or eukaryotic cells and by intracellular free radical scavengers? How is toxicity affected by these changes, and can it be prevented with specific pharmacological agents?

- *Electron Transfer:* Do bacteria exchange electrons with nanoparticles via either extracellular (e.g., phenazine type compounds) shuttles or cellular appendages, thus leading to their breakdown? What are the mechanisms and evidence for electron transfer between bacteria and nanoparticles, and what are the results? Does electron transfer happen either on, or in, the cytoplasmic membrane?

- *Variations from Nanoparticle Type:* How do transport, stability, and toxicity characteristics change with changing nanoparticle chemistry? Is there a "nanoparticle effect" common to nanoparticles of toxic (i.e., Cd-based) versus nontoxic (i.e., Ti-based) metals? Can nanoparticles be strategically designed to induce, and thereby study, specific cellular responses?

- *Natural Analogue Comparisons:* What are the effects of engineered QD nanoparticles relative to "natural," biotically synthesized nanoparticles of the same size? How do the differences in effects, if they exist, depend on the environment in which the nanoparticles reside? What is the relevance to this subject of natural nanoparticle formation, either geochemically or biogeochemically? Considering nanoparticles from minerals and even viral nanoparticles, what is the importance to bacterial evolution, physiology, and ecology? How can we use this information to extrapolate to engineered nanoparticle fates and effects in bacterial systems?

- *Toxicity End Points:* Are effects of QDs in the environment similar to that of Cd(II) ions? Do engineered QD nanoparticles become part of the soil matrix and effectively "blend in" to the soil mineralogy? Do they alter soil microbial function or structure quantifiably?

- *Biofilms:* How is the formation of bacterial biofilms altered in the presence of quantum dots? What is the toxicity of QDs to existing biofilms? How do QDs interact with bacterial biofilms? How do bacteria alter the fates of QDs in porous media?

- *Ecosystem Effects:* Other than breakdown, what other roles do bacteria have with regard to nanoparticles in the environment? What roles do nanoparticles, both natural and engineered, play in bacterial and natural ecosystem ecology? How can such roles best be understood?

9.8 CONCLUSIONS

Semiconductor nanocrystals (quantum dots or QDs) differ in important ways from bulk semiconductor materials. Their increased band gap means that they function as strong oxidizing and/or reducing agents, and their small size allows them to pass into living cells. Conjugation of biomolecules to the crystal surface can alter any or all of these properties. With their increased use in cellular labeling and in photovoltaics, quantum dots will enter the external environment. Consequences of releasing quantum dots into soil and water include obvious heavy metal toxicity and intact nanoparticle interactions with organisms. Bacteria factor importantly into the environmental fates and effects of quantum dots, because toxicity may affect their function within ecosystems; bacteria can also enhance heavy metal release through accelerating quantum dot breakdown. Interactions between bacteria and quantum dots in the environment could thus result in toxicological activation of the nanoparticles. Bacteria could also either limit transport by sequestering nanoparticles/metals, or enhance transport, for example, by either altering aggregate integrity or carrying biosorbed metals and nanoparticles through water and soil pores. In this chapter, quantum dots and potential physical associations with bacteria were introduced, biochemical interactions between quantum dots and bacteria were presented, environmental implications to quantum dot fate and transport were discussed, and outstanding research questions, some of which are applicable to the broader concern of nanoparticle interactions with environmental bacteria, were advanced. The overall message of this chapter is that quantum dots can have profound effects on bacteria; likewise, bacteria are likely to significantly influence quantum dot fates and effects in environmental systems.

ACKNOWLEDGMENTS

Research support was in part from the U.S. EPA STAR program (R831712 and R833323), the University of California Toxic Substances Research and Training Program: Lead Campus Program in Nanotoxicology, and by the Office of Science (BER), U.S. Department of Energy, Grant No. DE-FG02-05ER63949. XANES was performed by Dr. Sam Webb at the Stanford Synchrotron Radiation Laboratory, a national user facility operated by Stanford University on behalf of the U.S. Department of Energy, Office of Basic Energy Sciences. The SSRL Environmental Remediation Science Program is supported by the Department of Energy Office of Biological and Environmental Research. JLN acknowledges salary support from the Canada Research Chairs (CRC) program, the National Science and Engineering Research Council (NSERC) NanoIP and Discovery research grants, and the Canadian Space Agency CARN program. TCSPC data were collected with assistance from the Steve Bradforth Lab (USC). Several of the electron micrographs were acquired via the assistance of Randy Mielke (Jet Propulsion Laboratory) and Jose Saleta (MEIAF, UCSB). Peter Stoimenov (UCSB) contributed to portions of the research described in this chapter.

REFERENCES

1. Acosta-Martinez, V., Tabatabai, M.A. Arylamidase activity in soils: effect of trace elements and relationships to soil properties and activities of amidohydrolases. *Soil Bio Biochem* 2001;33:17–23.

2. Aizenberg, J., Lambert, G., Weiner, S., Addadi, L. Factors involved in the formation of amorphous and crystalline calcium carbonate: a study of an ascidian skeleton. *J Am Chem Soc* 2002;124:32–39.

3. Alexander, M. *Biodegradation and Bioremediation*. San Diego: Academic Press;1994.

4. Allison, S.D., Jastrow, J.D. Activities of extracellular enzymes in physically isolated fractions of restored grassland soils. *Soil Biol Biochem* 2006;38:3245–3256.

5. Anatasio, C., Martin, S.T. Atmospheric nanoparticles. In: Banfield, J.F., Navrotsky, A., editors. *Nanoparticles and the Environment*, vol. 44. Washington, D.C.: mineralogical Society of America; 2001. pp. 293–349.

6. Bae, W., Abdullah, R., Mehra, R.K. Cysteine-mediated synthesis of CdS bionanocrystallites. *Chemosphere* 1998;37:363–385.

7. Bae, W.O., Abdullah, R., Henderson, D., Mehra, R.K. Characteristics of glutathione-capped ZnS nanocrystallites. *Biochem Biophys Res Commun* 1997;237:16–23.

8. Ballou, B., Lagerholm, B.C., Ernst, L.A., Bruchez, M.P., Waggoner, A.S. Noninvasive imaging of quantum dots in mice. *Bioconjug Chem* 2004;15:79–86.

9. Banfield, J.F., Zhang, H.Z. Nanoparticles in the environment. In: Banfield, J.F., Navrotsky, A., editors. *Nanoparticles and the Environment*. Washington, D.C.: Mineralogical Society of America; 2001. pp. 1–58.

10. Bazylinski, D.A., Frankel, R.B. Magnetosome formation in prokaryotes. *Nature Rev Microbiol* 2004;2:217–230.

11. Bender, J., Lee, R.F., Phillips, P. Uptake and transformation of metals and metalloids by microbial mats and their use in bioremediation. *J Ind Microbiol* 1995;14:113–118.

12. Beveridge, T.J., Makin, S.A., Kadurugamuwa, J.L., Li, Z.S. Interactions between biofilms and the environment. *Fems Microbiol Revi* 1997;20:291–303.

13. Biffinger, J.C., Ray, R., Little, B., Ringeisen, B.R. Diversifying biological fuel cell designs by use of nanoporous filters. *Environ Sci Technol* 2007;41:1444–1449.

14. Bitton, G., Freihofer, V. Influence of extracellular polysaccharides on toxicity of copper and cadmium toward *Klebsiella aerogenes*. *Microb Ecol* 1978;4:119–125.

15. Blaney, L.M., Cinar, S., SenGupta, A.K. Hybrid anion exchanger for trace phosphate removal from water and wastewater. *Water Res* 2007;41:1603–1613.

16. Borm, P., Klaessig, F.C., Landry, T.D., Moudgil, B., Pauluhn, J., Thomas, K., Trottier, R., Wood, S. Research strategies for safety evaluation of nanomaterials, Part V: Role of dissolution in biological fate and effects of nanoscale particles. *Toxicol Sci* 2006;90:23–32.

17. Borrok, D., Turner, B.F., Fein, J.B. A universal surface complexation framework for modeling proton binding onto bacterial surfaces in geologic settings. *Am J Sci* 2005;305:826–853.

18. Boult, S., Hand, V.L., Vaughan, D.J. Microbial controls on metal mobility under the low nutrient fluxes found throughout the subsurface. *Sci Total Environ* 2006;372:299–305.

19. Brodowski, S., Amelung, W., Haumaier, L., Abetz, C., Zech, W. Morphological and chemical properties of black carbon in physical soil fractions as revealed by scanning electron microscopy and energy-dispersive X-ray spectroscopy. *Geoderma* 2005;128:116–129.

20. Brown, A.D. *Microbial Water Stress Physiology*. Chichester: John Wiley & Sons; 1990. p. 183.

21. Brown, G.E., Henrich, V.E., Casey, W.H., Clark, D.L., Eggleston, C., Felmy, A., Goodman, D.W., Gratzel, M., Maciel, G., McCarthy, M.I., Nealson, K.H., Sverjensky, D.A., Toney, M.F., Zachara, J.M. Metal oxide surfaces and their interactions with aqueous solutions and microbial organisms. *Chem Rev* 1999;99:77–174.

22. Brown, R.L., Bowman, R.S., Kieft, T.L. Microbial effects on nickel and cadmium sorption and transport in volcanic tuff. *J Environ Qual* 1994;23:723–729.

23. Bruchez, M., Moronne, M., Gin, P., Weiss, S., Alivisatos, A.P. Semiconductor nanocrystals as fluorescent biological labels. *Science* 1998;281:2013–2016.

24. Burda, C., Green, T.C., Link, S., El-Sayed, M.A. Electron shuttling across the interface of CdSe nanoparticles monitored by femtosecond laser spectroscopy. *J Phys Chem B* 1999;103:1783–1788.

25. Champion, J.T., Gilkey, J.C., Lamparski, H., Retterer, J., Miller, R.M. Electron microscopy of rhamnolipid (biosurfactant) morphology—effects of pH, cadmium, and octadecane. *J Colloid Interface Sci* 1995;170:569–574.

26. Chan, W.C.W., Maxwell, D.J., Gao, X.H., Bailey, R.E., Han, M.Y., Nie, S.M. Luminescent quantum dots for multiplexed biological detection and imaging. *Curr Opini Biotechnol* 2002;13:40–46.

27. Chan, W.C.W., Nie, S. Quantum dot bioconjugates for ultrasensitive nonisotopic detection. *Science* 1998;281:2016–2018.

28. Chandrasekhar, S., Satyanarayana, K.G., Pramada, P.N., Raghavan, P., Gupta, T.N. Processing, properties and applications of reactive silica from rice husk—an overview. *J Mater Science* 2003;38:3159–3168.

29. Chang, J.S., Chang, K.L.B., Hwang, D.F., Kong, Z.L. *In vitro* cytotoxicitiy of silica nanoparticles at high concentrations strongly depends on the metabolic activity type of the cell line. *Environ Sci & Technol* 2007;41:2064–2068.

30. Characklis, W.G., Marshall, K.C. Biofilms: a basis for an interdisciplinary approach, In: Characklis W.G., Marshall K.C., editors. *Biofilms*, New York: John Wiley & Sons, Inc.; 1990. pp. 3–15.

31. Charron, A., Harrison, R.M. Primary particle formation from vehicle emissions during exhaust dilution in the roadside atmosphere. *Atmospheric Environ* 2003;37: 4109–4119.

32. Chaudhuri, S.K., Lovley, D.R. Electricity generation by direct oxidation of glucose in mediatorless microbial fuel cells. *Nat Biotech* 2003;21:1229–1232.

33. Chen, J.-H., Lion, L.W., Ghiorse, W.C., Shuler, M.L. Mobilization of adsorbed cadmium and lead in aquifer material by bacterial extracellular polymers. *Water Res* 1995;29:421–430.

34. Chen, K.L., Mylon, S.E., Elimelech, M. Aggregation kinetics of alginate-coated hematite nanoparticles in monovalent and divalent electrolytes. *Environ Sci Technol* 2006;40: 1516–1523.

35. Claessens, J., Van Cappellen, P. Competitive binding of Cu2+ and Zn2+ to live cells of Shewanella putrefaciens. *Environ Sci Technol* 2007;41:909–914.

36. Clarke, S.J., Hollmann, C.A., Zhang, Z.J., Suffern, D., Bradforth, S.E., Dimitrijevic, N. M., Minarik, W.G., Nadeau, J.L. Photophysics of dopamine-modified quantum dots and effects on biological systems. *Nature Mater* 2006;5:409–417.

37. Clever, H.L., Derrick, E.M., Johnson, S.A. The solubility of some sparingly soluble salts of zinc and cadmium in water and in aqueous electrolyte solutions. *J Phys Chem Ref Data* 1992;21:941–1044.

38. Comolli, L.R., Kundmann, M., Downing, K.H. Characterization of intact subcellular bodies in whole bacteria by cryo-electron tomography and spectroscopic imaging. *J Microsc* 2006;223:40–52.

39. Costerton, J.W., Cheng, K.J., Geesey, G.G., Ladd, T.I., Nickel, J.C., Dasgupta, M., Marrie, T.J. *Bacterial Biofilms in Nature and Disease* 1987.

40. Costerton, J.W., Lewandowski, Z., De Beer, D., Caldwell, D., Korber, D., James, G. Biofilms, the customized microniche. *J Bacteriol* 1994;176:2137–2142.

41. Cubison, M.J., Alfarra, M.R., Allan, J., Bower, K.N., Coe, H., McFiggans, G.B., Whitehead, J.D., Williams, P.I., Zhang, Q., Jimenez, J.L., Hopkins, J., Lee, J. The characterisation of pollution aerosol in a changing photochemical environment. *Atmos Chem Phys* 2006;6:5573–5588.

42. Cunningham, A.B., Characklis, W.G., Abedeen, F., Crawford, D. Influence of biofilm accumulation on porous media hydrodynamics. *Environ Sci Technol* 1991;25:1305–1311.

43. Danese, P.N., Pratt, L.A., Kolter, R. Exopolysaccharide production is required for development of Escherichia coli K-12 biofilm architecture. *J Bacteriol* 2000;182:3593–3596.

44. Debeer, D., Stoodley, P., Lewandowski, Z. Liquid flow in heterogeneous biofilms. *Biotechnol Bioeng* 1994;44:636–641.

45. Deng, S.P., Tabatabai, M.A. Cellulase activity of soils-effect of trace elements. *Soil Biol Biochem* 1995;27:977–979.

46. Derfus, A.M., Chan, W.C.W., Bhatia, S.N. Probing the cytotoxicity of semiconductor quantum dots. *Nano Lett* 2004;4:11–18.

47. DiChristina, T.J., Fredrickson, J.K., Zachara, J.M. Enzymology of electron transport: energy generation with geochemical consequences. In: Banfield, J.F., Cervini-Silva, J., editors. *Molecular Geomicrobiology*, vol. 59. Washington, D.C: Mineralogical Society of America; 2005. pp. 27–52.

48. Falla, J., Block, J.C. Binding of Cd^{2+}, Ni^{2+}, Cu^{2+}, and Zn^{2+} by isolated enveloped of *Pseudomonas fluorescens*. *FEMS Microbiol Lett* 1993;108:347–352.

49. Ferrari, B.C., Binnerup, S.J., Gillings, M. Microcolony cultivation on a soil substrate membrane system selects for previously uncultured soil bacteria. *Appl Environ Microbiol* 2005;71:8714–8720.

50. Ferris, F.G., Schultze, S., Witten, T.C., Fyfe, W.S., Beveridge, T.J. Metal interactions with microbial biofilms in acidic and neutral pH environments. *Appl Environ Microbiol* 1989;55:1249–1257.

51. Fountaine, T.J., Wincovitch, S.M., Geho, D.H., Garfield, S.H., Pittaluga, S. Multispectral imaging of clinically relevant cellular targets in tonsil and lymphoid tissue using semiconductor quantum dots. *Modern Pathol* 2006;19:1–11.

52. Frankel, R.B., Bazylinski, D.A. Biologically induced mineralization by bacteria. In: Dove, P.M., de Yoreo, J.J., Weiner, S., editors. *Biomineralization*, Washington, D.C.: Mineralogical Society of America; 2003. pp. 95–114.

53. Frankel, R.B., Zhang, J.P., Bazylinski, D.A. Single magnetic domains in magnetotactic bacteria. *J Geophys Res-Solid Earth* 1998;103:30601–30604.

54. Gao, X.H., Chan, W.C.W., Nie, S.M. Quantum-dot nanocrystals for ultrasensitive biological labeling and multicolor optical encoding. *J Biomed Opt* 2002;7:532–537.

55. Gerion, D., Parak, W.J., Williams, S.C., Zanchet, D., Micheel, C.M., Alivisatos, A.P. Sorting fluorescent nanocrystals with DNA. *J Am Chem Soc* 2002;124:7070–7074.

56. Giere, R., Blackford, M., Smith, K. TEM study of PM2.5 emitted from coal and tire combustion in a thermal power station. *Environ Sci Technol* 2006;40:6235–6240.

57. Gilbert, B., Banfield, J.F. Molecular-scale processes involving nanoparticulate minerals in biogeochemical systems. In: Banfield, J.F., Cervini-Silva, J., Nealson, K.M., editors. *Molecular Geomicrobiology*, Washington, D.C.: Mineralogical Society of America; 2005. pp. 109–155.

58. Goertz, M.P., Houston, J.E., Zhu, X.Y. Hydrophilicity and the viscosity of interfacial water. *Langmuir* 2007;23:5491–5497.

59. Gu, H.W., Ho, P.L., Tong, E., Wang, L., Xu, B. Presenting vancomycin on nanoparticles to enhance antimicrobial activities. *Nano Lett* 2003;3:1261–1263.

60. Guinness, E.A., Arvidson, R.E., Jolliff, B.L., Seelos, K.D., Seelos, F.P., Ming, D.W., Morris, R.V., Graff, T.G. Hyperspectral reflectance mapping of cinder cones at the summit of Mauna Kea and implications for equivalent observations on Mars. *J Geophys Res-Planet* 2007;112:E08511.

61. Hahn, M.A., Tabb, J.S., Krauss, T.D. Detection of single bacterial pathogens with semiconductor quantum dots. *Analy Chem* 2005;77:4861–4869.

62. Han, M.Y., Gao, X.H., Su, J.Z., Nie, S. Quantum-dot-tagged microbeads for multiplexed optical coding of biomolecules. *Nature Biotechnol* 2001;19:631–635.

63. Hardman, R. A toxicologic review of quantum dots: toxicity depends on physicochemical and environmental factors. *Environ Health Perspect* 2006;114:165–172.

64. Harris, R.F. Effect of water potential on microbial growth and activity. In: Parr, J.F., Gardner, W.R., Elliott, L.F., editors. *Water Potential Relations in Soil Microbiology, SSSA Special Publication Number 9*, Madison: Soil Science Society of America; 1981. pp. 23–95.

65. He, F., Zhao, D.Y., Liu, J.C., Roberts, C.B. Stabilization of Fe-Pd nanoparticles with sodium carboxymethyl cellulose for enhanced transport and dechlorination of trichloroethylene in soil and groundwater. *Ind Eng Chem Res* 2007;46:29–34.

66. Herman, D.C., Artiola, J.F., Miller, R.M. Removal of cadmium, lead, and zinc from soil by a rhamnolipid biosurfactant. *Environ Sci Technol* 1995;29:2280–2285.

67. Hershkovitz, N., Oren, A., Cohen, Y. Accumulation of trehalose and sucrose in cyanobacteria exposed to matric water-stress. *Appl Environ Microbiol* 1991;57:645–648.

68. Hersman, L.E. The Role of siderophores in iron oxide dissolution. In: Lovley, D.R., editor. *Environmental Microbe-Metal Interactions*, Washington, D.C.: ASM Press; 2000.

69. Hockin, S.L., Gadd, G.M. Linked redox precipitation of sulfur and selenium under anaerobic conditions by sulfate-reducing bacterial biofilms. *Appl Environ Microbiol* 2003;69:7063–7072.

70. Hoffman, D.J. Role of selenium toxicity and oxidative stress in aquatic birds. *Aquat Toxicol* 2002;57:11–26.

71. Holden, P.A. Biofilms in unsaturated environments. In: Doyle, R.J., editor. Methods of Enzymology, vol. 337. San Diego: Academic Press; 2001. pp. 125–143.

72. Holden, P.A., Halverson, L.J., Firestone, M.K. Water stress effects on toluene biodegradation by *Pseudomonas putida*. *Biodegradation* 1997;8:143–151.

73. Holden, P.A., Hunt, J.R., Firestone, M.K. Toluene diffusion and reaction in unsaturated *Pseudomonas putida* mt-2 biofilms. *Biotechnol Bioeng* 1997;56:656–670.

74. Holden, P.A., LaMontagne, M.G., Bruce, A.K., Miller, W.G., Lindow, S.E. Assessing the role of *Pseudomonas aeruginosa* surface-active gene expression in hexadecane biodegradation in sand. *Appl Environ Microbiol* 2002;68:2509–2518.

75. Iler, R.K. *The Chemistry of Silica: Solubility, Polymerization, Colloid and Surface Properties, and Biochemistry*. New York: John Wiley & Sons; 1979.

76. Jaiswal, J.K., Mattoussi, H., Mauro, J.M., Simon, S.M. Long-term multiple color imaging of live cells using quantum dot bioconjugates. *Nature Biotechnol* 2003;21:47–51.

77. Jamba, L., Nehru, B., Bansal, M.P. Selenium supplementation during cadmium exposure: changes in antioxidant enzymes and the ultrastructure of the kidney. *J Trace Elem Experimen Med* 1997;10:233–242.

78. Jarup, L. Cadmium overload and toxicity. *Nephrology Dialysis Transplantation* 2002;17:35–39.

79. Jarup, L., Berglund, M., Elinder, C.G., Nordberg, G., Vahter, M. Health effects of cadmium exposure—a review of the literature and a risk estimate. *Scandinavian J Work Environ Health* 1998;24:240–240.

80. Jenkins, M.B., Chen, J.-H., Kadner, D.J., Lion, L.W. Methanotrophic bacteria and facilitated transport of pollutants in aquifer material. *Appl Environ Microbiol* 1994;60:3491–3498.

81. Jiang, C.W., Green, M.A. Silicon quantum dot superlattices: modeling of energy bands, densities of states, and mobilities for silicon tandem solar cell applications. *J Appl Phys* 2006;99:084506.

82. Kao, P.H., Huang, C.C., Hseu, Z.Y. Response of microbial activities to heavy metals in a neutral loamy soil treated with biosolid. *Chemosphere* 2006;64:63–70.

83. Kemmling, A., Kamper, M., Flies, C., Schieweck, O., Hoppert, M. Biofilms and extracellular matrices on geomaterials. *Environ Geol* 2004;46:429–435.

84. Kirchner, C., Liedl, T., Kudera, S., Pellegrino, T., Javier, A.M., Gaub, H.E., Stolzle, S., Fertig, N., Parak, W.J. Cytotoxicity of colloidal CdSe and CdSe/ZnS nanoparticles. *Nano Lett* 2005;5:331–338.

85. Kloepfer, J.A., Mielke, R.E., Nadeau, J.L. Uptake of CdSe and CdSe/ZnS quantum dots into bacteria via purine-dependent mechanisms. *Appl Environ Microbiol* 2005;71:2548–2557.

86. Kloepfer, J.A., Mielke, R.E., Wong, M.S., Nealson, K.H., Stucky, G., Nadeau, J.L. Quantum dots as strain- and metabolism-specific microbiological labels. *Appl Environ Microbiol* 2003;69:4205–4213.

87. Kowshik, M., Deshmukh, N., Vogel, W., Urban, J., Kulkarni, S.K., Paknikar, K.M. Microbial synthesis of semiconductor CdS nanoparticles, their characterization, and their use in the fabrication of an ideal diode. *Biotechnol Bioeng* 2002;78:583–588.

88. Kupiainen, K., Klimont, Z. Primary emissions of fine carbonaceous particles in Europe. *Atm Environ* 2007;41:2156–2170.

89. Kurek, E., Francis, A.J., Bollag, J.M. Immobilization of cadmium by microbial extracellular products. *Arch Environ Contam Toxicol* 1991;21:106–111.

90. Kyriacou, S.V., Brownlow, W.J., Xu, X.H.N. Using nanoparticle optics assay for direct observation of the function of antimicrobial agents in single live bacterial cells. *Biochemistry* 2004;43:140–147.

91. Lagerholm, B.C., Wang, M.M., Ernst, L.A., Ly, D.H., Liu, H.J., Bruchez, M.P., Waggoner, A.S. Multicolor coding of cells with cationic peptide coated quantum dots. *Nano Lett* 2004;4:2019–2022.

92. Lane, T.W., Morel, F.M.M. A biological function for cadmium in marine diatoms. *Proc Natl Acad Sci USA* 2000;97:4627–4631.

93. Larsen, L., Little, B., Nealson, K.H., Ray, R., Stone, A., Tian, J. Manganite reduction by *Shewanella putrefaciens* MR-4. *Am Mineral* 1988;83:1564–1572.

94. Larson, D.R., Zipfel, W.R., Williams, R.M., Clark, S.W., Bruchez, M.P., Wise, F.W., Webb, W.W. Water-soluble quantum dots for multiphoton fluorescence imaging *in vivo*. *Science* 2003;300:1434–1436.

95. Liermann, L.J., Guynn, R.L., Anbar, A., Brantley, S.L. Production of a molybdophore during metal-targeted dissolution of silicates by soil bacteria. *Chemi Geol* 2005;220:285–302.

96. Liu, H., Logan, B.E. Electricity generation using an air-cathode single chamber microbial fuel cell in the presence and absence of a proton exchange membrane. *Environ Sci Technol* 2004;38:4040–4046.

97. Logan, B.E., Regan, J.M. Electricity-producing bacterial communities in microbial fuel cells. *Trends Microbiol* 2006;14:512–518.

98. Lovric, J., Bazzi, H.S., Cuie, Y., Fortin, G.R.A., Winnik, F.M., Maysinger, D. Differences in subcellular distribution and toxicity of green and red emitting CdTe quantum dots. *J Mol Med-Jmm* 2005;83:377–385.

99. Lovric, J., Cho, S.J., Winnik, F.M., Maysinger, D. Unmodified cadmium telluride quantum dots induce reactive oxygen species formation leading to multiple organelle damage and cell death. *Chem Biol* 2005;12:1227–1234.

100. Lowenstam, H.A., Weiner, S. *On Biomineralization.* New York: Oxford University Press; 1989.

101. Mao, C., Solis, D.J., Reiss, B.D., Kottmann, S.T., Sweeney, R.Y., Hayhurst, A., Georgiou, G., Iverson, B., Belcher, A.M. Virus-based toolkit for the directed synthesis of magnetic and semiconducting nanowires. *Science* 2004;303:213–217.

102. Marx, M.C., Kandeler, E., Wood, M., Wermbter, N., Jarvis, S.C. Exploring the enzymatic landscape: distribution and kinetics of hydrolytic enzymes in soil particle-size fractions. *Soil Biol Biochem* 2005;37:35–48.

103. Mattoussi, H., Mauro, J.M., Goldman, E.R., Anderson, G.P., Sundar, V.C., Mikulec, F. V., Bawendi, M.G. Self-assembly of CdSe-ZnS quantum dot bioconjugates using an engineered recombinant protein. *J Am Chem Soc* 2000;122:12142–12150.

104. Miller, R.M. Biosurfactant-facilitated remediation of metal-contaminated soils. *Environ Health Perspect* 1995;103:59–62.

105. Moreau, J.W., Weber, P.K., Martin, M.C., Gilbert, B., Hutcheon, I.D., Banfield, J.F. Extracellular proteins limit the dispersal of biogenic nanoparticles. *Science* 2007;316:1600–1603.

106. Morris, R.V., Golden, D.C., Ming, D.W., Shelfer, T.D., Jorgensen, L.C., Bell, J.F., Graff, T.G., Mertzman, S.A. Phyllosilicate-poor palagonitic dust from Mauna Kea Volcano (Hawaii): a mineralogical analogue for magnetic Martian dust? *J Geophys Res-Planet* 2001;106:5057–5083.

107. Nadeau, J.L., Perreault, N.N., Niederberger, T.D., Whyte, L.G., Sun, H.J., Leon, R. Fluorescence microscopy as a tool for in situ life detection. *Astrobiology* 2008;8(3).

108. Nadtochenko, V., Denisov, N., Sarkisov, O., Gumy, D., Pulgarin, C., Kiwi, J. Laser kinetic spectroscopy of the interfacial charge transfer between membrane cell walls of *E-coli* and TiO$_2$. *J Photochem Photobiol A* 2006;181:401–407.

109. Nealson, K.H., Belz, A., McKee, B. Breathing metals as a way of life: geobiology in action. *Antonie Van Leeuwenhoek Int J Gen Mol Microbiol* 2002;81:215–222.

110. Nealson, K.H., Stahl, D.A. Microorganisms and biogeochemical cycles: what can we learn from layered microbial communities? In: Banfield, J.F., Nealson, K.H., editors. Geomicrobiology: Interactions Between Microbes and Minerals, vol. 35. Washington, D. C.: Mineralogical Society of America; 1997. pp. 5–34.

111. Nies, D.H. Resistance to cadmium, cobalt, zinc, and nickel in microbes. *Plasmid* 1992;27:17–28.

112. Nozik, A.J. Quantum dot solar cells. *Physica E* 2002;14:115–120.

113. Oberdorster, G., Maynard, A., Donaldson, K., Castranova, V., Fitzpatrick, J., Ausman, K., Carter, J., Karn, B., Kreyling, W., Lai, D., Olin, S., Monteiro-Riviere, N., Warheit, D., Yang, H. Principles for characterizing the potential human health effects from exposure to nanomaterials: elements of a screening strategy. *Parti Fibre Toxicol* 2005;2:8.

114. Oremland, R.S., Herbel, M.J., Blum, J.S., Langley, S., Beveridge, T.J., Ajayan, P.M., Sutto, T., Ellis, A.V., Curran, S. Structural and spectral features of selenium nanospheres produced by se-respiring bacteria. *Appl Environ Microbiol* 2004;70:52–60.

115. Papendick, R.I., Campbell, G.S. Theory and measurement of water potential. In: Parr, J. F., Gardner, W.R., Elliot, L.F., editors. Water Potential Relations in Soil Microbiology, SSSA Special Publication Number 9, Madison: *Soil Sci Soci Am*; 1981. pp. 1–22.

116. Parak, W.J., Pellegrino, T., Plank, C. Labeling of cells with quantum dots. *Nanotechnology* 2005;16:R9–R25.

117. Paunesku, T., Rajh, T., Wiederrecht, G., Maser, J., Vogt, S., Stojicevic, N., Protic, M., Lai, B., Oryhon, J., Thurnauer, M., Woloschak, G. Biology of TiO$_2$-oligonucleotide nanocomposites. *Nature Mat* 2003;2:343–346.

118. Posfai, M., Moskowitz, B.M., Arato, B., Schuler, D., Flies, C., Bazylinski, D.A., Frankel, R.B. Properties of intracellular magnetite crystals produced by desulfovibrio magneticus strain RS-1. *Earth Planet Sci Lett* 2006;249:444–455.

119. Potts, M. Desiccation tolerance of prokaryotes. *Microbiolo Rev* 1994;58:755–805.

120. Poulsen, N., Sumper, M., Kroger, N. Biosilica formation in diatoms: characterization of native silaffin-2 and its role in silica morphogenesis. *Proc Nat Acad Sci USA* 2003;100:12075–12080.

121. Priester, J.H., Horst, A.M., Van De Werfhorst, L.C., Saleta, J.L., Mertes, L.A.K., Holden, P.A. Enhanced visualization of microbial biofilms by staining and environmental scanning electron microscopy. *J Microbiol Methods* 2007;68:577–587.

122. Priester, J.H., Olson, S.G., Webb, S.M., Neu, M.P., Hersman, L.E., Holden, P.A. Enhanced exopolymer production and chromium stabilization in *Pseudomonas putida* unsaturated biofilms. *Appl Environ Microbiol* 2006;72:1988–1996.

123. Qu, L.H., Peng, X.G. Control of photoluminescence properties of CdSe nanocrystals in growth. *J Am Chem Soc* 2002;124:2049–2055.

124. Qu, L.H., Peng, Z.A., Peng, X.G. Alternative routes toward high quality CdSe nanocrystals. *Nano Lett* 2001;1:333–337.

125. Rajh, T., Saponjic, Z., Liu, J.Q., Dimitrijevic, N.M., Scherer, N.F., Vega-Arroyo, M., Zapol, P., Curtiss, L.A., Thurnauer, M.C. Charge transfer across the nanocrystalline-DNA interface: probing DNA recognition. *Nano Lett* 2004;4:1017–1023.

126. Ramos, A.S.F., Techert, S. Influence of the water structure on the acetylcholinesterase efficiency. *Biophy J* 2005;89:1990–2003.

127. Rancourt, D.G. Magnetism of earth, planetary, and environmental nanomaterials. In: Banfield, J.F., Navrotsky, A., editors. Nanoparticles and the Environment, vol. 44. Washington, D.C.: Mineralogical Society of America; 2001. pp. 217–292.

128. Reimers, C.E., Tender, L.M., Fertig, S., Wang, W. Harvesting energy from the marine sediment-water interface. *Environ Sci Technol* 2001;35:192–195.

129. Reith, F., Rogers, S.L., McPhail, D.C., Webb, D. Biomineralization of gold: biofilms on bacterioform gold. *Science* 2006;313:233–236.

130. Rogach, A.L., Nagesha, D., Ostrander, J.W., Giersig, M., Kotov, N.A. "Raisin bun"-type composite spheres of silica and semiconductor nanocrystals. *Chemistry of Materials* 2000;12:2676–2685.

131. Sambhy, V., MacBride, M.M., Peterson, B.R., Sen, A. Silver bromide nanoparticle/ polymer composites: dual action tunable antimicrobial materials. *J Am Chem Soc* 2006;128:9798–9808.

132. Schlesinger, W.H. *Biogeochemistry: An Analysis of Global Change.* San Diego: Academic Press; 1997.

133. Schnell, S., King, G.M. Responses of methanotrophic activity in soils and cultures to water stress. *Appl Environ Microbiol* 1996;62:3203–3209.

134. Schooling, S.R., Beveridge, T.J. Membrane vesicles: an overlooked component of the matrices of biofilms. *J Bacteriol* 2006;188:5945–5957.

135. Schultze-Lam, S., Fortin, D., Davis, B.S., Beveridge, T.J. Mineralization of bacterial surfaces. *Chem Geol* 1996;132:171–181.

136. Smith, A.M., Duan, H.W., Rhyner, M.N., Ruan, G., Nie, S.M. A systematic examination of surface coatings on the optical and chemical properties of semiconductor quantum dots. *Phys Chem Chem Phys* 2006;8:3895–3903.

137. Spath, R., Flemming, H.C., Wuertz, S. Sorption properties of biofilms. *Water Science and Technology* 1998;37:207–210.

138. Steinberger, R.E., Allen, A.R., Hansma, H.G., Holden, P.A. Elongation correlates with nutrient deprivation in unsaturated *Pseudomonas aeruginosa* biofilms. *Microb Ecol* 2002;43:416–423.

139. Steinberger, R.E., Holden, P.A. Extracellular DNA in single- and multiple-species unsaturated biofilms. *Appl Environ Microb* 2005;71:5404–5410.

140. Steinberger, R.E., Holden, P.A. Macromolecular composition of unsaturated *Pseudomonas aeruginosa* biofilms with time and carbon source. *Biofilms* 2004;1:37–47.

141. Stoodley, P., DeBeer, D., Lewandowski, Z. Liquid flow in biofilm systems. *Appl Environ Micro* 1994;60:2711–2716.

142. Su, B., Fermin, D.J., Abid, J.P., Eugster, N., Girault, H.H. Adsorption and photo-reactivity of CdSe nanoparticles at liquid|liquid interfaces. *J Electroanal Chem* 2005;583:241–247.

143. Sun, Y.H., Liu, Y.S., Vernier, P.T., Liang, C.H., Chong, S.Y., Marcu, L., Gundersen, M. A. Photostability and pH sensitivity of CdSe/ZnSe/ZnS quantum dots in living cells. *Nanotechnology* 2006;17:4469–4476.

144. Sweeney, R.Y., Mao, C.B., Gao, X.X., Burt, J.L., Belcher, A.M., Georgiou, G., Iverson, B.L. Bacterial biosynthesis of cadmium sulfide nanocrystals. *Chem Biol* 2004;11:1553–1559.

145. Tan, H., Champion, J.T., Artiola, J.F., Brusseau, M.L., Miller, R.M. Complexation of cadmium by a rhamnolipid biosurfactant. *Environ Sci Technol* 1994;28:2402–2406.

146. Tebo, B.M., Ghiorse, W.C., van Waasbergen, L.G., Siering, P.L., Caspi, R. Bacterially mediated mineral formation: insights into manganese(II) oxidation from molecular genetic and biochemical studies. In: Banfield, J.F., Nealson, K.H., editors. Geomicrobiology: Interactions Between Microbes and Minerals, vol. 35. Washington, D.C.: Mineralogical Society of America; 1997. pp. 225–266.

147. Tebo, B.M., Johnson, H.A., McCarthy, J.K., Templeton, A.S. Geomicrobiology of manganese(II) oxidation. *Trends Microbiol* 2005;13:421–428.

148. Teitzel, G.M., Parsek, M.R. Heavy metal resistance of biofilm and planktonic *Pseudomonas aeruginosa*. *Appl Environ Microbiol* 2003;69:2313–2320.

149. Tender, L.M., Reimers, C.E., Stecher, H.A., Holmes, D.E., Bond, D.R., Lowy, D.A., Pilobello, K., Fertig, S.J., Lovley, D.R. Harnessing microbially generated power on the seafloor. *Nature Biotechnol* 2002;20:821–825.

150. Thill, A., Zeyons, O., Spalla, O., Chauvat, F., Rose, J., Auffan, M., Flank, A.M. Cytotoxicity of CeO$_2$ nanoparticles for *Escherichia coli*. Physico-chemical insight of the cytotoxicity mechanism. *Environ Sci Technol* 2006;40:6151–6156.

151. Tsuruoka, T., Akamatsu, K., Nawafune, H. Synthesis, surface modification, and multilayer construction of mixed-monolayer-protected CdS nanoparticles. *Langmuir* 2004;20:11169–11174.

152. Tungittiplakorn, W., Cohen, C., Lion, L.W. Engineered polymeric nanoparticles for bioremediation of hydrophobic contaminants. *Environ Sci Technol* 2005;39:1354–1358.

153. U.S. Environmental Protection Agency, Nanotechnology White Paper, In: Council, S.P., editor. U.S. Environmental Protection Agency; 2007. p. 120.

154. Utsunomiya, S., Ewing, R.C. Application of high-angle annular dark field scanning transmission electron microscopy, scanning transmission electron microscopy-energy dispersive X-ray spectrometry, and energy-filtered transmission electron microscopy to the characterization of nanoparticles in the environment. *Environ Sci Technol* 2003;37:786–791.

155. Valentine, N.B., Bolton, H., Kingsley, M.T., Drake, G.R., Balkwill, D.L., Plymale, A.E. Biosorption of cadmium, cobalt, nickel, and strontium by a *Bacillus* simplex strain isolated from the vadose zone. *J Ind Microbiol* 1996;16:189–196.

156. Vecchio, A., Finoli, C., Di Simine, D., Andreoni, V. Heavy metal biosorption by bacterial cells. *Fresenius J Anal Chem* 1998;361:338–342.

157. Venugopalan, V.P., Kuehn, A., Hausner, M., Springael, D., Wilderer, P.A., Wuertz, S. Architecture of a nascent *Sphingomonas* sp biofilm under varied hydrodynamic conditions. *Appl Environ Microbiolo* 2005;71:2677–2686.

158. Volesky, B., Holan, Z.R. Biosorption of heavy metals. *Biotechnol Progr* 1995;11: 235–250.

159. Walsh, C.T., Sandstead, H.H., Prasad, A.S., Newberne, P.M., Fraker, P.J. Zinc: health effects and research priorities for the 1990s. *Environ Health Perspect* 1994;102:5–46.

160. Wang, C.L., Michels, P.C., Dawson, S.C., Kitisakkul, S., Baross, J.A., Keasling, J.D., Clark, D.S. Cadmium removal by a new strain of *Pseudomonas aeruginosa* in aerobic culture. *Appl Environ Microbiol* 1997;63:4075–4078.

161. Watson, S.K., Potok, R.M., Marcus, C.M., Umansky, V. Experimental realization of a quantum spin pump. *Phy Rev Lett* 2003;91:258–301.

162. Weiner, S., Dove, P.M. An overview of biomineralization processes and the problem of the vital effect. In: Dove, P.M., de Yoreo, J.J., Weiner, S., editors. *Biomineralization*, Washington, D.C.: Mineralogical Society of America; 2003. pp. 1–29.

163. Whitchurch, C.B., Tolker-Nielsen, T., Ragas, P.C., Mattick, J.S. Extracellular DNA required for bacterial biofilm formation. *Science (Washington D.C.)* 2002;295:1487.

164. White, C., Gadd, G.M. Accumulation and effects of cadmium on sulphate-reducing bacterial biofilms. *Microbiology-UK* 1998;144:1407–1415.

165. Whitman, W.B., Coleman, D.C., Wiebe, W.J. Prokaryotes: the unseen majority. *Proc Natl Acad Sci USA* 1998;95:6578–6583.

166. Willner, I., Baron, R., Willner, B. Integrated nanoparticle-biomolecule systems for biosensing and bioelectronics. *Biosens Bioelectron* 2007;22:1841–1852.

167. Wu, X.Y., Liu, H.J., Liu, J.Q., Haley, K.N., Treadway, J.A., Larson, J.P., Ge, N.F., Peale, F., Bruchez, M.P. Immunofluorescent labeling of cancer marker Her2 and other cellular targets with semiconductor quantum dots. *Nature Biotechnol* 2003;21:41–46.

168. Yee, N., Fein, J.B. Does metal adsorption onto bacterial surfaces inhibit or enhance aqueous metal transport? Column and batch reactor experiments on Cd-Bacillus subtilis-quartz systems. *Chem Geol* 2002;185:303–319.

169. Yu, K., Singh, S., Patrito, N., Chu, V. Effect of reaction media on the growth and photoluminescence of colloidal CdSe nanocrystals. *Langmuir* 2004;20:11161–11168.

170. Zhang, T.T., Stilwell, J.L., Gerion, D., Ding, L.H., Elboudwarej, O., Cooke, P.A., Gray, J.W., Alivisatos, A.P., Chen, F.F. Cellular effect of high doses of silica-coated quantum dot profiled with high throughput gene expression analysis and high content cellomics measurements. *Nano Lett* 2006;6:800–808.

171. Zhelev, Z., Jose, R., Nagase, T., Ohba, H., Bakalova, R., Ishikawa, M., Baba, Y. Enhancement of the photoluminescence of CdSe quantum dots during long-term UV-irradiation: privilege or fault in life science research? *J Photochem Photobiol B-Biol* 2004;75:99–105.

172. Zhelev, Z., Ohba, H., Bakalova, R. Single quantum dot-micelles coated with silica shell as potentially non-cytotoxic fluorescent cell tracers. *J Am Chem Soc* 2006;128:6324–6325.

173. Zheng, Z.L., Stewart, P.S. Penetration of rifampin through *Staphylococcus epidermidis* biofilms. *Antimicrob Agents Chemother* 2002;46:900–903.

PART III

TOXICITY AND HEALTH HAZARDS OF NANOMATERIALS

Potential Toxicity of Fullerenes and Molecular Modeling of Their Transport across Lipid Membranes

DMITRY I. KOPELEVICH[1], JEAN-CLAUDE BONZONGO[2], RYAN A. TASSEFF[3], JIE GAO[2], YOUNG-MIN BAN[1], and GABRIEL BITTON[2]

[1]Department of Chemical Engineering, P.O. Box 116005, University of Florida, Gainesville, FL 32611, USA
[2]Department of Environmental Engineering Sciences, P.O. Box 116450, University of Florida, Gainesville, FL 32611, USA
[3]School of Chemical and Biomolecular Engineering, Cornell University Ithaca, NY 14853, USA

10.1 INTRODUCTION

The current advances in nanotechnology present new challenges in ensuring that manufactured nanomaterials (MNs) do not become dangerous pollutants. This is because their anticipated widespread production and use could lead to new environmental problems, such as new classes of toxins and related environmental hazards (1). In addition to the nanometer size, the production of MNs involves chemical elements and combination of elements, which could either directly impact living cells or undergo environmental transformations that would produce secondary toxic derivatives. The literature on nanoscience and nanotechnology is quite abundant with studies dealing with both manufacturing and use of nano-materials in different industrial, environmental, and medical applications (2–9). However, the potential impacts of this new technology on both the environment and living organisms have received less attention. In the current high-throughput societies, one would anticipate the peak in MN production and use to be followed by either intentional (landfills) and/or nonintentional (diffuse) introduction of these materials into different environmental compartments. It is then obvious that a significant effort is needed for "upstream" determination of the potential impacts of this new and emerging technology on the environment and biota

Nanoscience and Nanotechnology   Edited by Vicki H. Grassian
Copyright © 2008 John Wiley & Sons, Inc.

before the products of nanoscience and nanotechnology become widespread. Therefore, besides the abundant research on creating new means of detecting pollutants, cleaning polluted waste streams, recovering materials before they become wastes, and expanding the available resources, there is presently a growing need of data on the assessment of the potential impacts of this new technology on the environment and living organisms. At the international level, the bell on future consequences of nanotechnology was rang first by the UK Government, which commissioned the Royal Society and the Royal Academy of Engineering to carry out an independent study into current and future developments in nanoscience and nanotechnology and their impacts. Their findings were published in a report entitled *Nanoscience and Nanotechnologies: Opportunities and Uncertainties*, released in July 2004 (www.royalsoc.ac.uk/policy). Among several key points highlighted, the study noted the lack of published data on the fate and impact of MNs and recommended research into the hazards and exposure pathways of MNs to reduce the uncertainties related to their potential impact on health, safety, and the environment. Several research programs on these aspects of nanotechnology are ongoing around the world and obtained experimental data are now appearing in peer-reviewed literature. In this chapter, we report on fullerenes' toxicity data obtained using a combination of laboratory microbiotests and a molecular modeling approach. The latter was used to investigate the interactions of carbonaceous MNs with model cell membranes and to assess the potential mechanisms of cytotoxicity of carbonaceous MNs.

10.1.1 Toxicity of Environmental Contaminants at the Organismal Level

Several toxicological tests that assess the toxicity of environmental contaminants at the organism or population levels are available. Scientific studies have argued that aquatic toxicity tests such as different fish assays used in regulatory frameworks are often impracticable for routine environmental screening (10, 11). They are expensive, laborious, and both time and space consuming. In view of these drawbacks, several toxicity tests have been developed and these tests include the so-called *small-scale* toxicity tests, also termed *microbiotests* (12). In general, microbiotests use small-sized test species and are relatively rapid, simple, and low cost. The *Ceriodaphnia dubia* test, the *Selenastrum capricornutum* toxicity test, and MetPLATE™ are three well-known small-scale toxicity tests used in this study to assess the toxicity of MNs, namely, fullerenes, and a description of each of the three tests is given in Section 10.2. As part of a much larger project, the objective of this effort is to use microbiotests as toxicity screening tools for different MNs and, ultimately, assist in the synthesis/design of less toxic nanomaterials.

Unlike the focus on aquatic biota and human cells, only a few studies have investigated the mobility of fullerenes and other nanoparticles in simulated porous media under laboratory conditions (13, 14). Additionally, Hyung et al. (15) reported that carbonaceous nanomaterials such as multiwalled carbon nanotubes (MWNTs) can be readily dispersed as an aqueous suspension in organic-rich natural river water.

These studies have shown that MNs could exhibit differing transport behaviors, and the findings are important with regard to the assessment of both efficacy (e.g., when used for remediation purposes) and environmental impacts of MN. Accordingly, the ultimate introduction of MNs into natural systems could lead to direct exposure of soil/ sediment microorganisms to these new materials. Testing the effect of C_{60} on two common soil microbial species (Gram-negative *Escherichia coli* and Gram-positive *Bacillus subtilis*) with basic differences in cell wall composition, Fortner et al. (16) showed that depending on the composition of culture medium and C_{60} speciation, the above microorganisms could be negatively impacted. Overall, these observations point to the possibility of MNs to be mobile in sedimentary and aquatic environments, with potential negative effects on basic ecological functions that are driven by microorganisms.

10.1.2   Molecular Modeling of Interaction of Carbon-based MN with Cell Membranes

In contrast to the growing number of studies using standardized toxicity tests, there are currently very few theoretical molecular-scale investigations of the effects of MNs on biosystems (17, 18). The computational component of this study is an ongoing investigation of the interaction of MNs with cell membranes and the possibility of permeation of MNs through cell membranes into cell interior and/or damage of the cell membranes. The model MNs considered here are spherical and nearly spherical carbon-based nanoparticles. The cell membrane is modeled as a phospholipid bilayer and other constituents of the membrane, such as receptors and ion channels, are neglected in this initial effort. Although transport of molecules into the cell interior is usually mediated by proteins, some relatively small molecules, such as water and oxygen, are known to cross the membrane through direct, or basal, permeation (19). Molecular dynamics studies (20, 21) have shown that basal transport of these small molecules through a lipid bilayer occurs via a series of jumps through voids between lipid molecules. Despite the fact that it is unlikely that larger particles (such as C_{60}) will find voids of sufficient size for their transport through a lipid membrane, recent studies have demonstrated that a variety of nanoparticles can permeate cellular membranes by a currently unknown mechanism different from phagocytosis and endocytosis (22, 23). This suggests that the basal permeation of the nanoparticles may be a dominant mechanism of the transport. Therefore, in the current study we investigate the basal permeation of the nanoparticles through a lipid bilayer, as illustrated in Fig. 10.1. Although there is an extensive literature on transport across cell membranes, most of this literature is focused on transport of small molecules, ions, and organic drug molecules (24–26), and there are a very few theoretical molecular-level studies of MN transport across the lipid bilayer.

The specific objectives of the current modeling study are (1) to assess the timescales of permeation of nanoparticles into the membrane and the cell interior and (2) to investigate the effects of the MNs, size and shape on their transport through and residence within cellular membranes. The small length scales of interest in this work

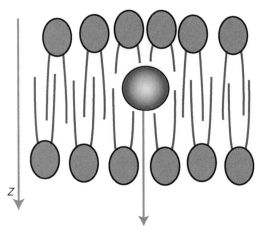

FIGURE 10.1   Schematic representation of basal transport of a nanoparticle through a lipid bilayer investigated in the current work.

motivate the use of molecular dynamic (MD) simulations, which enable investigations of transport processes at a molecular level. We anticipated and confirmed our results that the timescales of MN transport through the lipid membranes is beyond the reach of direct MD simulations, which can typically assess timescales limited by microseconds. Therefore, it is necessary to employ an indirect method of calculating transport properties of nanoparticles. In this work, we employ the potential of mean force (PMF) method (20, 21).

This study is performed using a coarse-grained molecular model (27). This model approximates small groups of atoms as coarse-grained beads, and allows one to explore larger time- and length scales while retaining realistic system dynamics. This simplified model allows us to understand the main principles of the mechanisms of MN transport through cell membranes and to perform a fast screening of multiple qualitatively different particles.

10.2   METHODS

10.2.1   Determination of the Potential Toxicity of Fullerenes (C_{60}) Using Microbiotests

10.2.1.1   *Ceriodaphnia dubia Acute Toxicity Assay*   *Ceriodaphnia dubia* is a freshwater cladoceran that naturally feeds on particles, including bacteria, algae, and yeast. It is used as test organism in the classical 48-h acute test to assess the toxicity of samples in freshwater (28). In this study, freshwater samples were replaced by C_{60} aqueous suspensions. Prior to the start of each assay, the neonates (<24 h) were separated from the adult daphnids and fed a mixture of 7 mL YCT (yeast, cereal leaves, and trout chow) and 7 mL algae (*S. capricornutum*) for every liter of invertebrate culture. After

feeding, 10 neonates were transferred to each test container (30-mL plastic cups) using a small wide-mouth plastic pipette to minimize the transfer of culture water. The sample dilutions, if necessary, were prepared with moderately hard water (MHW) and 20 mL of the sample or its dilutions were added to cups containing the neonates. MHW was used as the negative control and all test containers were then placed in a water bath at $20 \pm 2°C$ for 48 h while exposing the neonates to ambient lighting. All assays were conducted in triplicate with death/immobilization as an end point.

10.2.1.2 Selenastrum capricornutum (Renamed Pseudokirchneriella subcapitata)

The chronic toxicity of the C_{60} aqueous suspensions was evaluated according to the protocol of the 96-h *P. subcapitata* growth inhibition assay (28). Following dilution of C_{60} aqueous suspensions into algal growth medium to produce a C_{60} concentration gradient, each dilution was inoculated with a similar volume of algal culture. The samples and their dilutions as well as negative controls (algal medium without C_{60} sample) were incubated in triplicate for 96 h in a water bath at room temperature and under controlled light. The light intensity was set at $86 \pm 8.6 \, \mu E/m^2/s$. The algal growth was assessed by chlorophyll measurement. The EC_{50} was determined by plotting sample concentrations versus the percent inhibition determined using Equation 10.1, and performing a regression analysis in the linear portion of the graph, from which the EC_{50} was determined using Equation 10.2.

$$\% \text{ Inhibition} = \left(1 - \frac{[\text{chlrophyll } a]_{\text{sample}}}{[\text{chlrophyll } a]_{\text{control}}}\right) \times 100 \qquad (10.1)$$

$$EC_{50} = \frac{50 - Y_{\text{intercept}}}{X_{\text{variable}}} \qquad (10.2)$$

10.2.1.3 MetPLATE™ Test

This test is specific to heavy metal toxicity (29, 30). The test kit contains a bacterial reagent (an *E. coli* strain), a buffer, the chlorophenol red galactopyranoside (CPRG) that serves as the substrate for β-galactosidase, and moderately hard water that is used as diluent. The test is based on β-galactosidase activity, which is measured by the conversion of chlorophenol red galactopyranoside to chlorophenol red, which is determined by a microplate reader set at 570 nm. Briefly, the bacterial reagent is rehydrated with 5 mL of diluent and thoroughly mixed by vortexing. Next, 900 μL aliquot of the MN suspension or its dilution is added to a test tube containing 100 μL of the above-described bacterial reagent. In these assays, MHW serves as the negative control. The test kit also includes a positive control (Cu). The test tubes are vortexed and then incubated for 1.5 h at 35 °C. A 200-μL aliquot of the suspension is transferred to a 96-well microplate to which is added 100-μL of CPRG, the enzyme substrate, followed by shaking. The microplate is then incubated at 35 °C for color development. The response is quantified at 570 nm using a Multiscan microplate reader. The assay is performed in triplicate with four dilutions in a 96-well microplate, and the EC_{50} is determined following the approach described in Section 10.2.1.2.

10.2.2   Impacts of C_{60} on Microbial Degradation of Organic Matter in Sediment Slurries

Ecosystems accomplish numerous natural services and most, if not all, of them seem to have common main characteristics including the flow of energy, material, and information as well as the participation of biota and water. The ability to qualitatively and/or quantitatively characterize any of the above-listed natural services can be used to assess the impact of pollutants on ecosystem functions. Such abilities are provided by thermodynamics, which has been successfully applied in the description of some of the basic properties of ecosystems (e.g., flow of matter and energy). By combining the flow of material and energy to the participation of biota (microorganisms in this case), one can use a series of reactions involved in sedimentary cycling of organic carbon as a proxy for the detection of potential impact of MNs on ecosystem functions. Sediments used in this study were collected from two different locations: a polluted lake (Lake Alice) within the campus of the University of Florida (UF) and a pristine wetland (the Odum wetland), a protected aquatic system managed by the UF Center for Wetlands. In the field, surface sediments (top 10 cm) were collected using precleaned high-density polyethylene scoops, transferred into sieves (2 mm), and the sieved fraction collected into water precleaned plastic containers. Accordingly, obtained sediment samples were slurried during the sieving process on-site using water of the sampling site. In the laboratory, aliquots of the well-homogenized slurried sediments were transferred into 60-mL serum vials and spiked with a solution of sodium acetate to obtain a final concentration of about 75 mg CH_3COO^- per liter of slurry. To assess the effect of C_{60} on microbial-catalyzed degradation of acetate, one set of slurries was spiked with a C_{60} suspension, prepared using a method described by Degushi et al. (31), to a final concentration of 0.5 ppm (i.e., C_{60}-treated samples). Additionally, another set of sediment slurries received no C_{60} and served as control incubations. The mixtures were sealed and placed on a rota-shaker®. Using disposable syringes, samples were taken from the vials over time, filtered (0.45 μm), and analyzed for CH_3COO^- by ion chromatography (DX-320 IC System). Obtained results were then fit to existing kinetic models to help assess the impact of C_{60} on the rate of acetate oxidation by indigenous sediment's microorganisms.

10.2.3   Model Development for the Assessment of MNs' Ability to Cross Cell Membranes

As mentioned earlier, we investigate the nanoparticle transport using molecular dynamics simulations. Molecular dynamics simulations provide a research tool complementary to laboratory experiments and offer several advantages over the experimental approach. In particular, MD simulations enable investigations of events on a very small time- and length scales (down to 1 ps and 1 nm, respectively). Moreover, MD simulations provide tools to separately assess contributions of various factors, such as the different types of intermolecular interactions (e.g., electrostatic, van der Waals, hydrogen bonding, and others) and chemical reactions to the transport and toxicity of nanomaterials. On the other hand, the molecular simulations cannot adequately model complex systems and even simulations of an individual protein remain very challenging. In addition, the simulations cannot be extended to timescales much larger than 1 μs. However, one can overcome the latter limitation by using one of the several "indirect" simulation methods, which allow one to predict long-term behavior of a system based on a relatively short-scale simulation. In the current work, we use the constrained simulations method that enables predictions of slow transport of nanoparticles across cellular membranes based on relatively short-scale (1 μs) simulations.

Our studies are performed using a coarse-grained molecular dynamics (CGMD) model, which approximates small groups of atoms as a single united atom. Several such models have been introduced and applied to simulations of various complex molecular systems (27, 32–36). The development of coarse-grained molecular models and the improvement of the quantitative agreement with full atomistic models and experimental data are subjects of active ongoing research (27, 36). In this work, we use the model for water and phospholipids proposed by Marrink et al. (27). Although the coarse-grained model provides less details than an atomistically detailed model, simulations with this model are shown to yield good agreement with more detailed models, as well as with experimental data for several systems, including lipid membranes.

This model, on average, employs a four-to-one mapping approach to represent groups of atoms as a single spherical bead. The following four main types of beads are considered: polar (P), nonpolar (N), apolar (C), and charged (Q). For example, four water molecules are represented as a single polar bead and four methylene groups of the lipid tail are represented by a single hydrophobic apolar bead. This reduction allows one to perform simulations three to four orders of magnitude faster than simulations with the detailed atomistic model. This in turn enables fast screening of different nanoparticles and assessment of their interactions with lipid membranes.

Interactions between coarse-grained beads that are not connected by a chemical bond are modeled by the Lennard-Jones potential,

$$U_{LJ} = 4\varepsilon\left[\left(\frac{\sigma}{r}\right)^{12} - \left(\frac{\sigma}{r}\right)^{6}\right] \tag{10.3}$$

where the effective bead diameter σ is 0.47 nm, unless stated otherwise. The strength of the interaction ε is varied to mimic different types of interactions. Specifically, a large value of ε ($\varepsilon = 5$ kJ/mol) corresponds to strong attractions between the hydrophilic beads, whereas a small value of ε ($\varepsilon = 1.8$ kJ/mol) is used to model interactions between the hydrophobic and the hydrophilic beads, which is dominated by repulsion. Interaction between two hydrophobic particles is modeled using an intermediate interaction strength ($\varepsilon = 3.4$ kJ/mol). In addition, electrostatic interactions between charged beads are taken into account by a screened electrostatic potential.

Interactions between chemically bonded beads are modeled by harmonic potentials for the bond length and the bond angle vibrations,

$$U_{bond}(r) = \frac{1}{2} K_{bond}(r - r_0)^2 \qquad (10.4)$$

$$U_{angle}(\theta) = \frac{1}{2} K_{angle}[\cos(\theta) - \cos(\theta_0)]^2 \qquad (10.5)$$

Here, r is the instantaneous bond length, r_0 is the equilibrium bond length, θ is the instantaneous bond angle, and θ_0 is the equilibrium bond angle. In the current model, the force constants for the bond length and the bond angle vibrations are $K_{bond} = 1250 \, kJ/(mol \, nm)$ and $K_{angle} = 25 \, kJ/(mol \, rad^2)$, respectively; the equilibrium bond length is $r_0 = 0.47 \, nm$ and the choice of the equilibrium bond angle θ_0 is dictated by geometry of a molecule.

10.2.3.1 *Species Details*   The cell membrane is modeled by a dipalmitoylphosphatidylcholine (DPPC) lipid bilayer. The choice of the model system is dictated by the fact that phosphatidylcholines are major components of most biological membranes. The detailed atomistic structure of the DPPC molecule and the corresponding coarse-grained model are shown in Fig. 10.2a. The DPPC head group consists of choline (modeled as a positively charged bead Q^+), a phosphate group (modeled as a negatively charged bead Q^-), and a glycerol ester backbone (modeled by two nonpolar beads N). The two hydrocarbon tails of the lipid are modeled by two chains, each consisting of four apolar beads C. The lipid bilayer is surrounded on both sides by water. The coarse-grained model uses a single polar bead to represent four water molecules.

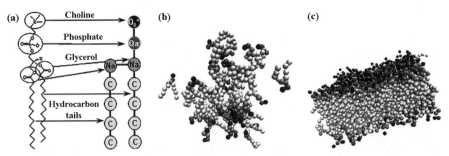

FIGURE 10.2   (a) Detailed atomistic structure and the corresponding coarse-grained model of a DPPC lipid. Mapping between different groups of atoms and the coarse-grained beads is also shown. The bead types are denoted as follows: C = apolar, P = polar, Qa (Na) = charged (nonpolar) groups acting as a hydrogen bond acceptor, and Q0 = charged groups with no hydrogen bonding capabilities (27); (b) Initial random dispersion of DPPC beads in water; (c) Self-assembled DPPC lipid bilayer. Head-group, glycerol, and tail beads are shown by black, dark gray, and light gray spheres, respectively. For clarity, water molecules are not shown. The molecular images in (b) and (c) are generated using VMD software package (Humphrey et al., (37)).

A computational model for the lipid bilayer is prepared through self-assembly. To prepare the membrane, we performed MD simulations of a DPPC lipid solution in water in a simulation cell of size $10 \times 10 \times 10\,nm^3$. The initial conditions for this simulation were chosen to be a random dispersion of lipids in water (see Fig. 10.2b). The lipids self-assemble into a bilayer such as that shown in Fig. 10.2c within 50 ns. For definiteness of the future discussion, we introduce the system of coordinates so that the z axis is perpendicular to the bilayer surface (see Fig. 10.1), and the x–y plane contains the bilayer center of mass.

The main focus of our study has been analysis of transport of spherical nanoparticles of various sizes. In addition, we have analyzed effects of deviations of a nanoparticle from the spherical shape on its transport properties. The primary goal of this work was to model carbon-based nanoparticles, such as fullerenes. Therefore, the effective diameters of the model particles are chosen to mimic those of fullerenes of the series C_{36}–C_{176} with the diameters ranging from 0.5 to 1.2 nm (38). However, in the spirit of coarse-grained simulations, we do not model explicitly all of the atoms of fullerenes but rather use a single sphere or an aggregate of several spheres to represent a nanoparticle, see Fig. 10.3. In the first stage of our analysis, the nanoparticles were modeled as hydrophobic spheres by assigning apolar to them the interaction strength, that is, the same interaction strength as that used for methylene groups. The effective Lennard-Jones diameter σ of the model nanoparticles varied between 0.47 and 1.1 nm.

Unfortunately, using a Lennard-Jones sphere as a model for a nanoparticle with diameter much larger than $\sigma \approx 1\,nm$ might lead to unrealistic results, since the Lennard-Jones parameter σ represents not only the effective diameter of the particle but also the effective range of interaction of this particle with other molecules of the system. Therefore, if σ is too large, one would end up with unrealistically long range of van der Waals interactions between the nanoparticle and its surroundings. A more realistic coarse-grained model for a larger nanoparticle should be based on a structure composed of several smaller beads connected together, rather than a single large bead. To check robustness of the results obtained for the spherical nanoparticles and to set the stage for investigations of more realistic (and more complex) models of nanoparticles, we initiated studies of nonspherical nanoparticles, such as cubical and tetrahedral

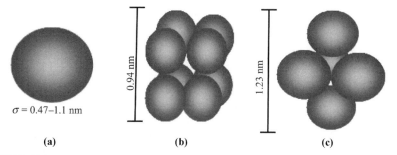

$\sigma = 0.47$–1.1 nm	0.94 nm	1.23 nm
(a)	**(b)**	**(c)**

FIGURE 10.3 Models of nanoparticles studied in the current work: (a) spherical beads with diameter σ ranging between 0.47 nm and 1.1 nm; beads of diameter $\sigma = 0.47$ nm arranged in (b) cubical and (c) tetrahedral structures.

arrangements of spherical beads of diameter 0.47 nm, as illustrated in Fig. 10.3b and c. The beads in these arrangements are connected by the harmonic bonds with the potential given by Equation 10.4.

10.2.3.2 Simulation Details

The CGMD simulations were preformed using the Groningen Machine for Chemical Simulations (GROMACS) software package (39). The model bilayer system contained 266 DPPC molecules and 4007 water beads. The periodic boundary conditions in all dimensions were assumed. Temperature and pressure were maintained at 323K and 1 bar using the Berendsen temperature and pressure coupling schemes (40). Anisotropic pressure coupling scheme was applied to maintain zero surface tension of the bilayer membrane. The time step for the numerical integration was 0.04 ps.

10.2.3.3 Constraint (Potential of Mean Force) Method

Transport of even small molecules across a lipid bilayer takes place over a long time. This transport is the so-called *rare event* and its timescale is out of reach of MD simulations. Several simulation techniques have been developed to investigate rare events, including umbrella sampling (41), coarse kinetic simulations (42, 43), and the potential of mean force method. In our investigation we employed the PMF method that has been successfully applied to similar systems (18, 20, 21).

The main idea of this method is to apply a force to constrain the nanoparticle at a certain location z_0 within the bilayer, as illustrated in Fig. 10.4. The nanoparticle is

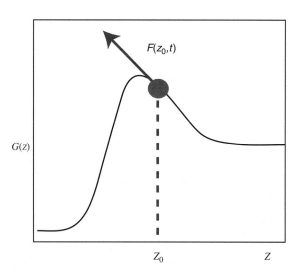

FIGURE 10.4 Schematic illustration of the potential of mean force method. The nanoparticle (shown by the circle) is constrained at a position z_0 inside the bilayer. The mean of the constraining force F (shown by the arrow) corresponds to the gradient of the free energy $G(z)$. If the constraining force is absent, the nanoparticle will move on average, toward a state with a lower free energy.

constrained in the direction z normal to the bilayer and is free to move in the directions x and y parallel to the bilayer surface. The constraint in our simulations is implemented using the SHAKE algorithm (44). The local environment around the particle is not uniform, which leads to a gradient in the free energy. If the particle is left unconstrained, it will move down the gradient of the free energy $G(z)$. Therefore, the applied constraining force $F(z_0, t)$ counteracts the free energy gradient and is equal (on average) to the gradient of the free energy. This force will also have a random component due to random fluctuations of molecules of the bilayer and the solvent. These fluctuations are directly related to the nanoparticle diffusivity within the bilayer, as will be discussed in Section 10.2.3.4. By repeating the constrained simulations for multiple locations z of the nanoparticle within the monolayer and performing the statistical analysis of $F(z_0, t)$ outlined in Section 10.2.3.4, one can fully reconstruct the parameters for a model describing the nanoparticle transport.

10.2.3.4 Calculation Details

The main assumption underlying our analysis is that the nanoparticle transport can be described by the Langevin equation (45):

$$m\ddot{z}(t) + \gamma(z)\dot{z}(t) + \frac{dG(z)}{dz} = \Gamma(z, t) \tag{10.6}$$

Here, m is the mass of the nanoparticle, γ is the friction coefficient, G is the free energy, Γ is the random force acting on the particle, and z is the particle's coordinate in the direction normal to the bilayer surface. The origin of the system of coordinates used here is shifted so that $z = 0$ corresponds to the bilayer center of mass (see Fig. 10.1). The random force Γ has a normal distribution with zero mean and the autocorrelation function obeying the fluctuation–dissipation theorem (46),

$$\langle \Gamma(z, t) \rangle = 0 \tag{10.7}$$

$$\langle \Gamma(z, t)\Gamma(z, t+\tau) \rangle = 2\gamma(z)\delta(\tau)k_B T \tag{10.8}$$

The brackets denote the ensemble average, k_B is the Boltzmann constant, T is the system temperature, and δ is the Dirac delta function. The fluctuation–dissipation theorem, Equation 10.8, provides a relationship between Γ and γ. As we shall see in section "Calculation of Friction Coefficient", this relationship allows us to compute the friction coefficient from the autocorrelation function of the constraining force, which is readily measurable from MD simulations. Moreover, the friction coefficient γ is directly related to particle diffusivity (45),

$$D(z) = k_B T / \gamma(z) \tag{10.9}$$

In what follows we discuss the computation of the terms of Equation 10.6 from the constrained MD simulations.

Calculation of Free Energy   If z is held constant by the application of a constraining force F, then the first and second terms of Equation 10.6 are zero. Further, the ensemble average of the random force term is zero, see Equation 10.7. Therefore, the ensemble average of Equation 10.6 yields the following relationship between the constraining force $\langle F \rangle$ and the free energy:

$$\langle F(z_0, t) \rangle = \frac{dG(z_0)}{dz} \tag{10.10}$$

In analysis of our simulations, we replace the ensemble averaging by the time averaging using the ergodic assumption.

Calculation of Friction Coefficient   To fully reconstruct Equation 10.6, it is necessary to obtain the friction coefficient γ. It can be shown (21, 47) that γ can be obtained from the autocorrelation function of the random force Γ,

$$\gamma(z) = \frac{1}{k_B T} \int_0^\infty \langle \Gamma(z, t) \Gamma(z, t + \tau) \rangle dt \tag{10.11}$$

The random force $\Gamma(z, t)$ is obtained by computing the deviation of the constraining force from its mean value,

$$\Gamma(z, t) = F(z, t) - \langle F(z, t) \rangle \tag{10.12}$$

Rate of Transport   One of the main goals of this study is to estimate timescales of (i) nanoparticle permeation into the membrane interior and (ii) nanoparticle motion across the membrane and permeation into the cellular interior. We will denote the average times required for these two processes by τ_1 and τ_2, respectively. It can be shown (45) that if a particle motion is described by Equation 10.6, then the average time τ of transport between two points with coordinates z_1 and z_2 is given by the following expression:

$$\tau(z_1 \rightarrow z_2) = \frac{1}{D} \int_{z_1}^{z_2} e^{G(y)/k_B T} dy \int_{z_1}^{y} e^{-G(x)/k_B T} dx \tag{10.13}$$

In the current work, point z_1 will be chosen to be a point of entrance into the bilayer and the following two choices for point z_2 will be considered: (i) to compute τ_1, point z_2 will be chosen in the interior of the bilayer and (ii) to compute τ_2, point z_2 will be chosen at the exit from the bilayer. Equation 10.13 is derived assuming constant diffusivity D; the validity of this assumption will be discussed in Section 10.3.2.

10.3 RESULTS

10.3.1 Experimental Assessment of the Toxicity of C_{60} Using Microbiotests and Sediment Indigenous Microorganisms

10.3.1.1 *Assessing the Toxicity of C_{60} Using Microbiotests* The concentrations of C_{60} that resulted in either 50% of growth inhibition in the algal test (i.e., EC_{50}) or 50% of dead organisms in *C. dubia* test (i.e., LC_{50}) are plotted in Fig. 10.5. These results show the following. First, not all microbiotests will respond positively to exposure to fullerenes. In this case, MetPLATE, a toxicity test based on the inhibition of the activity of a specific enzyme, β-galactosidase, which is inhibited primarily by heavy metals, was not responsive to tested carbonaceous nanomaterials. This observation confirms the specificity of this particular test to heavy metals. In contrast to MetPLATE, the second observation is that both the 48-h acute *C. dubia* and the 96-h chronic *S. capricornutum* tests were sensitive to C_{60}, with an average LC_{50} of 0.43 mg/L for *C. dubia* and an average EC_{50} of 0.13 mg/L for *S. capricornutum*. These two tests appear to be adequate for a rapid assessment of the potential toxicity of the tested carbonaceous MNs. Our results confirm previously observed trends based on laboratory studies using both macro- and microbiotests (48–52). However, more recent findings tend to suggest that the toxicity impact of C_{60} is affected by the type of pretreatments that the pristine C_{60} undergoes to produce aqueous suspensions (51, 52). For instance, the preparation of C_{60} aqueous suspension using THF as an intermediate step (31) enhances the toxicity impact of C_{60} on daphnia and fathead minnow (52). In spite of this last observation, our results as well as those published by other researchers point to the potential toxicity of fullerenes on biota. However, more relevant data could be obtained through laboratory experiments that mimic prevalent environmental conditions (e.g., use of C_{60} suspensions prepared directly into pore, river,

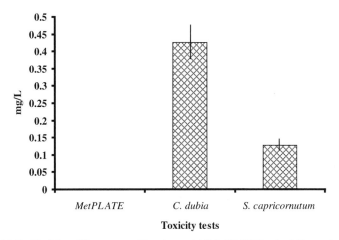

FIGURE 10.5   Toxicity of C_{60} measured by a bacterial (MetPLATE)-, algal (*S. capricornutum*)-, and invertebrate (*C. dubia*)- based microbiotests.

lake, or seawater). Nevertheless, current toxicity data provide a rationale for the investigation of different toxicity mechanisms associated with nanomaterials.

10.3.1.2 Impact on Sedimentary Microbial Degradation of Organic Matter   Although the redox potential (E_h) was not monitored in these experiments, the incubation of used sediment slurries in closed serum vials for 9–14 days could have led to the disappearance of molecular oxygen as initial terminal electron acceptor (TEA), followed by the reduction of any other thermodynamically favored TEAs as the acetate and naturally occurring organic matter (OM) present in tested sediments underwent decomposition. If so, measured reaction rates of acetate degradation would then reflect the activity of several indigenous microbial groups present in used sediment slurries. For mixtures such as sediments used in this study, complex reactions with consecutive steps involving more than one reactant are likely, and determining the exact order of reaction in such systems is prohibitively difficult. However, the experimental approach used in this study supports the use of a pseudo-first-order kinetic model, which allows the determination of the rates of reaction associated with each treatment and sediment types as shown in Fig. 10.6. These results show that the disappearance of acetate from the aqueous phase follows a pseudo-first-order kinetic model in both C_{60}-treated and C_{60}-nontreated sediment slurries. However, the apparent rate of reaction (k_{app}, determined as the slope of the regression of $\ln[C]$ versus time) is comparatively smaller in C_{60}-spiked slurries. In polluted sediments (Fig. 10.6a), the determined apparent reaction rate (k_{app}) was $0.20\,day^{-1}$ for non-treated (or control) sediment slurries and an order of magnitude lower in C_{60}-treated slurries ($k_{app} = 0.048\ day^{-1}$). Comparatively, much higher k_{app} were obtained with nonpolluted wetland sediments, with values of 0.415 and $0.196\,day^{-1}$ for the C_{60}-nontreated and C_{60}-treated slurries, respectively. Although the microbial and most physicochemical characteristics of used sediments were not determined (except for water content, percent organic matter determined as loss on ignition, and pH), our results suggest direct and/or indirect negative impacts of C_{60} on sediment microorganisms. It is likely that the slower degradation rates of acetate in C_{60}-treated slurries are the result of the bactericidal effect of fullerenes (16, 51), while the observed differences in k_{app} between the two types of sediments could be related to several factors, including but not limited to (i) the make up of the microbial communities in each sediment, (ii) the ability of microorganisms present to tolerate C_{60}-pollution, and (iii) physicochemical interactions between C_{60} and sediment/water chemical components and the resulting alteration or enhancement of the bactericidal activity.

   The literature on toxicity of MNs is fast growing. However, the toxicity mechanisms associated with most MNs remain poorly understood and mostly speculative. At the cell level, and as the laboratory data show, some of the possible toxicity mechanisms are physical damages associated with the permeation of cell membranes by MN as well as a destructive chemical effect due to lipid peroxidation caused by "MN". In this study, we use a modeling approach to theoretically predict the potential impacts and the toxicity mechanisms of carbonaceous MNs, namely, fullerenes, on cell membranes. The results of this modeling effort are presented below.

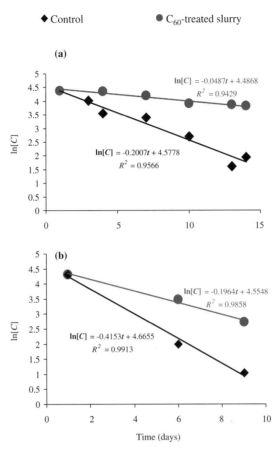

◆ Control   ● C$_{60}$-treated slurry

(a)

$\ln[C] = -0.0487t + 4.4868$
$R^2 = 0.9429$

$\ln[C] = -0.2007t + 4.5778$
$R^2 = 0.9566$

(b)

$\ln[C] = -0.1964t + 4.5548$
$R^2 = 0.9858$

$\ln[C] = -0.4153t + 4.6655$
$R^2 = 0.9913$

Time (days)

FIGURE 10.6 Kinetics of acetate disappearance from sediment slurries treated with C$_{60}$ as compared to nontreated control: (a) polluted lake sediments and (b) pristine wetland sediments.

10.3.2 Modeling Results

10.3.2.1 *Free Energy of Nanoparticles Inside a Lipid Membrane* To put the obtained results in perspective, in Fig. 10.7 we show the density profiles of lipid head- and tail groups and water molecules inside the bilayer and in surrounding areas. The free energy profiles obtained for the considered nanoparticles are shown in Fig. 10.8. The free energy is shifted so that its value approaches zero outside the bilayer. As expected, all considered hydrophobic nanoparticles exhibit a significant decrease of the free energy in the bilayer interior, that is, for -2 nm $< z < 2$ nm, which indicates a strong preference for these nanoparticles to be in the bilayer interior over the bulk water phase. This preference is caused by the stronger attraction of hydrophobic nanoparticles to the hydrophobic DPPC tails, as compared to the

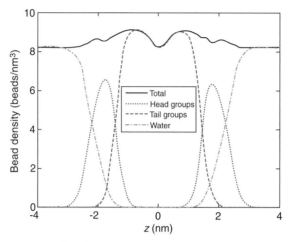

FIGURE 10.7   Density profile of the DPPC bilayer. The solid line shows the overall density of
the system, whereas the dotted, dashed, and dash-dotted lines show densities of the lipid head
groups, lipid tails, and water molecules, respectively. The density profiles are averaged in the
directions parallel to the bilayer surface, that is, in the x and y directions.

hydrophilic head groups or water molecules. These strong attractive interactions
provide a driving force for the nanoparticle to remain in the bilayer interior for a very
long time. Moreover, the depth of the free energy minimum increases with the
nanoparticle size due to a greater range of attractive interaction of larger nanoparticles

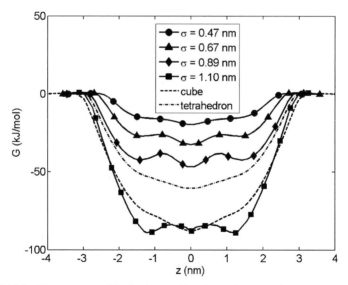

FIGURE 10.8   Free energy profiles for the considered model nanoparticles. The solid lines with
symbols, dashed lines, and dash-dotted lines show the free energy profiles for the spherical, cubical,
and tetrahedral particles, respectively. The effective (Lennard-Jones) diameters σ of the spherical
particles are shown in the plot legend. Note a very low energy barrier for entering the bilayer.

with lipid tails. Therefore, the time of residence of a nanoparticle in the membrane interior increases with the nanoparticle size.

As can be seen from Fig. 10.8, the energy barrier for nanoparticles to enter the DPPC bilayer is very small, $\sim$1 kJ/mol. The absence of a significant energy barrier implies that even relatively large nanoparticles can easily penetrate the membrane interior. This somewhat counterintuitive result can be explained by the following observations. First, the strength of nanoparticle interaction with hydrophilic lipid head group is very similar to that of the interaction with water. Further, the interactions of the hydrophobic nanoparticle with the glycerol ester backbone are slightly more attractive and the interactions with lipid tail beads are significantly more attractive than the interactions with water. Of course, the energy required to move apart the lipid head groups that are chemically bonded with relatively bulky tail groups is greater than the energy required to move apart small water molecules. However, as the density profile in Fig. 10.7 illustrates, the bilayer interface contains a large proportion of water beads, which mitigates the need, at least in this region, to move the DPPC beads apart. Once within the range of interaction with the glycerol ester backbone, the drive toward more favorable interactions becomes significant. This drive is further magnified as the nanoparticle approaches the tail region. Larger nanoparticles require more space and thus more energy to separate the lipid head groups, but the range and magnitude of attractive interactions with lipid tails are also increased and thus provide even more of a driving force to penetrate the membrane.

Another factor that reduces the resistance of the membrane to the nanoparticle entry is the difference in length scale of the cell membrane (with area of the membrane surface on the order of μm^2) and the considered nanoparticles (with the size limited to $\approx$1 nm). This difference allows the membrane to dissipate the mechanical stress caused by the nanoparticle entry throughout the membrane surface without any significant change of the membrane density in the neighborhood of the nanoparticle. In other words, it is expected that there is a relatively little resistance to the nanoparticle pushing the lipids apart. At the first sight, to observe this effect in the simulations, one would have to perform a simulation of a fairly large membrane segment, which would require prohibitively long simulation time. However, this effect is in fact reproduced in our simulations of a relatively small membrane patch (with area on the order of 100 nm^2) by applying the anisotropic pressure coupling scheme (40) to maintain zero surface tension of the bilayer membrane. This is achieved by adjusting the dimensions of the simulation box to maintain zero surface tension. In fact, we observe that the bilayer surface area changes by 2–3% upon the entry of the largest considered nanoparticle to maintain zero surface tension and to accommodate the nanoparticle in the membrane.

We observe that as the diameter of the spherical nanoparticle increases, the corresponding free energy profile develops additional free energy minima at 1.2 nm $< |z| <$ 1.5 nm, away from the membrane center at $z = 0$ nm. The origin in these minima is the soft polymer region of the lipid bilayer (21) characterized by a high density of ordered tail groups. As can be seen from Fig. 10.7, this region corresponds to $|z| <$ 1.1 nm. The increased density of the lipid tail region provides a resistance to the nanoparticle motion, which results in local energy barriers at 0.5 nm $< |z| <$ 0.8 nm (see Fig. 10.8). This barrier becomes more pronounced as the nanoparticle diameter increases, since more energy is required to create the necessary space for the larger particles. The local minima away from the bilayer center are developed between the bilayer entry and these local maxima. These minima are located at the points within the membrane that correspond to a balance between the attractive interactions between the lipid tails and the nanoparticles (due to their chemical nature) and the repulsive interactions between them (due to the increased lipid tail density). The existence of these minima implies that the spherical nanoparticles will be localized in different parts of the bilayer depending on their diameter and therefore will act differently on different groups of the lipid molecules.

Moreover, the preferred location of nanoparticles within the lipid membrane also depends on their shape. As can be seen from Fig. 10.8, the local free energy minima observed for the larger spherical nanoparticles do not exist for the considered nonspherical nanoparticles of comparable sizes. Therefore, the considered cubical and tetrahedral nanoparticles will be localized at the membrane center. Nevertheless, the general features of the free energy profile obtained with the nonspherical particles are the same as those for the spherical nanoparticles. Specifically, they still exhibit a very small energy barrier to enter the bilayer and a very large barrier to leave the barrier.

It is important to validate the coarse-grained model used in the current study by comparing the obtained results with the available molecular simulations data employing an atomistically detailed model. Since the coarse-grained models for DPPC lipids and water molecules have been validated by Marrink et al. (27), here we will focus on verification of the considered models for nanoparticles. Atomistic simulations of Qiao et al. (18) for C_{60} in a DPPC lipid bilayer predict a free energy profile that is qualitatively similar to that observed in our coarse-grained simulations for a spherical Lennard-Jones particle with diameter $\sigma = 1.1$ nm. The minimum in the free energy observed by Qiao et al. (18) is located at $z \approx 1.1$ nm, which compares favorably with the minimum location observed in our simulations at $z \approx 1.26$ nm. However, the free energy barrier between the local minimum and the bilayer center observed in the atomistic simulations (≈ 21.6 kJ/mol at $T = 325$ K) is significantly higher than the barrier observed in our simulations (≈ 4.5 kJ/mol at $T = 323$ K). Moreover, the difference in free energy between the bulk phase and the bilayer center obtained from the coarse grained simulations is twice as large as that observed in the atomistic simulations. Nevertheless, given the simplicity of our model, good qualitative agreement enables us to establish qualitative trends in the interactions between lipid membranes and nanoparticles.

10.3.2.2 *Diffusivity of Nanoparticles*
The diffusivities of the considered nanoparticles were computed using Equations 10.9, 10.11 and 10.12 for a range of particle locations z within the lipid membrane. Figure 10.9a demonstrates the obtained diffusivities for the considered spherical nanoparticles. As expected, the diffusivity decreases with particle size. Moreover, the diffusivity of the nanoparticles is position-dependent. In particular, all considered nanoparticles exhibit a sharp decrease in their diffusivity upon entering the bilayer. This decrease in diffusivity is probably due to the resistance of the lipids that are being pushed apart by the nanoparticles. In other words,

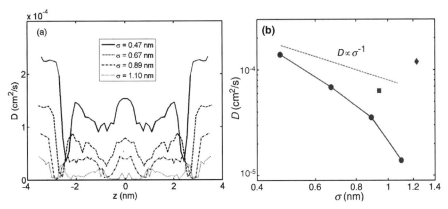

FIGURE 10.9   (a) Position-dependent diffusivity of spherical nanoparticles inside the lipid bilayer (σ = nanoparticle diameter). (b) Spatially averaged diffusivities of the spherical, cubical, and tetrahedral nanoparticles are shown by the circles, the square, and the diamond-shaped symbol, respectively. The dashed line shows the scaling of the diffusivity of a spherical particle expected from the Stokes–Einstein relationship.

the resistance of the lipids to the nanoparticle entry into the bilayer interior is manifested in the decreased diffusivity and not in the free energy barrier (see Fig. 10.8). This local decrease in diffusivity is expected to provide much less resistance to the nanoparticle transport than would be provided by a free energy barrier. Therefore, to obtain order of magnitude estimates of the transport rates, we will use the simplified model Eq. (10.13), which assumes that the diffusivity D of the nanoparticle within the bilayer is constant. The value of D will be chosen to be equal to the spatially averaged diffusivities shown in Fig. 10.9b. As a side note, Fig. 10.9b demonstrates that the scaling of diffusivity D of spherical nanoparticles with respect to their diameter σ does not obey the Stokes–Einstein relationship, $D \propto 1/\sigma$, which holds for Brownian diffusion of a spherical particle in viscous medium. This indicates that the diffusion within a bilayer cannot be described by this simple model.

In Section 10.3.2.1, we observed that the free energy profile obtained for the spherical nanoparticle with diameter $\sigma = 1.1$ nm most closely reproduces the free energy profile obtained in the atomistically detailed simulations of C_{60} in DPPC bilayer performed by Qiao et al. (18). The average diffusivity of this spherical nanoparticle obtained from our simulation is 1.4×10^{-5} cm^2/s, which compares reasonably well with the value obtained in the atomistic simulations $(2.7 \pm 0.4) \times 10^{-6}$ cm^2/s, given the simplicity of the coarse-grained model and the position dependence of the nanoparticle diffusivity inside the bilayer.

10.3.2.3 Timescales of Nanoparticle Transport   Based on the free energy profiles discussed in Section 10.3.2.1, we concluded that the nanoparticles can easily permeate the bilayer interior and will reside there for a long time. The measured diffusivities allow us to make this statement more quantitative and, using Equation 10.13,

compute the mean times τ_1 and τ_2 of the nanoparticle permeation into (i) the membrane interior and (ii) the cellular interior, respectively.

In order to apply Equation 10.13, it is necessary to specify the limit of integrations, that is the entry point z_1 and the exit point z_2. The point of entry into the bilayer was chosen to be $z_1 \approx -3.5$ nm and the point of exit was chosen as follows: (i) to compute τ_1, z_2 was set to correspond to the free energy minimum closest to the entry point z_1 and (ii) to compute τ_2, $z_2 \approx 3.5$ nm was chosen, which corresponds to the exit from the bilayer. In this calculation, we used the values of diffusivity averaged over all positions in the bilayer (see Fig. 10.9b). The computed values of τ_1 and τ_2 are shown in Fig. 10.10.

Since none of the considered nanoparticles experiences a significant energy barrier for entering the bilayer, the nanoparticles diffuse into the bilayer relatively quickly. This is confirmed by the small values of the computed permeation time τ_1 for all considered nanoparticles. However, the transport time τ_2 across the entire bilayer is several orders of magnitude larger, which demonstrates that the nanoparticles enter the bilayer very quickly and spend the majority of the transport time inside the bilayer interior. Moreover, the average time to cross the bilayer increases exponentially with the nanoparticle size. This increase in transport time is caused by the decreased diffusivity, the deeper free energy minimum, and the added free energy barrier in the bilayer interior. Moreover, the residence time depends not only on the particle size but also on the particle shape. During this long residence time, the nanoparticle may significantly disrupt the integrity of the bilayer either through chemical reactions or through physical effects. A possible chemical reaction leading to the disruption of cell

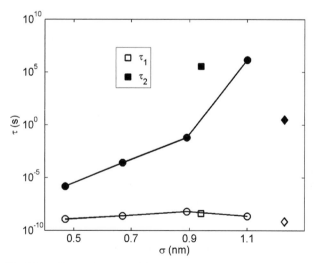

FIGURE 10.10 Mean times of nanoparticle permeation into (i) the membrane interior (τ_1) and (ii) the cellular interior (τ_2) are shown by the open and closed symbols, respectively. The transport times of the spherical, cubical, and tetrahedral nanoparticles are shown by the circles, the squares, and the diamond-shaped symbols, respectively.

membranes is lipid peroxidation, as proposed by Sayes et al. (49, 53). Physical effects of MNs that may lead to a local instability of the lipid bilayer could be similar to the nonspecific antimicrobial action of some peptides. These peptides act directly on the membrane of a target microorganism (54, 55) to produce a localized thinning of the membrane and/or formation of a nonlamellar lipid structure eventually leading to cell membrane lysis.

10.4   DISCUSSION AND CONCLUSIONS

Despite the anticipated safeguards that will be developed to limit environmental pollution with and human exposure to MNs, their use in activities such as the remediation of contaminated effluents (e.g., (56–58) and their introduction into different consumer products (59) are clear indications for future widespread occurrence in natural systems (60). This observation provides a rationale for studies focusing not only on toxicity but also on fate and transport on MNs in the environment. Although current findings on toxicity of MN are rather conflicting, they do raise relevant questions and concerns. The use of microbiotests in this study confirms the previously reported toxicity of C_{60} on bacteria and aquatic biota. Recently published toxicity studies have shown that the impact of C_{60} on bacteria and aquatic biota could vary significantly depending on the approach used to prepare aqueous C_{60} suspensions. Ideally, both short- and long-term toxicity experiments should closely mimic real-world scenarios to obtain data that could be more relevant for natural systems.

With regard to the identification of different toxicity mechanisms associated with MNs, one viable approach could be the development of predictive modeling tools that can be validated by laboratory toxicity experiments. The modeling component of this research was focused on investigation of the transport properties of carbonaceous nanoparticles across the lipid bilayer, a common feature of cell membranes. The nanoparticles are too large to fit within voids between the lipid molecules of the membrane and, therefore, their transport will not take place as a series of activated jumps between these voids observed by Marrink and Berendsen (20, 21) for small molecules. Instead, the nanoparticle diffusion through the membrane can be visualized as a Brownian motion through a continuous heterogeneous medium of the membrane and modeled by the Langevin equation 10.6.

This computational investigation was performed using the coarse-grained molecular dynamics model, which allowed us to relatively quickly assess transport properties of nanoparticles for a range of shapes and sizes. The coarse-grained model for lipids and water used in this work is taken directly from the work of Marrink et al. (27), where it has been thoroughly validated by comparison with the data from the atomistic simulations and experiments. One of the considered models for nanoparticles (a hydrophobic sphere with the Lennard-Jones diameter of 1.1 nm) allowed us to predict the free energy profile in a qualitative agreement with the profile obtained for C_{60} using the atomsitically detailed simulations (18).

The calculated free energy profiles demonstrate that there is no significant energy barrier to enter the bilayer for any of the nanoparticles studied. This suggests and is confirmed in Section 10.3.2.3 that these nanoparticles may enter the bilayer relatively quickly and that most of the transport time is spent inside the bilayer. Marrink and Berendsen (21) in their study of small molecules suggested that hydrophobic interior of a lipid bilayer acts as a trap for hydrophobic particles. Our study indicates further that the "trapping" effect is significantly increased with particle size. The free energy computed for the larger particles possesses a deeper minimum due to an increase in the number of possible hydrophobic interactions. This suggests that size plays a crucial role in the amount of time a nanoparticle will spend in the bilayer interior. In addition, our results demonstrate that the nanoparticle shape significantly affects its transport rate (see Fig. 10.10).

Since the hydrophobic nanoparticles will spend a significant amount of time within the cell membrane, they will have the opportunity to impact the membrane interior. As mentioned earlier, the specific mechanism(s) of action of fullerenes or other hydrophobic nanoparticles on cellular membranes is/are currently not clear. In several studies (49, 53, 61) it has been argued that C_{60} and its derivatives lead to the generation of reactive oxygen species and hence the peroxidation of cellular membranes that eventually leads to leaky membranes and cell death. On the other hand, other studies have demonstrated that C_{60} and some of its derivatives are not toxic and would actually act as powerful antioxidants (62, 63)

Even if fullerenes do not lead to a chemical damage of the membrane, it is possible that they cause damage due to physical interactions. Experiments of Jeng et al. (64) demonstrated that C_{60} can permeate into the lipid bilayer interior and induce changes in membrane morphology. Also, recent MD simulation studies have provided preliminary data supporting the possibility of lipid membrane damage by fullerenes and their derivatives (17, 18). In contrast to the above observations, and using a different experimental approach to that of Jeng et al. (64), Spurlin and Gewirth (65) did not observe the permeation of C_{60} into a lipid bilayer. Similar to the conflicting findings of toxicity studies of MN, the above discrepancy calls for further and more detailed investigations.

ACKNOWLEDGMENTS

This work was supported by the US EPA (STAR program grant # R832635.) and seed funds from the School of Natural Resources and the Environment (SNRE), University of Florida.

REFERENCES

1. Masciangioli, T., Zhang, X. Environmental technologies at nanoscale. *Environ Sci Technol* 2003;37:102A–108A.

2. Florence, A.T., Hillery, A.M., Hussain, N., Jani, P.U. Nanoparticles as carriers for oral peptide absorption: studies on particle uptake and fate. *J Control Release* 1995;36:39–46.

3. Jensen, A.W., Wilson, S.R., Schuster, D.I. Biological applications of fullerenes. *Bioorg Med Chem* 1996;6:767–779.

4. Wang, I.C., Tai, L.A., Lee, D.D., Kanakamma, P.P., Shen, C.K.-F., Luh, T.-Y., Wang, C.-B., Zhang, W.-X. Synthesizing nanoscale iron particles for rapid and complete dechlorination of TCE and PCBs. *Environ Sci Technol* 1997;31:2154–2156.

5. Davis, S.S. Biomedical applications of nanotechnology: implications for drug targeting and gene therapy. *Trends Biotechnol* 1997;6:217–224.

6. Zajtchuk, R. New technologies in medicine: biotechnology and nanotechnology. *Dis Mon* 1999;45:453–495.

7. Kotelnikova, R.A., Bogdanov, G.N., Frog, E.C., Kotelnikov, A.I., Shtolko, V.N., Romanova, V.S., Andreev, S.M., Kushch, A.A., Fedorova, N.E., Medzhidova, A.A. Nanobionics of pharmacologically active derivatives of fullerene C_{60}. *J Nanopart Res* 2003;5:561–566.

8. Eng, T.R. Population health technologies: emerging innovations for the health of the public. *Am J Prev Med* 2004;26:237–242.

9. Tungittiplakor, W., Lion, L.W., Cohen, C., Kim, J.-Y. Engineered polymeric nanoparticles for soil remediation. *Environ Sci Technol* 2004;38:1605–1610.

10. Blaise, C., Ferard, J.F., Vasseur, P. Microplate toxicity tests with microalgae: a review. In: Wells, P.G., Lee, K., Blaise, C., editors. *Microscale Toxicology, Advances, Techniques and Practice*, Boca Raton, USA: CRC Press; 1986.

11. Blaise, C. Microbiotests in aquatic ecotoxicology: characteristics, utility and prospects. *Environ Toxicol Water Qual Int J* 1991;6:145–156.

12. Janssen, C. Alternative assays for routine toxicity assessments: a review. In: Schuurmann, G., Markert, B. editors. *Ecotoxicology: Ecology Fundamentals, Chemical Exposure, and Biological Effects*, New York: Wiley & Sons, Inc.; 1997. pp. 813–839.

13. Lecoanet, H.F., Wiesner, M.K. Velocity effects on fullerene and oxide nanoparticle deposition in porous media. *Environ Sci Technol* 2004;38:4377–4382.

14. Lecoanet, H.F., Bottero, J.-Y., Wiesner, M.K. Laboratory assessment of the mobility of nanomaterials in porous media. *Environ Sci Technol* 2004;38:5164–5169.

15. Hyung, H., Fortner, J.D., Hughes, J.B., Kim, J.-H. Natural organic matter stabilizes carbon nanotubes in the aqueous phase. *Environ Sci Technol* 2007;41:179–184.

16. Fortner, J.D., Lyon, D.Y., Sayes, C.M., Boyd, A.M., Falkner, J.C., Hotze, E.M., Alemany, L.B., Tao, Y.J., Guo, W., Ausman, K.D., Colvin, V.L., Hughes, J.B. C_{60} in water: nanocrystal formation and microbial response. *Environ Sci Technol* 2005;39:4307–4316.

17. Chang, R., Violi, A. Insights into the effect of combustion-generated carbon nanoparticles on biological membranes: a computer simulation study. *J Phys Chem B* 2006;110:5073–5083.

18. Qiao, R., Roberts, A.P., Mount, A.S., Klaine, S.J., Ke, P.C. Translocation of C_{60} and its derivatives across a lipid bilayer. *Nano Lett* 2007;7:614–619.

19. Finkelstein, A. Water and nonelectrolyte permeability of lipid bilayer membranes. *J Gen Physiol* 1976;68:127–135.

20. Marrink, S.J., Berendsen, H.J.C. Simulation of water transport through a lipid membrane. *J Phys Chem* 1994;98:4155.

21. Marrink, S.J., Berendsen, H.J.C. Permeation process of small molecules across lipid membranes studied by molecular dynamics simulations. *J Phys Chem* 1996;100:16729.

22. Geiser, M., Rothen-Rutishauser, B., Kapp, N., Schürch, S., Kreyling, W., Schulz, H., Semmler, M., Hof, V.I., Heyder, J., Gehr, P. Ultrafine particles cross cellular membranes by nonphagocytic mechanisms in lungs and in cultured cells. *Environ Health Perspect* 2005;113:1555–1560.

23. Rothen-Rutishauser, B.M., Schürch, S., Haenni, B., Kapp, N., Gehr, P. Interaction of fine particles and nanoparticles with red blood cells visualized with advanced microscopic techniques. *Environ Sci Technol* 2006;40:4353–4359.

24. Wilson, M.A., Pohorille, A. Mechanism of unassisted ion transport across membrane bilayers. *J Am Chem Soc* 1996;118:6580.

25. Tieleman, D.P., Marrink, S.J., Berendsen, H.J.C. A computer perspective of membranes: molecular dynamics studies of lipid bilayer systems. *Biochem Biophys Acta* 1997;1331:235–270.

26. Ulander, J., Haymet, A.D.J. Permeation across hydrated DPPC lipid bilayers: simulation of the titrable amphiphilic drug valproic acid. *Biophys J* 2003;85:3475–3485.

27. Marrink, S.J., de Vries, A.H., Mark, A.E. Coarse grained model for semi-quantitative lipid simulations. *J Phys Chem B* 2004;108:750–760.

28. US EPA. *Short-Term Methods for Estimating the Chronic Toxicity of Effluents and Receiving Waters to Freshwater Organisms*, 3rd ed. EPA/600/4-91/002, Cincinnati, OH; 1994.

29. Bitton, G., Koopman, B. Bacterial and enzymatic bioassays for toxicity testing in the environment. *Rev Environ Contam Toxicol* 1992;125:1–22.

30. Bitton, G., Jung, K., Koopman, B. Evaluation of a microplate assay specific for heavy metal toxicity. *Arch Environ Contam Toxicol* 1994;27:25–28.

31. Deguchi, S., Rossitza, G.A., Tsujii, K. Stable dispersion of fullerenes, C60 and C70 in water: preparation and characterization. *Langmuir* 2001;17:6013–6017.

32. Smit, J. Attempts to put the standard model on the lattice. *Nucl Phys B* 1988;4:451.

33. Palmer, B.J., Liu, J. Simulations of micelle self-assembly in surfactant solutions. *Langmuir* 1996;12:746.

34. Goetz, R., Lipowsky, R. Computer simulations of bilayer membranes: self-assembly and interfacial tension. *J Chem Phys* 1998;108:7397.

35. Shelley, J.C., Shelley, M.Y., Reeder, R.C., Bandyopadhyay, S., Klein, M.L. A coarse grain model for phospholipid simulations. *J Phys Chem B* 2001;105:4464.

36. Izvekov, S., Voth, G.A. A multiscale coarse-graining method for biomolecular systems. *J Phys Chem B* 2005;109:2469–2473.

37. Humphrly, W., Dalke, A., Schulten, K. VMD–Visual molecular dynamics. *J Molec Graphics* 1996;14:33–38.

38. Goel, A., Howard, J.B., Vander Sande, J.B. Size analysis of single fullerene molecules by electron microscopy. *Carbon* 2004;42:1907–1915.

39. van der Spoel, D., Lindahl, E., Hess, B., Groenhof, G., Mark, A.E., Berendsen, H.J.C. GROMACS: fast, flexible, and free. *J Comput Chem* 2005;26:1701–1718.

40. Berendsen, H.J.C., Postma, J.P.M., DiNola, A., Haak, J.R. Molecular dynamics with coupling to an external bath. *J Chem Phys* 1984;81:3684.

41. Allen, M.P., Tildesley, D.J. *Computer Simulation of Liquids*. New York: Oxford University Press; 1987.

42. Kopelevich, D.I., Panagiotopoulos, A.Z., Kevrekidis, I.G. Coarse-grained kinetic computations for rare events: application to micelle formation. *J Chem Phys* 2005; 122:044908.

43. Hummer, G., Kevrekidis, I.G. Coarse molecular dynamics of a peptide fragment: free energy, kinetics, and long-time dynamics computations. *J Chem Phys* 2003;118:10762.

44. Ryckaert, J.P., Ciccotti, G., Berendsen, H.J.C. Numerical-integration of cartesian equations of motion of a system with constraints: molecular-dynamics of n-alkanes. *J Comput Phys* 1977;23:327.

45. Gardiner, C.W. *Handbook of Stochastic Methods for Physics, Chemistry, and the Natural Sciences.* Berlin: Springer-Verlag; 1983.

46. Zwanzig, R. *Nonequilibrium Statistical Mechanics.* New York, NY: Oxford University Press; 2001.

47. Roux, B., Karplus, M. Ion transport in a gramicidin-like channel: dynamics and mobility. *J Phys Chem* 1991;95:4856–4868.

48. Oberdoster, E. Manufactured nanomaterials (fullerenes, C_{60}) induce oxidative stress in the brain of juvenile Largemouth Bass. *Environ Health Perspect* 2004;112:1058–1062.

49. Sayes, C.M., Fortner, J.D., Guo, W., Lyon, D., Boyd, A.M., Ausman, K.D., Tao, Y.J., Sitharaman, B., Wilson, L.J., Hughes, J.B., West, J.L., Colvin, V.L. The differential cytotoxicity of water-soluble fullerenes. *Nano Lett* 2004;4:1881–1887.

50. Oberdoster, E., Zhu, S., Blickley, T.M., McClellan-Green, P., Haasch, M.L. Ecotoxicology of carbon-based engineered nanoparticles: effects of fullerenes (C60) on aquatic organisms. *Carbon* 2006;44:1112–1120.

51. Lyon, D.Y., Adams, L.K., Falkner, J.C., Alvarez, P.J.J. Antibacterial activity of fullerenes water suspensions: effects of preparation method and particle size. *Environ Sci Technol* 2006;40:4360–4366.

52. Zhu, S., Oberdörster, E., Haasch, M.L. Toxicity of an engineered nanoparticle (fullerene, C_{60}) in two aquatic species, *Daphnia* and fathead minnow. *Mar Environ Res* 2006;62(1): S5–S9.

53. Sayes, C.M., Gobin, A.M., Ausman, K.D., Mendez, J., West, J.L., Colvin, V.L. Nano-C_{60} cytotoxicity is due to lipid peroxidation. *Biomaterials* 2005;26:7587–7595.

54. Epand, R.M., Vogel, H.J. Diversity of antimicrobial peptides and their mechanisms of action. *Biochim Biophys Acta* 1999;1462:11–28.

55. Shai, Y., Oren, Z. From 'carpet' mechanism to de-novo designed diastereomeric cell-selective antimicrobial peptides. *Peptides* 2001;22:1629–1641.

56. Pitoniak, E., Wu, C.Y., Londeree, D., Mayzick, D., Bonzongo, J.C., Powers, K., Sigmund, W. Nanostructured silica-gel doped with TiO_2 for mercury vapor control. *J Nanopart Res* 2003;5:281–292.

57. Borderieux, S., Wu, C.Y., Bonzongo, J.C., Powers, K. Control of elemental mercury vapor in combustion systems using Fe_2O_3 nanoparticles. *Aerosol & Air Qual* 2004;4(1):74–90.

58. Puurunen, K., Vasara, P. Opportunities for utilising nanotechnology in reaching near-zero emissions in the paper industry. *J Clean Prod* 2007;15:1287–1294.

59. Melaiye, A., Sun, Z., Hindi, K., Milsted, A., Ely, D., Reneker, D.H., Tessier, C.A., Youngs, W.J. Silver(I)-imidazole cyclophane *gem*-diol complexes encapsulated by electrospun tecophilic nanofibers: formation of nanosilver particles and antimicrobial activity. *J Am Chem Soc* 2005;127(7):2285–2291.

60. Moore, M.N. Do nanoparticles present ecotoxicological risks for the health of the aquatic environment? *Environ Int* 2006;32:967–976.

61. Pickering, K.D., Wiesner, M.R. Fullerol-sensitized production of reactive oxygen species in aqueous solution. *Environ Sci Technol* 2005;39:1359–1365.

62. Wang, I.C., Tai, L.A., Lee, D.D., Kanakamma, P.P., Shen, C.K.F., Luh, T.Y., Cheng, C.H., Hwang, K.C. C-60 and water-soluble fullerene derivatives as antioxidants against radical-initiated lipid peroxidation. *J Med Chem* 1999;42(22):4614–4620.

63. Gharbi, N., Pressac, M., Hadchouel, M., Szwarc, H., Wilson, S.R., Moussa, F. [60] Fullerene is a powerful antioxidant *in vivo* with no acute or subacute toxicity. *Nano Lett* 2005;5:2578–2585.

64. Jeng, U.-S., Hsu, C.-H., Lin, T.-L., Wu, C.-M., Chen, H.-L., Tai, L.-A., Hwang, K.-C. Dispersion of fullerenes in phospholipid bilayers and the subsequent phase changes in the host bilayers. *Physica B* 2005;357:193–198.

65. Spurlin, T.A., Gewirth, A.A. Effect of C_{60} on solid supported lipid bilayers. *Nano Lett* 2007;7:531–535

CHAPTER 11

In Vitro Models for Nanoparticle Toxicology

JOHN M. VERANTH

Department of Pharmacology and Toxicology, University of Utah, 30 South 2000 East, Salt Lake City, UT 84112, USA

11.1 INTRODUCTION

This chapter discusses the use of *in vitro* cell culture models to study the responses of mammalian cells to environmental and occupational nanomaterials including fullerenes, carbon nanotubes, and oxides such as Al_2O_3, SiO_2, TiO_2, Fe_2O_3, and ZnO. The focus will be on the toxicology of manufactured carbon-based and metal oxide particles, but many of the issues are also applicable to incidental nanoparticles and to organic nanomaterials formulated for medical imaging and drug delivery applications.

Inhalation exposure of animals and humans to ambient ultrafine particles (diameter < 0.1 μm) has been associated with a range of adverse biological effects including pulmonary inflammation, alteration of cardiac rhythm, and increased blood coagulation (1, 2). Manufactured powders within this same size range are referred to as nanomaterials rather than as ultrafine particles, but this reflects common usage, not a real scientific difference between ultrafine and nanoparticles. The biochemical mechanisms by which ultrafine or nanosized solid particles induce cellular effects remain elusive. Studies using particles of the same nominal composition have produced contradictory results, suggesting uncontrolled variables. There is little understanding on how specific biological responses are induced by a real-world complex mixture of particles that have different sizes, chemical compositions, and surface modifications. Laboratory toxicology studies with environmental particles and manufactured nanoparticles have shown that insoluble inorganic particles induce pulmonary inflammation (3–6) but the proximal receptors that initiate the biochemical signaling have not been conclusively identified. Cell culture-based studies have been extensively used to study the toxicology of inhalable particles such as asbestos, silica,

---

Nanoscience and Nanotechnology   Edited by Vicki H. Grassian

and combustion emissions, and these methods are now being adapted to study nanomaterials.

Exposure to manufactured nanomaterials can occur by many different pathways. Particle inhalation can be the result of occupational dust from material processing or fugitive emissions to the atmosphere. Ingestion exposure can occur from hand-to-mouth contact in occupational settings, from environmental releases that eventually reach drinking water, and from nanosized powders that are used in formulating consumer products. Dermal exposure to nanomaterials can occur both at workplace and through the use of consumer products such as sunscreens. Systemic exposure can result from translocation of inhaled materials from the lung (7) by accidental skin puncture or by deliberate injection of materials used for medical imaging or as drug carriers. Since there are many exposure routes and target organs, it is logical that *in vitro* studies have used a wide range of cell types derived from lung, colon, vascular, liver, nerve, and skin sources.

Silicon dioxide and carbon black are examples of industrial chemicals that are already being produced in large quantities (many tons/hour) as nanosized powders. The catalogs of chemical suppliers and specialty nanomaterial synthesis companies include many novel particle compositions and grades. Suppliers are actively marketing new grades of these powders and suspensions with ever smaller primary particle size as they pursue the nanomaterials market. Production and the potential for human exposure are expected to increase as the novel properties of these nanomaterials find new applications.

Silica nanoparticles are commercially available with various surface functional group modifications including amine and carboxyl-functionalized particles. Inhalation of high concentrations of freshly ground crystalline silica has long been associated with occupational disease, but SiO_2 is also used in skin care and cosmetic formulations and is approved as a food additive where it can be present at levels up to 1–2% (8). When materials are approved for human contact and ingestion, they are often considered "safe" in industrial applications as well, and current environmental and occupational safety regulations do not make any distinction between nanoparticles and supermicron-sized particles of the same nominal chemical substance. Differences in surface functional groups are also rarely considered because these surface moieties are only a "trace" part of the bulk composition. Consumer Web sites are expressing concern regarding possible health effects of nanoparticles in cosmetics and topical skin products, but few studies of the toxicology of these nanomaterials have been published in the peer-reviewed literature. Human epidemiology only detects damage that has already been done, and animal studies are expensive. There is a great need for cost-effective, high-throughput assays for evaluating novel nanomaterials.

11.1.1 Benefits of *In Vitro* Testing for Particle Toxicology

Cell culture assays have many advantages for fundamental studies of particle toxicology. Cell cultures are easy to manipulate by molecular biology techniques, making *in vitro* models ideal for testing mechanistic biochemical hypotheses. Genetic manipulation techniques such as transfection to overexpress a specific receptor protein

(9) and gene silencing (10) are available to test the roles of specific biomolecules in mediating the response.

Another advantage is the low cost and high throughput of *in vitro* assays compared to animal studies. Animal exposures, controlled human exposures, and epidemiology will likely remain the definitive method for establishing exposure guidelines for occupational, household, and environmental airborne particles. However, due to the virtually unlimited possible combinations of particle size, bulk chemical composition, surface chemistry, and external mixtures with other particles, animal testing of all environmentally relevant particle types is not feasible. Multiwell cell culture plates allow testing many combinations of particle agonists, particle concentrations, exposure times, chemical cofactors, and inhibitor drugs in a few days.

Recent workshops on the environmental implications of nanotechnology have emphasized the role of *in vitro* assays especially in preliminary screening strategies to establish research priorities (11, 12). The advantages and limitations of cell culture assays for particle toxicology were comprehensively reviewed by Fubini et al. (13). This review sheds light on the current status of assays that are applicable to the toxicity, mutagenicity, and carcinogenicity of solid materials, discusses the particle surface and bulk properties that have been associated with adverse responses in animals, and addresses research needs. Although the emphasis of the review was on asbestos and mineral fibers, many of the issues are directly applicable to nanoparticles. The workshop recommended a testing strategy for physicochemical characterization of the test materials, acellular tests, and *in vitro* cellular tests prior to animal studies as a way to minimize animal use. There is increasing pressure from the public and government agencies for reducing, refining, and replacing animal use in scientific experimentation. Development of *in vitro* screening assays that can reliably predict which particle types are potent inducers of airway inflammation would be very valuable as a replacement for preliminary toxicological assessments using animal exposures. A large number of workshop and task force reports relevant to the development of nonanimal toxicology assays can be found on the Web site of the European Centre for the Validation of Alternative Methods.

A recommended screening strategy specifically for nanoparticles has also been proposed by the International Life Sciences Institute Research Foundation/Risk Science Institute (12). Key elements of this strategy include physicochemical characteristics, *in vitro* assays (acellular and cellular), and *in vivo* assays. Although public health standards are ultimately based on human data, *in vitro* toxicology is recognized as a powerful approach to elucidate mechanisms at the cellular, biochemical, and molecular level (14).

A major limitation of *in vitro* assays is that they lack the complex endocrine and neurological signaling interactions that take place between organs in an intact animal. Blood flow removes soluble components, brings macrophages and other mobile cells to the site of particle deposition, and carries biochemical signals between distant organs. Particle-induced neural stimuli affect voluntary and involuntary muscle action and can modulate endocrine signaling. Cocultures, such as experiments using both macrophages and epithelial cells, are a way to simulate *in vitro* the cytokine interactions between different cell types. To date, no consensus has developed regarding the most appropriate methods for *in vitro* particle toxicology, and investigators are using a wide range of model cells and assay methods.

11.1.2   Cells and Methods

Recent studies with nanoparticles have built on previous investigations of the toxicology of larger particles, and as a result many different cell types are being used for *in vitro* studies of nanoparticle toxicology. Typical human lung-derived cells include both immortalized lines such as BEAS-2B and A549 (available from American Type Culture Collection) and donor cells that have finite lifetime in culture such as normal human bronchial epithelial cells (NHBE) and small airway epithelial cells (SAEC) (both from Lonza). The immortalized cells are typically still in growth phase when treated with particles. In contrast, techniques have been developed to produce cultures containing pseudostratified and differentiated cells that more closely resemble the airway epithelium, for example, the EpiAirway™ model (MatTek). Rodent lung cells are also used. Examples include the mouse macrophage-like RAW264.7 cell line and primary macrophages harvested by bronchoalveolar lavage (BAL).

Cell models for other organs such as the colon and vascular system are also available. Currently, there are no procedures for the culture of primary colonocytes; therefore, colon cancer cell lines are the only human model systems for cell-based studies. RKO cells, while transformed, do not have mutations to genes that are commonly mutated in colon cancer, such as p53 or APC. CaCo-2 cells are unique in that they can be grown such that they show characteristics of organization/differentiation with a luminal and a basolateral surface. Studies of vascular cells have used commercial human aortic endothelial cells (Lonza) and human umbilical vein epithelial cells. Studies of neural signaling have used dorsal root ganglion (DRG) neuron explants (15).

Submerged cell culture, Fig. 11.1, is the most popular method. Cells are seeded in multiwell plates and typically given one to three days to attach to the plastic, assume a flattened shape, and continue proliferating. Mammalian cells are generally grown in an incubator at 37°C with 3–5% CO_2. The particles, suspended in cell culture media or buffer, are applied to the culture wells by replacing the media. Time from treatment to measurement of response varies. Only a few minutes are needed for calcium ion influx fluorescence-based assays. Harvesting cells for measurement of gene transcription by polymerase chain reaction (PCR) is typically done at 4–24 h. Assessing cell viability and harvesting the cell culture media for measurement of secreted cytokines is typically done at 24 h. Many cell culture protocol variations exist. Variables include the cell type, the cell seeding density, the culture media, the addition or withdrawal of serum from the media, the time between steps, and the biological end points measured.

In submerged culture, the cells are covered by 1–3 mm of cell culture media. Sufficient media volume must be available to supply nutrients and dilute waste products until the next media change, typically 1–2 days. The media must not be too deep since oxygen and carbon dioxide need to diffuse between the incubator

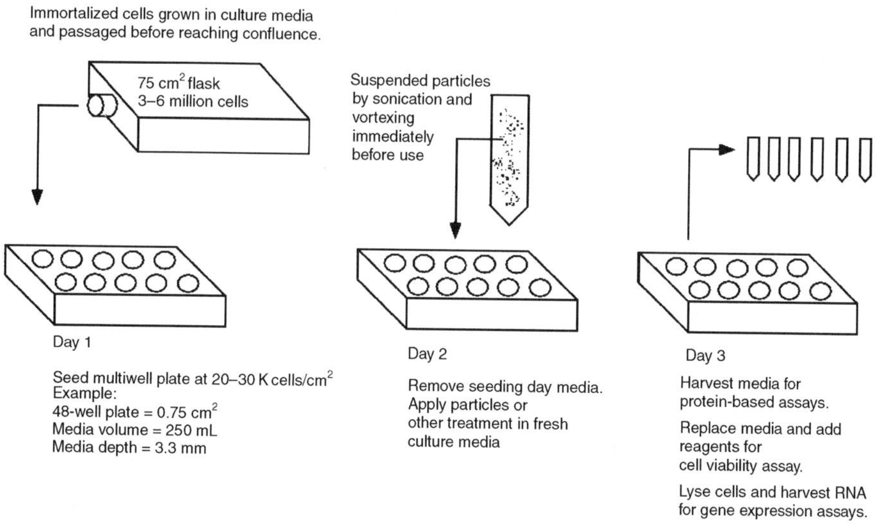

Immortalized cells grown in culture media
and passaged before reaching confluence.

75 cm² flask
3–6 million cells

Suspended particles
by sonication and
vortexing
immediately
before use

Day 1

Seed multiwell plate at 20–30 K cells/cm²
Example:
48-well plate = 0.75 cm²
Media volume = 250 mL
Media depth = 3.3 mm

Day 2

Remove seeding day media.
Apply particles or
other treatment in fresh
culture media

Day 3

Harvest media for
protein-based assays.

Replace media and add
reagents for
cell viability assay.

Lyse cells and harvest RNA
for gene expression assays.

FIGURE 11.1   Illustration of a typical *in vitro* cell culture particle toxicology experiment.

atmosphere and the cells. As discussed in Section 11.4, the geometry of submerged cell culture has a great effect on the effective nanoparticle dose and also causes dilution of secreted cytokines.

Cells grown on an air–liquid interface have been proposed as a more realistic lung cell model (16, 17). In this system, cells are grown on a porous membrane. During the particle exposure, the membrane is positioned at the top of the liquid so that cells are exposed to air while the liquid below maintains moisture and nutrient supply. Generally, the cells are grown in Transwells™ using submerged culture and are exposed at the air–liquid interface for only a few hours at a time since drying is a problem.

Another technique uses coculture of multiple cell types, often on the opposite sides of a Transwell™ membrane. The goal is to simulate short-range cytokine communication where one cell type stimulates a response in another. An example is the coculture of BEAS-2B human lung epithelial cells with THP-1 cells, a human peripheral blood monocyte-derived cell line (18).

11.1.3 Commonly Studied Toxicology End Points

Many different end points are measured by using *in vitro* assays depending on the laboratory capability and the biochemical hypotheses being tested. These include cytotoxicity, cell function, secretion of cytokines and other protein biomarkers, changes in gene regulation, DNA damage, and particle uptake.

Cytotoxicity can be quantified by assays that involve measurement of metabolism, changes in cell permeability, or leakage of cell contents. Metabolism-based assays such as the MTT [3-(4,5-dimethylthiazol-2-yl)-2,5-diphenyltetrazolium bromide] (19) or the WST-8 (2-(2-methoxy-4-nitrophenyl)-3-(4-nitrophenyl)-5-(2,4-disulfophenyl)-2H-tetrazolium, monosodium salt) assays are based on the rate at which live cells convert a tracer molecule to form a colored product that can be quantified using a plate reader. Many dyes such as propidium iodide and trypan blue show differential transport through intact versus damaged cell membranes, and this can be used to count total and dead cells either manually using a microscope or automatically using a flow cytometer. Lactate dehydrogenase (LDH) is a widely used cytotoxicity assay because it is present in all cell types and is rapidly released into the cell culture medium upon damage of the plasma membrane.

Apoptosis assays address the mechanism of cell death. Many different commercial kits are available that are based on differential labeling of live, apoptotic, and necrotic cells. Most commonly, these kits involve fluorescent markers and are suitable for flow cytometry cell counting.

Often, the research questions focus on particle-induced cell damage processes that stop short of death. These processes include alteration of cell growth rate, oxidative damage to DNA and macromolecules, and other reversible effects. These reversible effects can be quantified directly by detection of damage products or indirectly by upregulation of repair pathways. Examples of damage product assays include measurements of DNA strand breaks by the comet assay and measurement of lipid oxidation products such as malondialdehyde. Transepithelial cell resistance of confluent cell culture is another measure of cell function after particle exposure.

Many hypotheses regarding systemic effects of particle exposure involve tissue inflammation. Based on this, measurement of cytokines associated with the onset and resolution of inflammations is a common *in vitro* end point. Cytokines of interest include interleukin (IL)-1β, IL-6, IL-8, IL-10, and TNF-α, but many signaling molecules have also been measured in particle toxicology studies. References discuss the role of various cytokines in the lung (20, 21). Cytokines are secreted by the particle-treated cells into the culture medium that are then harvested for assay of the specific signaling protein, most commonly by enzyme-linked immunosorbent assay (ELISA). ELISA requires a good, species-specific antibody for the protein of interest. Commercial ELISA kits are available for many human and rodent cytokines.

Changes in proinflammatory signaling can also be measured at the gene regulation level by quantifying the level of messenger RNA for a protein of interest. These genes coding proteins of interest include the cytokines listed above, transcription factors regulating expression of other genes, and enzymes involved in various damage repair pathways. For gene expression analysis the cells are harvested, lysed, and processed to extract total RNA. The RNA is then transcribed into cDNA, which is much more stable in storage. Primers are synthesized for the genes of interest, and the cDNA is amplified using the PCR. Qualitative PCR involves running the amplified product on a gel and comparing relative size of the band corresponding to the gene of interest in treated and control cell samples. Quantitative real-time PCR (qRT-PCR) involves using a fluorescent marker to measure the amount of gene product after each amplification cycle. The relative amount of the starting mRNA for the gene of interest is determined by the number of amplification cycles needed for the signal from the amplified product to exceed a specified threshold amount (the C_t value). Advantages of qRT-PCR include high sensitivity and dynamic range spanning three or more orders of magnitude. Comparing the gene of interest to a "housekeeping gene" that transcribed at very constant levels allows correcting for differences between samples in the number of cells and the efficiency of the RNA processing. In addition to inflammation markers, gene regulation measurements often include oxidative stress response genes and DNA damage repair pathway genes.

Many hypotheses regarding nanoparticle toxicology involve the relative rate of particle uptake as a function of particle size. Measuring nanoparticle uptake is difficult unless the particles have been labeled with a fluorophore or radioisotope. Even with fluorescent particles, it is often difficult to distinguish between internalized particles and particles that have attached to the outside of the cell. Confocal microscopy can provide three-dimensional data proving that the particles are internalized. Semi-quantitative information on the extent of particle uptake can be obtained from image analysis.

11.2   CELL RESPONSES TO NANOMATERIALS

The *in vitro* responses of cells to nanomaterials is an area of active research, and no clear conclusions have emerged. Published results have suggested new hypotheses but have also raised questions as to whether the results can be generalized to whole animal and human responses. Owing to the number of papers being currently published, it is not possible to do a comprehensive and critical review of the literature. Rather, a selection of recent studies will be presented to illustrate the types of experiments that are being done and to highlight some of the results.

11.2.1   Effect of Particle Size

A widely stated hypothesis is that nanosized particles are more potent than larger particles of the same nominal substance because of their increased surface area per unit mass. This has been supported by animal studies with carbon black and titanium dioxide (22, 23). Both Donaldson (24) and Oberdörster (25) concluded in reviews that ultrafine particles of low-solubility and low-toxicity materials are more inflammogenic in the rat lung than larger particles from the same material, and they hypothesized that the effects are related to surface area and involve oxidative stress. *In vitro* results with a wider range of particle compositions have produced contradictory results.

Veranth et al. (26) measured the release of the proinflammatory cytokines IL-6 and IL-8 by BEAS-2B human bronchial epithelial cells treated with manufactured nano- and micron-sized particles of Al_2O_3, CeO_2, Fe_2O_3, NiO, SiO_2, and TiO_2, with soil-derived particles from fugitive dust sources, and with the positive controls LPS, TNF-α, and $VOSO_4$. Nanosized SiO_2 particles caused a statistically significant increase in IL-6 compared to both the untreated control and the cells treated with micron-sized SiO_2 particles in six consecutive experiments with BEAS-2B cells, indicating a reproducible pattern. For the other nano- and micron-sized particle pairs, there was no conclusive evidence for the nanoparticles being consistently more potent than an equal mass concentration of the micron-sized particles of the same nominal composition. Further, the manufactured metal and ceramic oxide particles were less potent for the induction of IL-6 release than an equal mass concentration of three different soil-derived dusts. The LPS, TNF-α, and soluble vanadium positive controls showed that the BEAS-2B cells are capable of producing IL-6 under the experimental conditions. The study concluded that metal oxide particles have low potency to induce IL-6 secretion in BEAS-2B cells.

A combined animal and *in vitro* study that compared crystalline silica (Min-U-Sil, 0.5 μm) with fine (300 nm) and nanosized (50 and 12 nm) quartz concluded that the pulmonary toxicities of α-quartz particles appeared to correlate better with surface activity than with particle size (6). In a hemolytic potential assay using human erythrocytes, the 50 nm quartz was less potent than either the 12-nm or the larger particles.

A study using guinea pig alveolar macrophages reported that the cytotoxicity apparently follows a sequence order on a mass basis: SWNTs > MWNT10 > quartz> C_{60} (27). Treatment concentrations were from 1.4 to 226 μg/cm^2. The trend in cytotoxicity for the carbonaceous materials did not correlate with the smallest dimension of the particle. The least cytotoxic material, C_{60}, was much small, about 1 nm diameter, compared to the diameter of the multiwalled nanotubes 10–20 nm. The authors concluded that carbon nanomaterials with different geometric structures exhibit quite different cytotoxicity and bioactivity *in vitro*.

The cytotoxic and genotoxic effects of nanoparticles and micron-sized particles of cobalt-chrome alloy were compared using human fibroblasts in tissue culture (28). Nanoparticles, which caused more free radicals in an acellular environment, induced more DNA damage than micron-sized particles using the alkaline comet assay. They induced more aneuploidy and more cytotoxicity at equivalent volumetric dose. Nanoparticles appeared to disintegrate within the cells faster than microparticles with the creation of electron-dense deposits in the cell, which were enriched in cobalt. The mechanism of cell damage appears to be different after exposure to nanoparticles and microparticles.

11.2.2 Comparisons of Different Nanoparticle Types and Toxicity Mechanisms

Identifying nanoparticle compositions and surface characteristics associated with biocompatibility versus toxicity has great interest for both environmental health and for design of nanoparticle-based products. Many *in vitro* studies have compared particle types, leading to an understanding of the biochemical mechanisms of action.

A study focusing on oxidative stress tested ambient ultrafine particles and manufactured titanium dioxide, carbon black, fullerol, and polystyrene nanoparticles in RAW264.7 cells (29). The ambient ultrafine particles and the cationic polystyrene nanospheres induced cellular uptake, reactive oxygen species (ROS) production, glutathione (GSH) depletion, and toxic oxidative stress through increased calcium uptake. The TiO_2 and fullerol particles did not induce oxidative stress in cells. The authors concluded that ROS generation and oxidative stress are important end points for comparisons of nanoparticle types.

In a study measuring IL-8 mRNA, IL8 protein release, and glutathione depletion A549 cells exposed to fine and ultrafine titanium dioxide and carbon black, the GSH assay confirmed that oxidative stress was involved in the response to all the particles (30). As expected, DQ12 quartz was more inflammatory than the low toxicity dusts, on a both mass and surface area basis.

Macrophage recruitment is often considered to be a marker of inflammation. An *in vitro* study of chemotactic substances capable of inducing macrophage migration exposed type II alveolar epithelial cells (cell line L-2) to a range of particles (31). Cells treated with carbon black nanoparticles showed significant release of macrophage chemoattractant compared to the negative control and to other dusts tested (fine carbon black and TiO_2 and nanoparticle TiO_2) as measured by macrophage migration toward type II cell-conditioned medium. Size fractionation of the chemotaxin-rich supernatant determined that the chemoattractants released from the epithelial cells were between 5 and 30 kDa in size.

Single- and multiwall carbon nanotubes were compared using A549 and NR8383 cells. A dose- and time-dependent increase of intracellular reactive oxygen species and a decrease of the mitochondrial membrane potential were observed after treatment of both cell types with commercial nanotubes, but the purified nanotubes had no effect (32). The authors concluded that metal traces associated with the commercial nanotubes are responsible for the biological effects.

11.2.3 Particle Uptake Studies

Studies on nanoparticle uptake by cells are motivated by both the seminal work on nanoparticle translocation from animal lungs to other organs (7) and by hypotheses related to particle interaction with intracellular biomolecules. Further, particles internalized by lysosomes experience a much lower pH than particles in the extracellular fluid and cytosol, and this may affect the dissolution and bioavailability of metals. Possible mechanisms of nanoparticle uptake include receptor-mediated phagocytosis that is activated by particle contact and passive uptake by normal pinocytosis that occurs because particles are small enough to be taken up along with the extracellular fluid.

Pulmonary macrophages and red blood cells were exposed to fluorescent polystyrene microspheres (1, 0.2, and 0.078 μm), and particle uptake was assessed by confocal laser scanning microscopy (33). Particle uptake *in vitro* into cells did not occur by any of the expected endocytic processes, but rather by diffusion or adhesive interactions. The authors observed that particles within cells are not membrane bound and hence have direct access to intracellular proteins, organelles, and DNA, which may greatly enhance their toxic potential.

A study using fluorescent-labeled silica particles ranging from 40 nm to 5 μm found that all particle sizes were taken up by the cytoplasm but that nanoparticles smaller than 70 nm were localized in the nucleus, where they interacted with proteins forming aberrant clusters (34).

An uptake study of polystyrene (0.078 μm), gold (0.025 μm), and titanium dioxide (0.02–0.03 μm) nanoparticles used a triple cell coculture model of the human airway wall composed of epithelial cells, macrophages, and dendritic cells (35). Titanium dioxide nanoparticles were detected as single particles without membranes as well as in membrane-bound agglomerations. Gold nanoparticles were found inside the cells as free particles only. Both particle size and material affected the cellular secretion of tumor necrosis factor-alpha.

11.2.4   Cell Model Differences

The term "cell model" describes a specific combination of cell source, culture conditions, and protocols. Typical choices of cells and techniques are described above, but no widely accepted and validated cell models exist for particle toxicology, and investigators have developed many variations. *In vitro* particle assays are assumed to be a simplified biological system that captures the essential biochemistry and cell signaling pathways relevant to the problem being studied. The choice of cell model will affect the observed data, and understanding cell model differences is important for interpretation of results.

Wottrich et al. (18) compared human epithelial (A549) and macrophage-like (THP-1 and Mono Mac 6) cell lines in mono- and coculture. The study used Fe_2O_3 (50–90 nm) and SiO_2 particles (60 and 100 nm). The end points included cytotoxicity by LDH release and the cytokines IL-6 and IL-8. The coculture showed increased cytokine response to the particles compared to individual cell types in monoculture. Data comparing multiple cell types treated with nanomaterials are limited, but additional insights can be gained from experiments with soluble components and micronsized particles.

Environmental nanoparticles can serve as carriers of endotoxin (lipopolysaccharide, LPS) due to their large surface area. A study compared LPS-induced activation of IL-6 and IL-8 secretion by bronchial epithelial cells (BEAS-2B) and type II-like pneumocytes (A549) (36). The A549 cells, but not the BEAS-2B cells, showed concentration-dependent response to soluble CD14, the cofactor for Toll-like receptor response to LPS. Further, the cytokine signaling response of A549 cells, but not BEAS-2B cells, could be suppressed by adding LPS binding protein. The authors concluded that distinct pathways exist for the LPS response in these two cell immortalized lung-derived cell lines.

Nanoparticles can be a source of soluble metals, especially when the particles are internalized by cells and exposed to the lower pH environment inside lysosomes. A comparison of cytotoxicity in three lung-derived cell lines exposed to metals indicated that rat NR8383 macrophages were most sensitive by an order of magnitude compared to rat RLE-6TN epithelial cells and human A549 cells (37). The three cell types ranked the LD_{50} concentrations for the metals similarly (V < Zn < Cu < Ni < Fe).

A study of cytotoxicity caused by polyvinyl chloride and silica particles used six different cell types (38). The sensitivity of the cell systems was A549 < human red blood cell < primary human type II pneumocytes < human alveolar macrophages < rat type II pneumocytes < rat alveolar macrophages.

An experiment comparing a suite of identical particle-relevant agonists in multiple cell culture models showed reproducible differences in the IL-6 cytokine secretion response between cell types and between the same cells grown in different cell culture media compositions (39). Cell types included A549, BEAS-2B, RAW 264.7, and primary macrophages. The identical treatments applied to all cells were soil-derived, diesel, coal fly ash, titanium dioxide, and kaolin particles along with soluble vanadium and lipopolysaccharide. The study also found statistically significant differences associated with protocol variations such as the size of the culture well used.

11.2.5 Comparisons of *In Vitro* and *In Vivo* Results

Use of *in vitro* assays for preliminary screening of particle types prior to animal and human testing requires the validation that the *in vitro* assays correlate with *in vivo* results. Developing cell culture models that predict whole animal responses is a more difficult problem than developing cell culture models that are suitable for testing specific biochemical mechanism hypotheses. Some studies have found good correlation between cell culture and animal results while others have reported contradictory results. Likely, the predictive value of *in vitro* assays will depend on selecting appropriate cell culture models, dose metrics, and biological end points.

The importance of surface area as a metric was supported by a study using A549 cells exposed to fine and ultrafine titanium dioxide and carbon black (30). The end points were IL-8 mRNA, IL8 protein release, and GSH depletion as markers of proinflammatory effects and oxidative stress. Dose–response relationships observed in the *in vitro* assays appeared to be directly comparable with those previously reported from *in vivo* experiments when the doses were similarly standardized. Both sets of data suggested a threshold in dose, measured as surface area of particles relative to that of the exposed cells, at around $1-10\,cm^2/cm^2$.

A study to assess the ability of *in vitro* screening studies to predict *in vivo* pulmonary toxicity of fine or nanoscale particle types in rats used carbonyl iron, crystalline silica, precipitated amorphous silica, nanosized zinc oxide, and fine-sized zinc oxide with particle size ranging from 90 to 500 nm (40). Animal end points were BAL, LDH, and neutrophil recruitment. Cell culture models were rat L2 lung epithelial cells, primary alveolar macrophages collected via BAL from unexposed rats, and alveolar macrophage—L2 lung epithelial cell cocultures. *In vitro* end points included cytotoxicity by LDH and MTT assays and measurement of inflammatory cytokines MIP-2, TNF-α, and IL-6. The study found little correlation of the *in vivo* and *in vitro* measurements and concluded that further assay development is needed.

A study of fullerenes analyzing cytotoxicity of underivitized C_{60} and fully soluble $C_{60}(OH)_{24}$ in human dermal fibroblasts and liver-derived cells (HepG2) concluded that the lethal dose varied over seven orders of magnitude with relatively a minor change in fullerene structure (41). The same research group subsequently did a rat exposure study and found little or no difference in lung toxicity effects between the two fullerene samples when compared to controls (42). Since these data were not consistent with the previously reported *in vitro* effects, they commented that these studies exemplify both the difficulty in interpreting and extrapolating *in vitro* toxicity measurements to *in vivo* effects and highlight the complexities associated with probing the relevant toxicological responses of fullerene nanoparticle systems.

A study using human lung epithelial cells and rat macrophages compared the relative toxicity of combined particulate and semivolatile fractions from gasoline and diesel engines. The authors reported that the rank order of potency from the *in vitro* assays in general did not correspond with the previous rankings from *in vivo* comparisons with the same samples (43).

11.2.6 Summary

The use of *in vitro* models to study nanoparticle toxicology is an active area of research. A wide range of questions regarding the biological effects of particle size, shape, surface chemistry, and bulk composition have been studied. Many hypotheses have been supported by experiments with *in vitro* models, but few general conclusions have been reached. Development of improved *in vitro* assays is an iterative process where the results of recent research are incorporated in the design of future studies.

11.3 QUANTITATIVE CONSIDERATIONS IN DESIGNING *IN VITRO* STUDIES

The recent *in vitro* nanoparticle toxicology studies reviewed above illustrate the wide range of particle types, cell culture models, and biological end points measured. This section presents some quantitative concepts relative to length scale, time, and concentration issues that may be important in both the design of experiments and the interpretation of results in the literature. Particle toxicology is an interdisciplinary problem, and it is very challenging to combine all the concepts from aerosol science, fluid flow, cell physiology, biochemistry, and physical chemistry that may apply to a single experiment. This discussion focuses primarily on the physical and chemical aspects of particle–cell interaction.

11.3.1 Length Scale Issues

Figure 11.2 is drawn to scale and attempts to illustrate a typical cell culture particle toxicology experiment on various length scales or magnifications. The cell culture scale (a) shows one corner of a typical plastic well with adherent cells treated with particles. After seeding, the cells settle to the bottom (A) within an hour or so, attach, develop a flattened morphology, and resume dividing. Some cells attach to the walls of the well (B) creating well-size-dependent edge effects (39). The culture media are replaced, typically one day after seeding, with fresh media that contain the particle suspension or other agonist and any cotreatments with cofactors or inhibitor drugs. The nanoparticles aggregate and settle to the bottom of the well (C), where they are in close proximity to the particle, but some particle fractions remain dispersed in the media and are far from the cells (D) (44). The distance from the liquid surface (E) to the cells presents both a diffusion barrier for gas exchange and a liquid volume that dilutes any soluble species dissolved from the particle or secreted by the cells. On the scale of this drawing, the liquid would be 3 m deep and the wells of a 48-well plate would be 13 m wide.

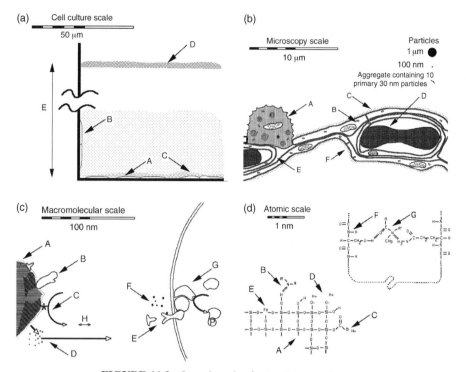

FIGURE 11.2 Length scales for *in vitro* experiments.

The microscopy scale (b) panel illustrates the conditions of the lung *in vivo* from the perspective of a histologist or pathologist. Alveolar type II cells (A) secrete lung surfactant and are involved in epithelial repair. The epithelial cells (B) cover most of the alveoli surface. Upper airway epithelial cells are ciliated, but the terminal bronchial and alveolar cells are not. The alveolar surface is covered by a thin layer of lung surfactant (C) that plays an important role in modifying surface tension and allowing the lung to inflate and contract properly. A red blood cell (D) is shown in a capillary formed from endothelial cells (E). Nanoparticles are known to translocate from the lung to systemic circulation, and the tight junctions (F) between cells are a possible route. Particles of 1000 (1 μm), 100, and 30 nm diameter are shown to scale.

The macromolecular scale (c) panel shows the cell from the perspective of a biochemist and a nanoparticle from the perspective of a material scientist. On this scale, a nanoparticle is typically an irregular and inhomogeneous solid, although one goal of nanoparticle synthesis is to achieve particles with controlled shape and crystalline phase. In cell culture media, the particle surface is expected to be coated with a layer of small molecules (A), but this coating is difficult to characterize since the adsorbed species are in dynamic equilibrium with the bulk liquid. Larger molecules, such as proteins (B), also adsorb to the particle surface. Many theories of nanoparticle toxicology speculate that particle surfaces have active sites that catalytically promote chemical reactions (C). Metal ions can dissolve from the particle surface and diffuse into the media (D). The cell on the right shows a typical biochemical textbook illustration of a receptor protein imbedded in the lipid bilayer and a matching molecular ligand (E). Sometimes, a molecule (F) interacts with a receptor protein at locations other than the normal ligand docking site. The binding of a ligand or exogenous agonist to the transmembrane receptor causes a conformational change in the receptor protein that in turn promotes a series of intracellular protein binding events leading to chemical changes such as activation of a protein catalyst by phosphorylation (G). The particle must be very close to the cell to interact with cell-surface receptor proteins unless the toxicity is mediated entirely by stable soluble molecules. For example, the short arrow (H) illustrates the typical distance an OH• free radical will diffuse in cell culture media before colliding with an organic molecule and reacting.

The atomic scale (d) panel illustrates the cell and a protein molecule from the perspective of a surface or physical chemist. A mineral particle is a crystalline lattice (A) that ultimately is interrupted by the surface. This is illustrated here by silicon dioxide but note that the actual tetrahedral arrangement of the 3D crystal is distorted by the 2D presentation. The illustration emphasizes that relatively large fractions of the atoms are on the surface, along edges, or at corners of a nanoparticle. These surface atoms are not coordinated the same way as atoms deeper in the crystal, and many surface atoms have reacted with, for example, hydroxyl groups. Surface hydroxyl groups can form hydrogen bonds with organic molecules (B). Silica surface sites may be reacted with other functional groups such as carboxyl (C) or amine (not shown). A portion of the surface functional groups are dissociated to form ions on the particle surface that are electrically balanced by a cloud of counterions in solution (D). This electrical double layer is the basis of the zeta potential. Heteroatoms, such as iron (E), may be loosely inserted into the crystal lattice or be on the particle surface as trace impurities. A small portion of a hypothetical receptor protein molecule is shown (F). The normal mechanism of receptor protein response to a biological ligand involves a small molecule (G) docking at a specific site in the folded protein and interacting with specific amino acid side chains by hydrogen bonds and ionic charges.

The illustrations in Fig. 11.2 cover almost five orders of magnitude in length scale. Thinking at these various scales raises many questions regarding how a solid nanoparticle might trigger biochemical responses. Any statement that a solid particle induced an intracellular biological response is a convenient shorthand that greatly oversimplifies the physical and chemical process taking place at the biochemical and atomic scale. Experiments to elucidate these fundamental processes is an excellent application for *in vitro* models since the isolated cells present a simplified and easily manipulated platform for the inorganic particle and cell biomolecules to interact.

11.3.2 Timescale Issues

The reported time from applying particles to cells *in vitro* to measuring biological changes ranges from less than a minute for calcium influx assays (15) to about 24 h for cytotoxicity and cytokine secretion. Upregulation of genes precedes protein expression and studies of gene expression generally harvest cells for RNA at 1–6 h. Since human-derived cells in culture typically double about once a day, experiments longer

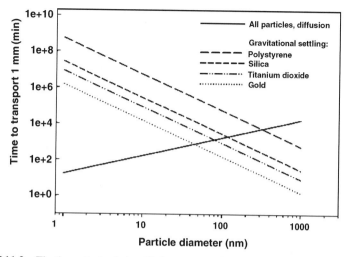

FIGURE 11.3   The theoretical relationship between particle diameter and time to diffuse (solid line) or settle (solid lines) 1 mm in water. Reproduced by permission from Teegaurden et al. (45).

than 24 h are seldom done since results could be confounded by relative rates of cell proliferation. However, nanoparticles are large objects compared to soluble molecules, and transport of particles in the culture dish may be an important factor affecting the effective dose at the particle surface.

A recent study introduced the concept of particokinetics (an analogy to pharmacokinetics) as a quantitative analysis of the particle transport in an *in vitro* experiment (44). The initial application of particles to the cell culture well by pipette is likely to result in a well-mixed suspension. Initiating the cell response requires that the particles come in contact with the cells adhering to the bottom of the well, unless the toxicity is solely caused by soluble molecules released from the particles. Particles in liquid move by gravitational settling and diffusion. Gravity settling depends on size and density, while diffusion depends on particle size only. Figure 11.3 shows the theoretical time needed for various particle sizes to travel 1 mm by settling and diffusion. Note that there is a minimum where neither settling nor diffusion is effective in moving the particles to the cells. This size for minimum transport rate occurs at about 40 nm for a dense gold particle and 400 nm for a polystyrene particle.

A consequence of the particle transport rate in aqueous media is that the number of particles in contact with the cells will start from near zero and linearly increase over the time of the experiment as particles from the upper part of the media settle. The cell surface is often sticky with secreted proteins, and particles colliding with the cell surface by diffusion are typically trapped as well. In addition, dispersed nanoparticles will collide in suspension and then settle as larger aggregate particles. As a result, the particles eventually become concentrated in close proximity of the cell.

The particokinetics study (44) presents the concept of a transport rate-adjusted particle dose. Although there are a number of simplifying assumptions in the adjustment, the graph, Fig. 11.4, clearly illustrates that there can be a difference in

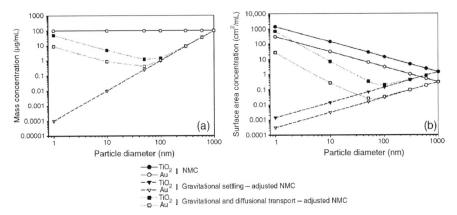

FIGURE 11.4 The relationship between the particle diameter and the transport rate-adjusted concentration for low-density and high-density particles. Cases: nominal concentration (solid line), gravitational settling transport only (short dashes), or gravitational settling plus diffusion (long dashes). The upper end of the nanoparticle range (70–100 nm) shows a minimum for rate-adjusted dose since neither settling nor diffusion is effective for rapidly bringing these particles into contact with the cells. Reproduced by permission from Teegaurden et al. (45).

effective dose depending on particle size. These authors reasoned that since the slowly transported particles do not reach the cells immediately, there is a difference in the time-integrated particle dose (equivalent to area under the curve, or AUC, in pharmacokinetics). For the thought experiment illustrated in Fig. 11.4, the nominal media concentration (mass of particles/total media volume) is independent of particle size, but the concentration in contact with the cells is size dependent. For nanoparticles in the 50–100 nm range, the effective concentration is reduced compared to larger or smaller particles since the transport in this size range is the slowest. The effect of different rates of particle transport from the media to the cell will confound experiments that attempt to measure biological response as a function of particle size and may lead to incorrect conclusions. The cited study (45) represents an initial attempt to quantitatively analyze the artifacts due to particle transport in a typical *in vitro* toxicology experiment. More sophisticated analysis can consider the effects of particle concentration on aggregate formation and the gradual accumulation of particles on the cell surface over the course of a 24-h experiment. A user-friendly computational model that incorporates these effects would be a great benefit for the design and analysis of *in vitro* particle toxicology experiments.

11.3.3 Dose and Concentration Issues

The concentration of particles in an *in vitro* experiment is typically reported as mass/volume of media or mass/surface area of cell culture well. Since typical media depth is 2–3 mm, the concentration expressed in $\mu g/cm^2$ is about 20–30% of the concentration in $\mu g/mL$. Expressing experimental results in terms of mass-based concentration is convenient since it can be unambiguously measured. At times concentrations are expressed in terms of particle surface area (cm^2 particles/cm^2 cells or cm^2 particles/mL of media) due to the hypothesis that particle surface area is a better dose metric. The particle surface area can be determined by gas adsorption (BET analysis) or estimated by geometry from an assumed particle diameter and density. Concentrations can also be reported as the number of particles per volume of fluid. For very small particles, the number concentration is sometimes expressed in molar units. For example, with 70-nm silica particles (nominally 3.6×10^{-16} g/particle) a mass concentration of 100 μg/mL corresponds to 0.5 nanomolar. Particle number and total particle surface area are greatly affected by the small-diameter tail of the size distribution. Assumptions made in converting data between mass, surface area, and particle count should be stated.

The majority of recent *in vitro* particle toxicology studies have used concentrations in the range 1–100 μg/cm^2 or up to 300 μg/mL. It is worthwhile to compare this concentration range with plausible concentrations *in vivo* from environmental, occupational, or medical exposures to nanomaterials. The basic analysis is straightforward. For a given exposure scenario, a plausible exposure concentration and an intake rate are assumed to establish an estimated human dose. This mass of particles is then divided by a target organ surface area or volume obtained from physiology texts to get an estimated concentration per square centimeter or milliliter in the body. Figure 11.5 compares the range of typical *in vitro* experiments with plausible concentrations for the lung, blood, colon, and skin. Simplifying assumptions include ignoring variation in body size, accumulation over multiple exposures, and clearance from the body. Although the estimates are subject to many uncertainties, as indicated by estimate range and error bars on the graphs, certain conclusions can be made.

The method of estimating plausible concentrations for inhalation will be presented in detail as an example. Inhalation is a likely exposure route and a person inhales about 10 m^3/day. A nanoparticle concentration of 10 μg/m^3 would be high for environmental exposure to manufactured particles but reasonable for occupational exposures and within the range of ambient ultrafine particle concentrations. The ICRP chart (45) suggests that the pulmonary deposition for nanoparticles in the lung is about 60% and the area of the lung is about 70 m^2 (46). Thus, a plausible nanoparticle dose is 0.08 ng/cm^2 based on total lung area. However, particles are not uniformly deposited in the lung. Phalen et al. (47) presented a detailed discussion of *in vivo* deposition versus *in vitro* experiments and indicated that deposition hot spots in the tracheobronchial region may be from three to four orders of magnitude higher than the average. However, the enhancement factors they reported vary with particle size since inertia is the dominant mechanism driving local deposition. Their calculated extreme case enhancement factor was 25,000 times the average for 5 μm particles but only 2900 for 1 μm particles, and the effect should be even smaller for nanoparticles in which inertial effects are small and diffusion is rapid. Oberdörster and Finkelstein (48) pointed out that Phalen's analysis was for the tracheobronchial region, not the alveolar region, and that the time course of exposure, gradual for *in vivo* versus bolus for *in vitro*, is also important. The estimate on the graph, Fig. 11.5, ranges from 0.08 nm/cm^2 to 80 ng/cm^2 because of uncertainties associated with deposition hot spots, uneven airflow in a diseased lung, increased ventilation during exercise, and inhaled particulate concentration. Particles in the lung are deposited on the alveolar surface, and the liquid layer of surfactant is very thin. Estimates of surfactant volume are around 10 mL, and

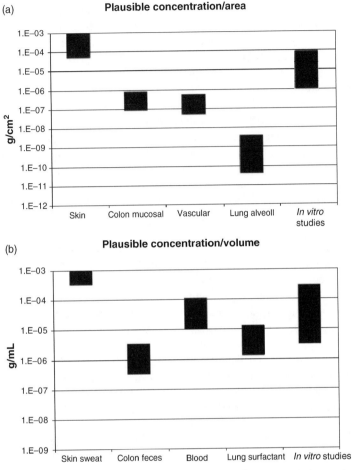

FIGURE 11.5   Plausible nanoparticle exposure concentrations compared to typical *in vitro* experiment conditions.

distributing the particles in surfactant volume gives a concentration of about 6 µg/mL, shown as a range from 1 to10 µg/mL to account for the uncertainty in ambient concentration, regional deposition, and surfactant thickness.

Nanoparticles in the blood are analyzed for two situations, translocation from the lung and deliberate injection. Although the translocation of nanoparticles from the lungs to other organs is documented, the amount is less than 1% of the lung deposition (7) and detection depends on the high sensitivity possible for a trace metal such as radiolabeled iridium. The estimated dose range for translocation is based on a small mass distributed over the 5–6 L of blood volume or the several hundred square meter of blood vessel internal surface area. However, nanoparticles deliberately injected as drug carriers or imaging agents could represent a much larger administered dose. Applications such as MRI contrast agents (49), chemotherapy drugs encapsulated in

PEG nanoparticles (50), or experimental use of quantum dots for imaging (51) involve doses on the order of a gram of particles dispersed in the human blood volume, which results in a very high exposure concentration. The concentration of nanoparticles reported for *in vitro* experiments with vascular endothelial cells (52) is within the range of experimental medical applications.

Ingestion is another possible pathway of nanoparticle exposure. Examples include children eating nanoparticle-containing products such as some zinc oxide sunscreens, nanoparticles entering the water supply by disposal of consumer products, and hand-to-mouth contact in occupational situations. Estimates of ingestion doses of particles vary widely. For soil ingestion by children, 50% take in less than 45 mg/day and 95% less than 208 mg/day (53). This analysis assumes an ingestion dose of 0.5–1 mg of nanoparticles by considering hand-to-mouth contact in an occupational setting and children eating nanoparticle-containing products such as sunscreen. The volume of ileal effluent entering the colon is about 1 L/day, so a nanoparticle concentration of 1 μg/mL is plausible. The mucosal area of the colon is about 2000 cm^2 (54, 55), so the assumed ingestion dose corresponds to 0.25–0.5 $μg/cm^2$.

Dermal exposure to nanomaterials is also considered in Fig. 11.5. Involuntary environmental nanomaterial dermal exposure is likely to be low, and occupational exposure can be controlled by simple methods such as gloves. However, application of nanoparticle sunscreen results in an extremely high exposure. Commercial sunscreens contain pigments such as TiO_2 or ZnO at 1–10% levels, and nanosized particles are being used since they block UV without leaving a white coating on the skin (56). The recommended application of sunscreen is 1.5–2 mg/cm^2 (57), giving a nanoparticle concentration of around 0.1 mg/cm^2. This is the highest concentration on the graph but can be misleading since the material is on the skin surface, a natural protective barrier. Most research on dermal exposure suggests that few nanoparticles from sunscreens or cosmetics penetrate through the stratum corneum into the living skin cells (56). However, other research suggests that skin is surprisingly permeable to nanomaterials with diverse physicochemical properties and may serve as a portal of entry for localized, and possibly systemic, exposure of humans to quantum dots and other engineered nanoscale materials (58).

The highly simplified analysis shown in Fig. 11.5 suggests that most *in vitro* experiments are at particle concentrations representing the high end of plausible *in vivo* exposures. However, the difference is not large for some situations such as colon cells studies of particles likely to be ingested. The issue of relevant dose needs to be considered in designing nanoparticle experiments and interpreting the human health significance of results. Using an increased dose to obtain statistically significant responses is an accepted technique in toxicology and is certainly appropriate when the goal is to use the cells to study specific biochemical pathways.

11.4   PARTICLE-INDUCED ARTIFACTS *IN VITRO*

Many *in vitro* assays that are being used for nanoparticles were originally developed for use with soluble agonists such as biomolecules and drugs. There is increasing evidence that assay interferences can occur due to high surface area and high particle number associated with typical mass concentrations of nanomaterials. This section highlights some recent publications that have reported assay artifacts with nanoparticle experiments. The problems appear to be most severe with carbon-based materials such as nanotubes. Validation of *in vitro* assays in the presence of specific nanoparticles being studied is important.

11.4.1 Adsorption of Assay Reagents

Metabolism-based measurements of viable cell count, or equivalently of cytotoxicity, involve adding the precursor of a light-absorbing indicator to the cell culture wells that contain both cells and particle treatments. If assay reagents are adsorbed on the particles, an incorrect result may be obtained. Monteiro-Riviere (59) reported that standard cytotoxicity assays such as MTT and neutral red that are well suited to assess chemical toxicity may generate conflicting results when carbon materials are assessed. The experiments were done with human keratinocyte cells using four sources of carbon black. Another study reported that the porous silica particles interfered with the MTT reagent through redox surface reactions (60). Loss of reagent will result in a false positive in the MTT cytotoxicity assay.

11.4.2 Adsorption of Cytokines

The secretion of cytokine proteins and other biomarkers is the basis of many *in vitro* studies of particle-induced proinflammatory signaling. Cytokines can be adsorbed from cell culture media onto the surface of nanomaterials. This adsorption of secreted protein will result in a false negative or reduced response for the nanoparticle-treated cells compared to cells treated with soluble agonists or supermicron-sized coarse particles. Adsorption of cytokines on high surface area particles was first shown by Seagrave et al. (17), who showed IL-8 adsorption on diesel particulate matter.

Veranth et al. (26) conducted experiments in which a known amount of recombinant IL-6 was added to suspensions of nanosized SiO_2 and TiO_2 particles in standard cell culture media and in the same culture media fortified with supplemental protein. The data were consistent with nonspecific binding of IL-6 to surfaces. Figure 11.6 shows the effect of increasing the exogenous protein supplementation using bovine serum albumin (BSA) or newborn calf serum on the measured IL-6 concentration in cell-free suspensions of kaolin, nanosized SiO_2, and TiO_2. Without particles, the measured IL-6 in cell culture media was close to the measurement for the standard prepared in serum-based assay diluent. The presence of 200 µg/mL of particles reduced the measured IL-6 concentration, but the measured IL-6 concentration increased with increasing amounts of supplemental protein. The kaolin, nano-SiO_2 and nano-TiO_2 all produced a statistically significant decrease in measured IL-6, suggesting that the typical *in vitro* cytokine induction experiment, Fig. 11.1, may be subject to false negatives due to adsorption of the secreted cytokine onto the surface of the particles.

Another study considered adsorption of multiple cytokines on a wide range of environmentally relevant particles, including wood smoke and diesel emissions

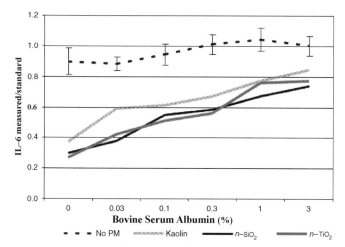

FIGURE 11.6   High surface area particles interfere with measurement of cytokines in cell-free media, but this artifact is ameliorated by adding exogenous protein to the media. The experiment shown involved addition of 200 μg/mL of kaolin, nano-SiO$_2$, or nano-TiO$_2$ to KGM media fortified with 0–3% bovine serum albumin (BSA). Measured IL-6 was compared to the assay standard prepared in 10% serum. Standard deviation is shown for only one treatment for clarity. Source paper provides full significance statistics (Veranth, 2007, #2147).

(incidental nanoparticles), ultrafine carbon black, and micron-sized quartz. The cytokines tested were TNF-α, IL-1β, IL-6, and IL-8. The authors concluded that cytokine binding was primarily with carbonaceous particles, not minerals, and confirmed that this effect could be blocked by adding serum proteins (61).

11.4.3   Adsorption of Nutrients

The cell culture media used for mammalian cells is a complex mixture containing many micronutrients, hormones, and growth factors essential for cell proliferation. Some of the essential factors such as vitamins and proteins contained in mixtures such a bovine pituitary extract may be at part per million levels. There have been reports that cell culture media components can be reduced below optimum levels by adsorption onto nanoparticle (59) (Robert Hurt, personal communication). Decreased cell proliferation rate caused by media depletion can be mistaken for particle-induced cytotoxicity. Also, any reduction in the number of viable cells in particle treatments compared to controls can cause a false low response in cytokine secretion assays.

11.5   *IN VITRO* ASSAY DEVELOPMENT AND VALIDATION

The rapidly increasing development of nanosized powders for industrial, chemical, medical, and consumer applications presents *in vitro* toxicologists with a challenge. Science is needed to design biocompatible nanomaterials for deliberate application to humans and to ensure that nanomaterials released to the environment are ecologically safe. There is a need for high-throughput screening assays that can prioritize materials for animal testing. There is also a need to use the tools of molecular biology to elucidate fundamental mechanisms of particle–cell interaction.

Current *in vitro* assays have provided much useful information. Indications that certain classes of nanomaterials *in vitro* can catalyze reactive oxygen species formation and cause DNA damage and oxidative stress are causes for human toxicology concern. Indication that metal oxide nanoparticles have potency comparable to other previously studied particle types but are not exceptionally (orders of magnitude) more potent than materials such as crystalline silica is reassuring to those concerned about the environmental implications of nanotechnology. Recent research suggests that the risks of human exposure to manufactured nanoparticles are real, but these risks are manageable with proper application of industrial hygiene and pollution control technology.

Development of improved *in vitro* assays for nanoparticle toxicology is both a challenge and an opportunity. Choice of the most appropriate cell types, protocols, and measured end points are all important questions. The ultimate goal needs to be to develop suites of *in vitro* assays that have high predictive power and good correlation with whole animal exposure outcomes.

REFERENCES

1. Pope, C.A., Dockery, D.W. Critical review: health effects of fine particulate air pollution: lines that connect. *J Air Waste Manage Assoc* 2006;56:709–742.

2. Dominici, F., McDermott, A., Daniels, M., Zeger, S.,Samet, J. Mortality among residents of 90 cities. In: revised analysis of time-series studies of air pollution and health. Special Report. Boston: Health Effects Institute, 2003.

3. Smith, K.R., Kim, S., Recendez, J.J., Teague, S.V., Menache, M., Grubbs, D.E., Sioutas, C., Pinkerton, K.E. Airborne particles of the California Central Valley alter the lungs of healthy adult rats. *Environ Health Perspect* 2003;111(7):902–908.

4. Smith, K.R., Veranth, J.M., Kodavanti, U.P., Aust, A.E.,Pinkerton, K.E. Acute pulmonary and systemic effects of inhaled coal fly ash in rats: comparison to ambient particles. *Toxicol Sci* 2006;93(2):390–399.

5. Evans, S.A., Al-Mosawi, A., Adams, R.A., Berube, K.A. Inflammation, edema, and peripheral blood changes in lung-compromised rats after instillation with combustion-derived and manufactured nanoparticles. *Exp Lung Res* 2006;32(8):363–378.

6. Warheit, D.B., Webb, T.R., Colvin, V.L., Reed, K.L., Sayes, C.M. Pulmonary bioassay studies with nanoscale and fine-quartz particles in rats: Toxicity is not dependent on particle size but on surface characteristics. *Toxicol Sci* 2007;95(1):270–280.

7. Kreyling, W.G., Semmler, M., Erbe, F., Mayer, P., Takenaka, S., Schultz, H., Oberdörster, G.,Ziesenis, A. Translocation of ultrafine insoluble iridium particles from lung epithelium to extrapulmonary organs is size dependent but very low. *J Toxicol Environ Health A* 2002;65(20):1513–1530.

8. Lomer, M.C.E., Hutchinson, C., Volkert, S., Greenfield, S.M., Catterall, A., Thompson, R. P.H., Powell, J.J. Dietary sources of inorganic microparticles and their intake in healthy subjects and patients with Crohn's disease. *Br J Nutr* 2004;92:947–955.

9. Reilly, C.A., Taylor, J.L., Lanza, D.L., Carr, B.A., Crouch, D.J.,Yost, G.S. Capsaicinoids cause inflammation and epithelial cell death through activation of vanilloid receptors. *Toxicol Sci* 2003;73:170–181.

10. Amara, N., Bachoual, R., Desmard, M., Golda, S., Guichard, C., Lanone, S., Aubier, M., Ogier-Denis, E.,Boczkowski, J. Diesel exhaust particles induce matrix metalloprotease-1 in human lung epithelial cells via a NADP(H) oxidase/NOX4 redox-dependent mechanism. *Am J Physiol Lung Cell Mol Physiol* 2007;294(1):L170–181.

11. Warheit, D.B., Borm, P.J., Hennes, C., Lademann, J. Testing strategies to establish the safety of nanomaterials: conclusions of an ECETOC workshop. *Inhal Toxicol* 2007;19 (8):631–643.

12. Oberdorster, G., Maynard, A., Donaldson, K., Castranova, V., Fitzpatrick, J., Ausman, K., Carter, J., Karn, B., Kreyling, W., Lai, D., Olin, S., Monteiro-Riviere, N., Warheit, D., Yang, H. Principles for characterizing the potential human health effects from exposure to nanomaterials: elements of a screening strategy. *Part Fibre Toxicol* 2005;2:8.

13. Fubini, B., Aust, A.E., Bolton, R.E., Borm, P.J., Bruch, J., Ciapett, G., Donaldson, K., Elias, Z., Gold, J., Jaurand, M.C., Kane, A.B., Lison, D., Muhle, H. Non-animal tests for evaluating the toxicity of solid xenobiotics. *Altern Lab Anim* 1998;26:579–615.

14. Devlin, R.B., Frampton, M.L., Ghio, A.J. *In vitro* studies: what is their role in toxicology? *Exp Toxicol Pathol* 2005;57(Suppl 1):183–188.

15. Oortgiesen, M., Veronesi, B., Eichenbaum, G., Kiser, P.F., Simon, S.A. Residual oil fly ash and charged polymers activate epithelial cells and nociceptive sensory neurons. *Am J Physiol Lung Cell Mol Physiol* 2000;278(4):L682–L695.

16. Knebel, J., Ritter, D., Aufderheide, M. Exposure of human lung cells to native diesel motor exhaust—development of an optimized *in vitro* test strategy. *Toxicol In Vitro* 2002;16:185–192.

17. Seagrave, J., McDonald, J.D., Mauderly, J.L. *Air–liquid interface culture: Towards more physiological in vitro toxicology of aerosols.* Society of Toxicology, Baltimore, MD, 2004. p. 1558. Abstract.

18. Wottrich, R., Diabate, S.,Krug, H.F. Biological effects of ultrafine model particles in human macrophages and epithelial cells in mono- and co-culture. *Int J Hyg Environ Health* 2004;207(4):353–361.

19. Mosmann, T. Rapid colorimetric assay for cellular growth and survival: application to proliferation and cytotoxicity assays. *J Immunol Methods* 1983;65(1–2):55–63.

20. Kelley, J. *Cytokines of the lung. Lung Biology in Health and Disease Series*, Vol. 61. New York: Marcel Dekker; 1993.

21. Driscoll, K.E. Cytokines and Regulation of Pulmonary Inflammation. In: Gardner, D.E., Crapo, J.D., McClellan, R.O., editors. *Toxicology of the Lung*, Ann Arbor: Taylor and Francis; 1999. pp. 149–172.

22. Li, X. Y., Brown, D., Smith, S., MacNee, W., Donaldson, K. Short-term inflammatory responses following intratracheal instillation of fine and ultrafine carbon black in rats. *Inhal Toxic* 1999;11(8): 709–731.

23. Churg, A., Gilks, B., Dai, J. Induction of fibrogenic mediators by fine and ultrafine titanium dioxide in rat tracheal explants. *Am. J. Physiol* 1999;277:L975–L982.

24. Donaldson, K., Brown, D., Coulter, A., Duffin, R., MacNee, W., Renwick, L., Tran, L., Stone V. The pulmonary toxicology of ultrafine particles. *J Aerosol Med* 2002;15(2): 213–220.

25. Oberdörster, G. Pulmonary effects of inhaled ultrafine particles. *Int Arch Occup Environ Health* 2001;74:1–8.

26. Veranth, J.M., Kaser, E.G., Veranth, M.M., Koch, M.,Yost, G.S. Cytokine responses of human lung cells (BEAS-2B) treated with oxide micron-sized and nanoparticles compared to soil dusts. *Part Fibre Toxicol* 2007;4:(2).

27. Jia, G., Wang, H., Yan, L., Wang, X., Pei, R., Yan, T., Zhao, Y.,Guo, X. Cytotoxicity of carbon nanomaterials: single-wall nanotube, multi-wall nanotube, and fullerene. *Environ Sci Technol* 2005;39(5):1378–1383.

28. Papageorgiou, I., Brown, C., Schins, R.P., Singh, S., Newson, R., Davis, S., Fisher, J., Ingham, E.,Case, C.P. The effect of nano- and micron-sized particles of cobalt-chromium alloy on human fibroblasts *in vitro*. *Biomaterials* 2007;28(19):2946–2958.

29. Xia, T., Kovochich, M., Brant, J., Hotze, M., Sempf, J., Oberley, T., Sioutas, C., Yeh, J.I., Wiesner, M.R., Nel, A.E. Comparison of the abilities of ambient and manufactured nanoparticles to induce cellular toxicity according to an oxidative stress paradigm. *Nano Lett* 2006;6(8):1794–1804.

30. Monteiller, C., Tran, L., MacNee, W., Faux, S., Jones, A.D., Miller, B.G.,Donaldson, K. The pro-inflammatory effects of low-toxicity low-solubility particles, nanoparticles and fine particles, on epithelial cells *in vitro*: the role of surface area. *Occup Environ Med* 2007;64(9):609–615.

31. Barlow, P.G., Coulter-Baker, A., Donaldson, K., MacCallum, J., Stone, V. Carbon black nanoparticles induce type II cells to release chemotaxins for alveolar macrophages. *Part Fibre Toxicol* 2005;2:11.

32. Pulskamp, K., Diabate, S., Krug, H. Carbon nanotubes show no sign of acute toxicity but induce intracellular reactive oxygen species in dependence on contaminants. *Toxicol Lett* 2007;168(1):58–74.

33. Geiser, M., Rothen-Rutishauser, B., Kapp, N., Schurch, S., Kreyling, W.G., Schultz, H., Im, H.V., Hyder, J., Gehr, P. Ultrafine particles can cross cellular membranes by nonphagocytic mechanisms in lungs and cultured cell. *Environ Health Perspect* 2005;113(11):1555–1560.

34. Chen, M., vonMikecz, A. Formation of nucleoplastic protein aggregates impairs nuclear function in response to SiO$_2$ nanoparticles. *Exp Cell Res* 2005;305:51–62.

35. Rothen-Rutishauser, B., Mühlfeld, C., Blank, F., Musso, C., Gehr, P. Translocation of particles and inflammatory responses after exposure to fine particles and nanoparticles in an epithelial airway model; *Part Fibre Toxicol* 2007;4:9.

36. Schulz, C., Farkas, L., Wolf, K., Krätzel, K., Eissner, G., Pfeifer, M. Differences in LPS-induced activation of bronchial epithelial cells (BEAS-2B) and type II-like pneumocytes (A-549) *Scand J Immunol* 2002;56:294–302.

37. Riley, M.R., Boesewetter, D.E., Turner, R.A., Kim, A.M., Collier, J.M., Hamilton, A. Comparison of the sensitivity of three lung derived cell lines to metals from combustion derived particulate matter. *Toxicol In Vitro* 2005;19:411–419.

38. Xu, H., Hoet, P.H.M., Nemery, B. *In vitro* toxicity assessment of polyvinyl chloride particles and comparisons of six cellular systems. *J Toxicol Environ Health A* 2002;65:1141–1159.

39. Veranth, J.M., Cutler, N.S., Kaser, E.G., Reilly, C.A.,Yost, G.S. Effects of cell type and culture media on interleukin-6 secretion in response to environmental particles. *Toxicol In Vitro* 2008;22(2):498–509.

40. Sayes, C.M., Reed, K.L.,Warheit, D.B. Assessing toxicity of fine and nanoparticles: comparing in vitro measurements to in vivo pulmonary toxicity profiles. *Toxicol Sci* 2007; 97(1):163–180.

41. Sayes, C.M., Fortner, J.D., Guo, W., Lyon, D., Boyd, A.M., Ausman, K.D., Tao, Y.J., Sitharaman, B., Wilson, L.J., Hughes, J.B., West, J.L., Colvin, V.L. The differential cytotoxicity of water-soluble fullerenes. *Nano Lett* 2004;4(10):1530–6984.

42. Sayes, C.M., Marchione, A.A., Reed, K.L., Warheit, D.B. Comparative pulmonary toxicity assessments of C_{60} water suspensions in rats: few differences in fullerene toxicity in vivo in contrast to in vitro profiles. *Nano Lett* 2007;7(8):2399–2406.

43. Seagrave, J., Mauderly, J.L., Seilkop, S.K. In vitro relative toxicity screening of combined particulate and semivolatile organic fractions of gasoline and diesel engine emissions. *J Toxicol Environ Health A* 2003;66(12):1113–1132.

44. Teeguarden, J.G., Hinderliter, P.M., Orr, G., Thrall, B.D., Pounds, J.G. Particokinetics in vitro: dosimetry considerations for in vitro nanoparticle toxicity assessments. *Toxicol Sci* 2007;95(2):300–312.

45. International Commission on Radiological Protection Task Force on Lung Dynamics. *Health Phys.* 1966;12:173–207.

46. Guyton, A.C., Hall, J.E. *Textbook of Medical Physiology.* Philadelphia, PA: Saunders; 1996.

47. Phalen, R.F., Oldham, M.J., Nel, A.E. Tracheobronchial particle dose considerations for in vitro toxicology studies. *Toxicol Sci* 2006;92(1):126–132.

48. Oberdörster, G., Finkelstein, J.N., To the editor. *Toxicol Sci* 2006;94(2):439.

49. Niendorf, H.P., Haustein, J., Cornelius, I., Alhassan, A.,Clauss, W. Safety of gadolinum-DTPA: extended clinical experience. *Magn Reson Med* 1991;22(2):222–228.

50. Dong, Y., Feng, S.S. In vitro and in vivo evaluation of methoxy polyethylene glycol–polyactide (MPEG-PLA) nanoparticles for small-molecule drug chemotherapy. *Biomaterials* 2007;28(28):4154–4160.

51. Ballou, B., Lagerholm, B.C., Lauren, A.E., Bruchez, M.P.,Waggoner, A.S. Noninvasive imaging of quantum dots in mice. *Bioconjug Chem* 2004;15:79–86.

52. Gojova, A., Gao, B., Kota, R.S., Rutledge, J.C., Kennedy, I.M.,Barakat, A.I. Induction of inflammation in vascular endothelial cells by metal oxide nanoparticles: effect of composition. *Environ Health Perspect* 2007;115(3):403–409.

53. Stanek, E.J., III,Calabrese, E.J. Daily estimates of soil ingestion in children. *Environ Health Perspect* 1995;103(3):276–285.

54. Debongnie, J.C., Phillips, S.F. Capacity of the human colon to absorb fluid. *Gastroenterology* 1978;74:698–703.

55. Sandle, G.I. Salt and water absorption in the human colon: a modern appraisal. *Gut* 2007;43:294–299.

56. Nohynek, G.J., Lademann, J., Ribaud, C., Roberts, M.S. Grey goo on the skin? nanotechnology, cosmetic and sunscreen safety. *Crit Rev Toxicol* 2007;37(3):251–277.

57. Bech-Thomsen, N., Wulf, H.C. Sunbathers' application of sunscreen is probably inadequate to obtain the sun protection factor assigned to the preparation. *Photodermatol Photoimmunol Photomed* 1992;9(6):242–244.

58. Ryman-Rasmussen, J.P.,Riviere, J.E.,Monteiro-Riviere, N.A. Penetration of intact skin by quantum dots with diverse physicochemical properties. *Toxicol Sci* 2006;91(1):159–165.

59. Monteiro-Riviere, N.A., Inman, A.O. Challenges for assessing carbon nanomaterial toxicity to the skin. *Carbon* 2006;44:1070–1078.

60. Laaksonen, T., Santos, H.I., Vihola, H., Salonen, J., Riikonen, J., Heikkila, T., Peltonen, L., Kumar, N., Murzin, D.Y., Lehto, V.-P., Hirvone, J. Failure of MTT as a toxicity testing agent for mesoporous silicon microparticles. *Chem Res Toxicol* 2007;20(12):1913-1918.

61. Kocbach, A., Totlandsdal, A.I., Låg, M., Refsnes, M., Schwartz, P.E. Differential binding of cytokines to environmentally relevant particles: a possible source for misinterpretation of in vitro results? *Toxicol Sci* 2007;176(2):131–137.

Biological Activity of Mineral Fibers and Carbon Particulates: Implications for Nanoparticle Toxicity and the Role of Surface Chemistry

PRABIR K. DUTTA[1], JOHN F. LONG[2], MARSHALL V. WILLIAMS[3], and W. JAMES WALDMAN[4]

[1]Department of Chemistry, The Ohio State University, Columbus, OH 43210, USA
[2]Department of Veterinary Biosciences, The Ohio State University, Columbus, OH 43210, USA
[3]Department of Molecular Virology, Immunology and Medical Genetics, The Ohio State University, Columbus, OH 43210, USA
[4]Department of Pathology, The Ohio State University, Columbus, OH 43210, USA

12.1 CORRELATION OF BIOLOGICAL PROPERTIES OF NATURAL MINERALS WITH STRUCTURE

Epidemiological data suggest that environmental and/or occupational exposure to minerals can cause lung disease (1–3). In particular, asbestos minerals can induce pulmonary inflammation, fibrosis of the lower respiratory tract (asbestosis) (1–4) and are a risk factor for developing bronchiogenic carcinoma and mesothelioma (5). These diseases develop over many years following exposure. Asbestos is a group of naturally occurring hydrated silicates and consist of six major fiber types with different chemical compositions, morphology, and durability (6). Several hypotheses for asbestos toxicity have been generated. Physicochemical properties such as shape, dimension, biopersistence, chemical composition, surface charge, and solubility are thought to be important parameters (7, 8). Phagocytosis of the fibers by macrophages results in the formation of reactive oxygen species (ROS) (1–4, 9, 10), where ROS is a collective term that includes radicals (superoxide anion, hydroxyl, peroxyl, and alkoxyl radicals) and hydrogen peroxide (H_2O_2). In addition, reactive nitrogen species

Nanoscience and Nanotechnology   Edited by Vicki H. Grassian
Copyright © 2008 John Wiley & Sons, Inc.

such as nitric oxide and peroxynitrite and reactive chloride species such as hypochlo-
rite are also involved in oxidative stress.

Iron on asbestos fibers can generate hydroxyl radicals via the Fenton reaction,
which can initiate lipid peroxidation and oxidation of intracellular proteins and DNA
(11, 12). Surface modification of amosite asbestos by hydrocarbon altered its
pathogenic properties (13), indicating that surface chemistry is relevant to toxicity.

Man-made mineral or vitreous fibers can also be bioactive, though their role in
respiratory disease in humans is not yet well established (14, 15). There are over 70
varieties of synthetic inorganic fibers, covering over 35,000 applications, with
different physicochemical and morphological characteristics. These include insula-
tion materials (e.g., glass wool, rock wool, and slag wool), glass filaments and
microfibers, and refractory ceramic fibers (14). As a consequence of the extensive
applications of these fibers, a significant fraction of the population is exposed,
necessitating that basis of toxicity of respirable fibers be understood. Besides asbestos,
iron on the surface of silica has shown enhanced inflammatory reaction in rats, and is
proposed to arise from Fenton chemistry. Iron on quartz has also shown to accelerate
the degradation of RNA, possibly through a hydroxyl-radical-mediated mechanism.

A recently discovered form of fibrous carbon, referred to as carbon nanotubes
(CNTs), is expected to find wide commercial use in composite materials, electronics, and even biomedical applications. These CNTs exhibit novel mechanical,
chemical, and electrical properties and is driving the considerable interest in these
materials. However, the parallel with asbestos fibers, at least from a morphological
point of view and biopersistence, has not gone unnoticed. Several studies have
focused on toxicity of CNTs, and considering that they are quite complex materials
(single-walled versus multiwalled, significant transition metal content), there are
varying reports on their toxicity in the literature. MWCNTs have been shown to
have inflammatory properties upon intratracheal instillation in the rat lungs (16).
ROS was observed in *in vitro* experiments with metal-containing SWCNTs,
whereas no ROS was observed with purified SWCNTs (17, 18). Unlike crystalline
silica, MWCNTs did not generate radical species upon interaction with H_2O_2 or
formate but were found to have a scavenging capacity (19). There are studies that
conclude that SWCNTs may be more toxic than quartz, especially pulmonary
injuries (20, 21). Cytotoxicity and impairment of phagocytic ability of alveolar
macrophages were more pronounced for SWCNTs as compared to MWCNTs,
quartz, and C_{60} (22).

Zeolites are well-defined crystalline aluminosilicates and can serve as model
systems for asbestos and other toxic minerals, primarily because the naturally
occurring zeolite erionite causes mesothelioma and is more carcinogenic than
crocidolite asbestos (23–25). On the other hand, mordenite with similar morphological and physical properties (26) is relatively benign (27–29). Pulmonary deposition of
fiber-shaped mordenite particles showed that substantial deep-lung deposition was
possible (15). Studies on *in vitro* superoxide generation induced by erionite and
mordenite indicated that mordenite was less active than erionite (9) and was attributed
to the fibrous shape of erionite. *In vivo* studies after intratracheal instillation of
mordenite showed acute and subacute inflammation of the lung, which was attributed

to the needle-shaped particles (10–12% of the mordenite sample) (30). *In vitro* studies also showed significant hemolysis with mordenite but moderate macrophage cell membrane damage, as compared to quartz. Fenton activity of erionite normalized to the iron content was reported to be about 200 times more reactive than crocidolite for DNA single-strand breaks (25, 27).

In this chapter, we focus primarily on our work on comparisons of the two zeolitic minerals, erionite and mordenite, to produce ROS intracellularly and extracelluarly by phagocytosis, as well as differences in the Fenton reactivity (creation of hydroxyl radicals) of the iron coordinated to their surfaces, and their mutagenic properties (31–35). Comparisons of the biological and chemical reactivity of erionite and mordenite should be helpful in developing correlations between structure and function of mineral fibers and other fibrous materials.

12.1.1 Characteristics of Zeolites

Zeolite minerals are characterized by their X-ray powder diffraction patterns (36). Figure 12.1 shows the powder diffraction patterns of naturally occurring erionite and mordenite (34). For erionite, the primary impurity phases are aluminum silicate hydrate, and for mordenite, the extraneous phase is quartz. Another characteristic feature is the morphology of zeolites. Figure 12.2 shows electron micrographs of the zeolite samples (33). The unfractionated erionite sample (Fig. 12.2, top left) includes long fibers, up to 45 μm, with a mode (most frequent length) of ~10 μm. These samples can be fractionated to provide populations ranging from 0.7 to 13 μm, with a mode of ~3 μm, the fiber morphology appearing intact (Fig. 12.2, top right). The unfractionated mordenite particles (Fig. 12.2, middle left) ranged from 0.1 to 10 μm, with a mode of ~3.7 μm. Size fractionation of mordenite allows for discrimination against the largest particles, as well as particles smaller than 1 μm (Fig. 12.2, middle right). Zeolite Y appears as octahedral crystals of a homogeneous size of slightly less than 1 μm (Fig. 12.2, bottom).

12.1.2 Cell–Fiber Interactions

12.1.2.1 Electron Microscopy   Electron microscopy was used to examine erionite uptake via phagocytosis by rat pulmonary alveolar macrophage-derived cells (NR8383) (32). Cells were fixed 20, 40, 60, and 90 min after fiber–cell contact. Within the 20-min period of contact between erionite and cells, the fibers were intracellular and within lysosomes (Figs. 12.3 and 12.4). The lysosomal membrane ranged from appearing to be completely intact morphologically (Fig. 12.3b) to having interruptions in its continuity (Fig. 12.4a). The uptake of silica has been noted to make membranes of the lysosomes unstable, releasing hydrolytic enzymes (37), and could be the cause for membrane damage. Frequently, multiple colloidal-size particles were seen within individual lysosomes (Fig. 12.4b), including coalesced particles. Aside from these lysosomal-associated changes, no specific morphologic change was noted in any other membranes or organelles during the time frame examined.

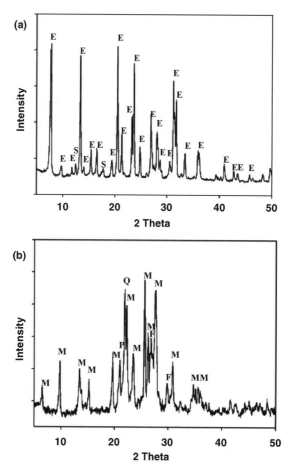

FIGURE 12.1   X-ray powder diffraction patterns of (a) Erionite (E), impurity peak S due to sodium aluminum silicate; (b) Mordenite (M), impurity peaks due to quartz (Q) and feldspar (F). (Adapted from Reference 34.)

12.1.2.2 Intracellular Oxidative Burst: Fluorescence Microscopy

Amounts of ROS within cells upon phagocytosis and changes of ROS with time upon exposure of macrophages to minerals have been reported. ROS has been measured in cultured neurons at the single-cell level by oxidation of $2',7'$-dichlorodihydrofluorescin and fluorescence detection, and this dye was used to examine the oxidative burst upon phagocytosis of erionite (31, 32). The reduced, esterified dye form (**I**) enters the cell and is deesterified by (nonspecific) cellular esterases to form **II** (H_2DCF). NR8383 cell line has been reported to exhibit high, nonspecific esterase activity (38). The oxidation of **II** to the fluorescent form **III** (DCF) by ROS can be readily detected. Figure 12.5 shows individual sections of fluorescence imaged from a control cell and an erionite-exposed cell (31). Clearly, the cell exposed to erionite exhibits intense fluorescence and is indicative of higher levels of intracellular peroxide

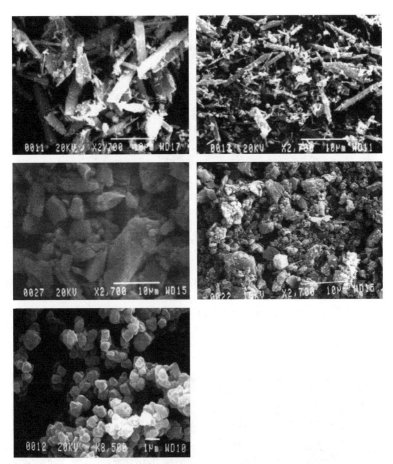

FIGURE 12.2   Scanning electron micrographs of zeolites. Top left and right are erionite before and after fractionation, respectively. Middle left and right are mordenite before and after fractionation, respectively. Bottom micrograph is that of zeolite Y. (Adapted from Reference 33.)

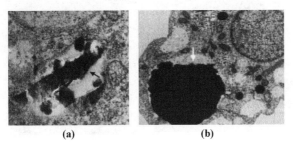

(a)   **(b)**

FIGURE 12.3   (a) Mineral fiber (arrow) plus fragments in a lysosome. The fiber morphologically resembles those seen by scanning electron microscopy in samples of erionite (18,000×). (b) Mineral particle (arrow) within a lysosome with a morphologically intact membrane (8700×). (Adapted from reference 32.)

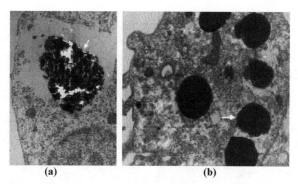

FIGURE 12.4 (a) Mineral particle (arrow) within a lysosome. The volume of the lysosome appears to have increased, the particle appears fragmented, and the membrane appears not to be intact (×15,100). (b) Macrophage with multiple particles, which are apparently within lysosomes (×12,000). The arrow points to a particle apparently within a lysosome. (Adapted from Reference 32.)

levels. The non zero background fluorescence in the control cell is attributed to oxidation resulting from mitochondrial respiration (39).

Figure 12.6 depicts the measurement of the entire intracellular fluorescence obtained by summing three confocal slices of fluorescence for erionite-exposed and control cells and plotted versus time (32). There is an overall increase of cellular

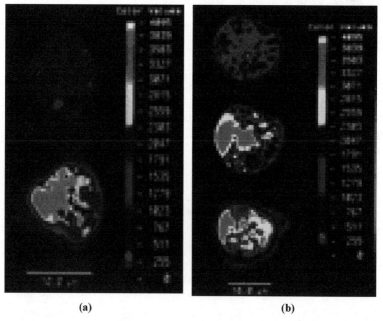

FIGURE 12.5 Confocal fluorescence images of a DCF-dye-loaded macrophage cell exposed to erionite. (a) Comparison of control cell (top) with erionite-exposed macrophage (30 min exposure to erionite). (b) Slice through an erionite-exposed cell. (Adapted from Reference 31.)

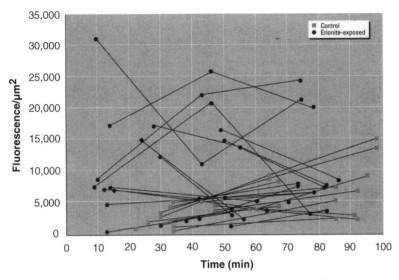

FIGURE 12.6 Integrated fluorescence intensity of erionite-exposed cells (•) versus control cells (■) over a 100 min period. Each line connects the data points for a single cell. (Adapted from Reference 32.)

fluorescence in the erionite-exposed cells compared to controls. Within 35 min after exposure to erionite, the mean cellular fluorescence was more than three times that of controls. Vilím et al. (40) suggested that the oxidative burst could be occurring during the first 5–10 min. The variability between the fiber-exposed cells can be related to variation in size and number of erionite particles in the cells. Some cells show five or six particles within lysosomes, while others have as few as one or even none. In previous studies on dust-induced macrophage chemiluminescence, the luminol signal depended on the ratio of fibers to cells (40). Using latex beads, Kobzik et al. (41) observed a rise in fluorescence in cells, roughly in proportion to the numbers of phagocytized beads. A similar effect was observed by Szejda et al. (42), with the amount of oxidative products formed by polymorphonuclear leukocytes upon phago-cytosis depending on the number of bacteria ingested by the cells.

The general decline in fluorescence signal for erionite-exposed cells beyond 40 min is consistent with previous studies. A study based on luminol fluorescence reported that the signal decreased after reaching a maximum at 10 min (43). Royall and Ischiropoulos (44) proposed that increased membrane permeability allows oxidized dye to diffuse out from the cell. Another possibility is intracellular acidification, which can quench the dye fluorescence (42, 45).

12.1.2.3 Extracellular Oxygen Species by Chemiluminescence Assay
Confocal microscopy studies discussed above provides information on intracellular ROS, whereas luminol chemiluminescence provides a measure of superoxide ions that leak out of the cell (46, 47). Figure 12.7 is a typical response of the luminol chemiluminescence induced by superoxide generated by mordenite-stimulated cells

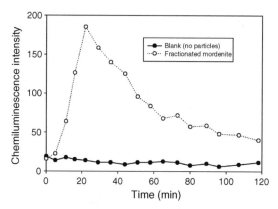

FIGURE 12.7   Measure of the oxidative burst using luiminol-based chemiluminescence data after exposure of macrophages to mordenite. (Adapted from Reference 33.)

as a function of time (33). Erionite, mordenite, and zeolite Y were examined, and in all cases, the peak luminescence occurred between 20 and 25 min postexposure. Table 12.1 summarizes the integrated chemiluminescence results generated by NR8383 cells treated with various concentrations of each of the three minerals. These data demonstrate dose-dependent response in all cases: the greatest luminescence signal was always associated with the highest mineral concentration. Data in Table 12.1 suggest that for comparable sizes of the particles, there is no significant difference in the oxidative burst, especially between the minerals erionite and mordenite.

Figure 12.8 presents the results obtained with three different size fractions of erionite, including a fraction of erionite called "fine erionite" (ranging from 0.2 to 4 μm, with a mode of 0.8 μm, as determined by scanning electron microscopy (SEM)) at a dose of 50 μg. It was noted that for exposure to the same mass of particulates, the chemiluminescence signal was inversely correlated to particle size for each category of zeolites, implying a positive correlation between the oxidative burst and the surface area of particles (33).

TABLE 12.1   Integrated Chemiluminescence Intensity Upon Macrophage–particle Interaction (Arbitrary Units)

	Median Size (μm)	10 μg	50 μg	250 μg
Mordenite	3.7	na[a]	503 ± 183	649 ± 303
Fractionated mordenite	1	946 ± 139	1846 ± 1134	8382 ± 1855
Erionite	10	na	695 ± 288	1166 ± 5 89
Fractionated erionite	3	na	965 ± 361	1110 ± 333
Fine erionite	0.8	na	1904	na
Zeolite Y	0.9	611	1359 ± 1133	3909 ± 1340

Deviations, when indicated, are the results of two or more measurements (data from Ref. 33).
[a]na: not analyzed.

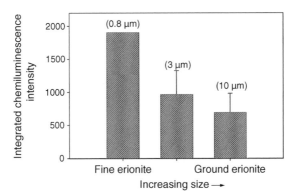

FIGURE 12.8   Comparison of chemiluminescence intensity (measure of oxidation burst) as a function of erionite particle size. (Adapted from Reference 33.)

12.1.3   Chemical Studies: Fenton Chemistry

12.1.3.1   Fe(II)-Mediated Fenton Chemistry   We have investigated the Fenton chemistry promoted by Fe(II) on the surface of different zeolite minerals (33, 34) as per the following reaction:

$$\text{Zeolite-Fe(II)} + H_2O_2 \rightarrow OH^{\bullet} + OH^{-} + \text{Zeolite-Fe(III)}$$

Fe(II) was introduced by ionexchange, and the surface Fe(II) concentration was measured after back exchange of the iron by a charged cationic polymer, Dab-4Br. The monomeric unit of Dab-4Br is shown below.

Dab-4Br is composed of cylindrically shaped units with a length of 8.7 Å and a diameter of 6.1 Å. Since erionite surface has a network of 8-membered rings with dimensions of 3.6×5.2 Å and mordenite has 4-, 8-, and 12-membered rings, with the largest having dimensions of 7.0×6.7 Å (48), Dab-4Br will only exchange with surface iron. Surface iron (Fe^{2+}) was found to increase with the concentration of the initial iron solution, ranging from 1.4 to $12.3 \times 10^2\ \mu g/g$ for zeolite Y, from 5 to $4.0 \times 10^2\ \mu g/g$ for mordenite, and from 1.6 to $3.3 \times 10^2\ \mu g/g$ for erionite.

Hydroxyl radical formation was measured by UV–Vis spectroscopy of methanesulfinic acid (MSA) formed by reaction of hydroxyl radicals with DMSO (49, 50). Figure 12.9 plots the amount of hydroxyl radical produced against the amount of surface iron (33). For mordenite and erionite, the production of hydroxyl radicals increases as surface iron increases. Increasing amounts of surface iron on zeolite Y does not give rise to increasing production of hydroxyl radicals. For similar amounts of

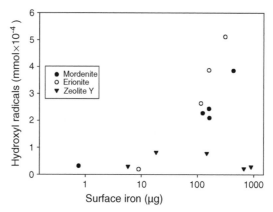

FIGURE 12.9 Production of hydroxyl radicals by iron-mediated Fenton chemistry for mordenite (●), erionite (○), and Zeolite Y (▼) as a function of surface iron (Fe^{2+}) content. (Adapted from Reference 33.)

surface iron, and especially at the higher surface iron levels, erionite induces a larger hydroxyl radical production than mordenite. The data obtained with free iron in solution showed a relatively small production of hydroxyl radicals (7.3×10^{-5} mmol for 350 μg iron).

12.1.3.2 Fe(III)-Mediated Fenton Chemistry

The iron acquired in biological systems by the minerals will be in the Fe(III) form (iron in serum ranges from 4 to 30 μM). However, the epithelial lung lining fluid (LLF) (51, 52) has ascorbic acid (AA) and glutathione (GSH), which can reduce the surface iron of inhaled particles (53, 54). Both AA and GSH can reduce iron; however, with their reduction potentials, AA should be able to more easily reduce ferric iron.

$$\text{AA (typically an anion)} + Fe(III) \rightarrow AA^{\bullet} + Fe(II), \quad E_{rxn} = \sim 0.4 \text{ V}$$

$$GSH + Fe(III) \rightarrow Fe(II) + GS^{\bullet}, \quad E_{rxn} = \sim -0.2 \text{ V}$$

Since Fe(III) ion exchange can lead to zeolite destruction, a different procedure (as compared to Fe^{2+}-zeolites discussed above) involving an ethereal solution of Fe $(SCN)_n^{(3-n)+}$ was used (55). The iron thiocyanate that phase transfers into water is hydrolyzed into $Fe(OH)^{2+}$, $Fe(OH)_2^{+}$, and $Fe_2(OH)_2^{4+}$, which ion exchange into the zeolite. The amount of surface iron for mordenite and erionite determined after exchange with Dab4-Br was found to be 1.8×10^1 and 8.8×10^{-1} μg iron per gram of zeolite, respectively (34). These numbers are different from the Fe(II) results above, which used ferrous sulfate solutions for ion exchange (33).

The Fenton activity of the Fe(III)-exchanged mineral fibers involved reaction with GSH or AA, followed by treatment with hydrogen peroxide in the presence of dimethyl sulfoxide. This method of analysis is also different from those used above in the Fe(II) chemistry (33). The methane sulfinic acid that is created is converted into a diazosulfone compound and quantitated by absorption spectroscopy. The overall process is outlined in reactions 12.1 and 12.2:

$$\text{dimethyl sulfoxide} + \text{OH}^\bullet \rightarrow \text{methane sulfinic acid (MSA)} + \text{CH}_3^\bullet \qquad (12.1)$$

$$\begin{aligned}\text{MSA (aq. phase)} + \text{diazonium salt (fast blue BB dye; aq. phase)} \rightarrow \\ \text{H}^+ + \text{diazosulfone (organic phase; yellow; 425 nm)} \qquad (12.2)\end{aligned}$$

The concentration of antioxidant was chosen to be 200 μM to simulate the typical concentration in LLF (56). Figure 12.10a shows the total hydroxyl radical generated, and Fig. 12.10b shows the hydroxyl radical production normalized to the amount of surface iron. The error bars in Fig. 12.10 are the result of two measurements. First, for both erionite and mordenite, the amount of hydroxyl radical follows the order: no antioxidant < glutathione < ascorbic acid. Note that in the absence of antioxidant, the Fe(III) can also be reduced to Fe(II) by H_2O_2. Second, the surface-iron-normalized data show that erionite produces significantly more hydroxyl radicals than mordenite. Third, the amount of hydroxyl radical relative to the surface iron varies from about

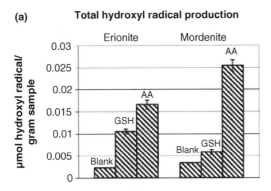

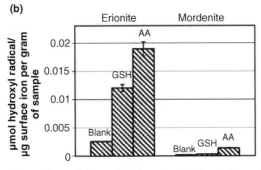

FIGURE 12.10 Comparison of (a) total hydroxyl radical and (b) surface-iron normalized hydroxyl radical generation between Fe^{3+}-erionite and Fe^{3+}-mordenite after reaction with ascorbic acid (AA) and gluthathione (GSH). (Adapted from Reference 34.)

~1% in the case of mordenite with no antioxidant added to ~106% with AA-treated erionite.

Erionite was about an order of magnitude more Fenton reactive than mordenite for all reductants, normalized to the surface iron levels. This is quite striking, especially considering that the Fenton reactivity of Fe(II) showed an increase by only a factor of 2 for erionite (Fig. 12.9).

12.1.4 Mutagenicity

We have studied the cytotoxic and genotoxic properties of erionite and mordenite using the AS52 cell line (35). To determine the effects of iron, erionite and mordenite were treated with either zeolite alone or in combination with iron, and the effects on cytotoxicity and the mutation frequency of the *gpt* recorder gene were determined and compared to values on nontreated controls.

Exposure of AS52 cells to either erionite or mordenite resulted in a dose-dependent cytotoxicity of the treated cells when compared to the nontreated control. A statistically significant ($p < 0.001$) increase in the cytotoxicity of cells was noted with mordenite at concentrations of $6 \, \mu g/cm^2$ or greater. Erionite was consistently more cytotoxic than mordenite, even at concentrations as low as $2 \, \mu g/cm^2$. The addition of ferrous ion to the mordenite-containing medium enhanced the cytotoxicity of mordenite ($p < 0.001$) (cells treated with $16 \, \mu g/cm^2$ mordenite and ferrous ion concentrations from 5 to 20 M). Likewise, cotreatment of AS52 cells with iron and erionite significantly ($p < 0.001$) enhanced the cytotoxicity as compared to cells treated with either iron or erionite (iron concentration of $20 \, \mu M$ and erionite concentrations greater than $6 \, \mu g/cm^2$).

Figure 12.11 shows the increase in the mutation frequency of cells treated with the fibers. In case of mordenite or with mordenite and ferrous ion, there was no increase in mutation frequency compared to the appropriate controls (nontreated and ferrous ion treated cells). In contrast to mordenite, there was a significant ($p < 0.01$–0.001) increase in the relative mutation frequency of cells treated with erionite when compared to nontreated cells (Fig. 12.11c) at concentrations of erionite $\geq 6 \, \mu g/cm^2$. The increase ranged from 1.6 to 4.8 for cells treated with 6 and $16 \, \mu g$ of erionite/cm^2, respectively. At concentrations of erionite below $8 \, \mu g/cm^2$, the mutation frequency of cells treated with erionite and ferrous ion was not significantly different from the mutation frequency obtained for cells treated with ferrous ion. However, at erionite concentrations $\geq 8 \, \mu g/cm^2$, there was a statistically significant ($p < 0.001$) increase in the relative mutation rate of cells treated with erionite and ferrous ion compared to cells treated with only ferrous ion. This ranged from a 1.2-fold increase in cells treated with $8 \, \mu g$ of erionite/cm^2 and $20 \, \mu M$ ferrous chloride to a 3.3-fold increase in cells treated with $16 \, \mu g$ of erionite/cm^2 and $20 \, \mu M$ ferrous chloride.

These data demonstrate that the cytotoxicities of the zeolites were enhanced in the presence of ferrous ions. Mordenite was not mutagenic and ferrous ion did not alter mordenite's mutagenic potential. Conversely, erionite was mutagenic and its mutagenic potential was significantly increased in the presence of iron.

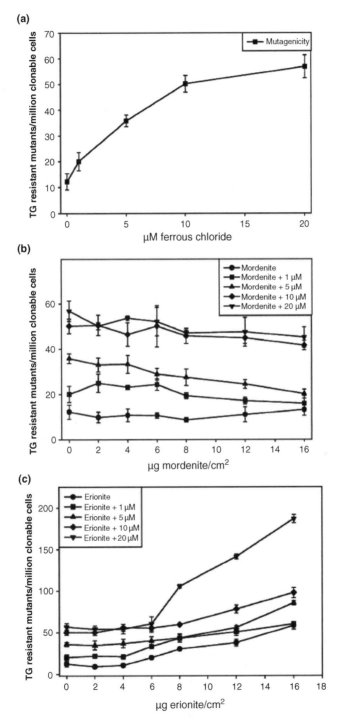

FIGURE 12.11 Iron- and zeolite-induced mutagenesis of AS52 cells. (a) Ferrous-chloride-treated cells. (b) Mordenite-treated cells with varying levels of Fe^{2+} in solution. (c) Erionite-treated cells with varying levels of Fe^{2+} in solution. (Adapted from Reference 35.)

12.1.5 Discussion: Correlation of Biological Activity with Structure

Summarizing the above studies, it can be concluded that erionite and mordenite exhibit considerable differences, as well as similarities. For example, there was no significant difference in the oxidative burst by macrophages between erionite and mordenite. Both exhibited ability to ion exchange iron; though, depending on the ion exchange method, the amount of exchange varied. Comparable levels of Fe(II) were found on the surface of both minerals (330–400 μg/g), but the Fe(III) level was significantly lower in erionite (0.88 μg/g), compared to mordenite (18 μg/g). Normalized to the surface iron levels, erionite exhibited a factor of 2–10-fold increase in Fenton chemistry (the latter number is for Fe(III) with added reductants) to produce hydroxyl radicals from hydrogen peroxide compared to mordenite. Erionite and mordenite also differed in their cytotoxic and mutagenic capabilities, a significant increase being noted for the mutagenicity of erionite in the presence of iron.

The differences in the biological and chemical effects of these two minerals are most profound in the presence of iron. However, physiological effects of these minerals cannot be solely a result of their ability to transport iron intracellularly since both mordenite and erionite can transport iron. Another hypothesis is that the chemical reactivity of iron on the two mineral surfaces is different. Figure 12.12 is a surface structure rendering of mordenite and erionite showing the different rings, 4-, 5-, 8-, and 12-membered rings for mordenite and 4-, 6-, and 8-membered rings for erionite. The a and b axes of erionite represent the fiber surface, and show a network of 8-rings (3.6×5.1 Å). Mordenite has channel openings made of 12-rings (7.0×6.5 Å), 8-rings (2.6×5.7 Å), and smaller 4-membered rings. Clearly, iron on the zeolite surface is coordinated to different aluminosilicate rings on these minerals. This coordination environment can modify the iron redox potential and influence the chemical reactivity that results in differences in the biological reactivity of these two minerals.

XPS data show that for Fe(III)-mordenite, the Fe $2p_{3/2}$ peak is shifted to higher binding energies by 0.8 eV, compared to erionite, suggesting that in mordenite the Fe(III) is coordinated by stronger electron-withdrawing ligands (34). Among the minerals, the closest parallel is the comparison between Fe_2O_3 and goethite, FeOOH. The presence of hydroxyl ligands in goethite leads to increased binding energy for both Fe 2p and Fe 3p peaks, for example, the Fe 2p by about 0.85 eV (57).

On the basis of the coordination hypothesis, we first address the differences in the Fenton reactivity of the two zeolite minerals. The coordination environment around the iron, especially the ability to form ternary ligand–metal–H_2O_2 complexes, is critical for promoting the Fenton reaction (58, 59); thus, the ligands around the iron are critical for participation in the Fenton reaction. It is known that iron complexes with low reduction potentials, for example, with EDTA (0.12 V), have higher rate constants ($10 \times 10^3 \, M^{-1} \, s^{-1})$ for the Fenton reaction than complexes with higher reduction potentials, for example, hydrated Fe(II), $76 \, M^{-1} \, s^{-1}$ (0.77 V) (60, 61). For an inner sphere mechanism, availability of coordination sites for H_2O_2 to Fe(II) is necessary. For example, with ligands such as EDTA and citrate, extra coordination sites are available and the complexes are Fenton active. Fe-EDTA has a higher Fenton rate

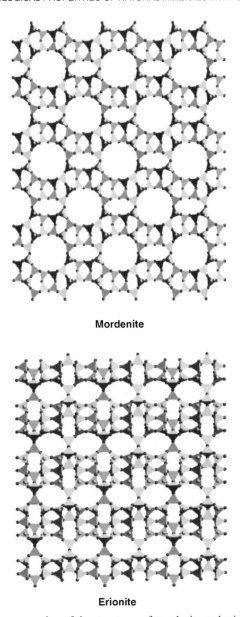

Mordenite

Erionite

FIGURE 12.12   Representation of the structures of mordenite and erionite. (Adapted from Reference 34.)

constant (factor of 10) than Fe-DTPA (E^0 value of 0.03 V) since extra coordination sites in DTPA (diethylenetriamine pentaacetic acid) are not available. On the other hand, strongly coordinating chelating ligands, such as desferrioxamine B and ferrozine, render the iron redox inactive.

Joshi and Limye (62) studied the catalytic decomposition of H_2O_2 by Fe(III)-mordenite, but did not conclude if the rate-determining step was the reduction of Fe(III) by the peroxide anion or the reduction of H_2O_2 by Fe(II) (Fenton reaction). The role of a mineral surface in modulating the reduction potential of Fe^{3+}/Fe^{2+} has been noted. The trend is to make the potential less positive, thus making Fe^{2+} more susceptible to oxidation. For example, the standard reduction potentials of Fe^{3+}/Fe^{2+} in aqueous solution, sorbed onto goethite and silica, are 0.77, 0.36, and 0.23–0.55 V, respectively (63). Other transition metals such as Mn^{2+} are more readily oxidized when sorbed onto alumina, quartz, and goethite (63). Chromium (IV), on the other hand, is reduced more readily by carboxylic acids in the presence of goethite, alumina, and TiO_2 (64). The reduction potential of iron in natural amphibole, magnetite, and biotite ranges from 0.27 to 0.52 V (65), values lower than the reduction potential of iron in water (0.77 V). Fubini et al. (66–68) have also argued that only few surface iron species on mineral samples are in the right redox and coordination state to be active in the hydroxyl radical generation. The reduction potential of Cu(II)-exchanged zeolites using cyclic voltammetry shows a 30 mV difference in $E_{1/2}$ between the cyclic voltammograms of copper-exchanged mordenite and copper-exchanged zeolite Y, suggesting that metal ions on different zeolites exhibit different reduction potentials (69). Thus, the increase in the amounts of radicals per surface iron for erionite compared to mordenite indicates that erionite surface sites provide a Fenton-enhancing coordination for iron.

The role of transition metals in the biological activity of carbon nanotubes is still undetermined. As far as the relation between metal ion content of CNTs and biological activity is concerned, Guo et al. (70) have shown that the bioavailability of iron depends on a host of parameters, including the vendor, purification processes, aging, sample handling including preparation, and possibly the reason for the variability in the published toxicity data. Even after purification of CNTs, transition metals such as Ni and Fe are still present (22). As-made SWCNTs can contain as much as 30 wt% iron, and such Fe-rich SWCNTs result in ultrastructural and morphological changes, oxidative stress, loss of cell viability (71), and changes in intracellular properties such as loss of glutathione and accumulation of lipid hydroperoxides (72). Upon purification, the biological effects of SWCNT including uptake by macrophages was limited (73). Such profound differences suggest that the metal content in CNTs as well as their aggregation characteristics modified by purification are playing a major role. Even without internalization, SWCNT bundles on surfaces of epithelial cells tend to increase the number of surfactant-storing lamellar bodies, presumably as a protective function (74). An interesting observation made with both Fe-rich and Fe-poor SWCNTs in that they are not readily phagocytized by macrophages *in vitro* unless modified by phosphatidylserine or detergent pluronic (75), in contrast to mineral fibers. Introduction of surface functionality, in the form of oxygen-containing groups (−OH, COOH, and C=O) is reported to lead to increase in toxicity of CNTs and C particulates (76). Thus, many of the issues with mineral fibers and asbestos have parallels in CNTs, especially the redox chemistry of surface bound transition metals on CNT, which has not yet been explored.

12.2 CORRELATIONS OF STRUCTURE WITH BIOLOGICAL RESPONSE OF PARTICULATES

Epidemiological studies have demonstrated an association between levels of airborne particulate matter of mass median aerodynamic diameter $\leq 2.5\,\mu m$ (PM2.5) and morbidity/mortality. Specific particulate-associated disease entities implicated by these studies include chronic bronchitis and interstitial fibrosis, exacerbation of asthmatic episodes and respiratory infectious diseases, as well as cardiovascular complications such as ischemic heart disease and stroke (77–80).

In support of the epidemiological observations cited above, a study of lung tissue acquired at autopsy from lifelong residents of Mexico City, a region of sustained high levels of airborne particulates, demonstrated extensive particulate deposition and greatly thickened, fibrotic airway walls compared to lung tissue of individuals from Vancouver, BC, a region of low ambient particulate levels (81). *In vitro* studies and animal experiments have documented a number of biological and pathological responses to particulate exposure, including enhancement of cytokine, chemokine, and intracellular oxidant production by macrophages and airway epithelial cells following phagocytosis of particulates (82–85), impairment of chemotactic motility of particulate-laden macrophages (86), acute and chronic pulmonary inflammation *in vivo* following experimental particulate administration or natural exposure in animals and humans (81, 87–89), and the subsequent development of alveolar epithelial hyperplasia and progressive pulmonary fibrosis (81, 87). Although data generated by these investigations provide some insight into possible etiologic roles of inhaled particulate matter, specific mechanisms by which particulate deposition in the alveoli might initiate or exacerbate cardiopulmonary disease remain to be fully resolved.

A major focus of uncertainty resides in the relationship between the physicochemical properties of particulates and their pathogenic potential. Ambient urban airborne particulate matter consists primarily of complex mixtures of fossil fuel combustion products such as coal fly ash (CFA), residual oil fly ash, and diesel exhaust particulates. The chemical composition of such particulates varies as a function of starting material and combustion process; however, common components include carbon aggregates, heavy metals, transition metals, acid salts, silicates, and a variety of organic compounds such as polycyclic aromatic hydrocarbons (90–92). Fullerenes (e.g., C_{60}) have been reported to be cytotoxic to human skin and liver cells, mediated by ROS generation (93).

Furthermore, the surface chemistry of particulates can be modified following their emission by reaction with atmospheric components. The etiologic significance of particulate chemistry has been suggested by several studies, which have demonstrated marked differences in biologic response to particulates as a function of chemical composition and/or surface characteristics (94–96). Hence, it is of major importance to define relationships between physicochemical properties of particulates and the mechanisms and intensity of biological responses induced following exposure.

Herein we review our studies of the impact of particulate surface chemistry and surface area upon particulate-induced oxidative stress and inflammatory responses in human macrophages. Since the endothelium is an integral component of the inflammatory cascade, we further summarize our studies of the relationship between particulate characteristics and macrophage-mediated endothelial inflammatory activation. To identify specific surface properties that determine biological activity, we have synthesized simplified model carbon particulates with or without surface metal or surface charge, and compared their biologic effects to those of environmental coal fly ash and diesel emission particulates (DEPs).

12.2.1 Internalization and Toxicity of Particulates

Human monocyte-derived macrophages (MDMs), isolated from human peripheral blood mononuclear cells and differentiated *in vitro* into the macrophage phenotype, were treated with $25\,\mu g/cm^2$ surface area of size-fractionated CFA ($<2.5\,\mu m$), DEPs, synthetic 1-μm-sized carbon particulate (μC), carbon spiked with iron (μC–Fe), or carbon–iron particle coated with fluoroaluminosilicate (μC–Fe/F–AlSi), or 100-nm-sized C or C–Fe. The synthesis of carbon particle replicas was done using an acid catalyzed condensation reaction of phenol and paraformaldehyde monomers using zeolite Y as a template for the reaction. The zeolite/polymer mix was pyrolyzed at $800°C$ to create the carbonaceous particulates and the zeolite template was removed by etching in hydrofluoric acid. Internalization of all particulates was verified by phase contrast microscopy or transmission electron microscopy. As shown in Fig. 12.13, all particulates were readily internalized by MDMs within 24 h (97, 98). This concentration of particulates was nontoxic as determined by trypan blue dye exclusion and absence of gross morphologic evidence of cell degeneration. However, as shown in Fig. 12.14, while μC particulates remained dispersed in the cytoplasm or formed small loose aggregates with no apparent disturbance of surrounding ultrastructure, μC–Fe sometimes formed large tightly associated clusters surrounded by a zone of organelle lysis (99).

12.2.2 Particulate-induced Macrophage Oxidative Burst

To assess the role of surface iron in the induction of oxidative stress in macrophages, MDMs in 96-well culture plates were exposed to $1\,\mu m$ C or C–Fe ($5\,\mu g/cm^2$) in the presence of luminol-supplemented culture medium. Luminescence, as a measure of superoxide liberation, was assayed at 10-min intervals, and luminescence indices were calculated by dividing the mean luminescence counts per minute (cpm) of four replicate particulate-treated wells by the mean cpm of four negative-control wells at each time point. As shown in Fig. 12.15, C–Fe-treated macrophages generated a curve with typical oxidative burst kinetics, with luminescence peaking at 20 min postexposure and gradually decreasing thereafter, while macrophages exposed to C generated little if any signal (99).

12.2.3 Macrophage-mediated Endothelial Activation by Particulates

To determine the impact of soluble factors elaborated by particulate-exposed MDMs upon proximal endothelia, human endothelial cells derived from umbilical vein (hUVEC), pulmonary artery (hPAEC), or pulmonary microvasculature (hPMVEC) were incubated with supernatants recovered from particulate-treated human MDM

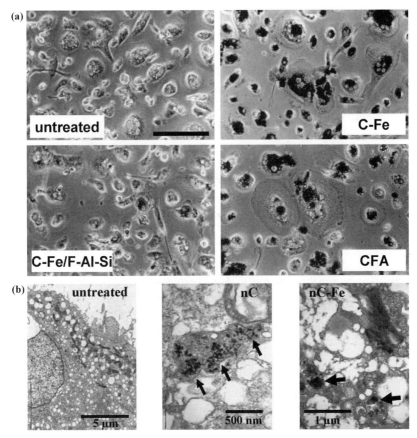

FIGURE 12.13   Internalization of particulates by macrophages. (a) Phase contrast micrographs of untreated MDMs or MDMs treated for 24 h with 1 μm synthetic C–Fe, C–Fe/F–AlSi, or size-fractionated (≤2.5 μm) CFA. (b) Transmission electron micrographs of untreated MDMs or MDMs treated for 24 h with nC or nC–Fe. (Adapted from References (97, 98).)

cultures, and then assayed by immunofluorescence flow cytometry for expression of endothelial–leukocyte adhesion molecules. Figure 12.16 presents data generated by immunofluorescence flow cytometric analysis of ICAM-1, VCAM-1, and E-selectin surface expression on hPMVEC following incubation with supernatants recovered from MDMs treated with various concentrations of each of the five types of 1 μm particulates (C, C–Fe, C–Fe/F–Al–Si, DEPs, CFA) (97). Mean fluorescence intensity (MFI) values shown in Fig. 12.16 were calculated by subtracting MFI generated by nonspecific staining with irrelevant isotype-matched control antibodies from MFI of specifically stained cells. These curves demonstrate dose-dependent induction of all three adhesion molecules upon hPMVEC incubated with supernatants recovered from MDMs treated with C–Fe, C–Fe/F–Al–Si, or DEPs. (Because toxic effects induced by these particulates upon MDMs began to occur at concentrations of 40–50 μg/cm^2,

(a)   **(b)**

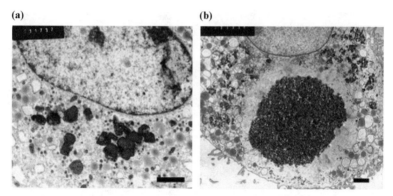

FIGURE 12.14   Ultrastructural disruption by iron-containing particulates. Transmission electron micrographs of MDM treated for 24 h with 1 μm synthetic C (a) or C–Fe (b). (Adapted from Reference 99.)

curves generated by these particulates terminate at 25 μg/cm^2.) In contrast, little or no endothelial activation was observed in response to supernatants recovered from MDMs treated with C or CFA, even at 2–4-fold greater concentrations (97). Data generated by these experiments suggest that the inflammatory potential of internalized particulates varies greatly as a function of particulate physicochemical characteristics. (Data are representative of experiments performed with MDMs isolated from each of five volunteer donors, 2–4 replicate experiments/donor, and with human endothelial cells derived from umbilical vein, pulmonary artery, and pulmonary microvasculature.)

To control for the possibility that normal MDM products released simply as a consequence of cell injury might induce this endothelial activation, ECs were also treated with clarified lysate of sonicated untreated MDMs. No adhesion molecule induction was observed in response to this treatment (97).

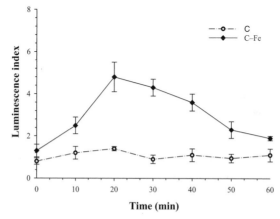

FIGURE 12.15   Induction of oxidative stress by iron-containing particulates. Luminol assay of oxidative burst kinetics in MDMs treated with 1 μm synthetic C or C–Fe. (Adapted from Reference 99.)

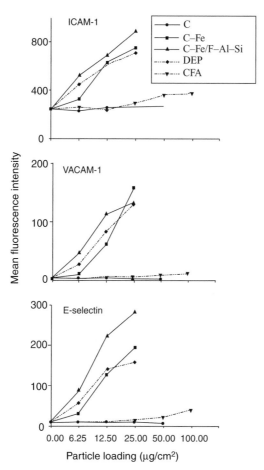

FIGURE 12.16 Differential induction of endothelial adhesion molecule expression by soluble factors elaborated by particulate-treated MDMs. Immunofluorescence flow cytometric analysis of hPMVEC following their incubation with supernatants recovered from MDMs treated with various concentrations of DEPs, CFA, or 1 μm synthetic C, C–Fe, or C–Fe/F–AlSi, and staining with FITC-conjugated antibodies specific for ICAM-1, VCAM-1, or E-selectin. (Adapted from Reference 97.)

To determine if MDM responses were elicited by endotoxin that may have been brought into cells on the particulates, a subset of the experiments described above was performed in the presence of 10 μg/mL polymyxin B, a known inactivator of endotoxin. While this concentration of polymyxin B in MDMs cultures completely neutralized endothelial adhesion molecule induction by supernatants recovered from MDMs stimulated by 100 pg/mL of LPS, it did not attenuate adhesion molecule induction by particulate-treated MDM supernatants (97).

To determine the impact of particulate surface area upon the magnitude of endothelial inflammatory activation, experiments were repeated with synthetic 100 nm carbon (nC) and carbon–iron (nC–Fe) particulates. Data generated by a

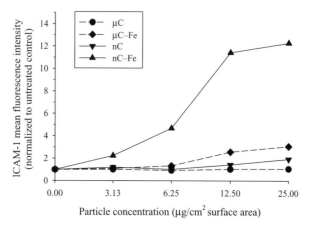

FIGURE 12.17 Enhanced macrophage-mediated endothelial inflammatory response induced by nanoparticulates. Immunofluorescence flow cytometric analysis of hPMVEC following their incubation with supernatants recovered from MDMs treated with various concentrations of 1 μm or 100 nm synthetic C or C–Fe, and staining with FITC-conjugated antibodies specific for ICAM-1. (Adapted from Reference 98.)

representative experiment (out of three replicates conducted with MDM isolated from three individual donors), expressed as mean fluorescence intensity normalized to untreated controls, are shown in Fig. 12.17 (98). As described above, supernatants recovered from MDMs treated with μC had no impact upon endothelial ICAM-1 expression. Similarly, supernatants recovered from nC-treated macrophages induced only a slight increase (approximately twofold) in endothelial ICAM-1. While supernatants recovered from MDMs treated with 25 μg/cm² of μC–Fe induced an approximately threefold enhancement of endothelial ICAM-1, supernatants recovered from macrophages treated with nC–Fe at all concentrations induced higher levels of endothelial ICAM-1 (up to 12-fold over untreated controls at 25 μg/cm²) (98).

The inducing agent elaborated by particulate-treated MDMs was identified as TNF-α by the complete attenuation of MDM-induced ICAM-1 by addition of blocking antibody specific for TNF-α to supernatants prior to their incubation with EC, as demonstrated by fluorescence flow cytometry data presented in Fig. 12.18 (98).

12.2.4 Quantitation of Particulate-induced TNF-α Production by Macrophages

To quantitate particulate-mediated TNF-α production by macrophages, supernatants of particulate-treated (or untreated) MDMs were analyzed by TNF-α-specific ELISA. As shown in Fig. 12.19 (representative of three replicate experiments conducted with MDMs isolated from three individual donors), while μC–Fe induced modest TNF-α production by MDMs (~0.15–0.2 ng/mL) only at the highest particulate concentration of 25 μg/cm², nC–Fe induced a dose-dependent elaboration of TNF-α by MDMs, reaching a level of >1.5 ng/mL in response to treatment with 12.5 μg/cm² (98). Data points represent mean values determined from duplicate microtiter wells. It should be

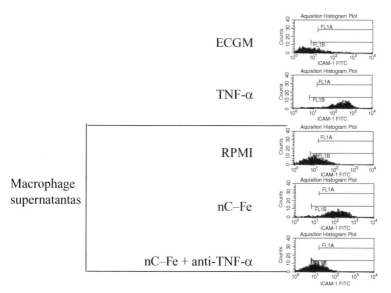

FIGURE 12.18 Attenuation of macrophage-mediated endothelial inflammatory activation by neutralization of TNF-α. hPMVEC ICAM-1 expression following their incubation with supernatants recovered from untreated MDMs (RPMI), supernatants recovered from MDMs treated with synthetic nC–Fe, or supernatants recovered from MDMs treated with synthetic nC–Fe in the presence of antibody with specific blocking activity against TNF-α. (Adapted from Reference 98.)

noted that TNF-α production induced by $25\,\mu g/cm^2$ nC–Fe exceeded the upper measurable limit of the assay. In contrast, TNF-α produced by nC- or μC-treated MDMs was below the detection limit of the assay at all particulate concentrations (up to $25\,\mu g/cm^2$) (98).

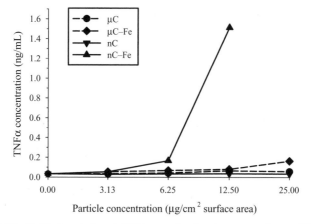

FIGURE 12.19 Particulate-induced TNF-α production by macrophages. TNF-α-specific ELISA of supernatants recovered from macrophages treated with various concentrations of 1 μm or 100 nm synthetic C or C–Fe. (Adapted from Reference 98.)

12.2.5 Determination of Fenton Activity of Particles

The ability of the particles to catalyze the decomposition of hydrogen peroxide to hydroxyl radicals through Fenton chemistry was measured by spin trapping with 5,5-dimethylpyroline-*N*-oxide (DMPO). The presence and magnitude of the four-line first derivative ESR signal (1:2:2:1 quartet) characteristic of the hydroxyl adduct, 2,2-dimethyl-5-hydroxy-1-pyrrolidinyloxyl (DMPO–OH), with a hyperfine splitting constant of 14.3 G correlated well with the inflammatory activities of each particulate (Fig. 12.20a and b). Specifically, μC, nC, and CFA, which elicited little if any inflammatory responses, generated little if any ESR signal. In contrast C–Fe, C–Fe/F–AlSi, and DEPs, all of which elicited substantial inflammatory responses, generated strong ESR signals, with nC–Fe generating stronger than its 1 μm counterpart (98).

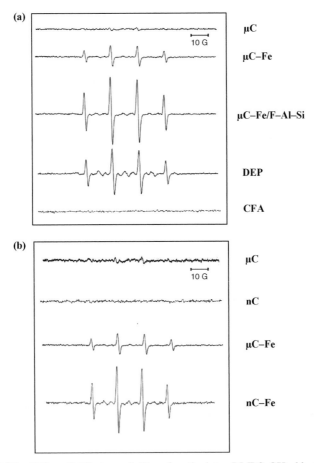

FIGURE 12.20 Differential Fenton activities of particulates. DMPO-OH adduct formation in solution as determined by electron paramagnetic resonance spectroscopy after exposure of 1 μm (a) or 100 nm (b) particulates. (Adapted from Reference 98.)

12.2.6 Discussion: Correlation of Biological Activity with Structure

In this chapter, we have summarized results of our studies that suggest a contributing role for surface Fenton-reactive metal ion in particulate-mediated inflammatory activation of macrophages and the consequent endothelial inflammatory activation by macrophage-elaborated soluble factors. Specifically, we have demonstrated that pure carbon particulates (C) show little or no Fenton reactivity with H_2O_2 as indicated by ESR of DMPO–OH adducts, and induce little if any TNF-α production by exposed macrophages. In contrast, equivalent doses of carbon-iron (C–Fe) particulates generate strong ESR DMPO–OH signals following reaction with H_2O_2 and induce high levels of TNF-α production by exposed macrophages in a dose-dependent manner. Furthermore, clarified culture supernatants recovered from C–Fe-treated macrophages raise vascular endothelial cells to an activated state as indicated by the induction of endothelial–leukocyte adhesion molecules ICAM-1, VCAM-1, and E-selectin (97). Finally, our observation that 100 nm particulates elicit much stronger responses than an equivalent mass of their 1-μm counterparts argues in favor of a role for surface chemistry in particulate-induced inflammation (98).

Various types of particulates have been demonstrated to induce the expression of a number of additional proinflammatory cytokines, chemokines, and receptors in exposed cultured cells, including interleukin (IL)-1β, IL-6, macrophage inflammatory protein-2 (MIP-2), histamine H1 receptor (H1R), IL-8, granulocyte macrophage-colony stimulating factor (GM-CSF), and monocyte chemoattractant protein-1 (MCP-1), effects which in some cases can be inhibited by antioxidants (82–85). Macrophages exposed to high concentrations (200 μg/mL) of ultrafine carbon black (14 nm) have been shown to produce a modest level of TNF-α, which can be blocked by introduction of the calcium channel blocker verapimil (100). It has been proposed that production of ROS leads to the oxidation of calcium pumps in the endoplasmic reticulum (100). The role of free radicals and reactive oxygen species created upon ultrafine particle inhalation in causing inflammation is being recognized for different types of particulates (101).

Iron can lead to the formation of hydroxyl radicals through the Fenton mechanism. Iron-containing particles have been proposed as a mechanism of introducing unregulated iron into cells (33, 97, 98, 102), and soluble iron has been shown to induce the expression of TNF-α in Kupffer cells, which was preceded by an increase in hydroxyl radicals (103).

Collectively, these findings suggest oxidative stress, induced by particulates in alveolar macrophages and possibly alveolar epithelia, as a potential contributing factor in the pathogenesis of pulmonary disease or exacerbation of existing disease associated with exposure to airborne particulates. Oxidative stress, in turn, triggers an inflammatory response in macrophages, which act as both phagocytes and sentinels in the lung and other tissues, resulting in their elaboration of TNF-α and other proinflammatory cytokines and chemokines. TNF-α then induces the expression of leukocyte adhesion molecules upon proximal septal capillary endothelium resulting in the recruitment of inflammatory leukocytes to the site. While in the case of acute injury or infection, inflammation is a beneficial response, required to clear pathogens or toxins and to initiate wound healing, chronic nonresolving inflammation, as would be expected under conditions of protracted inhalation exposure to airborne particulates, can lead to permanent tissue remodeling and fibrosis, as has been observed in individuals following long-term exposure.

Carbon-based nanomaterials including carbon nanoparticles are being considered for many applications, including optical devices, electronic applications, and biomedical use, for example, drug delivery. However, potential hazards stemming from exposure to carbon particulates are still being debated. In the context of the studies described here, as particles decrease in size with manifestation of novel properties, both the increased surface area and the functionality on the surface will play important roles in their biological activity.

ACKNOWLEDGMENTS

This work was supported in part by NIOSH Grant R01 OH009141 (PKD and WJW). We also acknowledge the contributions of the students involved in this project.

REFERENCES

1. Rom, W.N., Travis, W.D., Brody, A.R. Cellular and molecular basis of the asbestos-related diseases. *Am Rev Respir Dis* 1991;143:408–422.

2. Crouch, E. Pathobiology of pulmonary fibrosis. *Am J Physiol* 1990;259:L159–L184.

3. Heppleston, A.G. Minerals, fibrosis, and the lung. *Environ Health Perspect* 1991;94:149–168.

4. Quinlan, T.R., BeruBe, K.A., Marsh, J.P., Janssen, Y.M.W., Taishi, P., Leslie, K.O., Hemenway, D., O'Shaughnessy, P.T., Vacek, P., Mossman, B.T. Patterns of inflammation, cell proliferation, and related gene expression in lung after inhalation of chrysotile asbestos. *Am Pathol* 1995;147:728–739.

5. Maples, K.R., Johnson, N.F. Fiber-induced hydroxyl radical formation: correlation with mesothelioma induction in rats and humans. *Carcinogenesis* 1992;13:2035–2039.

6. Veblen, D.R., Wylie, A.G. Mineralogy of amphiboles and 1:1 layer silicates. In: Guthrie, G.D., Mossman, B.T., editors. *Reviews in Minerology*, 28. Chelsea, MI: Bookcrafters, Inc.; 1993. pp. 61–137.

7. Mossman, B.T., Gee, J.B.L. Asbestos-related diseases. *N Engl J Med* 1989;320(26): 1721–1730.

8. Mossman, B.T., Bignon, J., Corn, M., Seaton, A., Gee, J.B.L. Asbestos: scientific developments and implications for public policy. *Science* 1990;247:294–301.

9. Hansen, K., Mossman, B.T. Generation of superoxide O_2^- from alveolar macrophages exposed to asbestiform and nonfibrous particles. *Cancer Res* 1987;47:1681–1686.

10. Xu, A., Wu, L.J., Santella, R.M., Hei, T.K. Role of oxyradicals in mutagenicity and DNA damage induced by crocidolite asbestos in mammalian cells. *Cancer Res* 1999;59:5922–5926.

11. Janero, D.R. Malondialdehyde and thiobarhituric acid reactivity as diagnostic indices of lipid peroxidation and peroxidative tissue injury. *Free Radic Biol Med* 1990;9:515–540.

12. Refsgaard, H.H.F., Tsai, L., Stadtman, E.R. Modifications of proteins by polyunsaturated fatty acid peroxidation products. *Proc Natl Acad Sci USA* 2000;97:611–616.

13. Brown, R.C., Carthew, P., Hoskins, J.A., Sara, E., Simpson, C.F. Surface modification can affect the carcinogenicity of asbestos. *Carcinogenesis* 1990;11:1883–1885.

14. De Vuyst, P., Dumortier, P., Swaen, G.M.H., Pairon, J.C., Brochard, P. Respiratory health effects of man-made vitreous (mineral) fibers. *Eur Respir J* 1995;8:2149–2173.

15. Stephenson, D.J., Fairchild, C.I., Buchan, R.M., Dakins, M.E. A fiber characterization of the natural zeolite, mordenite: a potential inhalation health hazard. *Aerosol Sci Technol* 1999;30:467–476.

16. Muller, J., Huaux, F., Heilier, J.F., Arras, M., Delos, M., Nagy, B.J., Lison, D. Respiratory toxicity of carbon nanotubes. *Toxicol Appl Pharmacol* 2005;207:221–231.

17. Shvedova, A.A., Castranova, V., Kisin, E.R., Schwegler-Berry, D., Murray, A.R., Grandelsman, V.Z., Maynard, A., Baron, P. Exposure to carbon nanotube material: assessment of nanotube cytotoxicity using human keratinocyte cells. *J Toxicol Environ Health* 2003;66:1909–1926.

18. Shvedova, A.A., Kisin, E.R., Mercer, R., Murray, A.R., Johnson, V.J., Potapovich, A.I., Tyurina, Y.Y., Gorelik, O., Arepalli, S., Schwegler-Berry, D., Hubbs, A.F., Antonini, J., Evans, D.E., Ku, B.-K., Ramsey, D., Maynard, A., Kagan, V.E., Castranova, V., Baron, P. Unusual inflammatory and fibrogenic pulmonary responses to single walled carbon nanotubes in mice. *Am J Physiol Lung Cell Mol Physiol* 2005;289:L698–L708.

19. Fenoglio, I., Tomatis, M., Lison, D., Muller, J., Fonseca, A., Nagy, J.B., Fubini, B. Reactivity of carbon nanotubes: free radical generation or scavenging activity? *Free Radic Biol Med* 2006;40(7): 1227–1233.

20. Warheit, D.B., Laurence, B.R., Reed, K.L., Roach, D.H., Reynolds, G.A., Webb, T.R. Comparative pulmonary toxicity assessment of single-wall carbon nanotubes in rats. *Toxicol Sci* 2004;77:117–125.

21. Lam, C.W., James, J.T., McCluskey, R., Hunter, R.L. Pulmonary toxicity of single-wall carbon nanotubes in mice 7 and 90 days after intratracheal instillation. *Toxicol Sci* 2004;77:126–134.

22. Jia, G., Wang, H., Yan, L., Wang, X., Pei, R., Yan, T., Zhao, Y., Guo, X. Cytotoxicity of carbon nanomaterials: single-wall nanotube, multi-wall nanotube, and fullerene. *Environ Sci Technol* 2005;39(5): 1378–1383.

23. Wagner, J.C., Skidmore, J.W., Hill, R.J., Griffiths, D.M. Erionite exposure and esotheliomas in rats. *Br J Cancer* 1985;51:727–730.

24. Timblin, C.R., Guthrie, G.D., Janssen, Y.W.M., Walsh, E.S., Vacek, P., Mossman, B.T. Patterns of c-fos and c-jun proto-onocogene expression, apoptosis, and proliferation in rat pleural mesothelial cells exposed to erionite or asbestos fibers. *Toxicol Appl Pharamacol* 1998;151:88–97.

25. Hardy, J.A., Aust, A.E. Iron in asbestos chemistry and carcinogenicity. *Chem Rev* 1995;95:97–118.

26. Breck, D.W. *Zeolite Molecular Sieves*. New York: Wiley-Interscience Publication; 1974.

27. Eborn, S.K., Aust, A. Effect of iron acquisition on induction of DNA single-strand breaks by erionite, a carcinogenic mineral fiber. *Arch Biochem Biophys* 1995;316:507–514.

28. Guthrie, G.D. Biological effects of inhaled minerals. *Am Mineral* 1992;77; 225–243.

29. Palekar, L.D., Most, B.M., Coffin, D.L. Significance of mass and number of fibers in the correlation of V79 cytotoxicity with tumorigenic potential of mineral fibers. *Environ Res* 1988;46:142–152.

30. Adamis, Z., Tatrai, E., Honmas, K., Six, E., Ungvary, G. *In vitro* and *in vivo* tests for determination of the pathogenicity of quartz, diatomaceous earth, mordenite, and clinoptitolite. *Ann Occup Hyg* 2000;44:67–74.

31. Hogg, B.D., Dutta, P.K., Long, J.F. *In vitro* interaction of zeolite fibers with individual cells (macrophages NR8383): measurement of intracellar oxidative burst. *Anal Chem* 1996;68:2309–2312.

32. Long, J.F., Dutta, P.K., Hogg, B.D. Fluorescence imaging of reactive oxygen metabolites generated in single macrophage cells upon phagocytosis of natural zeolite fibers. *Environ Health Perspect* 1997;105:706–711.

33. Fach, E., Waldman, W.J., Williams, M., Long, J., Meister, R.K., Dutta, P.K. Analysis of the biological and chemical reactivity of zeolite-based aluminosilicate fibers and particulates. *Environ Health Perspect* 2002;110(11): 1087–1096.

34. Ruda, T.A., Dutta, P.K. Fenton chemistry of Fe(III)-exchanged zeolitic minerals treated with antioxidants. *Environ Sci Technol* 2005;39(16): 6147–6152.

35. Fach, E., Kristovich, R., Long, J.F., Waldman, W.J., Dutta, P.K., Marshall, M.V. The effect of iron on the biological activities of erionite and mordenite. *Environ Int* 2003;29:451–458.

36. Dutta, P.K., Zeolites: a primer. In: Auerbach, S.M., Carrado, K.A. editor. Dutta, P.K., coeditor, *Handbook of Zeolites and Layered Materials*. NY: Marcel Dekker, Inc.; 2003. pp. 1–19.

37. Allison, A.C., Harington, J.S., Birbeck, M. An examination of the cytotoxic effects on silica on macrophages. *J Exp Med* 1966;124:141–154.

38. (a) Helmke, R.J., Boyd, R.L., German, V.F., Mangos, J.A. *In Vitro* Cell. *Dev Biol* 1987;23:567–574 (b) Helmke, R.J., German, V.F., Mangos, J.A. *In Vitro* Cell. *Dev Biol* 1989;25:44–48.

39. (a) Rothe, G., Valet, G. *J Leukoc Biol* 1990;47:440–448. (b) Kobzik, L., Godleski, J.J., Brain, J.D. *J Leukoc Biol* 1990;47:295–303.

40. Vilím, V., Wilhelm, J., Brzák, P., Hurych, J. Stimulation of alveolar macrophages by mineral dusts *in vitro*: luminol-dependent chemiluminescence study. *Environ Res* 1987;42:246–256.

41. Kobzik, L., Godleski, J.J., Brain, J.D. Oxidative metabolism in the alveolar macrophage: analysis by flow cytometry. *J Leukoc Biol* 1990;47:295–303.

42. Szejda, P., Parce, J.W., Seeds, M.S., Bass, D.A. Flow cytometric quantitation of oxidative product formation by polymorphonuclear leukocytes during phagocytosis. *J Immunol* 1984;133:3303–3307.

43. Gormley, I.P., Kowolik, M.J., Cullen, R.T. The chemiluminescent response of human phagocytic cells to mineral dusts. *Br J Exp Pathol* 1985;66:409–416.

44. Royall, J.A., Ischiropoulos, H. Evaluation of 2′,7′-dichlorofluorescin and dihydrorhodamine 123 as fluorescent probes for intracellular H_2O_2 in cultured endothelial cells. *Arch Biochem Biophys* 1993;302:348–355.

45. Reynolds, I.J., Hastings, T.G. Glutamate induces the production of reactive oxygen species in cultured forebrain neurons following NMDA receptor activation. *J Neurosci* 1995;15:3318–3327.

46. Hempel, S.L., Buettner, G.R., O'Malley, Y.Q., Wessels, D.A., Flaherty, D.M. Dihydro-fluorescein diacetate is superior for detecting intracellular oxidants: comparison with 2',7'-dichlorodihydrofluorescein diacetate, 5(and 6)-carboxy-2',7'-dichlorodihydro-fluorescein diacetate, and dihydrorhodamine 123. *Free Radic Biol Med* 1989;27:146–159.

47. Ci, Y.X., Tie, J.K., Yao, J.F., Liu, Z.L., Lin, S., Zheng, W. Catalytic behavior of iron(II)-oxime complexes in the chemiluminescence reaction of luminol with hydrogen peroxide. *Anal Chim Acta* 1983;277:67–72.

48. von Ballmoos, R., Higgins, J.B. Collection of simulated x-ray powder patterns for zeolites. *Zeolites* 1990;10:313–520.

49. Babbs, C., Steiner, M.G. Detection and quantitation of hydroxyl radical using dimethyl sulfoxide as molecular probe. *Methods Enzymol* 1990;186:137–147.

50. Steiner, M.G., Babbs, C.F. Quantitation of the hydroxyl radical by reaction with dimethyl sulfoxide. *Arch Biochem Biophys* 1980;278:478–481.

51. Cross, C.E., van der Vliet, A., O'Neill, C.A., Louie, S., Halliwell, B. Oxidants, anti-oxidants, and respiratory tract lining fluids. *Environ Health Perspect* 1994;102(10): 185–191.

52. Slade, R., Crissman, K., Norwood, J., Hatch, G. Comparison of antioxidant substances on bronchoalveolar lavage cells and fluid from humans, guinea pigs, and rats. *Exp Lung Res* 1993;19:469–484.

53. Fubini, B., Arean, O. Chemical aspects of the toxicity of inhaled mineral dusts. *Chem Soc Rev* 1999;28:373–381.

54. Gutteridge, J.M., Mumby, S., Quinlan, G.J., Chung, K.F., Evans, T.W. Pro-oxidant iron is present in human pulmonary epithelial lining fluid: implications for oxidative stress in the lung. *Biochem Biophys Res Commun* 1996;220:1024–1027.

55. Evmiridis, N.P. Effect of crystal structure and percentage of ion exchange on ESR spectra of hydrated Fe(III) ion exchanged synthetic zeolites. *Inorg Chem* 1986;25:4362–4369.

56. Greenwell, L., Moreno, T., Jones, T., Richards, R. Particle-induced oxidative damage is ameliorated by pulmonary antioxidants. *Free Radic Biol Med* 2002;32(9): 898–905.

57. McIntyre, N.S., Zetaruk, D.G. X-ray photoelectron spectroscopic studies of iron oxides. *Anal Chem* 1977;49:1521–1529.

58. Addy, R.A., Gilbert, B.C. Iron–ligand interaction and the Fenton reaction. In: Berthon, G. editor. *Handbook of Metal–Ligand Interactions in Biological Fluids, vol. 2. Bioinorganic Chemistry.* New York: Marcel Dekker, Inc.; 1995. pp. 857–866.

59. Walling, C. Fenton's reagent revisited. *Acc Chem Res* 1975;8:125–131.

60. Graf, E., Mahoney, J.R., Bryant, R.G., Eaton, J.W. Iron-catalyzed hydroxyl radical formation. Stringent requirement for free iron coordination site. *J Biol Chem* 1984;259:3620–3624.

61. Singh, S., Hider, R.C. Colorimetric detection of the hydroxyl radical: comparison of the hydroxyl-radical-generating ability of various iron complexes. *Anal Biochem* 1998;171:47–54.

62. Joshi, R., Limye, S.N. Catalytic activities of transition metal complexes on zeolites for the decomposition of H_2O_2. *Oxid Commun* 1998;21:337–341.

63. Schoonen, M.A.A., Cohn, C.A., Roemer, E., Laffers, R., Simon, S.R., O'Riordan, T. Mineral-induced formation of reactive oxygen species. *Rev Mineral Geochem* 2006;64:179–221.

64. Deng, B., Stone, A.T. Surface catalyzed Cr(VI) reduction: reactivity comparisons of different organic reductants and different oxide surfaces. *Environ Sci Technol* 1996;30:2484–2494.

65. White, A.F., Peterson, M.L. Hochella M.F., Jr. Electrochemistry and dissolution kinetics of magnetite and ilmenite. *Geochim Cosmochim Acta* 1994;58:1859–1875.

66. Fubini, B., Mollo, L., Giamello, E. Free radical generation at the solid/liquid interface in iron containing minerals. *Free Radic Res* 1995;23:593–614.

67. Prandi, L., Bodoardo, S., Penazzi, N., Fubini, B. Redox state and mobility of iron at the asbestos surface: a voltammetric approach. *J Mater Chem* 2001;11:1495–1501.

68. Fubini, B., Mollo, L. Role of iron in the reactivity of mineral fibers. *Toxicol Lett* 1995; 82/83:951–960.

69. Senaratne, C., Zhang, J., Baker, M., Bessel, C.A., Rolison, D.R. Zeolite-modified electrodes: intra- versus extra-zeolite electron transfer. *J Phys Chem* 1996;100:5849–5862.

70. Guo, L., Morris, D.G., Liu, X., Vaslet, C., Hurt, R.H., Kane, A.B. Iron bioavailability and redox activity in diverse carbon nanotube samples. *Chem Mater* 2007;19(14): 3472–3478.

71. Shvedova, A.A., Castranova, V., Kisin, E.R., Schwegler-Berry, D., Murray, A.R., Gandelsman, V.Z., Maynard, A., Baron, P. Exposure to carbon nanotubes material: assessment of nanotubes cytotoxicity using human keratinocyte cells. *J Toxicol Environ Health A* 2003;66:1909–1926.

72. Kagan, V.E., Tyurina, Y.Y., Tyurin, V.A., Konduru, N.V., Potapovich, A.I., Osipov, A.N., Kisin, E.R., Schwegler-Berry, D., Mercer, R., Castranova, V., Shvedova, A.A. Direct and indirect effects of single walled carbon nanotubes on RAW 264.7 macrophages: role of iron. *Toxicol Lett* 2006;165:88–100.

73. Fiorito, S., Serafino, A., Andreola, F., Bernier, P. Effects of fullerenes and single-wall carbon nanotubes on murine and human macrophages. *Carbon* 2006;44:1100–1105.

74. Davoren, M., Herzog, E., Casey, A., Cottineau, B., Chambers, G., Byrne, J.H., Lyng, F.M. *In vitro* toxicity evaluation of single walled carbon nanotubes on human A549 lung cells. *Toxicol In Vitro* 2007;21(3): 438–448.

75. Kagan, V.E., Tyurina, Y.Y., Tyurin, V.A., Konduru, N.V., Potapovich, A.I., Osipov, A.N., Kisin, E.R., Schwegler-Berry, D., Mercer, R., Castranova, V., Shvedova, A.A. Direct and indirect effects of single walled carbon nanotubes on RAW 264.7 macrophages: role of iron. *Toxicol Lett* 2006;165(1): 88–100.

76. Magrez, A., Kasas, S., Salicio, V., Pasquier, N., Seo, J.W., Celio, M., Catsicas, S., Schwaller, B., Forro, L. Cellular toxicity of carbon-based nanomaterials. *Nano Lett* 2006;6(6): 1121–1125.

77. Dockery, D.W., Pope, C.A., III , Xu, X., Spengler, J.D., Ware, J.H., Fay, M.E., Ferris, B. G., Jr, Speizer, F.E. An association between air pollution and mortality in six U.S. cities. *N Engl J Med* 1993;329:1753–1759.

78. Koenig, J.Q., Larson, T.V., Hanley, Q.S., Rebolledo, V., Dumler, K., Checkoway, H., Wang, S.Z., Lin, D., Pierson, W.E. Pulmonary function changes in children associated with fine particulate matter. *Environ Res* 1993;63:26–38.

79. Ilabaca, M., Olaeta, I., Campos, E., Villaire, J., Tellez-Rojo, M.M., Romieu, I. Association between levels of fine particulate and emergency visits for pneumonia and other respiratory illnesses among children in Santiago. *Chile Air Waste Manag Assoc* 1999;49:154–163.

80. Pope, C.A., III, Burnett, R.T., Thun, M.J., Calle, E.E., Krewski, D., Ito, K., Thurston, G. D. Lung cancer, cardiopulmonary mortality, and long-term exposure to fine particulate air pollution. *JAMA* 2002;287:1132–1141.

81. Calderón-Garcidueñas, L., Mora-Tiscareño, A., Fordham, L.A., Chung, C.J., García, R., Osnaya, N., Hernández, J., Acuña, H., Gambliaz, T.M., Villarreal-Calderón, A., Carson, J., Koren, H.S., Devlin, R.B. Canines as sentinel species for assessing chronic exposures to air pollutants: Part 1. Respiratory pathology. *Toxicol Sci* 2001;61:342–355.

82. Finkelstein, J.N., Johnston, C., Barrett, T., Oberdörster, G. Particulate–cell interactions and pulmonary cytokine expression. *Environ Health Perspect* 1997;105:1179–1182.

83. Goldsmith, C.A., Imrich, A., Danaee, H., Ning, Y.Y., Kobzik, L. Analysis of air pollution particulate-mediated oxidant stress in alveolar macrophages. *J Toxicol Environ Health A* 1998;54:529–545.

84. Terada, N., Hamano, N., Maesako, K.I., Hiruma, K., Hohki, G., Suzuki, K., Ishikawa, K., Konno, A. Diesel exhaust particulates upregulate histamine receptor mRNA and increase histamine-induced IL-8 and GM-CSF production in nasal epithelial cells and endothelial cells. *Clin Exp Allergy* 1999;29:4–8.

85. Barrett, E.G., Johnston, C., Oberdörster, G., Finkelstein, J.N. Silica-induced chemokine expression in alveolar type II cells is mediated by TNF-alpha-induced oxidant stress. *Am J Physiol* 1999;276:L979–L988.

86. Donaldson, K., Brown, G.M., Brown, D.M., Slight, J., Robertson, M.D., Davis, J.M. Impaired chemotactic responses of bronchoalveolar leukocytes in experimental pneumoconiosis. *J Pathol* 1990;160:63–69.

87. Callis, A.H., Sohnle, P.G., Mandel, G.S., Wiessner, J., Mandel, N.S. Kinetics of inflammatory and fibrotic pulmonary changes in a murine model of silicosis. *J Lab Clin Med* 1995;105:547–553.

88. Clarke, R.W., Catalano, P.J., Koutrakis, P., Murphy, G.G., Sioutas, C., Paulauskis, J., Coull, B., Ferguson, S., Godleski, J.J. Urban air particulate inhalation alters pulmonary function and induces pulmonary inflammation in a rodent model of chronic bronchitis. *Inhal Toxicol* 1999;11:637–656.

89. Nightingale, J.A., Maggs, R., Cullinan, P., Donnelly, L.E., Rogers, D.F., Kinnersley, R., Chung, K.F., Barnes, P.J., Ashmore, M., Newman-Taylor, A. Airway inflammation after controlled exposure to diesel exhaust particulates. *Am J Respir Crit Care Med* 2000; 162:161–166.

90. Lies, K.H., Hartung, A., Postulka, A., Gring, H., Schulze, J. Composition of diesel exhaust with particular reference to particle bound organics including formation of artifacts. *Dev Toxicol Environ Sci* 1986;13:65–82.

91. Borm, P.J.A. Toxicity and occupational health hazards of coal fly ash (CFA). A review of data and comparison to coal mine dust. *Ann Occup Hyg* 1997;41:659–676.

92. Lee, S.W. Source profiles of particulate matter emissions from a pilot-scale boiler burning North American coal blends. *J Air Waste Manag Assoc* 2001;11:1568–1578.

93. Sayes, C.M., Fortner, J.D., Guo, W., Lyon, D., Boyd, A.M., Ausman, K.D., Tao, Y.J., Sitharaman, B., Wilson, L.J., Hughes, J.B., West, J.L., Colvin, B.L. The differential cytotoxicity of water-soluble fullerenes. *Nano Lett* 2004;4:1881–1887.

94. Warheit, D.B., Webb, T.R., Colvin, V.L., Reed, K.L. Sayes CM Pulmonary bioassay studies with nanoscale and fine-quartz particles in rats: toxicity is not dependent upon particle size but on surface characteristics. *Toxicol Sci* 2007;95:270–280.

95. Warheit, D.B., Webb, T.R., Reed, K.L., Frerichs, S., Sayes, C.M. Pulmonary toxicity study in rats with three forms of ultrafine-TiO$_2$ particles: Differential responses related to surface properties. *Toxicology* 2007;230:90–104.

96. Grassian, V.H., O'Shaughnessy, P.T., Adamcakova-Dodd, A., Pettibone, J.M., Thorne, P. S. Inhalation exposure study of titanium dioxide nanoparticles with a primary particle size of 2 to 5 nm. *Environ Health Perspect* 2007;115:397–402.

97. Kristovich, R., Knight, D.A., Long, J.F., Williams, M.V., Dutta, P.K., Waldman, W.J. Macrophage-mediated endothelial inflammatory responses to airborne particulates: impact of particulate physicochemical properties. *Chem Res Toxicol* 2004;17:1303–1312.

98. Waldman, W.J., Kristovich, R., Knight, D.A., Dutta, P.K. Inflammatory properties of iron-containing carbon nanoparticles. *Chem Res Toxicol* 2007;20:1149–1154.

99. Long, J.F., Waldman, W.J., Kristovich, R., Williams, M., Knight, D., Dutta, P.K. Comparison of ultrastructural cytotoxic effects of carbon and carbon/iron particulates on human monocyte-derived macrophages. *Environ Health Perspect* 2005;113:170–174.

100. Brown, D.M., Donaldson, K., Borm, P.J., Schins, R.P., Dehnhardt, M., Gilmour, P., Jimenez, L.A., Stone, V. Calcium and ROS-mediated activation of transcription factors and TNF-alpha cytokine gene expression in macrophages exposed to ultrafine particles. *Am J Physiol Lung Cell Mol Physiol* 2004;286:L344–L353.

101. Dick, C.A., Brown, D.M., Donaldson, K., Stone, V. The role of free radicals in the toxic and inflammatory effects of four different ultrafine particle types. *Inhal Toxicol* 2003;15:39–52.

102. Baldys, A., Aust, A.E. Role of iron in inactivation of epidermal growth factor receptor after asbestos treatment of human lung and pleural target cells. *Am J Respir Cell Mol Biol* 2005;32:436–442.

103. She, H., Xiong, S., Lin, M., Zandi, E., Giulivi, C., Tsukamoto, H. Iron activates NF-kappaB in Kupffer cells. *Am J Physiol Gastrointest Liver Physiol* 2002;283:G719–G726.

CHAPTER 13

Growth and Some Enzymatic Responses of *E. Coli* to Photocatalytic TiO$_2$

AYCA ERDEM, DAN CHA, and CHIN PAO HUANG

University of Delaware, Newark, DE 19716, USA

13.1 INTRODUCTION

Industrial pollution and excessive usage of water in the past 30 years have created severe stress on bodies of water throughout the world, prompting government agencies to implement water pollution management programs aimed at the development of treatment technologies including physical, chemical, and biological processes. The photocatalytic process is an advanced chemical oxidation technology that has received much attention in the field. Photocatalysis using titanium dioxide (TiO$_2$) as the photocatalyst has dominated both the scientific literature and practice. Thus, it is inevitable that photocatalytic TiO$_2$ particles will escape into the environment. The presence of photocatalytic TiO$_2$, especially that in the nanoscale, in the environment can have significant ecological and health consequences. This chapter reviews the recent work on the inactivation and survival of waterborne organisms such as bacteria, viruses, and human cells in response to photocatalytic nanoparticle TiO$_2$ particles.

Since the discovery of the photocatalytic splitting of water molecules on the TiO$_2$ electrode by Fujishima and Honda in 1972 (1) and the biocidal ability of TiO$_2$ by Matsunaga et al. (2) in 1985, TiO$_2$ has been studied extensively for its efficiency in the inactivation of microorganisms, the destruction of organic compounds in air and water, and human health-related concerns. The photocatalytic chemistry of TiO$_2$ has been studied extensively over the past 25 years for various purposes, including the removal of organic and inorganic compounds and the inactivation of harmful microorganism from contaminated water and air. Inactivation of bacteria with TiO$_2$ nanoparticles

Nanoscience and Nanotechnology Edited by Vicki H. Grassian

offers a number of advantages. First, TiO$_2$ is nonhazardous and inexpensive (3). Second, the UV content of solar light (about 5%) is sufficient to activate the TiO$_2$ photocatalyst to produce reactive oxygen species (ROS) and OH radicals, which are the primary factors in the inactivation of bacteria.

TiO$_2$ has three distinct crystalline structures, that is, anatase (tetragonal), brookite, (orthorhombic), and rutile (tetragonal). Most studies on photocatalytic reactivity have been conducted on anatase and/or rutile. The indirect band gap energies are $3.2\,eV$ ($\lambda_{bg} = 387\,nm$) and $3.0\,eV$ ($\lambda_{bg} = 413\,nm$), respectively, for anatase and rutile. Although the rutile shows lower band gap energy than anatase, anatase has been found to exhibit better photocatalytic activity and thus is more effective in inactivation of bacteria. The photocatalytic activity, photocatalytic systems, and the applications of photocatalytic chemistry to water and wastewater disinfection have been discussed (1–6).

13.2   FACTORS AFFECTING THE PHOTOCATALYTIC ACTIVITY OF TIO$_2$ WITH RESPECT TO BACTERIAL INACTIVATION

Several factors can affect the photocatalytic activity of TiO$_2$. Crystal forms, impurity, surface area or particle size, and density of surface hydroxyl groups are but some of the obvious parameters. Environmental parameters such as the intensity and wavelength of the light source and illumination time (continuous and intermittent) can also affect the inactivation of microorganisms, for example, bacteria. Photocatalysis occurs upon the irradiation of TiO$_2$ with a light source whose wavelength is smaller than its band gap energy when hydroxyl (OH$^{\bullet}$) radicals and ROS are generated.

13.2.1   Light Intensity

Both the light source and the light intensity are important factors affecting the inactivation of bacteria. Wei et al. (7) studied the effect of light intensity on the disinfection of bacteria using photocatalytic TiO$_2$ radiated with UV light below long wavelength (>380 nm). The light intensity was increased from 1800 to 16600 W/m^2 during the experiments. The researchers reported that the time to reach 90% mortality decreased from 100 to 30 min (or the half-life increased from 25 to 50 min) when the light intensity increased from 1800 to 16,000 W/m^2. They also found that the presence of strong electron acceptors such as O$_2$ or H$_2$O$_2$ enhanced the killing efficiency of the photocatalytic TiO$_2$ due to minimization of the recombination of h$_{vb}^{+}$ and e$_{cb}^{-}$. The formation of ROS and OH affects DNA replication and damages the cell membrane, which results in cell death. Bekbolet (8) studied the effect of light intensity on the inactivation of *Escherichia coli* at light intensity in the range from 24.9 to 67.9 μE (s m^2) (or 249 to 679 W/m^2) and in the presence of 1 mg/mL of TiO$_2$. They reported a linear increase in the inactivation rate with an increase in light intensity. The effect of light intensity on the die-off of *E. coli* was also investigated by Huang et al. (9). They reported that the viability of cells decreased from 45% to 3% when the light intensity was increased from 0.78 to 2.03 mW/cm^2 or (or 7.8 to 20.3 W/m^2) respectively. Chai et al. (10) also reported that the increase in UV light intensity decreased the survival ratio of the *E. coli*. The disinfection of the bacteria (e.g., 90% kills) was observed within 4 h with a 10 W/m^2 UV lamp, but when the light intensity was increased to 70 W/m^2, the disinfection was finished within 40 min. Rincon and Pulgarin (11) studied the inactivation of *E. coli* using an artificial sunlight source in the absence and the presence of a TiO$_2$ catalyst. They demonstrated that when the light intensity was increased from 400 to 1000 W/m^2, the bacterial inactivation rate was increased. In particular, the higher die-off of bacteria was observed when TiO$_2$ particles were present. It is interesting to note that when the reaction was interrupted, the bacterial self-defense mechanism kicked in and the disinfection time increased. Figure 13.1 clearly shows the exponential half-life decrease with light intensity on the disinfection of bacteria exemplified by *E. coli* (3, 7–10, 12).

13.2.2   Light Source and Wavelength

The earliest studies employed only UV light sources because the main objective was to disinfect all bacteria in the system. During the last decade, researchers have given more attention to solar light (i.e., photolysis) and light system design in the inactivation of bacteria.

Results show that visible violet and blue light have little disinfection capability, while the other components of sunlight, for example, UV-A, UV-B, and UV-C, are able to inactivate organisms. UV-C radiation (i.e., 260 nm) has the greatest potency because it corresponds to maximum absorption by DNA. Because of its germicidal ability to initiate changes in nucleic acids and other structures such as enzymes and immunogenic antigens, UV-C (i.e., 254 nm) has been a popular agent for the disinfection of drinking water and secondary wastewater effluents. However, results of recent studies have shown that near-ultraviolet (UV-A) light has the greatest effect on the inactivation of microorganisms. Wegelin et al. (13) studied the killing of bacteria and viruses under different light sources and reported that the light sources between 300 and 370 nm have a biocidal action on bacteria and viruses and that UV-C light is less effective than UV-A at inactivating micro-organisms. According to Wegelin et al. (13), it was able to achieve a 3-log inactivation of bacteria when *E. coli* were exposed to midday summer sun (solar light intensity = ~2000 kJ/m^2) for 5 h. However, much larger solar light intensity (approximately 34,300 kJ/m^2) was needed to inactivate viruses to the same level and time span. Davies and Evison (14) found that solar light with wavelength lower than 370 nm was the most effective in the inactivation of *E. coli* in water. They reported that 10 h of sunlight exposure resulted in 1-log inactivation of *E. coli* and 4 h of exposure yielded a 4-log inactivation of *Salmonella typhimurium*. Acra et al. (15) compared the germicidal effects of different wavelengths between 260 and 850 nm on the inactivation of coliform bacteria. Their results showed that the decrease in survival rate was the most significant when the bacteria were exposed to

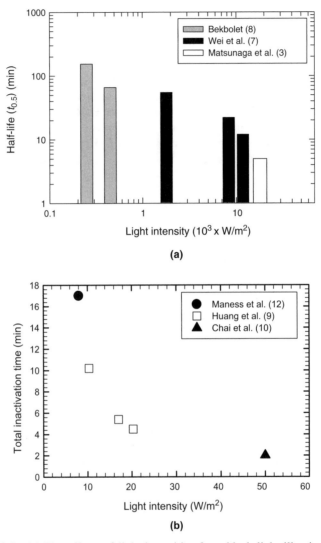

FIGURE 13.1   (a) The effects of light intensities from black light illumination on the half-life of *E. coli* (3, 7, 8). TiO$_2$ with a concentration of 1 g/L was used in each experiment. The cell concentrations were 10^3 CFU/mL in the gray bars, 10^6 CFU/mL in the black bars, and 10^2 CFU/mL in the open bar. (b) The effects of light intensities from UV light illumination on the half-lives of *E. coli* (9, 10, 12). Experimental conditions are as follows: •, 1 g/L TiO$_2$ and 2 × 10^9 CFU/mL; □, 1 g/L TiO$_2$ and 5.5 × 10^5 CFU/mL; ▲, 0.1 g/L TiO$_2$ and 1.36 h^{-1} growth rate.

the wavelengths between 260 and 350 nm. Since the UV-C component of sunlight does not reach the Earth, they concluded that 315–400 nm are the most bactericidal wavelengths. Figure 13.2 shows the effect of light source on the photocatalytic inactivation of *E. coli*. The results show that the most effective light source is UV

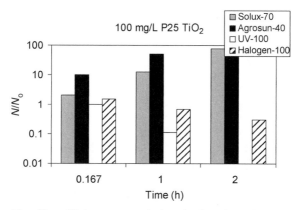

FIGURE 13.2 The effect of light sources on the inactivation of *E. coli*. Particle concentration, 100 mg/L P25 TiO$_2$.

light, which confers 99% disinfection of *E. coli* within 2 h. The halogen light with a power output of 100 W was also effective on the inactivation of *E. coli* cells when compared to simulated solar lights, for example, Agrosun (40 W) and Solux (70 W) light sources.

13.2.3 pH

The pH limits the activity of enzymes that enable an organism to synthesize new protoplasm. The optimum growth pH of *E. coli* in a culture is 6.0–7.0 at 37°C. A minimum pH of 4.4 and maximum pH of 9.0 are required for bacteria growth. Bacteria obtain their nutrients for growth and division from their environment; thus, any change in the concentration of these nutrients would cause a change in the growth rate (16).

E. coli, with a pH$_{zpc}$ of ~ 2–3, has a negative cell surface charge and the TiO$_2$ particle, with a pH$_{zpc}$ of ~ 5–6, will be neutral or negatively charged under ambient conditions. At alkaline pH, it is expected that electrostatic repulsion between *E. coli* and TiO$_2$ would be significant because both surfaces are negatively charged. Cho et al. (17) noted that in photocatalytic degradation of charged substrates, the electrostatic interaction at the TiO$_2$/water interface could play an important role, although not a sole decisive one; other factors can affect the interaction as well. However, the presence of diffusing OH radicals in the TiO$_2$ suspension could change this behavior. The role of electrostatic interaction between the charged cell and the TiO$_2$ surface could be insignificant in determining the overall photocatalytic activity. The role of surface interaction on the inactivation of microorganisms remains a subject that needs further investigation.

Cho et al. (18) investigated the effects of pH (5.6, 7.1, and 8.2) on the inactivation of *E. coli*. They observed no significant pH effect during the 120 min of microbial inactivation experiments (Fig. 13.3). They further pointed out that electrostatic interaction of bacterial cells and the TiO$_2$ particles had little effect on the photocatalytic inactivation of bacteria. Rincon and Pulgarin concluded that the charge of the TiO$_2$ and organic substances could be different at different pH values due to possible surface modification

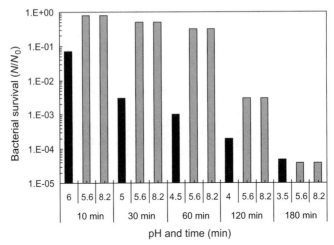

FIGURE 13.3 Effect of pH on the survival of *E. coli* (18, 19). Cell concentrations of 10^6 CFU/mL were used in each experiment. The TiO$_2$ concentrations in black bars, 1000 mg/L; in gray bars, 500 mg/L.

of both the TiO$_2$ particles and the cells. The results shown in Fig. 13.3 clearly indicate that the degree of bacterial inactivation was much pronounced under alkaline conditions, suggesting pH-dependent photocatalytic disinfection (19). Two distinct trends are observed in **Table 13.2**. According to Rincon and Pulgarin (19) (black columns), there was a linear relationship between bacterial survival and pH; there was a decrease in survival fraction with pH as the exposure time increased from 10 to 180 min. However, Cho et al. (18) demonstrated that the change in pH from 5.6 to 8.2 did not have any apparent effect on the die-off of bacteria. It must be noted that both experiments were conducted under light condition in the presence of TiO$_2$ nanoparticles.

13.2.4 Temperature

It has been reported that temperature can affect the photocatalytic inactivation of bacteria. Moreover, it is well known that as the temperature decreases, the microbial metabolism slows, in turn decreasing the inactivation of the bacteria in question (20). It has been shown that the proportions of fatty acids in the lipids of *E. coli* can be altered by temperature variations. As the temperature for growth increases, the unsaturated fatty acids composition decreases. Cells in the stationary phase tend to have larger amounts of cyclopropane fatty acids than those in the exponential phase when the growth temperature is decreased (21). Rincon and Pulgarin (11) studied the effect of temperature on the survival of mixed cultures of Gram-negative (*E. coli*) and Gram-positive (enterococci) organisms and reported that different groups of bacteria responded differently to temperature; the Gram-positive bacteria were less resistant to temperature increase than the Gram-negative bacteria mainly due to morphological differences. According to these authors, the inactivation rate of Gram-positive bacteria was faster than that of the Gram-negative ones when the

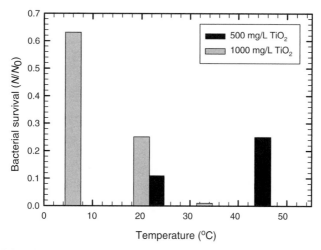

FIGURE 13.4   Effect of temperature and TiO$_2$ nanoparticles on the survival of Gram-negative bacteria (11, 18). Cell concentrations of 10^6 CFU/mL were used in each experiment.

temperature was increased from 23 to 45°C. Both groups of bacteria responded in a similar way to temperature; that is, the inactivation rate increased with temperature (Fig. 13.4) (18). Cho et al. (18) investigated the inactivation characteristics of *E. coli* at various temperatures, that is, 6, 20, and 33°C, and observed that the survival rate of *E. coli* decreased from 62% to <2% as the temperature increased from 6 to 33°C in the presence of 1000 mg/L of TiO$_2$. However, the survival fraction increased from about 10% to 26% when the temperature increased from 20 to 45°C in the presence of 500 mg/L of TiO$_2$.

13.2.5   O$_2$/N$_2$ Atmosphere

Since the ROS and OH$^\bullet$ species that formed during the application of photocatalytic TiO$_2$ nanoparticles may play an important role in bacterial inactivation, the addition of an electron donor can control the levels of these species in the system. The use of nitrogen as a purging gas limits the generation of superoxide radicals in the system, provides a less oxidizing environment for the bacteria, and ultimately leads to reduced bacterial inactivation. However, the O$_2$ sparging increases the electron scavenging capability and subsequently results in more cell death. This is because the conduction band (CB) electrons reduce the oxygen to a superoxide radical (O$_2$) during photocatalysis (22). Wei et al. (7) investigated the effect of gas composition on the photocatalytic inactivation of *E. coli*. Five different gas compositions—namely, 100% N$_2$, 25% O$_2$–75% N$_2$, 50% O$_2$–50% N$_2$, 75% O$_2$–25% N$_2$, and 100% O$_2$—were studied. The survival of the bacteria under different gas compositions changed dramatically when the gas flow was switched from 100% N$_2$ to 100% O$_2$. Dunlop et al. (23) treated *E. coli* cultures with air and oxygen-free nitrogen gas for 60 min under dark and light conditions and reported that the maximum die-off rate was when the samples were sparged with air under UV-A irradiation.

TABLE 13.1   Effect of Air Purging with Nanoparticles and Light Intensity on the Survival Rate of Bacteria

Nanophotocatalysts	Wavelength	Survival rate (%)	References
TiO$_2$, 1000 mg/L	300–400 nm	56	Wei et al. (7)
Immobilized P25 TiO$_2$	<300 nm	49	Dunlop et al. (23)
TiO$_2$, 2000 mg/L	365 nm	37	Liu et al. (22)

Table 13.1 shows an example of the bacterial survival rate when the bacteria were treated with air purging and TiO$_2$ nanoparticles under UV light. The air used in the experiments was composed of 25% O$_2$ and 75% N$_2$.

13.3   TARGET ORGANISM

13.3.1   Type of Bacteria

Bacteria are single-cell microorganisms with prokaryotic cellular components and they do not contain nucleus or membrane-bound organelles, which are the definitive characteristic of eukaryotic cells such as the cells of plants and animals. They range between 0.5 and 5 μm in size and have many shapes including spheres (cocci), rods (bacilli), or spirals (spirilla) (24). Bacteria are classified into two group, Gram-positive and Gram-negative, according to the reaction of the bacterial cell walls to Gram stain. Gram-positive bacteria have a thick cell wall (20–80 nm) containing many layers of peptidoglycan (80%), teichoic acids (10–20%), and small amounts of lipids, proteins, and lipopolysaccharides. In contrast, Gram-negative bacteria have a very thin cell wall (2–6 nm) consisting of peptidoglycan layers surrounded by an outer lipid membrane (6–18 nm thickness) that contains 50% lipopolysaccharides, 35% phospholipids, and 15% lipoproteins. The main role of the outer membrane is to protect the bacteria from substances that would damage the inner membrane or cell wall such as antibiotics, dyes, and detergents. Therefore, Gram-negative bacteria are more resistant than Gram-positive bacteria toward external stresses (24, 25). Results of a recent study have found Gram-positive bacteria to be more sensitive than Gram-negative bacteria to the antibacterial effects of TiO$_2$ (26).

As early as 1985, Matsunaga et al. (2) investigated the effect of photocatalytic inactivation of several microorganisms such as Gram-positive bacteria (*Lactobacillus acidophilus*), yeast (*Saccharomyces cerevisiae*), Gram-negative bacteria (*E. coli*), and green algae (*Chlorella vulgaris*) in water. The sterilization was done with Pt–TiO$_2$ particles under irradiation for 60–120 min. This has created new methods and approaches for sterilization and the use of photocatalytic technology for water and wastewater treatments. The technology has sparked great interest among researchers who have investigated the disinfection capability of photocatalytic TiO$_2$ using Gram-positive (2, 4, 12, 22) and Gram-negative bacteria (2–4, 6, 9, 10, 18, 23, 27–29), cancer cells (4, 30), human cells (31, 32), viruses (33, 34), and other types of cells (35, 36). Among those microorganisms, *E. coli*, which is considered an ideal indicator of fecal contamination, has been the most studied organism.

13.3.2 Growth Phase

The age or growth phase of cells is another important factor governing the photo-catalytic disinfection of bacteria using TiO_2. In most of the studies, bacteria at the stationary growth phase were employed most frequently (37, 38). With the appropriate media, the *E. coli* were cultured at 37°C for 16–20 h in a rotary shaker that rotates at a speed of 200 rpm. Some studies have employed cells from the exponential phase (3–5 h) (39–41), whereas others have defined the time of growth as "overnight" (42, 43) or the "late-exponential" phase (44) under similar conditions. Walker et al. (39) studied the responses of *E. coli* of different cell ages (i.e., mid-exponential at 3 h and stationary at 18 h) on adhesion to various organic surfaces. They reported that 16% of the 3-h cells were partitioned into dodecane, 34% of the 18-h cells were partitioned into hydrocarbon, and the outer membrane proteins of *E. coli* were mostly hydrophilic (acidic). Huisman et al. (45) and Nikaido (46) have found that when the culture is younger (i.e., mid-exponential phase), the hydrophobicity increases. Other researchers have also reported similar findings that the change in bacterial hydrophobicity was due to the presence of proteins (38, 40, 47) (Fig. 13.5).

13.3.3 Growth Media

The composition of the medium affects the sensitivity of the bacteria to the photo-treatment. Nutrient broth (11, 28, 48, 49), LB plate (50), LB broth (7, 9, 29, 38), and

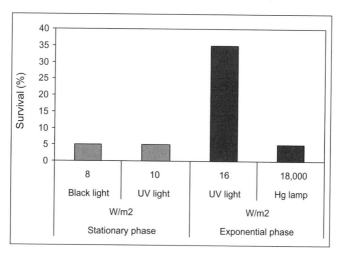

FIGURE 13.5 Effect of cell age on the photocatalytic inactivation of *E. coli* by TiO_2 (38, 40, 47).

TABLE 13.2 Components of Bacterial Growth Media

Ingredients	Nutrient Broth	LB Plate	LB Broth	Tryptic Soy Broth
Peptones (g/L)	15			20
Tryptone (g/L)		10	10	
Yeast extract (g/L)	3	5	5	
NaCl (g/L)	6	5	10	5
Dextrose/glucose (g/L)	1	15		2.5
K$_2$HPO$_4$ (g/L)				2.5

tryptic soy broth (18) are among the most commonly used culture media. Besides these media, some studies have employed wastewater from treatment plants, surface water, and even bottled water (51, 52). Table 13.2 shows examples of the ingredients of the most frequently used media.

Among all these media, LB broth is the most suitable media for growing the *E. coli* cultures. The high salt concentration enables an ionic strength of 150–200 mM; however, the aggregation of the nanoparticles in the LB broth medium occurs more rapidly because of this high salt concentration. Due to the electrostatic attraction, large aggregated nanoparticles were readily formed in the media (53).

Rincon and Pulgarin (51) studied the effect of ionic strength on the inactivation of *E. coli* and showed that the ion concentration of the media could affect the inactivation of the bacteria. They indicated that the fastest detrimental effect was when the bacteria were exposed in surface water under sunlight within 30 min in the presence of TiO$_2$. Due to the different chemical compositions of the media used in the systems, it is expected that the effect on bacterial inactivation would vary according to the specific interactions between the cells and the characteristics of the systems, including the reduction of oxidation rates by anions, adsorption and aggregation of organic substances and anions onto the nanoparticles, competition of the anions for the photogenerated holes, and competitive anion uptake by bacteria. The mode and extent of the interactions also depend on the type of medium (e.g., ionic strength), the presence or the absence of light, and the light source. However, Otto et al. (54) studied the effect of ionic strengths on the survival of *E. coli* (e.g., ionic strength of 60, 122.5, 185, 310, and 560 mM) and reported that there was no significant difference in the die-off of *E. coli* in the range of ionic strength studied in the absence of TiO$_2$ particles.

Figure 13.6 shows the effects of growth media on the photocatalytic inactivation of *E. coli*. In terms of ionic strength alone, the LB broth, M9 media, and DI water had ionic strengths of 170, 70, and 10 mM, respectively. Results indicate that the initial die-off rate constant increases in the following order: DI water (5.26×10^{-2} min^{-1}) > M9 media (3.70×10^{-2} min^{-1}) > LB broth (2.22×10^{-2} min^{-1}). The bacterial survival at 45 min was 2-log, 3-log, and 4-log for DI, M9, and LB broth, respectively. In general, as the ionic strength of the media decreases, the photocatalytic inactivation of *E. coli* increases.

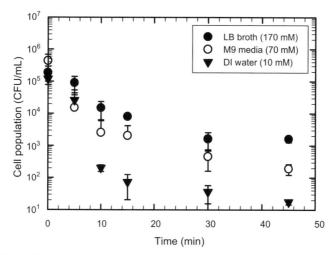

FIGURE 13.6   The effects of media ionic strength on the inactivation of *E. coli*. Solar light simulated lamps were employed to inactivate initial cell concentration of $7 \times 10^6 \, \text{CFU/mL}$ during the experiments.

13.4 TOXICOLOGICAL EFFECTS OF NANOPARTICLES

13.4.1 Ecotoxicity

The earliest toxicological studies showed that the UV light has an adverse effect on the bacteria. Bacterial inactivation can be brought about by numerous means, including damage to the cell membrane, oxidative stress, and DNA damage. The ROS and OH• produced by the illuminated TiO_2 nanoparticles promote a more destructive effect on the cell membrane and the DNA of the cell, thereby inducing photodynamic cell death in the bacteria. In 1985, when Matsunaga et al. (2) demonstrated the biocidal actions of TiO_2 on bacteria, a new area of photocatalytic toxicology was opened. Many studies have also confirmed the inactivation of bacteria by photocatalytic reactions, but the mechanism underlying the death of the bacteria remains unknown. Nakagawa et al. (55) investigated the photogenotoxicity of TiO_2 particles on rats and reviewed the toxicity studies published from 1990 to 1994. In their study, they used the Chinese hamster cell line CHLrIU and the mouse lymphoma cell line L5178Y/tk$^{+/-}$ clone 3.7.2C(L5178Y) for the single-cell gel (SCG) and mammalian cell mutation assays. Their results showed that under dark conditions TiO_2 particles exhibited no or weak genotoxicity to the cells. The genotoxic effects were observed when TiO_2 particles were irradiated with UV/vis light. They also showed that pulmonary inflammation in rats was stimulated by the inhalation of TiO_2 particles and that chronic exposure resulted in tumors in rats. However, when the rats were orally exposed to TiO_2 particles, no toxicological or carcinogenic affects were observed. Results of *in vitro* studies using chromosomal aberration assay, single-cell electrophoresis assay, microbial mutation assay, micronucleus assay, and mouse lymphoma assay showed the toxicity effects of TiO$_2$ particles; however, no negative effects were observed in the absence of UV/vis light illumination.

13.4.2 Toxicity on Human Health

Warheit et al. (56) investigated the lung toxicity of intratracheally instilled nano-TiO$_2$ particles using the pulmonary bioassay. The pulmonary toxicity of instilled rutile-type ultrafine-TiO$_2$ particles was compared with α-quartz (positive control), fine-TiO$_2$ particles (negative control), ultrafine uf-3 TiO$_2$ particle control, and vehicle controls (PBS). The results show that ultrafine-TiO$_2$ or fine-TiO$_2$ did not produce any adverse pulmonary effects in any of the end points (i.e., BAL inflammatory indicators, cell proliferation, or histopathology). At a higher dose of 5 mg/kg continued through the 3-month postexposure study period, quartz particles showed significant effects in pulmonary inflammation, cytotoxicity, and lung parenchymal cell proliferation end points. They concluded that the most hazardous particles were quartz particles and inflammation was increased due to the ultrafine particles. Table 13.3 summarizes the results of the toxicity of nanoparticles on lung cells (60, 61).

The toxicity of other nanoparticles has also been investigated. Lam et al. (58) studied the pulmonary toxicity of single-wall carbon nanotubes (SWCNTs), carbon black, and quartz particles. They used male mice (B6C3F, 2 months old) intratracheally instilled with nanoparticles at various concentrations (e.g., 0, 0.1, or 0.5 mg of particles in 50 μL of mouse serum) (58). They examined mice exposed to the nanoparticles for either 7 or 90 days; the mice were then euthanized for histopathological examination. The results showed that these three nanoparticles stimulated dose-dependent lung lesions, characterized by interstitial granulomas. They concluded that carbon black and quartz were less toxic than SWCNTs and CNTs containing Ni with SWCNTs containing Ni exhibiting the highest mortality rate.

Warheit et al. (59) investigated the toxicological effects of SWCNTs in rats by applying intratracheal instillation. Eight-week-old male rats were exposed to 1 or 5 mg/kg of unrefined SWCNTs (55–65% SWCNTs, 30–40% amorphous carbon, and 5% Ni and Co), quartz (as a positive control), carbonyl iron particles (as a negative

TABLE 13.3 Comparison of Attributes of Lung Overload in Rat and Human Species[a]

Attributes	Rat	Human
Chronic pulmonary inflammation	Yes	Not certain
Hyperplasia of macrophages and epithelial cells	Yes	Not certain
Altered pulmonary clearance	Yes	Probably not
Large pulmonary burdens of particles	Yes	Probably not
Increased interstitilization of deposited particles	Yes	Yes
Increased translocation of particles from lung to thoracic lymph nodes	Probably	Probably
Interstitial lung disease (fibrosis)	Yes	Yes, but less severe
Production of lung tumors	Yes	No

[a]Ref. 57.

control), graphite particles (5% Ni/Co as SWCNTs), and phosphate-buffered saline solution. BAL fluid markers, cell proliferation assays, and histopathological examination at 24 h, 1 week, 1 month, and 3 months after instillation were examined. When the rats were exposed to 5 mg/kg SWCNTs, 15% of the population died. They observed transient inflammation and identified a series of non-dose-dependent multifocal granulomas in rats exposed to SWCNTs. The granulomas were nonuniform in distribution, not progressive after 1 month.

Muller et al. (60) studied the effect of intratracheally instilled MWCNTs on the pulmonary function of rats. Female Sprague–Dawley rats were exposed to MWCNTs at different concentrations, that is, 0.5, 2, and 5 mg, and monitored for up to 60 days. To compare the effects of particles, asbestos and carbon black were also used. Rats were examined using BAL, inflammatory and fibrotic markers, biopersistence tests, and histopathological assay. The results showed that dose-dependent inflammation and granuloma formation were observed. Asbestos fibers were found to be the most inflammatory, whereas carbon black was the least inflammatory. The authors urged the introduction of appropriate safety measures for handling CNTs while calling for more studies to accurately establish the toxicology of CNTs.

Lin et al. (31) investigated the toxicity effect of CeO_2 nanoparticles (i.e., 20 nm) on the human bronchoalveolar carcinoma-derived cell line (A549). The results demonstrated that nanoparticles at dosage levels of 3.5–23.3 mg/L can induce significant oxidative stress, as revealed by elevated ROS, reduced GSH, increased lipid peroxidation, and increased membrane damage. There is a strong correlation between decreased cell viability and increased ROS levels after 24 h of exposure. A significant correlation was also observed between decreased cell viability and increased LDH activity, indicating that cell death was the primary cause of the cell number reduction. Together, these results indicate that the reduction of cell viability is due to increased cellular stress, which results in cell mortality, as indicated by elevated LDH activity.

Figure 13.7 summarizes the effect of nanoparticles, for example, CNTs and CeO_2, on the mortality of colon and lung cells in mammals (31, 61). Results show

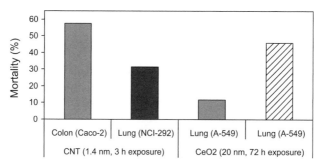

FIGURE 13.7   The lethal effects of the nanoparticles on the human cell lines (31, 61). CNTs with a concentration of 100 mg/L were used in colon (Caco-2) and lung (NCI-292) toxicity experiments. CeO_2 with concentrations of 3.5 and 23.3 mg/L were used in lung (A-549) toxicity experiments.

that in the presence of CNTs at 100 mg/L mortality was 58% and 32% for colon and lung cells, respectively, whereas the mortality of lung cells increased from 12% to 46% when the concentration of CeO$_2$ nanoparticles increased from 3.5 to 23.3 mg/L, respectively.

13.5 KILLING MECHANISMS

13.5.1 TiO$_2$/UV Process

The inactivation of bacteria present in wastewaters using irradiated TiO$_2$ has received much attention. As early as 1985, Matsunaga et al. (2) discovered that bacteria can be completely (99%) inactivated when incubated with Pt–TiO$_2$ particles for 60–120 min and irradiated with a metal halide lamp. This early finding raised the attention of many researchers. To improve the mechanism of semiconductor-catalyzed degradation of bacteria in aqueous media, the properties of the nanoparticles should be well understood. The spectral absorption characteristics of TiO$_2$ allow its excitation in the UV-A, UV-B, and UV-C regions. Irradiated TiO$_2$ in water generates electrons in the conductance band (e_{cb}^-) and positive holes in the valence band (h_{vb}^+) through irradiation with photonic energy equal to or greater than its band gap energy of 3.2 eV. These generated e^-/h^+ pairs mostly recombine within 10 ns; after 250 ns of formation, the remaining photogenerated e^-/h^+ pairs are trapped at TiO$_2$ surfaces. Studies also show that the electrons are trapped either at Ti^{+4} sites or at/near surfaces of TiO$_2$ nanoparticles (36).

As described in the equations below, the electron is excited from the valence band (VB) to the conduction band (Eq. 13.1). The CB electrons reduce O$_2$ to the superoxide ion (Eq. 13.2), and this reaction prevents the e^-/h^+ recombination. The further reduction of superoxide ions produces H$_2$O$_2$ (Eq. 13.3).

$$TiO_2 \xrightarrow{uv} h_{vb}^+ + e_{cb}^- \qquad (13.1)$$

$$O_2 + e_{cb}^- \rightarrow O_2^{\cdot -} \qquad (13.2)$$

$$O_2^{\cdot -} + e_{cb}^- + 2H^+ \rightarrow H_2O_2 \qquad (13.3)$$

It has been shown that when H$_2$O$_2$ reacts with superoxide ions, the photodegradation rate is increased and OH$^\cdot$ are formed via the Harber–Weiss reaction (62); alternatively, CB electrons reduce the H$_2$O$_2$ to generate OH radicals (Eqs. 13.4 and 13.5).

$$O_2^{\cdot -} + H_2O_2 \rightarrow OH^\cdot + OH^- + O_2 \qquad (13.4)$$

$$e_{cb}^- + H_2O_2 \rightarrow OH^\cdot + OH^- \qquad (13.5)$$

The positive hole at the VB reacts with hydroxyl ions (OH$^-$) and forms OH$^\cdot$ (Eq. 13.6). It has also been documented that the generation of OH$^\cdot$ can be attributed to the oxidative decomposition of water resulting from the VB h$^+$ (Eq. 13.7). The recombination of OH$^{\bullet}$ can lead to the production of H_2O_2 under aerobic conditions (Eq. 13.8).

$$h_{vb}^{+} + OH^{-} \rightarrow OH^{\bullet} \qquad (13.6)$$

$$h_{vb}^{+} + H_2O \rightarrow H^{+} + OH^{\bullet} \qquad (13.7)$$

$$2\,OH^{\bullet} \rightarrow H_2O_2 \qquad (13.8)$$

Among all the ROS, OH has the most bactericidal role in the inactivation processes (6, 63).

13.5.2 Cell Membrane Damage

13.5.2.1 Lipid Peroxidation: MDA Formation
The bacterial cell membrane is composed of layers of lipids (20–30%) and proteins (80–70%). The composition changes in different species of bacteria. For example, Gram-negative bacteria such as *E. coli* have an additional outer membrane that is also composed of lipids and proteins. The differences between outer and inner membranes are the enzymes that do the electron transfer and other material transports (64). When irradiated TiO_2 nanoparticles are present with bacteria, the cell membrane is the primary target of the initial oxidative attack. The polyunsaturated phospholipid component of the cell membrane is oxidized by the ROS and hydroxyl radicals generated by TiO_2 nanoparticles and forms malondialdehyde (MDA). MDA is an index measuring the degree of lipid peroxidation of the membrane. The method is based on the reaction between MDA and thiobarbituric acid (TBA) to form a pink MDA–TBA mixture. It has been very well documented in the literature that MDA increases with the extent of membrane oxidation. Figure 13.8 shows that under light conditions, TiO_2 nanoparticles have detrimental effects on the cell membrane (12, 27). Results clearly indicate that the presence of irradiated TiO_2 nanoparticles induced lipid peroxidation of membrane. It is also interested to note that the presence of TiO_2 at nanosize, that is, 25 nm, also brought about lipid peroxidation.

13.5.2.2 Respiration Activity
When the cell membrane is damaged by the lipid peroxidation, further alterations may occur in the membrane system. For example, membrane-bound proteins and electron mediators may be affected, and functional changes can occur. The ROS attacks from the TiO_2 nanoparticles can affect the semipermeability of the cell membrane and the leakage of ions such as K^{+} and Ca^{2+}. As reported by Matsunaga et al. (2, 3) and Maness et al. (12), the cell membrane is also responsible for respiration activity. When the membrane is damaged by the ROS and other radical attacks, it loses that function, leading to cell death. Cellular respiration can be monitored by determining the uptake of O_2 and/or by studying the reduction of 2,3,5-triphenyltetrazolium chloride (TTC) to its reduced product, 2,3,5-triphenyltetrazolium formazan (TTF), a red precipitate. Succinate or glucose can be used as the electron donor in the system. Apparently, the electron transport pathway from carbon source to TTC might be short-circuited because of the electron mediator disruption in the cell membrane. During respiration, the bacteria reduce the

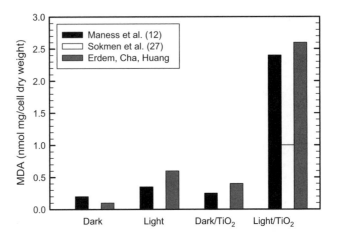

FIGURE 13.8 Effects of different conditions on the lipid peroxidation of *E. coli* cells. In each experiment, 0.1 g/L TiO$_2$ concentrations were used. Maness et al. [12] used 2.5×10^8 CFU/mL of *E. coli* under dark and UV light (8 W/m^2) for 30 min. Sokmen et al. [27] employed 10^8 CFU/mL of *E. coli* under only UV light (5.8 W/m^2) for 10 min. Erdem et al. used 10^6 CFU/mL of *E. coli* under dark and solar light for 60 min.

TTC by their cytochrome systems, and this reduced TTC has been used to maintain metabolic activities. Therefore, if bacterial cell membranes cannot reduce TTC, they cannot generate negative redox potential. This situation has been observed when bacteria are inactivated by TiO$_2$ nanoparticles under light conditions (12).

Figure 13.9 shows the effect of TiO$_2$ nanoparticle concentration on bacterial respiratory activity under dark and light conditions. The results show that smaller

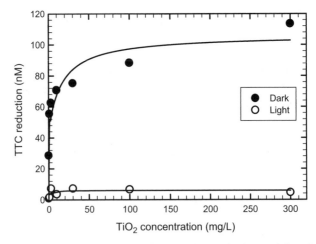

FIGURE 13.9 Effects of different conditions on the respiration activity of *E. coli* cells. Experimental conditions: 10^6 CFU/mL of *E. coli* with 0.1 g/L of 30-nm TiO$_2$ under dark and solar light for 60 min.

TiO$_2$ concentrations affect the cellular respiration more than higher concentrations under the dark condition. However, under the light condition, because of the disruption of the cell membrane by superoxides and free radicals, the reduction of TTC is much decreased.

13.5.3 Enzymatic Response: Glutathione S-Transferase Activity

Glutathione S-Transferases (GSTs) are a multigene family of isoenzymes, which are defense mechanisms developed by an organism exposed to contaminants. Three and six GSTs have been identified in humans and rats, respectively (65). These GSTs have been grouped into seven classes (alpha, mu, pi, theta, sigma, kappa, zeta, and omega) based upon sequence homology and ability to catalyze the conjugation of glutathione to a broad range of electrophilic substrates in organisms. Some of the bacteria, mostly aerobic eukaryotes and in some prokaryotes, express a beta class GST and GST family (66). Glutathione conjugation involves the addition of the tripeptide (GSH) to an electrophilic site on the substrate, catalyzed by GST. This forms the initial step in the formation of *N*-acetylcysteine (mercapturic acid). The rate-limiting step in GSH synthesis is catalyzed by glutamyl-cysteine synthetase (GCS). Oxidized glutathione can be reduced by glutathione reductase (GR). GST, GSH, GCS, and GR have all been investigated as potential enzymatic markers. GST activity often increases when the bacteria are exposed to contaminants.

Figure 13.10 shows the GST activity of the *E. coli JM105* (human GST p1-1 encoded) toward TiO$_2$ and solar light illumination for 60 min. The isopropyl β-D-1-thiogalactopyranoside (IPTG) induced *E. coli* were cultured overnight in a buffer medium; the enzymatic response toward 2,4-dinitrophenol, glutathione, and 1-chloro-2,4-dinitrobenzene (CDNB) was then measured to show the GST activity of the cells. Different concentrations of TiO$_2$ (0.1, 1, 10, and 100 mg/L) were used during the experiments. Results clearly show a significant increase in GST in the presence of the TiO$_2$ photocatalyst.

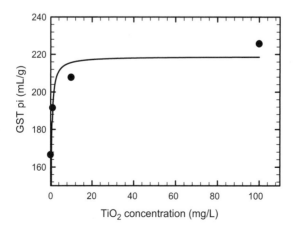

FIGURE 13.10   GST activity of JM105 *E. coli* toward 30 nm P25 under solar light.

13.5.4   Genetic Response: DNA Damage

UV exposure has cellular and molecular effects on biological systems. It is a well-known phenomenon that UV radiation results in photochemical modification of the genetic material (DNA), and alterations can remain as permanent mutations in the DNA if the amount and the type of UV radiation are too high. UV radiation directly affects the inhibition of cell division due to the DNA damage, and it also damages the essential enzymes or proteins, which are involved in membrane transport processes. However, most of the damage can be efficiently repaired by the cell itself (67). Studies have shown that UV-A, UV-B, and UV-C radiation have more disinfection capability than the visible violet and blue light. UV-A radiation, also known as long wave and black light, has spectra values from 320 to 400 nm and has been found to be the most effective wavelength for the inactivation of microorganisms. UV-B, or medium wave, ranges between 280 and 320 nm. The most germicidal wavelength of UV-C, short wave, is an effective radiation below 280 nm, which coincides with that of the maximum absorption of DNA. As a result, UV-C exhibits the greatest potency in germicidal capability that initiates changes in nucleic acids and enzymes (13). Depending upon the dose absorbed by the organism, any portion of UV or visible radiation may have either harmful or beneficial effects. For instance, UV-B radiation is often known as "harmful radiation," whereas UV-A is considered "beneficial radiation" because of its photorepair mechanisms (68).

During UV radiation, the photochemical reaction starts with the absorption of a single photon by a molecule; then, an electron is raised from the excited site to a higher energy level. This photoreaction depends mainly upon the molecular structure, the UV wavelength, and the reaction conditions (69). The initial products of the UV radiation (ROS-induced damage) are DNA lesions, including base damage, DNA single- and double-strand breaks (SSB and DSB) (70). Figure 13.11 shows

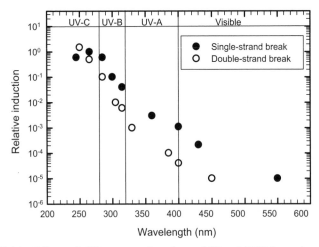

FIGURE 13.11   Effects of different wavelengths on SSB and DSB formations in cultured cells [69].

the effects of different wavelengths on SSB and DSB formations in cultured cells. These radiation damages can occur either individually or together, creating a clustered damage. This clustered damage may involve a SSB or DSB combined with base damage but can also involve more complex associations including multiple closely spaced DSB. It has been noted that as the complexity of the damage increases, it becomes more difficult to repair and more likely to lead to biological consequences (71).

One of the most frequent ways in which the biological activity of DNA is altered by UV radiation is *thymine dimerization*, a reaction where two thymine molecules are fused together to form a dimer, as shown below (72):

These thymine residues cross the chains and form an interchain dimerization. Dimerization results in disruption of hydrogen bonding between bases in the DNA molecule. UV irradiation can also cause cytosine hydration, which may also result in hydrogen bond disruption as shown in the following reaction (72):

The methods to detect DNA damage are based on the chromatographic separation of bases, enzymatic or biochemical incision of DNA by gel electrophoresis, and antibody binding. More specifically, single gel electrophoresis (SGE or comet assay), which is a very sensitive and fast assay for individual cells, has been used to detect DNA strand breaks (73). To unwind the cellular DNA, individual cells are applied in an agarose gel and then lysed under alkaline conditions. Comet assay is then carried out to distinguish the DNA fragments of the cells because smaller DNA fragments migrate to the anode faster and any strand breaks in DNA would result in a comet-like pattern of DNA fragments (comet tail). The intensity and length of the comet tail relative to the head, which is undamaged DNA, reflect the number of DNA breaks (74). It has been reported that the single-cell electrophoresis assay measures only DNA strand breaks and cannot identify or detect modifications to DNA bases (75). Other techniques, such as immunoassays using capillary electrophoresis with laser induced fluorescence (76), are particularly useful for detecting specific DNA adducts.

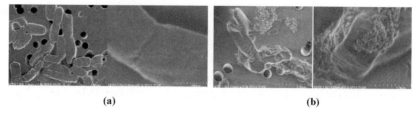

(a)   (b)

FIGURE 13.12   SEM images of cells in TiO₂ suspension in dark (a) and under irradiation (b). The right panels both upper and lower, are cells in the absence of TiO₂ particles.

13.5.5 Overall Killing Mechanisms

While attack by ROS is key to the survival of microorganisms, particle–particle interactions between bacteria and photocatalytic TiO₂ nanoparticles can also play a role in the inactivation of microorganisms. Figure 13.12 illustrates vividly that in the absence of light the attachment of nanosized TiO₂ particles onto the *E. coli* cell surface takes place. The figure compares the images of *E. coli* cells in the absence (left) and the presence (right) of TiO₂ nanoparticles. It is clear that the nanosized TiO₂ particles accumulate on the cell–water interface. Attachment of the nanosized TiO₂ particles onto the cell wall can have far-reaching significance, for example, blocking the transport of nutrients and metabolites across the cell membrane. Figure 13.12 also shows the effect of photocatalytic nano-TiO₂ particles on *E. coli* cells. In the presence of photocatalytic TiO₂ particles, the *E. coli* cells display vivid cell wall damage. A similar observation was reported by Nadtochenko et al. (77), who studied the morphological alteration of *E. coli* cells in the presence of TiO₂ particles under illumination. Figure 13.13 shows the AFM images of *E. coli* cells in suspensions of illuminated TiO₂ (P25) particles, before exposure in dark (a), after exposure to TiO₂ slurry in dark for 1 h (b), before illumination in TiO₂ suspension, (c) and 1 h after exposing to TiO₂ suspension under illumination (d). Results also show the physical interactions between the nano-TiO₂ particles and the cells in dark. Illumination of the TiO₂ suspensions brought severe damage to the cells. In another study, Sunada et al. (78) exposed *E. coli* cells to illuminated TiO₂ thin film at a light intensity of 1.0 mW/cm².

(a)   (b)   (c)   (d)

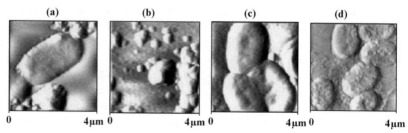

0   4 μm 0   4 μm 0   4 μm 0   4 μm

FIGURE 13.13   AFM images of *E. coli* cells in TiO₂ suspension. (a) *E. coli* from TiO₂ slurry before exposure; (b) *E. coli* cells from TiO₂ slurry after 1-h exposure in dark; (c) *E. coli* cells before illumination; (d) *E. coli* cells after illumination for 1 h in the presence of TiO₂ particles (P25) [77].

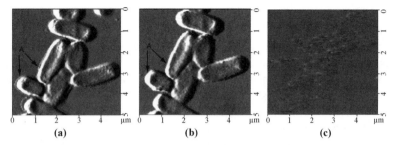

FIGURE 13.14   AFM images of *E. coli* in the presence of TiO$_2$ thin film: (a) in dark; after (b) illumination for 1 day; (c) after illumination for 6 days. Light intensity = 1 mW/cm^2 [78].

The AFM images clearly show that extended exposure to illuminated TiO$_2$ thin films brought severe damages to the *E. coli* cells (Fig. 13.14).

Sunada et al. (78) reported a two-stage killing mechanism of *E. coli* in the presence of photocatalytic TiO$_2$ nanoparticles. The first stage involves disordering of the outer membrane of *E. coli* cells on illuminated TiO$_2$. This stage is necessary for further penetration of ROS into the inner membrane. The arrival of ROS inside the cell plasma brings about disordering of the inner membrane, that is, cytoplasmic membrane. Figure 13.15 depicts the steps by which *E. coli* cells are photocatalytically killed (78). Combination of the ROS attack on the cell membrane and the coating of photocatalytic TiO$_2$ nanoparticles on the membrane might be the total killing mechanisms.

13.6 SUMMARY

Nanomaterials are receiving growing attention as the result of their increased use in engineering and biomedical industries. The excessive deployment of nanoparticles has brought about a new environmental concern, that is, nanotoxicity. Despite the huge potential and benefits of nanoparticles and nanotechnology, there is little doubt,

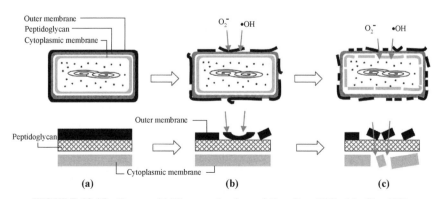

FIGURE 13.15   Proposed killing mechanism of *E. coli* on TiO$_2$ thin film [78].

despite limited studies to date, that nanoparticles can pose severe threats to environmental health.

Results of recent investigation have shown that several pertinent parameters can affect the photocatalytic inactivation of organisms, that is, bacteria. The crystal form of TiO_2, wavelength and light intensity, pH, growth media, concentration of nanoparticles, and type and physiological state of the bacteria are some of the most obvious factors. On a mass basis, nanosized particles are more potent than fine particles in creating environmental and health stress.

The combination of physical and chemical forces exerted on microbial cells is responsible for the inactivation of microorganisms. Attachment of nanosized particles on the cell surface might create a barrier for the transport of nutrients and metabolites across the cell boundary. Further attack on the outer membrane by ROS brings down the cell structure, in turn further exposing the inner cellular matters to attack by ROS. Results obtained by various researchers have shown consistently that photocatalytic TiO_2 is effective in inactivating bacteria such as *E. coli*.

ACKNOWLEDGMENTS

This work was supported by EPA STAR grant # R-83172101. Dr Nora Savage was the Project Manager. The Industrial Technology Research Institute in Hsin Chu, Taiwan, provides additional support to this study. The first author wishes to thank the Akdeniz University for the award of fellowship. We wish to thank Diane Kukich for her editorial assistance.

REFERENCES

1. Fujishima, A., Honda, K. Electrochemical photolysis of water at a semiconductor electrode. *Nature* 1972;238:37–38.

2. Matsunaga, T., Tomoda, R., Nakajima, T., Wake, H. Photoelectrochemical sterilization of microbial cells by semiconductor powders. *FEMS Microbiol Lett* 1985;29(1–2): 211–214.

3. Matsunaga, T., Tomoda, R., Nakajima, T., Wake, H., Nakamura, N., Komine, T. Continuous-sterilization system that uses photosemiconductor powders. *Appl Environ Microbiol* 1988;54 (6):1330–1333.

4. Blake, D.M., Maness, P.C., Huang, Z., Wolfrum, E.J., Huang, J., Jacoby, W.A. Application of the photocatalytic chemistry of titanium dioxide to disinfection and the killing of cancer cells. *Sep Purif Methods* 1999;28(1):1–50.

5. Fujishima, A., R.T., Tryk, D.A. Titanium dioxide photocatalysis. *J Photochem Photobiol C* 2000;1:1–21.

6. Srinivasan, C., Somasundaram, N. Bactericidal and detoxification effects of irradiated semiconductor catalyst, TiO_2. *Curr Sci* 2003;85(10):1431–1438.

7. Wei, C., Lin, W.Y., Zainal, Z., Williams, N.E., Zhu, K., Kruzic, A.P., Smith, R.L., Rajeshwar, K. Bactericidal activity of TiO_2 photocatalyst in aqueous media: toward a solar-assisted water disinfection system. *Environ Sci Technol* 1994;28(5):934–938.

8. Bekbolet, M. Photocatalytic bactericidal activity of TiO_2 in aqueous suspensions of *E. coli*. *Water Sci Technol* 1997;35(11–12):95–100.

9. Huang, N., Xiao, Z., Huang, D., Yuan, C. Photochemical disinfection of *Escherichia coli* with a TiO_2 colloid solution and a self-assembled TiO_2 thin film. *Supramol Sci* 1998;5:559–564.

10. Chai, Y.S., Lee, J.C., Kim, B.W. Photocatalytic disinfection of *E. coli* in a suspended TiO_2/UV reactor. *Korean J Chem Eng* 2000;17(6):633–637.

11. Rincon, A.G., Pulgarin, C. Photocatalytical inactivation of *E. coli*: effect of (continuous-intermittent) light intensity and of (suspended-fixed) TiO_2 concentration. *Appl Catal B* 2003;44:263–284.

12. Maness, P.C., Smolinski, S., Blake, D.M., Huang, Z., Wolfrum, E.J., Jacoby, W.A. Bactericidal activity of photocatalytic TiO_2 reaction: toward an understanding of its killing mechanism. *Appl Environ Microbiol* 1999;65(9):4094–4098.

13. Wegelin, M., Canonica, S., Mechsner, K., Fleischmann, T., Pesaro, F., Metzler, A. Solar water disinfection: scope of the process and analysis of radiation experiments. *J Water Supply Res Technol* 1994;43(3):154–169.

14. Davies, C., Evison, L. Sunlight and the survival of enteric bacteria in natural waters. *J Appl Bacteriol* 1991;70:265–274.

15. Acra, A., Raffoul, Z., Karahagopian, Y. Solar disinfection of drinking water and oral rehydration solutions: guidelines for household application in developing countries. Department of Environmental Health, Beirut, 1984.

16. Atlas, R.M. *Principles of Microbiology*. St. Louis: W.C. Brown; 1995. p. 1298.

17. Cho, M., Chung, H.M., Choi, W.Y., Yoon, J.Y. Different inactivation behaviors of MS-2 phage and *Escherichia coli* in TiO_2 photocatalytic disinfection. *Appl Environ Microbiol* 2005;71(1):270–275.

18. Cho, M., Chung, H., Choi, W., Yoon, J. Linear correlation between inactivation of *E. coli* and OH radical concentration in TiO_2 photocatalytic disinfection. *Water Res* 2004;38(4):1069–1077.

19. Rincon, A.G., Pulgarin, C. Bactericidal action of illuminated TiO_2 on pure *Escherichia coli* and natural bacterial consortia: post-irradiation events in the dark and assessment of the effective disinfection time. *Appl Catal B* 2004;49(2):99–112.

20. Kim, S., Choi, W. Kinetics and mechanisms of photocatalytic degradation of $(CH_3)_nNH_{4-}{}^{n+}$ $(0 \leq n \leq 4)$ in TiO_2 suspension: the role of OH radicals. *Environ Sci Technol* 2002;36:2019–2025.

21. Marr, A.G., Ingraham J.L. Effect of temperature on the composition of fatty acids in *Escherichia coli*. *J Bacteriol* 1962;84:1260–1267.

22. Liu, H.L., Yang, T.C.K. Photocatalytic inactivation of *Escherichia coli* and *Lactobacillus helveticus* by ZnO and TiO_2 activated ultraviolet light. *Process Biochem* 2003;39:475–481.

23. Dunlop, P.S.M., Byrne, J.A., Manga, N., Eggins, B.R. The photocatalytic removal of bacterial pollutants from drinking water. *J Photochem Photobiol A* 2002;148(1–3):355–363.

24. Cabeen, M.T., Jacobs-Wagner, C. Bacterial cell shape. *Nat Rev Microbiol* 2005;3(8):601–610.

25. Woese, C., Kandler, O., Wheelis, M. Towards a natural system of organisms: proposal for the domains Archaea, Bacteria, and Eucarya. *Proc Natl Acad Sci USA* 1990;87(12):4576–4579.

26. Fu, G., Vary, P.S., Lin, C.-T. Anatase TiO$_2$ nanocomposites for antimicrobial coatings. *J Phys Chem B* 2005;109:8889–8898.

27. Sokmen, M., Candan, F., Sumer, Z. Disinfection of *E. coli* by the Ag–TiO$_2$/UV system: lipidperoxidation. *J Photochem Photobiol A* 2001;143(2–3):241–244.

28. Sunada, K., Kikuchi, Y., Hashimoto, K., Fujishima, A. Bactericidal and detoxification effects of TiO$_2$ thin film photocatalysts. *Environ Sci Technol* 1998;32:726.

29. Yu, J.C., Tang, H.Y., Yu, J., Chan, H.C., Zhang, L., Xie, Y., Wang, H., Wong, S.P. Bactericidal and photocatalytic activities of TiO$_2$ thin films prepared by sol-gel and reverse micelle methods. *J Photochem Photobiol A* 2003;153:211–214.

30. Cai, R., Hashimoto, K., Itoh, K., Kubota, Y., Fujishima, A. Photokilling of malignant cells with ultrafine TiO$_2$ powder. *Bull Chem Soc Jpn* 1991;64:1268–1275.

31. Lin, W., Huang, Y.W., Zhou, X.D., Ma, Y. Toxicity of cerium oxide nanoparticles in human lung cancer cells. *Int J Toxicol* 2006;25:451–457.

32. Long, T.C., Saleh, N., Tilton, R.D., Lowry, G.V., Veronesi, B., Titanium dioxide (P25) produces reactive oxygen species in immortalized brain microglia (BV2): implications for nanoparticle neurotoxicity. *Environ Sci Technol* 2006;40(14):4346–4352.

33. Watts, R.J., Kong S.H., Orr, M.P., Miller, G.C., Henry, B.E., Photocatalytic inactivation of coliform bacteria and virus in secondary waste-water effluent. *Water Res* 1995;29(1): 95–100.

34. Sjogren, J.C. Sierka, R.A., Inactivation of phage MS2 by iron-aided titanium dioxide Photocatalysis. *Appl and Environ Microbiol* 1994;60(1):344–347.

35. Bernard, B.K., Osheroff, M.R., Hofmann, A. and Mennear, J.H., Toxicology and carcinogensis studies of dietary titanium dioxide-coated mica in male and female Fischer 344 rats. *J. Toxicol Environ Health* 1990;29(4):417–429.

36. Wamer, W.G., Yin, J.J., Wei, R.R. Oxidative damage to nucleic acids photosensitized by titanium dioxide. *Free Radic Biol Med* 1997;23(6):851–858.

37. Kikuchi, Y., Sunada, K., Iyoda, T., Hashimoto, K., Fujishima, A. Photocatalytic bactericidal effect of TiO$_2$ thin films: dynamic view of the active oxygen species responsible for the effect. *J Photochem Photobiol A* 1997;106(1–3):51–56.

38. Huang, Z., Maness, P.-C., Blake, D.M., Wolfrum, E.J., Smolinski, S.L., Jacoby, W.A. Bactericidal mode of titanium dioxide photocatalysis. *J Photochem Photobiol A* 2000;130:163–170.

39. Walker, S.L., Hill, J.E., Redman, J.A., Elimelech, M. Influence of growth phase on adhesion kinetics of *Escherichia coli* D21g. *Appl Environ Microbiol* 2005;71(6):3093–3099.

40. Xu, M., Ma, J., Gu, J., Lu, Z. Photocatalytic TiO$_2$ nanoparticles damage to cellular membranes and genetic supramolecules. *Supramol Sci* 1998;5:511–513.

41. Walker, S.L., Redman, J., Elimelech, M. Role of cell surface lipopolysaccharides (LPS) in *Escherichia coli* K12 adhesion and transport. *Langmuir* 2004;20:7736–7746.

42. Daughney, C.J., Fowle, D.A., Fortin, D.E. The effect of growth phase on proton and metal adsorption by *Bacillus subtilis*. *Geochim Cosmochim Acta* 2001;61:1025–1035.

43. Christensen, P.A., Curtis, T.P., Egerton, T.A., Kosa, S.A.M., Tinlin, J.R. Photoelectrocatalytic and photocatalytic disinfection of *E. coli* suspensions by titanium dioxide. *Appl Catal B* 2003;41:376–386.

44. Abu-Lail, N.I., Camesano, T.A. Role of lipopolysaccharides in the adhesion, retention, and transport of *Escherichia coli* JM109. *Environ Sci Technol* 2003;37:2173–2183.

45. Huisman, G.W., Siegele, D.A., Zambrano, M.M., Kolter, R., Morphological and physiological changes during stationary phase. In: Neidhardt, F.C., editor. *Escherichia coli and Salmonella: Cellular and Molecular Biology*, Washington, DC: ASM Press; 1996. pp. 1672–1682.

46. Nikaido, H. Molecular basis of bacterial outer membrane permeability revisited. *Microbiol Mol Biol Rev* 2003;67:593–656.

47. Jana, T.K., Srivastava, A.K., Csery, K., Arora, D.K. Influence of growth and environmental conditions on cell surface hydrophobicity of *Pseudomonas fluorescens* in non-specific adhesion. *Can J Microbiol* 2000;46:28–37.

48. Nadtochenko, V., Denisov, N., Sarkisov, O., Gumy, D., Pulgarin, C., Kiwi, J. Laser kinetic spectroscopy of the interfacial charge transfer between membrane cell walls of *E. coli* and TiO_2. *J Photochem Photobiol A* 2006;181(2–3):401–407.

49. Christensen, P.A., Curtis, T.P., Egerton, T.A., Kosa, S.A.M.,Tinlin, J.R. Photoelectrocatalytic and photocatalytic disinfection of *E. coli* suspensions by titanium dioxide. *Appl Catal B* 2003;41(4):371–386.

50. Dadjour, M.F., Ogino, C., Matsumura, S., Shimizu, N. Kinetics of disinfection of *Escherichia coli* by catalytic ultrasonic irradiation with TiO_2. *Biochem Eng J* 2005;25:243–248.

51. Rincon, A.G., Pulgarin, C. Solar photolytic and photocatalytic disinfection of water at laboratory and field scale. Effect of the chemical composition of water and study of the postirradiation events. *J Sol Energy Eng* 2007;129:100–110.

52. Meichtry, J.M., Lin, H.J., De La Fuente, L., Levy, I.K., Gautier, E.A., Blesa, M.A., Litter, M.I. Low-cost TiO_2 photocatalytic technology of water potabilization in plastic bottles for isolated regions. Photocatalyst fixation. *J Sol Energy Eng* 2007;129:119–126.

53. Hiemenz, P.C., Rajagopalan, R. *Principles of Colloid and Surface Chemistry*, New York: Marcel Dekker, Inc.; 1997. pp. 355–404.

54. Otto, K., Elwing, H., Hermansson, M. Effect of ionic strength on initial interactions of *Escherichia coli* with surfaces, studied on-line by a novel quartz crystal microbalance technique. *J Bacteriol* 1999;181(17):5210–5218.

55. Nakagawa, Y., Wakuri, S., Sakamoto, K., Tanaka, R. The photogenotoxicity of titanium dioxide particles. *Mutat Res* 1997;394:125–132.

56. Warheit, D., Webb, T.R., Reed, K.L., Frerichs, S., Sayes, C.M. Pulmonary toxicity study in rats with three forms of ultrafine-TiO_2 particles: differential responses related to surface properties. *Toxicology* 2007;230:90–104.

57. Warheit, D.B. Nanoparticles: health. *Mater Today* 2004;7(2):32–35.

58. Lam, C., James, J.T., McCluskey, R., Hunter, R. Pulmonary toxicity of single-wall carbon nanotubes in mice 7 and 90 days after intratracheal instillation. *Toxicol Sci* 2004; 77(1):126–134.

59. Warheit, D.B., Laurence, B.R., Reed, K.L., Roach, D.H., Reynolds, G.A.M., Webb, T.R. Comparative pulmonary toxicity assessment of single-wall carbon nanotubes in rats. *Toxicol Sci* 2004;77(1):117–125.

60. Muller, J., Huaux, F., Moreau, N., Misson, P., Heiler, J.F., Delos, M., Arrasa, M., Fonsecab, A., Nagyb, J.B., Lison, D. Respiratory toxicity of multi-wall carbon nanotubes. *Toxicol Appl Pharmacol* 2005;207(3):221–231.

61. Panessa-Warren, B.J., Warren, J.B., Wong, S.S., Misewich, J.A. Biological cellular response to carbon nanoparticle toxicity. *J Phys: Condens Matter* 2006;18:S2185–S2201.

62. Kehrer, J.P. The Haber–Weiss reaction and mechanism of toxicity. *Toxicology* 2000;149:43–50.

63. Legrini, O., Oliveros, E., Braun, A.M. Photochemical processes for water treatment. *Chem Rev* 1993;93(2):671–698.

64. Mindich, L. Synthesis and assembly of bacterial membranes. In: Leive, L., editor. *Bacterial Membranes and Walls*. New York: Marcel Dekker, Inc.; 1973. p. 495.

65. Habig, W.H., Jakoby, W.B. Assays for the differentiation of glutathione S-transferases. *Methods Enzymol* 1981;77:398–405.

66. Hoarau, P., Garello, G., Gnassia-Barelli, M., Romeo, M., Girard, J.P. Purification and partial characterization of seven glutathione S-transferase isoforms from the clam *Ruditapes decussates*. *Eur J Biochem* 2002;269:4359–4366.

67. Holm-Hansen, O., Lubin, D., Helbling, E.W. Ultraviolet radiation and its effects on organisms in aquatic environments. In: Young, A.R., Bjom, O., Moan, J., Nultsch, W., editors. *Environmental UV Photobiology*. New York: Plenum Press; 1993. pp. 379–425.

68. Mitchell, D.L., Kanentz, D. The induction and repair of DNA photodamage in the environment. In: Young, A.R., Bjom, O., Moan, J., Nultsch, W., editors. *Environmental UV Photobiology*. New York: Plenum Press; 1993. pp. 345–377.

69. World Health Organization (WHO). Environmental health criteria 160: ultraviolet radiation. Finland; 1994. p. 352.

70. Ward, J.F. Radiation mutagenesis: the initial DNA lesion responsible. *Radiat Res* 1995;142:362–368.

71. Nikjoo, H., O'Neill, P., Terrissol, M., Goodhead, D.T. Quantitative modelling of DNA damage using Monte Carlo track structure method. *Radiat Environ Biophys* 1999;38:31–38.

72. Yu, M.H. Environmental toxicology. *Biological and Health Effects of Pollution*. Boca Raton, FL: CRC Press; 2005. p. 339.

73. Singh, N.P. Technologies for detection of DNA damage and mutations. In: Pfeifer, G.P., editor. *Microgel Electrophoresis of DNA from Individual Cells: Principles and Methodology*. New York: Plenum Press; 1996. pp. 3–24.

74. Collins, A.R. The comet assay for DNA damage and repair. *Mol Biotechnol* 2004;26:249–261.

75. Kleparnik, K., Horky, M. (review) Detection of DNA fragmentation in a single apoptotic cardiomyocyte by electrophoresis on a microfluidic device. *Electrophoresis* 2003;24(21):3778–3783.

76. Wang, H., Lu, M., Mei, N., Lee, J., Weinfeld, M., Le, X.C. Immunoassays using capillary electrophoresis laser induced fluorescence detection for DNA adducts. *Anal Chim Acta* 2003;500(1):13–20.

77. Nadtochenko, V.A., Rincon, A.G., Stanca, S.E., Kiwi, J. Dynamics of *E. coli* membrane cell peroxidation during TiO₂ photocatalysis studied by ATR–FTIR spectroscopy and AFM microscopy. *J Photochem Photobiol A* 2005;169(2):131.

78. Sunada, K., Watanabe, T., Hashimoto, K. Studies on photokilling of bacteria on TiO₂ thin film. *J Photochem Photobiol A* 2003;156(1–3):227–233.

Bioavailability, Trophic Transfer, and Toxicity of Manufactured Metal and Metal Oxide Nanoparticles in Terrestrial Environments

JASON UNRINE, PAUL BERTSCH, and SIMONA HUNYADI

Department of Plant and Soil Sciences, University of Kentucky, Lexington, KY 40546, USA

14.1 INTRODUCTION

Rapid progress in nanotechnology will likely benefit nearly every sector of science and industry, particularly in computing and microelectronics, medicine, biological sciences, electronic sensors, environmental controls and remediation, transportation, energy production, chemical manufacturing, agriculture, and consumer products (1–6). Nanomaterials are currently produced in metric tons per year, and this quantity is expected to increase rapidly over the next decade (2, 3, 7). At present, there are nearly 600 consumer products and other items that use nanotechnology according to the Project on Emerging Nanotechnologies (http://www.nanotechproject.org). There may be parallels between the nanotechnology revolution and past technological advances. For example, immediately following World War II, rapid progress in chemical synthesis led to the development of new, more efficacious pesticides and other persistent organic chemicals (8). These new materials were released to the environment under the assumption that the quantities involved would be diluted in the environment and pose little risk to humans. However, by the 1960s the caveat to this assumption became painfully clear. The new, more persistent organic chemicals were present in high concentrations in biota due to food chain accumulation and had caused widespread adverse effects on humans and wildlife (9–12). As was true for technological advances in materials of the past, such as improvements in the synthesis of organic chemicals, the benefits of nanotechnology may have associated risks. We are compelled by the lessons that have been learned after releasing novel organic chemicals to carefully consider the potential impacts of nanotechnology prior to their large-scale release into the environment.

It is unavoidable that nanomaterials will be released into the environment, either intentionally or accidentally, and it is very likely that they will occur in terrestrial environments. Yet, despite extensive research on potential applications of nanotechnology in recent years, comparatively little has been done to evaluate their potential hazards (13). This is particularly true for the fate and transport of nanomaterials in the environment as well as their potential ecological effects and risks (7, 14–19). Very little attention has been focused thus far on the fate of nanomaterials in terrestrial environments. It is particularly important to consider food chain accumulation of nanomaterials because diet is a key contaminant exposure pathway for humans and other ecological receptor species. Some nanoparticulate materials can be absorbed gastrointestinally and penetrate cell membranes (19–22). Gastrointestinal absorption of nanoparticle-based drug delivery systems has been intensively researched in recent years and has been demonstrated as a potential route of uptake for some nanomaterials (23, 24). Since nanomaterials can be absorbed from the diet, it is natural to ask the question whether or not they can be transmitted through food webs. Several studies exist on the uptake and toxicity of nanoparticles to microbes, cell cultures, and mammals; however, few studies have investigated uptake in ecological receptor species and have not focused on the modes of uptake of the particles and their stability within tissues (7, 14–19). This chapter focuses on factors that may be important in determining bioavailability, trophic transfer, and potential toxicity of metal and metal oxide nanomaterials in terrestrial environments.

Entry of nanomaterials into terrestrial food webs and subsequent transfer to humans and other higher trophic-level ecological receptors critically depends upon their uptake by plants, microorganisms, and detritivores (Fig. 14.1) as they serve as the base of terrestrial food webs. Bioavailability is likely to be governed by interactions between factors intrinsic to the nanomaterials themselves (particle size, shape, composition, charge, etc.) and factors extrinsic to the particles, such as soil geochemistry, which may determine partitioning of particles between aqueous and solid phases in the soil. The surfaces of the particles could also be modified by biotic and abiotic ecosystem components. For example, microbial transformations of the particles or modification of particle surfaces by plant root exudates may occur. The charge of metal oxides is variable and depends upon pH, the nature and concentration of cations and anions, the presence of organic matter, and temperature. Thus, site-specific environmental conditions may be important determinants of bioavailability of nanomaterials. Further, environmental conditions may partly determine the stability of the materials in terms of oxidation or dissolution.

Limited information is available on the transport of manufactured nanoparticles through porous media (25, 26), but transport of naturally occurring colloids through soil and unconsolidated geological media has been widely researched, and many of the principles developed regarding generation and transport of colloids through porous media should be readily applicable to manufactured nanoparticles (27–33). A key question is whether or not manufactured nanomaterials, which are much more

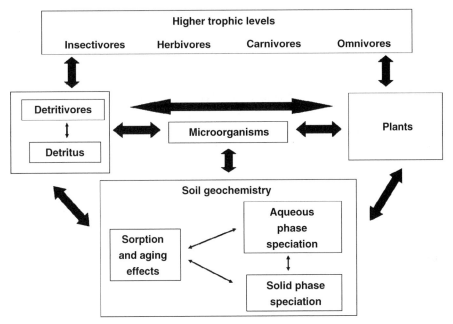

FIGURE 14.1 Schematic illustrating pathways and processes potentially involved in controlling bioavailability and trophic transfer of nanomaterials in terrestrial environments.

monodisperse and engineered (e.g., with surfactants, polymers, coatings, and other modifications) to possess specific properties, will behave in the environment similarly to naturally occurring nanoparticles.

Although our understanding of the environmental aspects of nanotechnology is extremely limited at this time, the field is rapidly expanding. Thus, our intention with this chapter is to discuss the importance of considering trophic transfer of nanomaterials and suggest possible factors that may influence their bioavailability and should be investigated. We begin by describing the properties and uses of common metal and metal oxide nanomaterials as well as a brief description of their toxicity, focusing on Au, Cu, Ag, TiO_2, and ZnO. We consider the role of chemical speciation in determining bioavailability and toxicity of metals and how this might relate to the properties of nanoscale metals and metal oxides. We discuss how the structure and properties of nanoparticles may relate to their bioavailability and how these properties may be altered in the environment. Finally, we summarize areas that need to be investigated to understand the processes governing terrestrial bioavailability.

14.2 METAL AND METAL OXIDE NANOPARTICLES, THEIR USES, AND PROPERTIES

Considerable attention has been focused on development of nanomaterials with tunable structural, catalytic, optical, and surface properties. Of the nanoscale metal and metal oxide materials, Au, Cu, Ag, ZnO, and TiO$_2$ are a few of the basic materials that are already routinely produced for which there are some, although very limited, toxicity data available. Innumerable applications for nanoscale metals and metal oxides are under development and include advanced coatings, biomedical applications such as drug delivery, imaging, disinfection, and cancer treatment, emission control in diesel engines, catalysts for fuel cells, environmental remediation, and in the chemical industry. Along with uses in optical and chemical sensing, medical imaging, and drug delivery, new uses for gold nanoparticles include self-assembling nanowires and conductive inks for printing flexible electronics (34–36). Silver nanoparticles have been used as coatings and confer antimicrobial properties to materials such as fabrics and plastics (35–38). Copper nanoparticles can be used in pesticides, microelectronics, advanced coatings, heat transfer fluids (engine coolant), sorbants in environmental remediation, fuel cells, antimicrobials, and fungicides (39–41). Nanoparticulate metals are relatively inexpensive and simple to manufacture. For example, Au nanoparticles can be easily synthesized in a wide range of sizes simply by reducing Au ions with a mild reductant such as sodium citrate. The resulting Au nanoparticles can be readily functionalized with a wide array of polymers and biomolecules to alter their surface properties and confer special functions. These basic nanomaterials will be key intermediates for more complex nanomaterials and nanocomposites as well as building blocks for nanodevices (36). Given their diversity of applications and roles as intermediates for more complex nanomaterials and nanocomposites, it is likely that nanoscale metals and metal oxides will be released into the environment in significant quantities.

Extensive evidence exists for toxicity caused by Ag nanoparticles to bacteria, and one of the primary applications of Ag nanoparticles and composites is as a germicide. Fabrics treated with Ag nanoparticles have been shown to be highly toxic to bacteria (38, 42). Xu et al. (43) reported that Ag nanoparticles up to 80 nm (geometric diameter) were transported across the outer and inner membranes of *Pseudomonas aeruginosa*, and they provided strong evidence for the involvement of the MexAB-OprM efflux pump in transmembrane transport. They also reported that concentrations of Ag nanoparticles up to 1.3 pM did not cause significant toxicity, implicitly suggesting that concentrations exceeding this level inhibited growth of *P. aeruginosa*. Toxicity has also been observed in eukaryotic cells *in vitro*. Exposure to 10–50 mg/L of either 15 or 100 nm Ag particles resulted in significant shrinkage in BRL 3A rat liver cells. Shrinkage was accompanied by decreased mitochondrial potential and function, depletion of reduced glutathione (GSH) levels, and increases in reactive oxygen species (ROS), suggesting oxidative stress as the mode of action (44). In another study, Ag nanoparticles were the most toxic of the several nanomaterials studied to mammalian germ line stem cells (45). Gold nanoparticles have been used extensively in research on medical imaging and drug delivery for cancer treatment. The Au$_{55}$ nanocluster has the ability to bind to DNA and cause cell death in a variety of human cancer cell lines. Gold nanoparticles are thought to be more toxic if they are conjugated with cationic side chains as opposed to anion side chains, perhaps due to differences in bioavalability (46). Composites containing copper nanoparticles have also been shown to have antimicrobial and antifungal properties. The mode of action may involve the oxidation of copper and subsequent release of copper ions through dissolution (40).

Intense interest in uses of nanoscale metal oxides such as ZnO and TiO_2 partially stems from the unique optical properties of these materials when particle sizes are less than 100 nm. These properties include high absorbance of ultraviolet (UV) light while remaining transparent in the visible spectrum, making them useful as UV protectants and sunscreens. They also exhibit semiconductor and photocatalytic properties causing concern over potential toxicity and phototoxicity. In addition to the use as a UV protectant, TiO_2 nanomaterials are also being explored for use as antimicrobial agents (47) and catalysts for environmental remediation (48, 49). ZnO is currently receiving much attention as a major candidate for nanoparticle applications (50–54). Manufactured nanosized ZnO structures, composites, and substrates are currently used for a broad range of applications, including pigments, rubber additives, sunscreens, personal care products, biological and chemical sensors, varistors, transducers, photoelectrodes, and catalysts, leading to ZnO being categorized as the richest family of nanostructures among all materials including carbon nanotubes, both in the variety of structures and properties (55).

Toxicity from TiO_2 has been observed in aquatic organisms such as *Daphnia magna* and in algae, and the potential for phototoxicity has been demonstrated (56, 57). Studies in human fibroblasts have suggested that crystal structure of TiO_2 is an important determinant of toxicity, with anatase-structured TiO_2 being about 100 times more toxic than rutile-structured TiO_2 (58). Toxicities of various forms of TiO_2 were well correlated with their ability to generate ROS under illumination (58), although photoactivation is not necessarily required for TiO_2 to produce ROS (59). Both TiO_2 and ZnO were shown to cause toxicity in *Escherichia coli*, where illumination was also found to be a significant factor influencing toxicity (60). In another study, 13 nm diameter. ZnO was shown to be much more toxic to microorganisms (*E. coli* and *Staphylococcus aureus*) than to human T lymphocytes, suggesting different mechanisms of toxicity in eukaryotes and prokaryotes (61). In a study of metal oxide toxicity in Neuro 2A cells, it was found that ZnO was much more toxic than TiO_2, Fe_3O_4, Al_2O_3, or CrO_3 particles of a similar size (30–45 nm). Other studies have suggested that ZnO is more toxic than TiO_2 or SiO_2 as well (60). Given the diversity of applications of metal and metal oxide nanomaterials combined with their demonstrated potential for toxicity, they are an important class of nanomaterials to consider in terms of ecological risks and effects.

14.3 CHEMICAL SPECIATION, BIOAVAILABILITY, AND TOXICITY OF METALS

Consideration of the bioavailability and toxicity of metal and metal oxide nanoparticles might best begin with what is currently known about metals in the environment. An extensive body of research has been conducted on the relationship between chemical speciation, bioavailability, and toxicity of metals, much of which is based on the premise that the free ion is the primary chemical species that elicits toxicity (i.e., the free ion activity model, FIAM) (62). The biotic ligand model (BLM) takes this idea one step further and relates speciation of metals to toxicity by taking into account competition between environmental and biotic ligands and has emerged as a dominant paradigm in metal ecotoxicology (63–65). It is likely that some metal-containing nanomaterials can elicit toxicity through the release of free ions (40, 41) consistent with the FIAM and the BLM (Fig. 14.2). While the BLM and FIAM have typically not accounted for dietary exposure (66, 67), they have been extended to earthworm–soil models (68–71). However, the assumption that toxicity is mainly related to the free ion fails to consider unique properties of other forms of metals as they relate to particle size or the possibility that they may penetrate cell membranes intact or initiate toxicity through contact with cell membranes.

Endocytosis may be an important route of entry not only in animals but also in plants (72). The endocytotic route of particle entry into cells has long been known, but this route of exposure has, in general, not been considered by ecotoxicologists (73). For example, low-density lipoprotein particles ($\sim$22 nm diameter) enter the cell through receptor-mediated endocytosis as the primary means by which cells acquire cholesterol (74). This route of entry may be particularly important for nanoparticles if they become coated with macromolecules that are recognized by receptor sites on the outer surface of the cell membrane (see Section 14.6). Other possibilities are that the particles may nonspecifically enter cells during pinocytocis or phagocytosis along

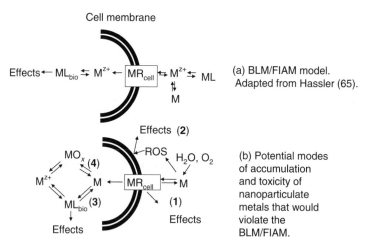

FIGURE 14.2 Schematic illustrating mechanisms governing bioavailability and toxicity of metals according to the biotic ligand model (BLM; panel (a); based on Hassler et al. (65)). (M, zerovalent metal; L, ligand; R_{cell}, cell surface receptor; ROS, reactive oxygen species; M^{z+}, metal ion; ML, metal–ligand complex; L_{bio}, biotic ligand; FIAM, free ion activity model). Panel (b) illustrates potential mechanisms of nanomaterial toxicity that would violate the BLM including (1) effects caused by direct binding of nanomaterials to cell surface receptors, (2) generation of reactive oxygen species, either intracellularly or intercellularly, (3) direct binding of nanomaterials to sensitive biotic ligands, or (4) uptake of intact metal particles into cells with subsequent intercellular oxidation and release of free metal ions.

with the targeted materials or that they can penetrate membranes when they become compromised or damaged (75).

The paradigm of metal toxicology is that metallic or uncharged species are relatively nontoxic or inert; however, as materials reach nanometer dimensions, structural defects and discontinuous crystal planes that alter the electronic structure arise and these create reactive surface sites that can, for example, generate ROS (17, 76). Surface reactive sites that generate ROS can cause damage to cell and organelle membranes or upset the redox balance of cells (17, 76). The prevalence of these changes and the resulting emergent surface properties are highly dependent on chemical composition of the material (77). Nanoparticles and biomolecules can also bind causing adverse effects in ways that are difficult to predict and deviate from the FIAM. For example, the Au_{55} cluster fits precisely into the major groove of B-DNA where it is irreversibly bound. In one study, this caused nearly 100% mortality in a variety of cancer cell lines at concentrations exceeding 0.2–2.5 μM depending on cell line. In a study in mammalian germ line stem cells, several inorganic nanoparticles were found to be toxic, while equivalent concentrations of the corresponding inorganic salts were not toxic (45). High concentrations of positively charged nanoparticles have also been shown to disrupt the blood brain barrier, altering its permeability to drugs (78). Previous studies have also demonstrated that the FIAM often fails to predict Al toxicity. The Al nanocluster, the tredecameric Al (Al_{13}) (formula $AlO_4Al_{12}(OH)_{24}$ $(H_2O)_{12}^{7+}$), is several orders of magnitude more toxic to plants than the hexaaquo Al^{3+} species, and other studies have shown similarly enhanced toxicity of the Al_{13} nanocluster to algae and fish (79–84). While the environmental relevance of Al_{13} species remains controversial, a recent report has demonstrated the importance of Al_{13}, containing phases in streams receiving acid drainage from mines (84). The Al_{13} nanocluster is also a manufactured component of a number of personal care products, including antiperspirants and precursor used in the synthesis of alumina nanoparticles (51, 84, 85).

Many organisms produce nanoparticles intracellularly; for example, the magnetic magnetite particles (Fe_3O_4) are produced by many single and multicellular organisms with crystallinity and morphological characteristics that cannot, as yet, be replicated by laboratory-manufactured pathways (33). Other nanoparticles are produced by microorganisms extracellularly, including uraninite (UO_2), pyromorphite, and sphalerite (ZnS) as well as a variety of iron oxohydroxides (33, 86, 87). The observation that some microorganisms readily biomineralize nanocrystalline phases of varying composition might argue against nanoparticles generally having a high intrinsic toxicity; however, it is unclear how nanoparticles biomineralized via controlled processes differ from manufactured nanoparticles and how these differences might ultimately influence regulatory pathways and toxicity.

Chemical composition may be important in determining the relative importance of the free ion versus the metal or metal oxide particle in terms of bioaccumulation and toxicity. For example, the group 11 elements (which all have useful properties on the nanoscale) differ in the thermodynamic favorability of oxidation. The heats of formation (ΔH_0) of metal oxides for the group 11 elements are −2160, −7740, and −43880 g cal/mol in Au, Ag, and Cu, respectively. As a result, Au nanoparticles are highly stable in aqueous dispersion, while Ag particles are much less so, and Cu particles are very susceptible to oxidation and not stable in oxygenated suspensions. On the other hand, the aqueous solubility of Ag_2O and Au_2O_3 are 0.0013 and 5.7×10^{-11}, respectively, whereas Cu_2O is soluble only under acidic conditions (88). Smaller particles with greater surface area may be more susceptible to oxidation and dissolution as oxidation in larger particles is slowed down once a protective surface layer of oxide forms, creating a barrier to the diffusion of oxygen (89). This suggests that, for example, the importance of free Au ions would be much less than that of Ag or Cu ions in terms of bioavailability and that oxidation of metals may be more important for smaller particle sizes.

It is clear that current models of metal ecotoxicity may serve as a reasonable starting point for understanding the behavior of nanomaterials, but these paradigms are likely incomplete as they apply to nanomaterials and will most likely have to be revised. Understanding the factors controlling binding of nanomaterials to the cell membrane is essential if the goal is to develop a model that relates their chemical speciation to bioavailability and, ultimately, toxicity. The BLM relates to the free ion activity of metals and competition between external and biotic ligands to their toxicity. A key parameter in this model is the partitioning of the ion between external ligands and receptors on the surface of the cell membrane (63). It is conceivable that a similar approach may be taken to develop an explanatory framework for nanomaterials based in part on their charge, although the factors influencing binding to cell membranes and other ligands in the environment may prove to be equally or more complex than those that influence ions.

14.4   FACTORS LIKELY TO INFLUENCE BIOACCUMULATION AND TROPHIC TRANSFER OF NANOPARTICLES

The bioavailability of nanoparticles may depend on a number of inherent properties of the materials. Particle size may be a primary factor influencing bioavailability (75). Therefore, bioavailability may be influenced by the tendency of nanoparticles to form aggregates in pore waters or in living cells. In animals, the available evidence suggests that only a small fraction of orally applied nanoparticle doses are translocated across the gut. Particles that are smaller than 10 nm in radius may be absorbed in the gut and retained in the body to a greater extent than larger nanoparticles that are near 100 nm in radius. The extent of particle translocation once absorbed is also likely to be partially determined by particle size. Experiments with Ir particles administered by inhalation in rats showed that, although nano-particles can be quickly translocated from pulmonary to extrapulmonary tissues, elimination and clearance of these particles is rapid preventing the particles from accumulating over time (90, 91). Other evidence suggests that while smaller nanoparticles can be redistributed to a number of organs, larger nanoparticles are transported either in the hepatic portal circulation to the liver or in the lymphatic system to the spleen where they are rapidly filtered, in analogy to first-pass metabolism of orally administered xenobiotics (90, 91). The primary filaments of the spleen are spaced approximately 100 nm apart, which would tend to trap larger nanoparticles or aggregates while letting small nanoparticles to pass (21).

Few studies have examined the influence of particle size on bioavailability of nanoparticles to soil organisms. We investigated the role of particle size in determining bioavailability of citrate-capped Au nanoparticles in earthworms. Electrophoretic mobility measurements indicate that these particles have a slight negative charge at circumneutral pH. We exposed earthworms to either 4 or 18 nm diameter. Au particles, which were characterized with respect to size by transmission electron microscopy (TEM), ultraviolet–visible spectroscopy (UV-Vis), and dynamic light scattering (DLS). The earthworms were exposed to artificial soils consisting of 70% silica sand, 10% peat moss, and 20% kaolin, which were brought to a total moisture content of 30% using suspensions of Au nanoparticles containing 10 mg/L Au or a solution containing an equivalent concentration of $HAuCl_4$ (92). After 1 week of exposure, the earthworms were sacrificed, embedded in hydrophilic melamine resin, and sectioned to approximately 200 μm. The sections were then analyzed to determine the content and distribution of Au in tissues using laser ablation-inductively coupled plasma mass spectrometry (LA-ICP-MS). The results suggested that Au uptake was dependent on particle size, with uptake of Au ions being the highest followed by 4 and 18 nm particles (see Fig. 14.3 for example data). The regions of the highest Au intensity were the gut followed by the skin. Interestingly, earthworms exposed dermally for 48 h to the same suspensions applied to filter paper showed a similar distribution of particles in the cross section of the worms, suggesting that the particles could penetrate the skin and be translocated to tissues that are remote from the portal of entry. Although LA-ICP-MS is not capable of determining the speciation of Au in the tissues, in another experiment with *Caenorhabditis elegans*, we determined that Au particles are present in tissues as metallic Au using synchrotron-based X-ray micro-fluorescence spectroscopy (μSXRF) and spatially resolved X-ray absorption near-edge spectroscopy (μXANES; Fig. 14.4). We have also observed similar patterns of particle size-dependent uptake in plants (*Nicotiana xanthi*) using μSXRF and μXANES (93).

Bioaccumulation and biomagnification are intimately related processes. The basic requirements for biomagnification are that a substance is readily taken up through ingestion and that the substance is slowly eliminated from the body causing enrichment with each trophic transfer. As outlined above, only a small fraction of nanoparticles are expected to be readily translocated across the epithelium of the gut. There is indirect evidence that metal nanoparticles may be more bioaccumulative through trophic exposure than through direct exposure. For example, surface modification of gold nanoparticles with biocompatible coatings such peptides or polyethylene glycol (PEGylation) enhances absorption by cells (20, 94). It is possible that metal nanoparticles are coated in cells by lipids, peptides, or carbohydrates and would therefore be more available for uptake after biotransformation than the primary core particle. Some nanoparticles may bind to the pore of molecular chaperones (95). Another possibility is that metal nanoparticles adsorbed to the surface of the gastrointestinal epithelium would be coated by surfactants and other substances released by digestive glands such that the nanoparticles would then become more bioavailable to consumer

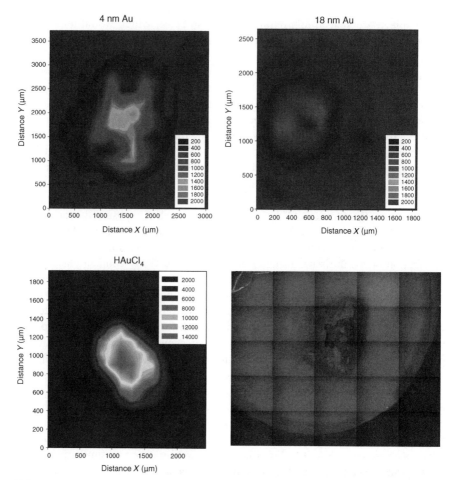

FIGURE 14.3   Distribution of Au in cross sections of earthworms (example data) exposed to artificial soils containing 3 mg/kg of either 4 nm or 18 nm diameter citrate-capped Au spheres or an equivalent concentration of $HAuCl_4$. A light microscopy image of a representative cross section is also shown (lower right). The elemental distribution plots were produced by scanning the earthworm sections using laser ablation coupled to inductively coupled plasma mass spectrometry (LA-ICP-MS). The units of the color scale are counts per second of Au^{197}. Please note the difference in scale between the $HAuCl_4$-treated individual and the 4 or 18 nm sphere-treated individuals.

organisms than the primary core particle even in the absence of intestinal absorption by the consumer organism.

Particle surface charge is another potentially important factor controlling bioavailability of metal and metal oxide nanoparticles. Both positively charged and negatively charged spherical Au nanoparticles have been shown to be taken up into human cells (96), but some studies have suggested that positively charged nanoparticles are more readily taken up into cells than negatively charged particles. This likely relates to the

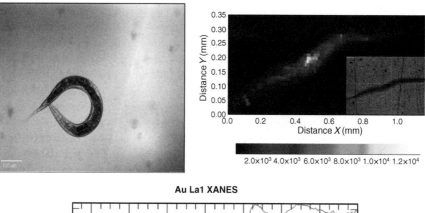

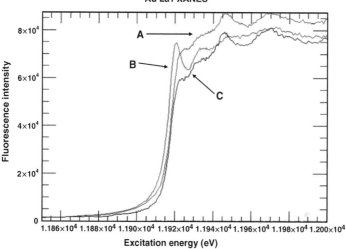

FIGURE 14.4   Light microscopy image of a nematode (*Caenorhabditis elegans*) exposed to 10 mg/L of 4 nm diameter citrate-capped Au spheres (top left). The distribution of Au in an exposed nematode as determined by scanning the Au $L_{\alpha 1}$ fluorescence using a synchrotron-based X-ray microprobe (µSXRF; top right). Areas of high Au intensity within the nematode are composed of metallic Au^0 as determined using X-ray absorption near-edge microspectroscopy (µXANES; bottom; A, Au^0 standard; B, $HAuCl_4$ standard; C, a pixel of high Au abundance within the nematode). Light microscopy image courtesy of A. Neal.

attractive or repulsive forces between the particles and the negatively charged exterior surface of the cell membrane. For example, a study in human oral carcinoma cells showed that removal of the cationic surfactant cetyl trimethylammonium bromide (CTAB) from Au nanorods and replacement with the anionic polymer polyethylene glycol (PEG) reduced cellular uptake (97). Another study demonstrated that Au spheres conjugated with cationic thiols were more toxic than those conjugated with anionic thiols in green monkey kidney cells, bovine erythrocytes, and *E. coli* (46). Evidence exists for a role of receptor-mediated endocytosis in the uptake of the particles based on recognition of extracellular proteins bound to the nanoparticle surface (77). Therefore, charge may play a critical role in both binding of proteins to the surface and binding of the nanoparticle–protein complex to the cell membrane.

Finally, as more diverse nanostructures need to be researched, such as rods, wires, cubes, triangles, stars, dog bones, and other forms, shape should be considered as a factor potentially influencing bioavailability. A few studies have investigated the relative bioavailability of Au spheres and rods of varying shape and aspect ratio (77). In general, spherical particles were taken up by cells more readily than rods with the either same length or diameter as the spheres. It is thought that the difference in angle of contact with the cell membrane might be an important factor. Spherical particles contact the cell membrane with a greater surface area due to their curvature and thus may contact a greater number of cell surface receptors involved in endocytosis.

14.5   THE SURFACE AND ENVIRONMENTAL MODIFICATIONS OF THE SURFACE

It is widely accepted that natural humic materials and metal oxohydroxides often act as armoring agents or surface coatings on fine-grained mineral surfaces altering or even controlling the surface chemistry (98). For example, it has been demonstrated that organic constituents coassociated with variable charge minerals significantly alter the point of zero net charge (pznc) by shifting it to significantly lower pH values while reducing the overall sensitivity of surface charge to pH (69, 99–102). Conversely, Fe and Al oxohydroxides coassociated with quartz, permanent charge phyllosilicate, and metal phosphate minerals have been found to shift the pznc to higher pH values while increasing the overall sensitivity of surface charge to pH (30, 98, 103–105). It has also been observed in contaminated groundwater systems that Fe-oxohydroxides associated with nanosized colloidal phases containing Pu and other transuranics influence or control surface chemical properties (106–108).

Given the high surface area and reactivity of manufactured nanoparticles, it is certain that once released into the environment they will interact with abundant organic ligands as well as metal oxohydroxide components. As described, there is ample evidence from the investigation of naturally occurring nanosized colloidal mineral phases that the surface charge properties are controlled by the natural organic ligands and metal oxohydroxides "coating" or "armoring" the nanosized colloids. The interaction of nanomaterials released into the environment with naturally occurring humic substances and Fe-oxohydroxides will likely result in the surface modification of the nanoparticles that will influence their reactivity, stability, mobility, and bioavailability to ecological receptors. We expect that metal and metal oxide nanoparticles released into the environment will rapidly react with naturally occurring soluble humic substances and naturally occurring Fe-oxohydroxides.

Our preliminary studies suggest that citrate-capped Au nanoparticles can become coated with both Fe-oxohydroxides and humic substances. We added 0.6 mM aqueous $FeCl_2$ to a suspension of the Au nanoparticles and simultaneously increased the pH from 2.6 to 5.4 (to oxidize the Fe^{2+}). We analyzed the suspension using size exclusion chromatography (SEC) coupled to ICP-MS and observed that Fe and Au coeluted with

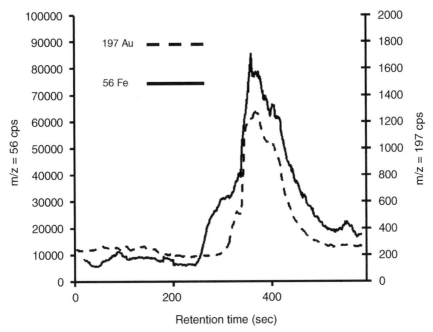

FIGURE 14.5 Size exclusion chromatogram showing the elution of a particle containing both Fe and Au. The particle was prepared by exposing 18 nm diameter citrate-capped Au spheres to a solution containing Fe^{2+} and increasing pH from 3 to 5 to affect precipitation of Fe-oxohydroxides onto the particle surface. Both Fe and Au were detected using inductively coupled plasma mass spectrometry in dynamic reaction cell mode using H_2 as the reaction gas.

similar peak shapes (Fig. 14.5). This suggests that when Fe is oxidized in the presence of Au particles, Fe-oxohydroxides can coat the nanoparticle surface. We have also successfully synthesized 20 nm Au spheres using technical-grade humic acid (HA; 0.025% mass/volume in the reaction mixture) as the sole reductant and capping agent. The resulting nanoparticulate suspensions were stable, spherical, and monodisperse as revealed by TEM, DLS, and UV–Vis (TEM; Fig. 14.6), suggesting that humic acid acts as a surfactant on the particle surface. Humic substances coating nanosized colloidal U particles in contaminated riparian sediments have been observed previously (109). Humic acid is also known to readily sorb to the surface of nanosized Fe^0 particles and significantly alter their surface chemistry (110, 111).

Given that alteration of the surface may change the charge of the particles, bioavailability of the particles may therefore be altered. For example, positively charged particles, which are attracted to the cell membrane, may be rendered less bioavailable after coating with natural organic matter (NOM), which will tend to confer a net negative charge over a wide pH range. Conversely, negatively charged particles may sorb Fe^{2+} ions under anoxic conditions, but after oxidation Fe-oxohydroxides may coat the surface. The particles could then become net positively

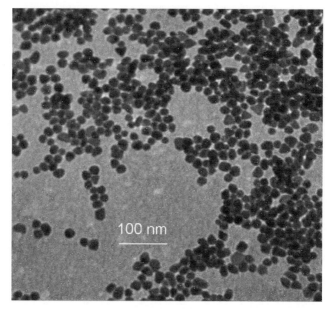

FIGURE 14.6   Transmission electron micrograph showing the size, shape, and polydispersity of 20 nm diameter humic acid-capped Au spheres. Humic acid (technical grade) was used as the sole reductant and surfactant during synthesis.

charged. This may render the particles more bioavailable. Thus, environmental modification may actually reverse the propensities for each of these materials to bioaccumulate.

Sorption of Fe-oxohydroxides and NOM may alter the aggregation/dispersion behavior of particles, which would relate to mobility and bioavailability in the environment. It is expected that the positively charged nanoparticles will be more readily coated with humic substances because of electrostatic attraction. Previous studies in other manufactured nanomaterials such as buckminsterfullerene (C_{60}) have demonstrated that humic substances tend to stabilize aqueous dispersions (112). Coating of Au particles with Fe-oxohydroxides, which have a charge that is more sensitive to pH, may decrease the stability of dispersions because the zpnc for amorphous Fe-oxohydroxides is expected to be circumneutral. Therefore, under many environmental conditions, forces from electrostatic repulsion between the particles may be weak and result in destabilization of the dispersions. In a natural system, the Au particles may be coated with both Fe-oxohydroxides and NOM. The NOM may partially buffer the pH sensitivity of the metal oxohydroxides. Sorption of HA to hematite particles results in a decrease in the zpnc and an increase in the tolerance of dispersion to a wide range of pH values and salt concentrations (113). Adsorption of NOM to C_{60} also stabilizes aqueous suspensions at high salt concentrations, suggesting steric repulsion as the stabilization mechanism (114).

14.6 SUMMARY AND RESEARCH NEEDS

Although little is known about the potential for uptake and trophic transfer of nanomaterials in food webs, some basic questions are beginning to emerge that should be the basis for future investigations. Interactions between factors that are intrinsic and extrinsic to the particles are likely to yield complex, species-specific relationships between the structure of particles and the bioavailability. The chemistry of the particle surface and in particular particle surface charge may be critically important in determining uptake and toxicity by influencing the competition between charged mineral and organic surfaces in soil environments and cell membranes, which are generally negatively charged. The aggregation and dispersion behavior of nanomaterials may also be an important determinant of uptake through influence on particle size and stability in soil pore waters. Once nanomaterials enter the environment, it is very likely that their surfaces will be modified through biotic or abiotic processes. These transformations are likely to alter the surface chemistry of the particles and change their stability under a given set of environmental conditions.

Given these complexities, there is a great need for extensive characterization of nanomaterials with respect to particle size, shape, charge, crystallinity, and sensitivity of charge to pH, cations, and anions. The materials also need to be evaluated with respect to their reactions with naturally occurring organic and mineral substances under varying environmental conditions that may lead to transformations in the environment. Studies should also focus on their interactions with proteins and other macromolecules inside or outside of living organisms and how this might affect their fate and bioavailability. To avoid confusion over mechanisms, bioaccumulation studies should focus on differentiation between metal accumulation as the intact particles and accumulation of ions released from the particles. Stability both under environmental and physiological conditions is a key issue. Rather than solely tracing the uptake of metals or metal oxides by measuring metal content, studies should employ methods that differentiate between metal ions and intact particles. Such an approach requires a variety of complementary techniques that include imaging techniques such as TEM, spectroscopic techniques such as X-ray nanospectroscopy, and separation techniques such as field-flow fractionation (FFF) coupled to ICP-MS. For metals such as Zn, which are abundant and highly regulated in cells, the difficulty in differentiating Zn-containing nanomaterials from naturally occurring Zn-binding proteins (which have sizes that are similar to nanoparticles) is particularly challenging. Likewise, toxicity studies should be referenced to the free metal ion to help uncover mechanisms of nanoparticle toxicity that are unrelated to the release of free metal ions as well as to put the observed effects into context.

Although multiple trophic levels and levels of biological organization should be studied simultaneously, there should be considerable effort focused on the entry of nanoparticles into food webs at the lowest levels. Uptake by plants, microorganisms, and detritivores could serve as a route of entry into food webs, and uptake in these organisms is probably the prerequisite for exposure in higher trophic level organisms including humans. The diet has proved to be one of the most important routes of exposure to environmental contaminants such as methylmercury, selenium, DDT, and PCBs (115). These contaminants have caused widespread adverse effects on ecosystems and human populations. To evaluate the safety of manufactured nanomaterials, it is essential to know if and how they will enter food webs, how efficiently they are transferred in food webs, what their fate is in organisms, and what the resulting adverse effects may be. Understanding the mechanisms of uptake, transfer, and toxicity in food webs is essential and may contribute toward the minimization of risks to humans and ecosystems that may be associated with nanotechnology.

ACKNOWLEDGMENTS

Portions of this work were funded by the United States Environmental Protection Agency (EPA), National Center for Environmental Research (NCER), through STAR grants R83333501 and RD832530. Portions of this work were performed at Beamline X26A, National Synchrotron Light Source (NSLS), and Brookhaven National Laboratory. X26A is supported by the United States Department of Energy (DOE)—Geosciences (DE-FG02-92ER14244 to The University of Chicago—CARS). Use of the NSLS was supported by DOE under Contract No. DE-AC02-98CH10886. The authors gratefully acknowledge A. Willis, W. Rao, A. Lanzirotti, D. Karapatakis, A. Neal, and O. Tsyusko-Unrine.

REFERENCES

1. LaVan, D., McGuire, T., Langer, R. Small-scale systems for *in vivo* drug delivery. *Nat Biotechnol* 2003;21:1184–1191.

2. Masciangioli, T., Zhang, W.X. Environmental technologies at the nanoscale. *Environ Sci Technol* 2003;37:102A–108A.

3. Mazzola, L. Commercializing nanotechnology. *Nat Biotechnol* 2003;21:1137–1143.

4. Service, R.F. Nanotoxicology: nanotechnology grows up. *Science* 2004;304:1732–1734.

5. Whitesides, G. The "right" size in nanobiotechnology. *Nat Biotechnol* 2003;21:1161–1165.

6. Nanotechnology white paper. EPA 100/B-07/001. Nanotechnology Work Group, Science Policy Council. United States Environmental Protection Agency, Washington, DC, USA; 2007.

7. Colvin, V.L. The potential environmental impact of engineered nanomaterials. *Nat Biotechnol* 2003;21:1166–1170.

8. Ecobichon, D. Toxic effects of pesticides. In: Klaassen, C.D., editor. *Casarett and Doulls Toxicology: The Basic Science of Poisons.* New York: McGraw-Hill Health Professions Division; 1995 pp. 643–690.

9. Woodwell, G.M. Toxic substances and ecological cycles. *Sci Am* 1967;216:24.

10. Woodwell, G.M., Wurster, C.F., Isaacson, P.A. DDT residues in an east coast estuary: a case of biological concentration of a persistent insecticide. *Science* 1967;156:821.

11. Carson, R. *Silent Spring* Boston: Houghton Mifflin; 1962. p. 368.

12. Colborn, T., Dumanoski, D., Meyers, J. *Our Stolen Future: Are We Threatening Our Fertility, Intelligence, and Survival?—A Scientific Detective Story.* New York: Dutton; 1996. p. 306.

13. Maynard, A. Nanotechnology: a research strategy for addressing risk. Project on Emerging Nanotechnologies, Woodrow Wilson International Center for Scholars, Washington, DC USA; 2006.

14. Dagani, R. Nanomaterials: safe or unsafe? *Chem Eng News* 2003;81:30–33.

15. Warheit, D. Nanoparticles: health impacts? *Mater Today* 2004;7:32–35.

16. Maynard, A.D. Addressing the potential environmental and human health impact of engineered nanomaterials. *Epidemiology* 2005;16:S154–S154.

17. Oberdorster, E. Manufactured nanomaterials (fullerenes, C-60) induce oxidative stress in the brain of juvenile largemouth bass. *Environ Health Perspect* 2004; 112: 1058–1062.

18. Oberdorster, E. Toxicity of nC-60 fullerenes to two aquatic species: Daphnia and largemouth bass. *Abstr Pap Am Chem Soc* 2004;227:U1233–U1233.

19. Oberdorster, G., Oberdorster, E., Oberdorster, J. Nanotoxicology: an emerging discipline evolving from studies of ultrafine particles. *Environ Health Perspect* 2005;113:823–839.

20. McNeil, S.E. Nanotechnology for the biologist. *J Leukoc Biol* 2005;78:585–594.

21. Jani, P.U., Mccarthy, D.E., Florence, A.T. Titanium-dioxide (rutile) particle uptake from the rat GI tract and translocation to systemic organs after oral-administration. *Int J Pharm* 1994;105:157–168.

22. Geiser, M., Rothen-Rutishauser, B., Kapp, N., Schurch, S., Kreyling, W., Schulz, H., Semmler, M., Hof, V.I., Heyder, J., Gehr, P. Ultrafine particles cross cellular membranes by nonphagocytic mechanisms in lungs and in cultured cells. *Environ Health Perspect* 2005;113:1555–1560.

23. Kreuter, J. Nanoparticle-based drug delivery systems. *J Control Release* 1991;16:169–176.

24. Kreuter, J. Peroral administration of nanoparticles. *Adv Drug Deliver Rev* 1991;7:71–86.

25. Lecoanet, H.F., Bottero, J.Y., Wiesner, M.R. Laboratory assessment of the mobility of nanomaterials in porous media. *Environ Sci Technol* 2004;38:5164–5169.

26. Lecoanet, H.F., Wiesner, M.R. Velocity effects on fullerene and oxide nanoparticle deposition in porous media. *Environ Sci Technol* 2004;38:4377–4382.

27. Seaman, J.C., Bertsch, P.M., Miller W.P. Chemical controls on colloid generation and transport in a sandy aquifer. *Environ Sci Technol* 1995;29:1808–1815.

28. Kretzschmar, R., Sticher, H. Transport of humic-coated iron oxide colloids in a sandy soil: influence of Ca^{2+} and trace metals. *Environ Sci Technol* 1997;31:3497–3504.

29. Ranville, J.F., Chittleborough, D.J., Shanks, F., Morrison, R.J.S., Harris, T., Doss, F., Beckett, R. Development of sedimentation field-flow fractionation-inductively coupled plasma mass-spectrometry for the characterization of environmental colloids. *Anal Chim Acta* 1999;381:315–329.

30. Seaman, J.C., Bertsch, P.M. Selective colloid mobilization through surface-charge manipulation. *Environ Sci Technol* 2000;34:3749–3755.

31. Buffle, J., Wilkinson, K.J., Stoll, S., Filella, M., Zhang, J.W. A generalized description of aquatic colloidal interactions: the three-colloidal component approach. *Environ Sci Technol* 1998;32:2887–2899.

32. Ryan, J.N., Elimelech, M. Colloid mobilization and transport in groundwater. *Colloids Surf A* 1996;107:1–56.

33. Banfield, J.F., Zhang, H.Z. Nanoparticles in the environment. In: Banfield, J.F., Navrotsky, A., editors. *Nanoparticles and the Environment*. Washington, DC: The Minerological Society of America; 2001. pp. 1–58.

34. Corti, C., Holliday, R. Commercial aspects of gold applications: from materials science to chemical science. *Gold Bull* 2004;37:20–26.

35. Nawafune, H., Akamatsu, K. Minute wiring technology using nano-sized particles. *J Jpn Inst Met* 2005;69:179–189.

36. Brust, M., Kiely, C.J. Some recent advances in nanostructure preparation from gold and silver particles: a short topical review. *Colloids Surf A* 2002;202:175–186.

37. Lee, B.I., Qi, L., Copeland, T. Nanoparticles for materials design: present & future. *J Ceram Process Res* 2005;6:31–40.

38. Lee, H.J., Yeo, S.Y., Jeong, S.H. Antibacterial effect of nanosized silver colloidal solution on textile fabrics. *J Mater Sci* 2003;38:2199–2204.

39. Eastman, J.A., Choi, S.U.S., Li, S., Yu, W., Thompson, L.J. Anomalously increased effective thermal conductivities of ethylene glycol-based nanofluids containing copper nanoparticles. *App Phys Lett* 2001;78:718–720.

40. Cioffi, N., Torsi, L., Ditaranto, N., Tantillo, G., Ghibelli, L., Sabbatini, L., Bleve-Zacheo, T., D'Alessio, M., Zambonin, P.G., Traversa, E. Copper nanoparticle/polymer composites with antifungal and bacteriostatic properties. *Chem Mater* 2005;17:5255–5262.

41. Cioffi, N., Torsi, L., Ditaranto, N., Sabbatini, L., Zambonin, P.G., Tantillo, G., Ghibelli, L., D'Alessio, M., Bleve-Zacheo, T., Traversa, E. Antifungal activity of polymer-based copper nanocomposite coatings. *App Phys Lett* 2004;85:2417–2419.

42. Lee, H.J., Jeong, S.H. Bacteriostasis and skin innoxiousness of nanosize silver colloids on textile fabrics. *Text Res J* 2005;75:551–556.

43. Xu, X.H.N., Brownlow, W.J., Kyriacou, S.V., Wan, Q., Viola, J.J. Real-time probing of membrane transport in living microbial cells using single nanoparticle optics and living cell imaging. *Biochemistry* 2004;43:10400–10413.

44. Hussain, S.M., Hess, K.L., Gearhart, J.M., Geiss, K.T., Schlager, J.J. In vitro toxicity of nanoparticles in BRL 3A rat liver cells. *Toxicol In Vitro* 2005;19:975–983.

45. Braydich-Stolle, L., Hussain, S., Schlager, J.J., Hofmann, M.C. In vitro cytotoxicity of nanoparticles in mammalian germline stem cells. *Toxicol Sci* 2005;88:412–419.

46. Goodman, C.M., McCusker, C.D., Yilmaz, T., Rotello, V.M. Toxicity of gold nanoparticles functionalized with cationic and anionic side chains. *Bioconjuz Chem* 2004;15:897–900.

47. Fu, G.F., Vary, P.S., Lin, C.T. Anatase TiO_2 nanocomposites for antimicrobial coatings. *J Phys Chem B* 2005;109:8889–8898.

48. Narr, J., Viraraghavan, T., Jin, Y.C. Applications of nanotechnology in water/wastewater treatment: a review. *Fres Environ Bull* 2007;16:320–329.

49. Gao, Z.Q., Yang, S.G., Na, T., Sun, C. Microwave assisted rapid and complete degradation of atrazine using TiO_2 nanotube photocatalyst suspensions. *J Hazard Mater* 2007;145:424–430.

50. Tian, Z., Voight, J., Liu, J., McKenzie, B., McDermott, M., Rodgrigues, M., Konishi, H., Xu, H. Complex and oriented ZnO nanostructures. *Nature* 2003;2:821–826.

51. Wang, L., Muhammed, M. Synthesis of zinc oxide nanoparticles with controlled morphology. *J Mater Chem* 1999;9:2871–2878.

52. Jing, L., Xu, Z., Sun, X., Shang, J., Cai, W. The surface properties and photocatalytic activities of ZnO ultrafine particles. *App Surf Sci* 2001;180:308–314.

53. Zheng, Z.X., Xi, Y., Huang, H., Wu, L., Lin, Z. Preparation and optical properties of nano-sized ZnO colloidal particles using NH_3 gas as a volatile catalyst. *J Chem Eng Jpn* 2001;34:15–21.

54. Yu, W., Li, X., Gao, X. Self-catalytic synthesis and photoluminesence of ZnO nanostructures on ZnO nanocrystal substrates. *Appl Phys Lett* 2004;84:2658–2660.

55. Wang, Z. Nanostructures of zinc oxide. *Mater Today* 2004;7:26–33.

56. Lovern, S.B., Klaper, R. Daphnia magna mortality when exposed to titanium dioxide and fullerene (C-60) nanoparticles. *Environ Toxicol Chem* 2006;25:1132–1137.

57. Hund-Rinke, K., Simon, M. Ecotoxic effect of photocatalytic active nanoparticles TiO_2 on algae and daphnids. *Environ Sci Pollut Res Int* 2006;13:225–232.

58. Sayes, C.M., Wahi, R., Kurian, P.A., Liu, Y.P., West, J.L., Ausman, K.D., Warheit, D.B., Colvin, V.L. Correlating nanoscale titania structure with toxicity: a cytotoxicity and inflammatory response study with human dermal fibroblasts and human lung epithelial cells. *Toxicol Sci* 2006;92:174–185.

59. Gurr, J.R., Wang, A.S.S., Chen, C.H., Jan, K.Y. Ultrafine titanium dioxide particles in the absence of photoactivation can induce oxidative damage to human bronchial epithelial cells. *Toxicology* 2005;213:66–73.

60. Adams, L.K., Lyon, D.Y., Alvarez, P.J.J. Comparative eco-toxicity of nanoscale TiO_2, SiO_2, and ZnO water suspensions. *Water Res* 2006;40:3527–3532.

61. Reddy, K.M., Feris, K., Bell, J., Wingett, D.G., Hanley, C., Punnoose, A. Selective toxicity of zinc oxide nanoparticles to prokaryotic and eukaryotic systems. *Appl Phys Lett* 2007;90:213902.

62. Morell, F.M.M. *Principles of Aquatic Chemistry.* New York: Wiley-Interscience; 1990. pp. 300–309.

63. Di Toro, D., Allen, H., Bergman, H., Meyer, J., Paquin, P., Santore, R. Biotic ligand model of the acute toxicity of metals. 1. Technical basis. *Environ Toxicol Chem* 2001;20:2283–2396.

64. Bell, R.A., Ogden, N., Kramer, J. The biotic ligand model and a cellular approach to class B metal aquatic toxicity. *Comp Biochem Physiol C* 2002;133:175–188.

65. Hassler, C.S., Slaveykova, V.I., Wilkinson, K.J. Some fundamental (and often overlooked) considerations underlying the free ion activity and biotic ligand models. *Environ Toxicol Chem* 2004;23:283–291.

66. Szebedinszky, C., McGeer, J., McDonald, D., Wood, C. Effects of chronic Cd exposure via the diet or water on internal organ-specific distribution and subsequent gill Cd uptake kinetics in juvenile rainbow trout (*Oncorhynchus mykiss*). *Environ Toxicol Chem* 2001; 20:597–607.

67. De Schamphelaere, K., C.R.J. Effects of chronic dietary copper exposure on growth and reproduction of *Daphnia magna. Environ Toxicol Chem* 2004;23:2038–2047.

68. Ritchie, J., Cresser, M., Cotter-Howells, J. Toxicological response of a bioluminescent microbial assay to Zn, Pb and Cu in an artificial soil solution: relationship with total metal concentrations and free ion activities. *Environ Pollut* 2001;114:129–136.

69. Lofts, S., Spurgeon, D.J., Svendsen, C., Tipping, E. Deriving soil critical limits for Cu, Zn, Cd and Pb: a method based on free ion concentrations. *Environ Sci Technol* 2004; 38: 3623–3631.

70. Cotter-Howells, J., Charnock, J., Winters, C., Kille, P., Fry, J., Morgan, J. Metal compartmentation and speciation in a soil sentinel: the earthworm, *Dendrodrilus rubidus*. *Environ Sci Technol* 2005;39:7731–7740.

71. Steenbergen, N., Iaccino, F., De Winkel, M., Reijnders, L., Peijnenburg, W. Development of a biotic ligand model and a regression model predicting acute copper toxicity to the earthworm *Aporrectodea caliginosa*. *Environ Sci Technol* 2005;39:5694–5702.

72. Low, P.S., Chandra, S. Endocytosis in plants. *Annu Rev Plant Physiol Plant Mol Biol* 1994;45:609–631.

73. Simkiss, K. Ecotoxicants at the cell-membrane barrier. In: Newman, M.C., Jagoe, C.H., editors. *Ecotoxicology: A Hierarchical Treatment*. Boca Raton, FL: CRC Press; 1996. pp. 59–84.

74. Alberts, B., Bray, D., Lewis, J., Raff, M., Roberts, K., Watson, J. Vesicular traffic in the secretory and endocytic pathways. Molecular Biology of the Cell. New York: Garland; 1994. pp. 600–647.

75. Hillyer, J.F., Albrecht, R.M. Gastrointestinal persorption and tissue distribution of differently sized colloidal gold nanoparticles. *J Pharm Sci* 2001;90:1927–1936.

76. Nel, A., Xia, T., Madler, L., Li, N. Toxic potential of materials at the nanolevel. *Science* 2006;311:622–627.

77. Chithrani, B.D., Ghazani, A.A., Chan, W.C.W. Determining the size and shape dependence of gold nanoparticle uptake into mammalian cells. *Nano Lett* 2006;6:662–668.

78. Lockman, P.R., Koziara, J.M., Mumper, R.J., Allen, D.D. Nanoparticle surface charges alter blood–brain barrier integrity and permeability. *J Drug Target* 2004;12:635–641.

79. Parker, D.R., Kinraide, T., Zelazny, L. On the phytotoxicity of polynuclear hydroxy-aluminum complexes. *Soil Sci Soc Am J* 1989;53:789–796.

80. Parker, D.R., Kinraide, T., Zelazny, L. Aluminum speciation and phytotoxicity in dilute hydroxy aluminum solutions. *Soil Sci Soc Am J* 1988;52:438–444.

81. Kinraide, T. Identity of the rhizotoxic aluminum species. *Plant Soil* 1991;134:167–178.

82. Shann, J., Bertsch, P.M. Differential cultivar response to polynuclear hyroxo-aluminum complexes. *Soil Sci Soc Am J* 1993;57:116–120.

83. Maison, A., Bertsch, P.M. Aluminium speciation in the presence of wheat root cell walls: a wet chemical study. *Plant Cell Environ* 1997;20:504–512.

84. Bertsch, P.M., Parker, D.R. Aqueous polynuclear aluminium species. In: Sposito, G., editor. *The Environmental Chemistry of Aluminium*. Boca Raton, FL: CRC Press; 1996. pp. 117–168.

85. Casey, W., Phillips, B., Furrer, G. Aqueous aluminium polynuclear complexes and nanoclusters: a review. In: Banfield, J.F., Navrotsky, A., editors. *Nanoparticles and the Environment*, Washington, DC: The Minerologial Society of America; 2001. pp. 167–190.

86. Templeton, A.S., Trainor, T., Traina, S., Spormann, A., Brown, G. Pb(II) distribution at biofilm-metal oxide interfaces. *Proc Natl Acad Sci USA* 2001;98:11897–11902.

87. Rancort, D. Magnetism of the earth, planetary and environmental nanomaterials. In: Banfield, J.F., Navtrotsky, A., editors. *Nanoparticles and the Environment*. Washington, DC: The Mineralogical Society of America; 2001: p. 349.

88. Lide, D., editor. *CRC Handbook of Chemistry and Physics: A Ready-reference Book of Chemical and Physical Data*, Boca Raton, FL: CRC Press; 1994.

89. Phung, X., Groza, J., Stach, E., Williams, L., Ritchey, S. Surface characterization of metal nanoparticles. *Mater Eng A* 2003; A359: 261–268.

90. Semmler, M., Seitz, J., Erbe, F., Mayer, P., Heyder, J., Oberdorster, G., Kreyling, W.G. Long-term clearance kinetics of inhaled ultrafine insoluble iridium particles from the rat lung, including transient translocation into secondary organs. *Inhal Toxicol* 2004; 16: 453–459.

91. Kreyling, W.G., Semmler, M., Erbe, F., Mayer, P., Takenaka, S., Schulz, H., Oberdorster, G., Ziesenis, A. Translocation of ultrafine insoluble iridium particles from lung epithelium to extrapulmonary organs is size dependent but very low. *J Toxicol Environ Health A* 2002;65:1513–1530.

92. Ecological effects test guidelines: OPPTS 850.6200 earthworm subchronic toxicity test. EPA 712-C-96-167. Office of Prevention, Pesticides and Toxic Substances, United States Environmental Protection Agency, Washington, DC, USA; 1996.

93. Newman, L., Unrine, J., Murphy, C.J., Sabo-Attwood, T. Uptake of gold nanoparticles in the tobacco *Nicotiana Xanthi*. (In preparation).

94. Tkachenko, A.G., Xie, H., Coleman, D., Glomm, W., Ryan, J., Anderson, M.F., Franzen, S., Feldheim, D.L. Multifunctional gold nanoparticle-peptide complexes for nuclear targeting. *J Am Chem Soc* 2003;125:4700–4701.

95. Ishll, D., Kinbara, K., Ishida, Y., Ishll, N., Okochi, M., Yohda, M., Alda, T. Chaperonin-mediated stabilization and ATP-triggered release of semiconductor nanoparticles. *Nature*, 2003;423,628–632.

96. Connor, E.E., Mwamuka, J., Gole, A., Murphy, C.J., Wyatt, M.D. Gold nanoparticles are taken up by human cells but do not cause acute cytotoxicity. *Small* 2005;1:325–327.

97. Huff, T.B., Hansen, M.N., Zhao, Y., Cheng, J.X., Wei, A. Controlling the cellular uptake of gold nanorods. *Langmuir* 2007;23:1596–1599.

98. Bertsch, P.M., Seaman, J.C. Characterization of complex mineral assemblages: implications for contaminant transport and environmental remediation. *Proc Natl Acad Sci USA* 1999;96:3350–3357.

99. Kaplan, D.I., Bertsch, P.M., Adriano, D.C. Mineralogical and physicochemical differences between mobile and nonmobile colloid phases in reconstructed pedons. *Soil Sci Soc Am J* 1997;61:641–649.

100. Kaplan, D.I., Bertsch, P.M., Adriano, D.C., Miller, W.P. Soil-borne mobile colloids as influenced by water flow and organic carbon. *Environ Sci Technol* 1993;27:1193–1200.

101. Kretzschmar, R., Robarge, W.P., Amoozegar, A. Filter efficiency of three saprolites for natural clay and iron oxide colloids. *Environ Sci Technol* 1993;28:1907–1915.

102. Kretzschmar, R., Sticher, H. Transport of humic coated iron oxide colloids in sandy soil: influence of Ca^{2+} and trace metals. *Environ Sci Technol* 1997;31:3497–3504.

103. Coston, J.A., Fuller, C., Davis, J. Pb^{2+} and Zn^{2+} adsorption by a natural aluminum- and iron-bearing surface coating on an aquifer sand. *Geochim Cosmochim Acta* 1995;59: 3535–3547.

104. Rengasamy, P., Oades, J.M. Interaction of monomeric and polymeric species of metal ions with clay surfaces: changes in surface properties of clays after addition of Fe(III). *Aus J Soil Res* 1977;15:235–242.

105. Seaman, J.C., Bertsch, P.M., Strom, R.N. Characterization of colloids mobilized from southeaster coastal plain sediments. *Environ Sci Technol* 1997;31:2782–2790.

106. Novikov, A., Kalmykov, S.N., Utsunomiya, S., Ewing, R.C., Horreard, F., Merkulov, A., Clark, S.B., Tkachev, V.V., Myasoedov, B.F. Colloid transport of plutonium in the far-field of the Mayak production association, Russia. *Science* 2006;314:638–641.

107. Kaplan, D.I., Bertsch, P.M., Adriano, D.C. Facilitated transport of contaminant metals through an acidified aquifer. *Ground Water* 1995;33:708–717.

108. Kaplan, D.I., Bertsch, P.M., Adriano, D.C., Orlandini, K.A. Actinide association with groundwater colloids in a Coastal Plain aquifer. *Radiochim Acta* 1994;66/67:101–187.

109. Jackson, B.P., Ranville, J.F., Bertsch, P.M., Sowder, A.G. Characterization of colloidal and humic-bound Ni and U in the "dissolved" fraction of contaminated sediment extracts. *Environ Sci Technol* 2005;39:2478–2485.

110. Lee, P.F., Sun, D.D., Leckie, J.O. Adsorption and photodegradation of humic acids by nano-structured TiO_2 for water treatment. *J Adv Oxid Technol* 2007;10:72–78.

111. Giasuddin, A.B.M., Kanel, S.R., Choi, H. Adsorption of humic acid onto nanoscale zerovalent iron and its effect on arsenic removal. *Environ Sci Technol* 2007;41:2022–2027.

112. Terashima, M., Nagao, S. Solubilization of (60)fullerene in water by aquatic humic substances. *Chem Lett* 2007;36:302–303.

113. Illes, E., Tombacz, E. The effect of humic acid adsorption on pH-dependent surface charging and aggregation of magnetite nanoparticles. *J Colloid Interface Sci* 2006; 295:115–123.

114. Chen, L.C., Elimelech, M. Influence of humic acid on the aggregation kinetics of fullerene (C60) nanoparticles in monovalent and divalent electrolyte solutions. *J Colloid Interface Sci* 2007;309:126–134.

115. Dourson, M., Price, P., Unrine, J. Health risks from eating contaminated fish. *Comments Toxicol* 2002;8:399–419.

CHAPTER 15

Health Effects of Inhaled Engineered Nanoscale Materials

AMY K. MADL[1,2] and KENT E. PINKERTON[1]

[1]Center for Health and the Environment, University of California, Davis, One Shields Avenue, Davis, CA 95616, USA
[2]ChemRisk, Inc., 25 Jessie Street at Ecker Square, Suite 1800, San Francisco, CA 94105, USA

15.1 INTRODUCTION

The emergence of engineered nanoscale materials provides tremendous promise of significant advancements in the fields of imaging, electronics, and therapeutics (1–7). While many nanotechnology applications are not yet on the market, there are a number of products used today that do contain nanomaterials. These products include, but are not limited to, computer chips, sporting goods, clothing, cosmetics, and dietary supplements (8). Engineered nanomaterials thought to have the greatest promise for commercial application include carbon nanotubes, nanowires, quantum dots, superparamagnetic nanoparticles, and small molecules (polymers, dendrimers, and micelles). Although there exists much enthusiasm for the potential societal benefits of engineered nanomaterials, concerns have also been raised by scientists, regulators, industry, and insurers about whether our knowledge of possible health risks of these engineered nanomaterials is keeping pace with products going to market.

15.1.1 What is a Nanoparticle?

Nanotechnology may be a newly emerging field, but the study of particles less than 100 nm in diameter (also known as ultrafine particles) has been around for decades. Nanoparticles come from many different sources; they exist naturally in the environment (forest fires, viruses, and volcanoes), are produced as byproducts of industrial or combustion processes (engines, power plants, and incinerators), and are intentionally

Nanoscience and Nanotechnology   Edited by Vicki H. Grassian
Copyright © 2008 John Wiley & Sons, Inc.

made for various industrial or consumer product applications (pigments and chemical catalysts). The emergence of nanotechnology has added a new type nanoparticle to this list.

Particles less than 100 nm in diameter, whether generated from combustion, manufacturing, or engineered processes, are considered nanoparticles; researchers have tried to find terminology to classify nanoparticles generated from these different processes. The ASTM (American Society for Testing and Materials) International, for example, in partnership with a number of organizations (i.e., the American Institute of Chemical Engineers (AIChE), American Society of Mechanical Engineers (ASME), and NSF International), has published standard terminology for nanotechnology (9). The ASTM defines nanotechnology as a field that "involves a wide range of technologies that measure, manipulate, or incorporate materials and/or features with at least one dimension between approximately 1 and 100 nanometers (nm). Such applications exploit the properties, distinct from bulk/macroscopic systems, of nanoscale components" (9). Additionally, the ASTM defines nanoparticles as a subclassification of ultrafine particles with lengths in two or three dimensions greater than 1 nm and smaller than about 100 nm (9).

15.2 CONSIDERATIONS FOR STUDYING INHALED ENGINEERED NANOMATERIALS

Although nanoscale particles (<100 nm diameter) resulting from manufacturing (i.e., ultrafine titanium dioxide, TiO_2 or carbon black, CB) or combustion processes (i.e., vehicle exhaust and air pollution) have been studied extensively, the potential biocompatibility and toxicity of engineered nanomaterials have only recently received attention from the scientific community. Many of the nanosized materials being developed today are produced in a variety of compositions (metal, elemental semi-conductor, compound semiconductor, and metal oxide), shapes (spiral, wire, belt, spring, pillar, helix, etc.), structures (core/shell and single composition), and sizes, further complicating the health implications of these materials. Research during the last few years has begun to assess which parameters of different nanoscale materials might contribute to toxicity; a tremendous amount of information still needs to be gathered to better understand the circumstances under which these materials might be compatible or toxic to biological systems.

15.2.1 Potential Inhalation Exposure of Nanoparticles

What makes understanding the human health risks associated with engineered nanomaterials particularly challenging is that their potential applications could result in a wide range of plausible chemical exposure scenarios (i.e., worker, environmental, and consumer product), resulting in very different toxicities depending on production method, chemical and physical properties, or ultimate end use. While workers, consumers, or the general public may potentially be exposed to nanoparticles through a number of pathways (e.g., dermal, ingestion, and ocular), inhalation, at least from an occupational standpoint, is likely to be one of the most significant routes of exposure, and will therefore be the focus of this chapter.

To appropriately evaluate the potential health effects of any airborne material, including nanoparticles, it is necessary to understand the (1) conditions under which the materials will become aerosolized, (2) characteristics which most influence particle uptake, distribution, and retention in the body, and (3) dose metrics that best correlate with observed health effects from particle exposure. In most cases, workers handling materials in industrial settings will have the greater potential for inhalation exposure and health risk to airborne particles compared to persons exposed to the same material in environmental or consumer product scenarios. With the development of nanoparticles for therapeutic and diagnostic applications though, as well as current and future use in thousands of consumer products, assuming the industrial worker is the highest exposed or the only population at risk may not be well founded.

The extent to which a material may become airborne, whether occurring in an industrial environment or during the use of a consumer product, depends on how the material is being handled and if it can be readily aerosolized, broken into small fragments, or reduced to powder (i.e., friable). Even if a nanomaterial does become airborne, a number of other factors should be taken into account when evaluating the extent and significance of human exposure. Particle size and air concentration, for example, are obvious factors that will influence inhalation exposure. Other factors include the composition and stability, pathway of exposure (i.e., direct versus incidental), rigor and method of manipulation, use of dust control measures (i.e., engineering controls and personal protective equipment), and the dispersion and fate of particles within a given environment.

Some of the challenges to understanding the health effects of nanoparticles, which represent an incredible range of materials of varying chemistry, size, shape, and reactivity, are that these materials may not only act very differently from the parent material, but also may have significantly different physicochemical characteristics in the manufacturing environment compared to those generated from the final product present in a household or environmental setting. An additional difficulty to evaluating the toxicology of these materials, particularly in experimental animal studies, is being able to administer the materials in a manner comparable to that which is experienced in the real world.

15.2.2 General Concepts of Pulmonary Deposition and Clearance of Nanoparticles

Although the respirability and probability of particle deposition in the respiratory tract depend on many factors (e.g., size, solubility, shape, and density), the parameter that best determines by what mechanism and to what extent airborne particles will enter and deposit in the nasopharyngeal, tracheobronchial, and alveolar regions of the lung is the particle size. Larger diameter particles (i.e., >2–3 μm diameter) act primarily by inertial mechanisms, and will preferentially deposit in the upper respiratory tract, whereas smaller particles (i.e., <100 nm diameter) act by diffusion, and will deposit

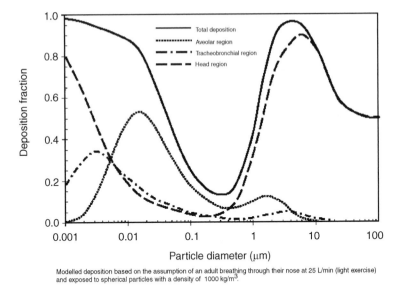

Modelled deposition based on the assumption of an adult breathing through their nose at 25 L/min (light exercise) and exposed to spherical particles with a density of 1000 kg/m³.

FIGURE 15.1   Particle deposition probability in different regions of the respiratory tract [10].

both in the nasopharyngeal and tracheobronchial regions, and, to a greater extent, in the alveolar or gas exchange region of the lungs (Fig. 15.1). Particle deposition models suggest that up to 90% or more of inhaled particles of 100 nm diameter or less will deposit in the respiratory tract with approximately 50% depositing in the alveoli (10). Agglomeration of individual particles can change the size and aerodynamic properties of inhaled nanoscale materials; therefore, while subunits of nanomaterials may be of the nanoscale, agglomeration or aggregation may cause airborne particles to act similar to particles of larger size.

Once particles are deposited within the respiratory tract, the mechanism by which they are cleared from the lungs depends not only on the site of deposition, but also on the size of the particle itself. Solid particles are cleared from the lungs through a variety of mechanisms: (1) sneezing, coughing, and removing mucus from the nasopharyngeal region, (2) direct or macrophage-mediated transport along the mucociliary escalator and subsequent elimination by the gastrointestinal tract, (3) direct or macrophage-mediated transport across the bronchiolar or alveolar epithelium and subsequent clearance by the systemic circulation or interstitial lymphatics, and (4) physicochemical processes, including dissolution, leaching, and physical breakdown of particles.

Comparisons of transport patterns of different sized particles suggest that nanosized particles are retained in the lungs to a greater extent than their fine-sized counterparts (up to 2.5µm). It has been shown, for example, that 80% of fine-sized particles were retrieved with macrophages in bronchoalveolar lavage 24 h following intratracheal instillation, compared to 20% of nanosized particles (11, 12). In addition, it has been reported that nanosized particles lack the rapid phase clearance typically observed in the first 24 h of exposure, and corresponding to tracheobronchial deposition of larger sized particles (13–15). Research suggests that nanosized particles clear slower than larger particles because they are more readily taken up by epithelial cells and, therefore, less likely to be phagocytized by macrophages (16, 17). The fact that nanosized particles have a potentially high efficiency for deposition, target both the upper and lower regions of the respiratory tract, are retained in the lungs for a long period of time, induce more oxidative stress, and cause greater inflammatory effects than their fine-sized equivalents all suggest a need to better understand the true impact of these particles on the body.

15.2.3   Instillation Versus Inhalation Studies

Because there are so many parameters that may ultimately influence the toxicity or biocompatibility of inhaled nanoparticles, researchers have relied on a combination of particle delivery techniques (intratracheal instillation/aspiration/inhalation or nose-only/whole body inhalation) as a means to study the pulmonary and systemic effects of nanoparticles. Intratracheal instillation involves injection of particles suspended in saline through a catheter inserted in the trachea of the animal, whereas intratracheal aspiration entails administration of a suspension as droplets in a puff of air, and intratracheal inhalation involves cannulating the trachea, attaching the open end of the cannula to a port of an aerosolization system, and ventilating the animal at a known rate and pressure (18–20). For these intratracheal techniques, animals are anesthetized during particle exposures; however, with nose-only or whole-body inhalation exposures, animals are not sedated, and are either constrained in a tube or allowed to roam freely within the cage during the exposure.

While these methods of particle administration to the respiratory tract have their limitations, they do have unique advantages (20–22) (Table 15.1). The greatest benefit that intratracheal instillation provides, for example, is a method of delivery of a known amount of material so that the effects of different doses and formulations (e.g., particle size, shape, and chemistry) can be compared easily and within a reasonable time frame. This method also circumvents the need for specialized equipment and expertise that are usually required for inhalation studies. When deciding on a method of particle delivery to the lungs, it is important to understand that patterns of particle deposition, translocation, and retention following intratracheal instillation may not accurately reflect the physiologic patterns observed with particle inhalation. Scientists have attempted to address this issue and also to design a system that would allow the delivery of a known amount of test material to the lungs by using intratracheal inhalation. As with intratracheal instillation, intratracheal inhalation bypasses the normal scrubbing mechanisms of the nasal turbinates and, therefore, particles administered in this manner are delivered to a region of the lungs that would not otherwise be accessible. Despite this shortcoming, intratracheal or nose-only inhalation studies offer the advantage over whole-body inhalation studies because particles are delivered by only one route. Recent work evaluating the extrapulmonary transport of ultrafine carbon particles delivered by whole-body inhalation demonstrates that particle accumulation in the liver could not be explained solely by particles deposited in the lungs and that particle ingestion through external contamination of animal pelt likely contributed to the liver dose (23).

TABLE 15.1   Potential Advantages and Limitations of Different Methods of Particles Administration to the Respiratory Tract

	Intratracheal instillation	Inhalation
Advantages	Inexpensive	Provides a natural way for delivery of toxicants
	Actual delivered dose is known	Deposition and clearance patterns comparable to that in a real-world setting
	Minimal risks to workers administering the material	Evaluate effects at all levels of the respiratory tract
	Administration of multiple doses within a short period of time	Results in even distribution of delivered toxicant
	Comparison of responses to different toxicant formulations (size, shape, chemistry)	
	Avoids exposure to animal skin or pelt	
	Localize exposure to specific lung lobes in larger animals	
	Useful as a screening tool for dosing and toxicity ranking	
Limitations	Administration of material in a nonphysiologic invasive manner	Expensive
	Dose rate is greater than by inhalation	Requires specialized expertise and equipment for system development and design
	Distribution in the lungs will differ compared to inhalation exposures	Sufficient test material required for duration of testing
	Avoids natural scrubbing mechanisms of nasal passages	Dermal/fur contamination with whole-body exposures
	Bypasses upper respiratory tract that could normally be a target	Proper handling of hazardous aerosols
	Suspension may not represent nature of the material in a real-word setting	Delivered dose can be estimated or measured through sophisticated labeling
	Unnatural pattern of deposition may translate to unnatural clearance, retention, and response patterns	Variability of particle burden can be great
	Reproducibility of material delivery dependent on technician experience	
	Clearance mechanisms and kinetics differ may not be comparable to inhalation	
	Particle clumping, local inflammation and irregular particle retention may be unique to method of administration	

Adapted from Refs. (20–22).

Despite the differences in particle fate and transport observed between intratracheal instillation and inhalation studies, general patterns of toxicity, such as markers of inflammation (e.g., immune cell profile, interleukin-1, tumor necrosis factor-α, and macrophage inflammatory protein-2) and lung injury (e.g., lactate dehydrogenase and total protein) show similar trends that vary in timing and severity depending on the administration method (18, 19, 22, 24, 25). For this reason, intratracheal instillation studies are generally viewed as a useful way to screen the potential health effects of different materials. Intratracheal instillation also provides an efficient and cost effective way to compare the relative toxicity of different materials over a range of doses, as well as to evaluate the potential mechanisms by which different materials, including nanoparticles, elicit different biological responses in the lungs and extrapulmonary organs. Instillation methods, though, cannot be used to evaluate particle deposition patterns. Caution should also be taken when interpreting patterns of particle translocation and retention and pulmonary histopathology because particle clumping, local inflammation, and irregular particle retention may be a reflection of the administration method, and not the inherent nature of the administered material.

15.2.4 Aerosol Characterization for Inhalation Studies

Additional factors for evaluating the effects of nanoparticles include the appropriate characterization of exposure/dose metrics and understanding differences between exposures in the workplace (or another human exposure setting) and those delivered in experimental animal studies. While it is known that some materials or processes generate nanosized particles (e.g., polytetrafluoroethylene and welding fumes) in an industrial setting, airborne particles are generally controlled for and regulated on a total or respirable (i.e., $<10 \mu m$ diameter) mass basis and, as a result, methods for characterizing airborne particles are not unique for ultrafine or nanosized particles. In fact, for many of the approaches currently being used to measure airborne particles (e. g., personal respirable dust samplers or impactors), no distinction can be made between particles in the fine or ultrafine size range. However, such a distinction may be important because recent studies suggest that particle mass may not be the best exposure metric to correlate with biological responses, such as pulmonary inflammation.

The term nanoparticle is one that is very broad, and only describes the nature of the particle in terms of size. Nanoparticles represent an incredible range of materials with respect to chemical and molecular form (e.g., metal alloy versus oxide, carbon particle versus nanotube), surface area (e.g., single versus agglomerate), solubility (e.g., manganese oxide versus soluble salts), charge, and biopersistence. Many of these properties appear to have some influence on biological effects. In addition, for a given particle type, particle number and surface area will increase as the particle diameter decreases for a given mass. This effect has been illustrated by Finlayson-Pitts (26) in size distribution studies of urban ambient particulates (26–28) (Fig. 15.2). Because pulmonary responses to nanosized particles are not dictated by particle mass alone, researchers have tried to better characterize the physicochemical characteristics of particles being delivered in experimental studies and to determine how these parameters impact the biological effects observed *in vivo*.

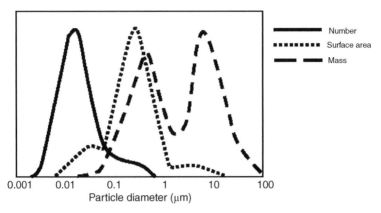

Adapted from Refs (26,28) and also presented in Ref. (27).

FIGURE 15.2 Size distribution according to number, surface, and mass of particles found in urban ambient air.

There are a number of methods available to assess the different characteristics of airborne particles, and several comprehensive reviews describe the advantages and limitations of these available analytical methods (29–34). While a critical analysis of each of these techniques is beyond the scope of this chapter, a brief summary of the approaches important to toxicological investigations are summarized in Table 15.2. With a greater understanding of these exposure/dose–response relationships and proper characterization of particles in experimental studies and human exposure settings, regulators will be better equipped to make policy decisions and establish standards to protect human health.

15.3 RESPIRATORY TOXICOLOGY STUDIES OF ENGINEERED NANOMATERIALS

Carbon nanotubes, nanowires, and quantum dots include a few of the nanomaterials that have recently received a great deal of attention due to their superior electronic, optical, mechanical, chemical, and even biological properties (35–37). Other engineered nanomaterials that have been investigated for their use in biology and medicine include superparamagnetic, polymeric, colloidal gold, metallic and composite nanoparticles, as well as semiconductor nanocrystals, nanoshells, and nanoeggs (38–42). These novel materials will certainly help foster technological advances in virtually every major industry. While many industries already utilize nanoscale materials in a variety of applications (e.g., dispersive agents in cosmetics, composites in automotive parts, tubes in flat panel displays, and fibers in textiles), as new substances and applications move from research and development into high-volume industrial production, understanding toxicological and biocompatible properties of these materials within a variety of settings (e.g., occupational, environmental, or consumer product) will become critical.

TABLE 15.2 Methods for Characterization of Ultrafine Particles

Property and Method	Basis of Measurement	Considerations
Morphology		
Scanning electron microscopy (SEM)	Electron scatter producing high-resolution image of sample surface with three-dimensional appearance	Offline analyses. Good for surface morphology. Can be coupled with instruments (e.g., energy dispersive X-ray analysis [EDX]) for compositional analysis
Transmission electron microscopy (TEM)	Transmission of electrons through sample to give high-resolution image	Offline analyses. Considered gold standard for structure and morphology analysis. Can be coupled with instruments (e.g., energy dispersive X-ray analysis) for compositional analysis
Concentration		
Filter measurement	Mass–gravimetric difference	Useful for correlation of historical measurements or established methods with new monitoring approaches. Can be limited to certain particle size ranges with particle size separator, but not specific to nanoparticles
Condensation particle counter (CPC)	Particle number based on detection of condensed particles	Generally not size specific unless equipped with a particle size separator (up to 100 nm)
Optical particle counter	Particle mass and number. Light scattering of particles assuming certain particle geometry	Particles smaller than 300 nm not detected
Size distribution		
Scanning mobility particle sizer (SMPS)	Size selection based on mobility of charged particles and analysis with a differential mobility analyzer	Real-time measurement of number weighted distributions
Electrical low pressure impactor (EPI)	Size selection based on aerodynamic diameter and charge and impaction on low pressure impactor stages	Real-time measurement of number weighted distributions

(continued)

TABLE 15.2 (*Continued*)

Property and Method	Basis of Measurement	Considerations
Cascade impactors	Size selection based on aerodynamic diameter and impaction on impactor stages	Real-time measurement of aerodynamic mass weighted distributions. Andersen impactors are suitable for personal sampling, whereas berner LPl or micro-office cascade impactors [MOUDl] are not
Surface area		
Epiphanometer	Radioactive tagging based on Fuchs surface area	Online measurement. Good for particles<100 nm and not for particles >1 μm
Diffusion charger	Detection of charged aerosol using an electrometer	Online measurement. Good for particles <100 nm but underestimates surface area for particles >100 nm
Scanning mobility particle sizer (SMPS) coupled with CPC or ELPI	Projected area equivalent diameter or fractal dimensions	Indirect measure using aerosol size distribution. Estimates surface area-weighted distribution assuming certain particle geometry
Brunaeur, Emmett, & Teller (BET)	Adsorption of nitrogen or carbon dioxide	Offline measurement. Accounts for surface pores but requires large sample sizes
Composition and surface chemistry		
Zeta potential	Surface charge based on light-scattering electrophoresis or electroacoustophoresis	Function of surface charge, adsorbed species, and composition and ionic strength of surrounding solution
Surface reactivity	Use of probe molecules to measure changes in oxidative state	Can potentially be used to monitor reactivity in biological fluids
Surface energy/wettability	Measured through immersion microcalorimetry studies or contact angle measurements with various liquids	Important for understanding aggregation, dissolution and bioaccumulation
SEM/TEM coupled with EDX	Unique X-ray spectrum based on chemical composition	Useful in analysis of composition and trace contamination

Adapted from Refs. (29–34).

Research on nanomaterials, including carbon nanotubes, semiconductor crystals, nanocomposites, and other ultrafine particles like TiO_2, will be examined to illustrate what is currently known and what remains to be known about how particle characteristics (e.g., size, agglomeration, morphology, solubility, and surface chemistry), exposure/dose metrics (e.g., mass, size, and surface area), and route of exposure (e.g., inhalation, ingestion, and dermal) influence the biological fate and toxicity of nanosized particles. Aggressive research programs are currently investigating these issues and have evolved, in part from our understanding of the dose–response relationships seen with other industrial long-utilized nanosized particles, specifically TiO_2 and carbon particles.

15.3.1 Nanoscale Titanium Dioxide Particles

Although the potential biocompatibility and toxicity of engineered nanomaterials has only recently received attention from the scientific community, our understanding of well-studied particles like TiO_2 provides a framework for evaluating the characteristics of the engineered nanoparticles that have the greatest influence on biological fate and toxicity. TiO_2 particle research is useful to illustrate how particle characteristics such as size, crystalline form, aggregation state, and surface chemistry, as well as dose metrics (e.g., particle mass, number, and surface area concentrations) correlate with pulmonary toxicity.

While studies of nanosized TiO_2 particles suggest that particle surface area is an important parameter for predicting inflammatory effects in the lungs (43–45), other studies have shown that surface area alone may not dictate toxicity (46–50). Investigations of TiO_2 particles in the 1990s found that ultrafine TiO_2 particles are more toxic than similarly composed but larger sized particles (51–57). With some of these studies (51, 52), however, the crystalline structure of the fine and ultrafine TiO_2 particles differed, suggesting that the differential effects observed between the particle types might be more complicated, and may not be due to particle size alone (49).

Ultrafine TiO_2 particles have several industrial applications, and, as such, also have different sizes, shapes, chemistry, and crystalline structures. Fine or ultrafine TiO_2 particles are used as white pigments in paints, cosmetics, plastics, and paper, as well as in food as anticaking or whitening agents (58). These particles can physically exist in a single particle form or as agglomerates, morphologically occur as spheres, rods, and dots, chemically coated with alumina and amorphous silica, and arrange in different crystalline forms (rutile or anatase). Recently, TiO_2 has been engineered into nanowires for potential applications as dye-sensitized solar cells (59). Because TiO_2 particles were historically viewed as having low toxicities and were generally classified as a nuisance dust, for many years these particles were used as negative controls in inhalation and intratracheal instillation animal studies investigating the toxicity of other particles (45, 60). It was not until reports of lung tumors in rodents exposed to extremely high doses of TiO_2 particles (i.e., particle overload conditions) (61, 62) and different potencies of lung effects for fine particles versus ultrafine particles were published that the mechanisms by which these particles induced adverse health effects were further evaluated (63–65).

A significant amount of research has been conducted over the last few years to better understand how the different physicochemical forms of TiO_2 particles can influence their potential toxicity. Recent investigations of the physicochemical aspects of TiO_2 particles show that particle coating, size, agglomeration state, and crystalline form influence pulmonary toxicity; however, particle surface area showed no effect on the cytotoxic responses of pulmonary cells. More specifically, *in vitro* culture of human epithelial cells and intratracheal instillation studies of nanosized TiO_2 particles showed consistently greater cytotoxic effects with anatase compared to rutile TiO_2 particles (46, 49). Anatase and rutile TiO_2 particles, delivered at similar surface area doses, increased release of lactate dehydrogenase, interleukin-8, and reactive oxygen species, as well as depressed mitochondrial activity in dissimilar patterns in cultured human epithelial cells (46). Based on these biologic parameters, researchers concluded that anatase TiO_2 is 100 times more cytotoxic than rutile TiO_2 particles at a similar particle size and surface area. *In vivo* it was observed that ultrafine anatase TiO_2 particles caused greater changes in bronchoalveolar lavage inflammatory indicators, cell proliferation, and histopathology compared to ultrafine rutile TiO_2 particles; however, with both crystalline forms of TiO_2 particles pulmonary effects were transient and resolved by 1 week postexposure (49). Inhalation and intratracheal instillation studies of ultrafine pigment-grade TiO_2 particles covered with different surface coatings (e.g., alumina or amorphous silica) demonstrate that the TiO_2 particle formulations with the greatest amount of alumina or amorphous silica only produced mild pulmonary effects compared to the reference TiO_2 control particles (47). In addition, intratracheal studies of different shaped (particles, rods, and dots) and sized (fine and ultrafine) TiO_2 show that particles with a greater surface area (TiO_2 rods and dots) induced only transient inflammatory and cellular effects, similar to those observed in larger sized TiO_2 particles. These results thus run counter to current hypotheses indicating that particle surface area might be a better predictor of inflammatory responses and disease in the lungs.

The study of TiO_2 particles have also contributed to the proposed frameworks for safety testing and future understanding of the toxicity and biocompatibility of engineered nanoparticles (47, 49). In particular, comparisons of the administration of intratracheal instilled versus inhaled TiO_2 particles demonstrate that pulmonary responses, as measured by bronchoalveolar lavage inflammatory markers and histopathology, are similar (47). Through this study, a testing scheme has been proposed to "bridge" the results of testing different particles to a reference material. The goal of this testing regimen is to rank the relative pulmonary toxicity of different materials through intratracheal instillation studies and to correlate the findings to effects observed with the reference material via inhalation. Thus, this "bridging" of effects of known materials in instillation and inhalation studies serves as a basis for understanding the potential toxicity of new substances (47).

In summary, the study of TiO_2 particles illustrates that no single particle characteristic or parameter is a hallmark indicator of toxicity. Recent studies show that the crystalline form of nanosized TiO_2 particles has the greatest impact on pulmonary responses, whereas particle surface area, alumina or amorphous silica coating, shape, and, to some extent, particle size (fine versus ultrafine) in contrast appear to only have lesser influences on toxicity. These findings may merely reflect the low inherent toxicity of these materials; however, as the database of health studies on engineered nanoparticles develops, our understanding of which physicochemical properties of different classes of nanoparticles most influence biocompatibility or adverse health effects will also mature.

15.3.2 Nanotubes

Carbon-based engineered nanoparticles, such as single-walled carbon nanotubes (SWCNTs), multi-walled carbon nanotubes (MWCNTs), and fullerenes, have received notable attention due to their superior electronic, optical, mechanical, chemical, and even biological properties. Questions have been raised as to whether the unique properties of these materials may exert biological effects distinct from their parent material. Carbon nanotubes are hollow graphite tubes that can be visualized as a single sheet of graphite rolled to form a cylinder. Carbon nanotubes are composed of either a single layer (e.g., SWCNT) or of multiple layers of individual SWCNTs stacked within one another (e.g., MWCNT), and are manufactured either by electrical arc discharge, laser ablation, or chemical vapor deposition processes. These materials show promise for a wide array of applications, including substrates for neuronal cell growth (66–68), supports for liposaccharides to mimic cell membranes (69), ion channel blockers (70, 71), and drug delivery systems (72–75). In addition to their medical applications, carbon nanotubes are expected to advance electronics, within energy storage devices, thermal insulators, conducive fillers, and molecular electronic devices (72–74, 76–82).

Recent laboratory and field investigations suggest that further research is needed to identify the best methods for sampling and characterization of nanomaterials, to understand the mechanisms of release from their bulk material, and to evaluate the nature of the particles with respect to their size, shape, morphology, and composition (83). Laboratory studies have indicated that agitation SWCNT bulk material can release fine particles into the air, while aerosol concentrations generated during the handling of unrefined material in the field are very low, with airborne nanotube concentrations of 53 $\mu g/m^3$ or less (83). The implications of this monitoring for larger-scale operations is unknown.

Recent research efforts on the health effects of engineered nanoparticles have primarily focused on carbon nanotubes (84–91). The majority of these studies were conducted using *in vitro* methods, while only a few studies have evaluated health effects *in vivo*. Of the majority of studies conducted to date, all have delivered carbon nanotubes to the respiratory tract via intratracheal instillation or pharyngeal aspiration, and none has involved the aerosolization of carbon nanotubes and administration by inhalation with one recent exception (92). The type of carbon nanoparticle (i.e., SWCNT, MWCNT, and fullerene), method of processing (i.e., refined or unrefined), the presence of residual transition metal catalysts, and functionality of different reactive groups are a few of the parameters researchers have tested in cultured cells to better understand which physicochemical characteristics influence toxicity (93–96). SWCNTs, for example, appear to have greater toxic effects on cultured human fibroblasts than MWCNT, active carbon, carbon black, and graphite carbon. In addition, acid treatment (a method of carbon nanotube refinement and removal of residual metal catalysts) of SWCNT produced more toxicity than its unrefined counterpart (96). These findings are supported by other studies that show that acid treatment and subsequent functionalization of SWCNTs or fullerenes influence the extent of toxicity on human lung-tumor cell lines and primary immune cells (93–95). The addition of carbonyl, carboxyl, or hydroxyl groups on the surface of carbon nanotubes induces cell death in lung tumor cells (94). Functionalization of SWCNTs with water-soluble functional groups appears to influence cellular-specific uptake and tolerance by primary immune cells, whereas nonfunctionalized carbon nanotubes induce oxidative stress and apoptosis in a variety of cell systems (97–102).

While clear toxicological differences between carbon nanoparticles functionalized with different chemical moieties have been observed with *in vitro* cell systems, these same responses are not always seen when administering the same material *in vivo* (95, 103). Of the *in vivo* studies conducted on carbon nanotubes, all report inflammation, progressive fibrosis, and granulomas in rodents exposed to carbon nanotubes via intratracheal installation or pharyngeal aspiration. More specifically, as a result of these exposures, acute dose-dependent changes in alveolar wall thickness, immune cell recruitment, and indicators of cellular damage and oxidative stress (measured by levels of inflammatory cells, cytokines, and protein in bronchoalveolar lavage) were observed (87, 88, 90, 91, 104). Carbon nanotubes also produce pulmonary function deficits, impairment of bacterial clearance, aortic plaques, and atherosclerotic lesions (90, 105–107).

In an attempt to understand how different physical and chemical parameters contribute to these toxicological effects, researchers have evaluated the impact of the method of carbon nanotube production, as well as the influence of milling carbon nanotubes or altering the content and type of metal catalyst on the toxicity in animals. Results suggest, however, that carbon nanotubes themselves induce cellular changes, because all of the various formulations of carbon nanotubes produced pulmonary lesions (87). A recently published study shows that these effects can be exacerbated by feeding animals a vitamin E-deficient diet, and thereby reducing the anti-oxidant capacity (glutathione, ascorbate, and α-tocopheral) of the lungs while enhancing acute inflammation and fibrotic responses (104).

Only one study to date has evaluated the effects of carbon nanotubes via inhalation (92). In this study, MWCNTs were delivered to mice at concentrations of 0.3, 1, or $5 \, mg/m^3$ for 7 or 14 days and were followed for 7 and 14 days postexposure. Inflammatory responses in the spleen were more sensitive to particle exposures than that observed in the lungs. Alveolar macrophages in bronchoalveolar lavage and in lung tissue sections contained black particles, however white blood cell counts in bronchoalveolar lavage were not elevated and no lung damage was observed. Changes in immunosuppression markers (e.g., T-cell antibody and proliferative response) and cytokine gene expression of IL-10 and NAD(P)H oxidoreductase were observed in the spleen, whereas no changes in oxidant stress markers were seen in the lungs (92). Although this is the first inhalation study conducted with carbon nanotubes, additional studies will be required to compare these effects with those associated with SWCNTs, a nanomaterial expected to have a greater market share than MWCNTs. Further, the biological significance and human health implications of these findings will need to be further investigated.

One aspect of toxicity testing of nanoparticles that is worth noting is that cellular toxicity may actually be a desired outcome and intended use of the nanoparticle, particularly if the target is tumor cells. Certainly, future research efforts will need to explore and understand the similarities and discrepancies of *in vitro* and *in vivo* biological responses as different functionalized carbon nanotubes, as well as other nanomaterials, are being developed. While research in the area of biocompatible carbon nanotube design is still in its infancy, researchers are aggressively pursuing ways to functionalize these nanoparticles to have sufficient biocompatibility, functionality, distribution, retention, and specificity in the hope that these nanomaterials can someday be utilized as carriers of biologic and therapeutic molecules.

Despite early findings, limited studies have evaluated the effects of carbon nanotubes via inhalation exposure, which would model an exposure route more comparable to real-world environmental or occupational conditions. Inhalation studies are likely to confirm or disprove the pattern, extent, and timing of pulmonary changes observed in intratracheal instillation and aspiration studies. Without inhalation studies, uncertainties and questions about correlations between the existing body of experimental animal literature and the experience in human exposure settings, as well as the applicability of developing occupational and environmental standards, will continue to remain.

15.3.3 Nanowires

Nanowires are most simply defined as wires of a nanosize range. Although they are tube-shaped like carbon nanotubes, they are generally not hollow in nature. Nanowires can be metallic, semiconducting, or insulating and consist of different elemental compositions, such as silicon, germanium, gallium, nitride, or cadmium sulphide (108). Due to their regular crystal structures and unique and uniform electronic properties, these engineered nanomaterials show significant promise for a number of industries (4, 109). Additionally, the greater control of electronic properties of nanowires compared to nanotubes make these materials an attractive candidate to researchers developing electronic nanosensors (110).

Future applications of nanowires are anticipated to be similar to those of carbon nanotubes (e.g., sensors and transistors); however, nanowires are being produced from a greater array of elements (i.e., silicon, silica, germanium, zinc oxide, and TiO_2) compared to carbon nanotubes (i.e., carbon and metal catalysts). More recently, nanowires of various forms and compositions have gained applicability for future markets. TiO_2 nanotubes (111) and silicon nanowires (4, 112), for example, have been proposed for use in future photovoltaic cells. Silica nanowires can be used as high surface area supports or as templates to make other nanomaterials (113, 114). The use of nanowires has also been proposed in solar cells and other energy devices. Considering that a broad range of different metals are being developed as nanowires, that the applications of these nanomaterials are expected to exceed carbon nanotubes, and that no toxicological studies on nanowires have been conducted, there is a compelling need to understand the potential health effects of nanowires before they are produced on a large industrial scale and ultimately reach consumers.

15.3.4 Nanosized Model Particles

Nanosized particles, such as those specifically engineered in the nanotechnology field or those existing as a result of combustion processes, have received considerable interest due in part to their large functional surface area and their ability to adsorb relatively large quantities of compounds compared to particles of larger diameter. This property has been explored in applications such as drug delivery and in understanding the dynamics of copollutants in particulate air pollution. Model particles, such as quantum dots, iridium, gold, and manganese oxide, have been used to assess how surface properties of nanosized particles delivered via the respiratory tract may influence the disposition, fate, and transport, as well as the biological responses in pulmonary and extrapulmonary tissues. These studies provide some initial insight as to physicochemical properties, such as particle surface coatings, size, and chemistry, which influence particle transport patterns in the lungs and to other organs (i.e., brain, liver, spleen, and heart). As this information evolves, scientists will gain a better understanding of the dynamics existing between particle physicochemistry and the cellular milieu of the respiratory tract, thus providing a basis for developing nano-material safety testing schemes.

15.3.4.1 Quantum Dots Quantum dots are semiconductor nanocrystals that have unique optical and electronic properties. Although the characteristics of these materials were first described in 1982 (115), it was not until the last decade that the methods of high-quality production and chemical conjugation were developed, thus prompting exploration of quantum dot use in biodetection and imaging applications. Quantum dots consist of a metalloid crystalline core, made from a variety of materials (e.g., ZnS, CdSe, CdTe, InAs, and GaAs), and covered by a shell (e.g., proteins, peptides, nucleic acids, carbohydrates polymers, and small molecules), depending on the desired application and physicochemical properties (e.g., antibody detection, receptor–ligand response) (116–125). The stability, resistance against photobleaching, multiple color excitation fluorescence, and general fluorescent brightness of quantum dots, as well as improvements in their synthesis, surface chemistry, and conjugation have proved them useful for cellular, organ, and tumor imaging and molecular profiling (116, 126). Quantum dots have also been explored for delivering bioactive molecules, such as antibodies and receptor ligands, to specific sites, including neoplastic cells, peroxisomes, DNA, and cell membrane receptors (121, 127–132).

Aside from their potential applications in human medicine, quantum dots, thanks to their unique imaging properties, also provide an exceptional opportunity to understand the mechanisms by which different physicochemical properties (i.e., surface coatings, size, and chemistry) of particles delivered via various routes (dermal, inhalation, and ocular) influence the disposition, fate, and transport, as well as the biological responses within both target tissues and those distant from the administration site (133–135). This feature is particularly important given that ultrafine particles can enter the systemic circulation and translocate to sites (i.e., heart, liver, central nervous system, and brain) distant from the port of entry (i.e., lungs) (136, 137). Much of the work that has been performed on quantum dots has focused on intracellular targeting of cultured cells with bioactive moieties, and very little has been published regarding the kinetics and toxicity of these particles *in vivo* (121).

Because of their metalloid crystalline core, quantum dots can be toxic. A limited number of studies have shown that surface coatings can alter the deterioration of the CdSe core, which can ultimately influence the cellular toxicity and biocompatibility of quantum dots (138–140). These surface coatings can also influence the distribution of these particles in the body (117, 121, 141, 142). Data from the few published animal studies suggest that quantum dots can accumulate in certain organs depending on the chemical characteristics and bioactive nature of their coatings. Findings that conjugation of quantum dots with bioactive moieties show specific labeling to cultured cell membranes are encouraging for potential applications of tissue targeting (i.e., tumors) (130); however, preliminary *in vivo* research shows that factors other than particle surface coatings may play a role in specific cellular targeting. Quantum dots, for example, accumulate in the liver and spleen regardless of the peptide used for quantum dot conjugation (117). Administration of quantum dots coated with different molecular weights of polyethelene glycol (PEG) indicates that differential tissue and organ deposition in mice are time and molecular weight dependent, with larger molecular weight PEG-coated quantum dots clearing more slowly and accumulating more in lymph nodes, liver, and bone marrow than those coated with low molecular weight PEG (141). More recently, quantum dots have been administered to the respiratory tract of rodents to understand how different functional groups and surface coatings influence the deposition, transport, and retention of ultrafine particles in pulmonary and extrapulmonary tissues delivered via the respiratory tract. Preliminary findings demonstrate that nanosized quantum dots can translocate from olfactory epithelium of the nasal cavity to the olfactory bulb region of the brain (133).

15.3.4.2 *Other Model Particles (Iridium, Gold, and Carbon)*   A number of different types of particles have been used to evaluate how different particle physicochemical parameters affect the kinetics of nanosized particles in the lungs of humans and experimental animals. Some of these model particles include the isotopes iridium-192 and technetium-99m, as well as nanosized gold and carbon particles. A significant advantage of using radiolabeled ultrafine particles and administering them to animals by intratracheal inhalation is that the dose of particles delivered to the lungs is known, and therefore a complete balance of radiolabeled particle distribution and excretion can be measured. Much of this type of work has been conducted with radiolabeled iridium particles (136, 143–146).

Research on radiolabeled particles confirms that ultrafine particles are cleared more slowly from the lungs than particles of equivalent type, but larger size. In addition, it has been observed that clearance of particles via phagocytosis by alveolar macrophage occurs to a lesser extent with ultrafine particles compared to fine-sized particles. These findings were initially reported in rats exposed by inhalation to TiO_2 particles, where ultrafine particles (20 nm diameter) showed greater pulmonary retention, translocation to the pulmonary interstitium, persistent inflammation, and impaired alveolar macrophage clearance of test particles compared to exposures from larger sized (250 nm diameter) TiO_2 particles (63–65). Subsequent studies using radiolabeled ultrafine particles with iridium show that only 20% of retained ultrafine (15–20 nm diameter) particles are lavaged from the lungs 24 h after exposure, whereas more than 80% of fine-sized polystyrene particles (0.5, 2, and 10 μm) are retrieved by bronchoalveolar lavage at the same postexposure time period (136–144).

In addition, transport of ultrafine particles to the epithelium appears to be a major route for particles of this size to be cleared from the airway or alveolar lumen. Immediately after inhalation, iridium-radiolabeled nanoparticles were predominately free or associated with macropahges in the lavage, whereas 3 weeks after exposure, the lavageable fraction of nanoparticles decreased to 0.06 of the initial lung burden, with 80% of the retained nanoparticles translocated into the epithelium and interstitium (17). Although macrophage uptake of ultrafine particles is less than fine-sized particles, the major route of elimination for both particle size ranges is the gastrointestinal tract, thus suggesting that even after the initial clearance by macrophages via the mucociliary escalator, some fraction of ultrafine particles continue to be reentrained from the epithelium and interstitium to the airway or alveolar lumen (17, 63, 145, 147). The uptake of ultrafine particles by epithelial cells and subsequent translocation to the interstitial space is thus rather efficient, with nonlavageable nanoparticle fractions retained in the lungs longer than 3 days being internalized in either of these compartments (17). These results are confirmed by transmission electron microscopy of type I, endothelial, and alveolar septal cells in the lungs following exposure to ultrafine gold particles (148). In summary, then, research demonstrates that nanosized particles will be retained in the lungs and have direct interactions with the epithelium and interstitium to a greater extent and over a longer period of time than their fine-sized counterparts.

Researchers have also utilized gold and other types of metal particles to understand how different particle surface coatings, solubility, and shapes influence particle fate and biological responses in the body. Surface coatings of gold nanoparticles with PEG, serum albumin, or citrate following intratracheal aspiration to rats, for example, have affected the extent of uptake by mononuclear phagocytic cells, as well as produced changes in markers for cellular toxicity (149). Whereas in this study PEG appears to have the greatest impact on pulmonary responses to gold-labeled nanoparticles; in another study of ovalbumin-coated carbon nanoparticles (14 or 56 nm diameter) administered intratracheally, nanoparticles exhibit adjuvant activity and aggravated antigen-related airway inflammation (bronchoalveolar lavage cellular, cytokine, and chemokine profile) and immunoglobulin production, with the greatest effects observed with the smallest nanosized particles (150). In addition, the formation of 8-hydroxy-2'-deoxyguanosine (8-OHdG), a marker for oxidant stress, was markedly induced by antigen plus nanoparticle, compared to nanoparticle or antigen alone (150).

While particle size significantly influences particle fate, transport, and toxicity, it is not the sole determinant of these responses. *In vitro* assays of ceramic particles (TiO_2, Al_2O_3, ZrO_2, Si_3N_4, and SiC) of different sizes and shapes with murine fibroblasts and macrophages show that cellular cytotoxicity corresponds with the release of soluble metal ions and monomers, and is affected minimally by the size (90–1600 nm diameter) or the shape (spheric, spindle, or dendritic) of the nanoparticle (151). Overall, the study of the health effects of nanomaterials is still so immature that a clear pattern regarding which physicochemical characteristics are likely to predict cellular toxicity or biocompatibility has yet to appear. Some key parameters, however, such as particle antigenicity, size, persistence, solubility, charge, reactivity, crystalline, or chemical form do seem to emerge as important factors, and are mindfully being considered in current research testing regimens.

15.3.5   Extrapulmonary Transport of Inhaled Nanosized Particles

A number of recent publications have shown that nanosized particles can translocate to extrapulmonary organs following inhalation exposure (23, 136, 137, 145, 152–159). While nanoparticles do undergo classic mechanisms of clearance from the lungs, there are transport pathways that are unique to particles within the nanosize range (Fig. 15.3). Classic mechanisms of nanoparticle clearance include macrophage phagocytosis, transport along the mucociliary escalator, and physicochemical processes (i.e., dissolution and physical breakdown). Unique clearance mechanisms include uptake and transport along sensory neurons as observed in the nasal olfactory epithelium, as well as direct uptake of particles by the epithelium of the respiratory tract and subsequent transport to the interstitium, lymphatics, and systemic circulation (12, 137). Transepithelial movement of particles occurs to a greater extent, and macrophage phagocytosis occurs to a lesser degree for nanosized particles compared to those within the fine size range (11, 12, 16, 17). These observations explain the enhanced access to the systemic circulation and extrapulmonary distribution that appears to occur in some circumstances with nanosized particles (12, 160). Particle conjugation with albumin and/or phospholipids may be important processes to facilitate the transport of nanoparticles from the pulmonary lumen to the interstitial space (161). Particle size, surface chemistry (e.g., solubility, reactivity, and charge), and other physiochemical characteristics (e.g., shape and agglomeration) are also likely parameters that will impact how nanoparticles are internalized by the pulmonary epithelium.

Some of the attention on extrapulmonary transport of particles, particularly with transport mechanisms of particles to the brain, stemmed from observations that inhalation exposure to manganese-containing dusts or fumes among workers, who are involved with steel manufacturing, welding, and mining and processing of manganese ores, show progressive functional deficits in the central nervous system (162, 163). Researchers have thus observed that metals, such as manganese, can be effectively transported from the nasal olfactory epithelium along the axons of the olfactory neurons to the olfactory bulb of the brain (155, 164–166). Particle transport via neurons along the ganglion nodosum of the neck has also been documented following intratracheal instillation of nanosized polystyrene microspheres (167). Since these observations were made, researchers have evaluated the extrapulmonary transport of ultrafine particles (e.g., carbon, iridium, zinc, and manganese oxide) and polystyrene beads to all major organs of the body (23, 136, 137, 145, 152, 159).

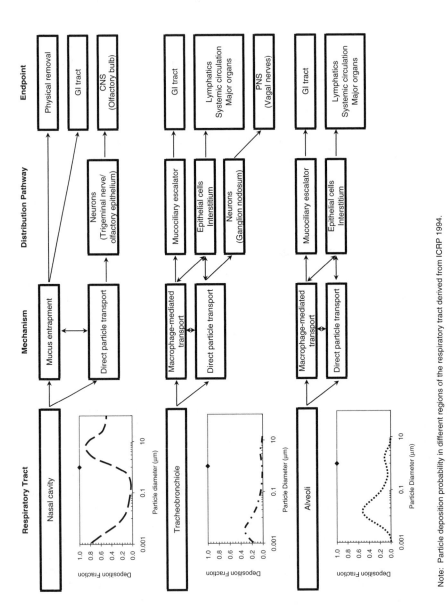

Note: Particle deposition probability in different regions of the respiratory tract derived from ICRP 1994.

FIGURE 15.3 Transport pathways of insoluble nanoparticles in different regions of the respiratory tract.

As experimentally shown in rats and humans, translocated fractions of inhaled ultrafine particles to extrapulmonary organs after a single exposure are generally below 5% (136, 143–146). In a 6 month translocation kinetics study involving a single 1 h inhalation exposure to 15–20 nm-sized radiolabeled iridium particles, maximum accumulation of iridium ultrafine particles occurred in liver, spleen, kidneys, heart, and brain about 1 week after inhalation, with a maximum deposited mass of 0.1–0.5% of the pulmonary dose. Thereafter, the load in each target organ declined to less than 0.01–0.05%, and remained clearly detectable (144). The significance of such a small mass fraction of particles translocating to other organs is unknown; however, if these results were expressed as a particle number or surface area measurement, a different pattern of extrapulmonary transport might be observed. This trend of extrapulmonary transport appears to be more significant (i.e., greater particle fraction is translocated) with smaller nanosized particles (e.g., 15 nm diameter) than larger ones (e.g., 80 nm diameter) (136). These observations, then, lend themselves to the hypothesis that transport via the systemic circulation following inhalation particle exposure is an inverse particle-size dependent phenomenon (136).

For obvious reasons, the research community has paid significant interest to particle transport to the brain resulting from inhalation and particle deposition in the olfactory mucosa of the nasal epithelium. Despite its recent attention, this pathway initially was described over a half century ago in studies of poliovirus (30 nm diameter) and silver-coated gold (50 nm diameter) particle transport in nonhuman primates (168–170). Today, studies have focused on the extent, timing, and duration (e.g., kinetics) exhibited by solid nanosized particles as they deposit in the nasopharyneal region and translocate and accumulate in different regions of the brain (137, 171). In an inhalation study of ultrafine elemental carbon particles (36 nm count median diameter), for example, significant increases in particle concentrations 7 days after exposure were observed in the olfactory bulb of the brain with increased but variable concentrations seen in the cerebrum and cerebellum (137). Similarly, airborne manganese oxide particles (30 nm diameter) administered to rats via occluded nostrils show a 3.5-fold increased concentration of manganese in the olfactory bulb of the brain following 12 days postexposure, and also measurable increases in the striatum, frontal cortex, and cerebellum (152). While the fractional transport of ultrafine particles from the nasal epithelium to the brain has been proposed to be on the order of 20% (137), the toxicological and biological significance of these doses to the brain yet remains to be elucidated. The extrapulmonary transport of inhaled ultrafine particles therefore will certainly continue to be an active area of research.

15.4 BRIDGING THE KNOWLEDGE GAP BETWEEN EXPERIMENTAL STUDIES AND HUMAN EXPOSURES

15.4.1 Human Clinical Studies of Nanoparticles

Over the last two decades, researchers have intensely studied how the fine-sized fraction of particulate air pollution (PM2.5) might exacerbate cardiopulmonary effects among susceptible individuals in the general population. While the mechanisms by which ambient particulate matter (PM2.5) causes increased morbidity and mortality due to respiratory and cardiovascular diseases is not well understood, investigators have recently proposed that the ultrafine fraction of ambient particulate matter may be responsible for, or at least contribute to, these effects. Controlled human studies reveal effects on blood leukocyte adhesion molecular expression, cardiac repolarization, and heart rate variability following inhalation exposure to ultrafine (<100 nm diameter) or nanosized particles (153, 172, 173).

To understand the potential effects of particulate air pollution, over the last decade researchers have developed and utilized sophisticated methods both to deliver airborne particles via inhalation and to detect particle deposition, transport, and retention in controlled clinical studies. Generally, ultrafine carbon particles, either labeled or unlabeled with technetium-99m, have been administered in a single short-term exposure to healthy or compromised individuals (i.e., asthma, emphysema, or chronic obstructive pulmonary disease) at rest or undergoing light to moderate exercise. These studies show that deposition of ultrafine particles is increased in individuals with moderate to severe airway obstruction compared to healthy subjects, as well as in exercising individuals compared to those at rest (143, 174, 175). Human studies have also confirmed a high deposition efficiency of ultrafine particles in the respiratory tract as originally predicted by the International Commission on Radiological Protection (175).

Evaluating particle kinetics and translocation in humans using inhaled technetium-99m-labeled particles has somewhat been challenging. Nemmar (160) has detected activity in the liver, thyroid, salivary glands, and stomach within an hour of exposure (160). Some researchers, however, have suggested that leaching and contamination may contribute to observations of rapid transport of technetium-99m in the blood and liver in human inhalation studies rather than to the actual direct transport of particles from the lungs to the systemic circulation (143, 160). Several studies, though, have not been able to confirm these observations of rapid transport of ultrafine particles to the blood (143, 146, 176, 177). In addition, in clearance studies able to account for particle leaching or contamination, technetium-99m-labeled carbon particles show no differences in the rate of removal of particles from the lungs of healthy and diseased individuals (143, 146). Although effects on the cardiovascular system have been detected in controlled human experiments involving inhalation exposure to nanoparticles, it remains unknown whether these effects occur as a result of direct distribution of particles to the systemic circulation, or are instead a result of some indirect response occurring from particle-induced release of cellular mediators. Moreover, doses required to detect extrapulmonary transport of ultrafine particles with these methods may be too high for application in human clinical studies; future research is needed to resolve the discrepancies in extrapulmonary transport observed in human and animal studies.

15.4.2 Correlating Nanoparticle Exposure, Dosimetry, and Health Effects

Very little epidemiologic data are available concerning the human health impact of the ultrafine fraction of ambient particulate pollution. The cardiovascular and pulmonary

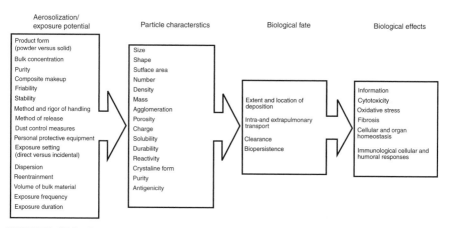

FIGURE 15.4 Potential important factors in correlating exposure, dosimetry, and health effects of inhaled nanoparticles.

effects shown in human clinical studies (153, 172, 173, 178), however, as well as observations in experimental animal studies showing that ultrafine particles can translocate from the respiratory tract to extrapulmonary organs (i.e., heart, liver, spleen, and brain), prove that this pathway will be a significant and complicated area of ongoing research. There are many factors that are likely to influence the aerosolization and potential inhalation exposure of nanoparticles, many of which are also likely to impact the physicochemical characteristics of the airborne nanoparticles, in turn, affecting the biological fate and response (Fig. 15.4).

Identifying the exposure or dose metrics that correlate to the most sensitive health effects will be important for conducting risk assessments and establishing standards for protecting human health. Owing the emerging views that particle mass may not be the most predictive metric of the health effects of inhaled particles, researchers are developing parallels between other exposure/dose parameters and disease. Several studies show that as particles decrease in size, alveolar macrophages have a decreased capacity to phagocytize and clear particles from the lungs (43–45). One hypothesis advocated for this phenomenon is that the surface area per unit mass posits critical to the impairment of alveolar macrophage-mediated particle clearance and to the elicitation of cellular toxicity. The work of Donaldson and colleagues (43, 45) indicate that the difference between different-sized dusts in the level of inflammation and translocation to the lymph nodes can be explained most simply when the lung burden is expressed as total particle surface area. In addition, Donaldson et al. (179) show that ultrafine TiO_2 and ultrafine carbon black have a greater ability to generate free radicals and to induce oxidative stress compared to their fine-sized particle equivalents. In considering the mechanisms by which engineered nanoparticles induce fibrosis, it is thus imperative that experimental studies characterize the delivered aerosol in a manner that allows occupational and environmental standards to be derived from data best reflective of the most sensitive health outcomes. Further research is required to better understand the relationship between different exposure metrics (e.g., particle mass, number, and surface area) and the development of disease from nanomaterials, particularly given the wide array of chemical, physical, and morphological forms of engineered nanoparticles.

Fortunately for current researchers, though, several of these issues have already been raised and investigated while studying other materials such as asbestos; as such, researchers can draw on many well-established methods and tools to help them better understand the dynamics between the environment and biological systems. In case of asbestos, for example, it is well known that the formulation of the bulk material, its potential friability, the type of asbestos (serpentine versus amphibole), and its size and shape (fiber diameter and length) will influence exposure and biopersistence and the ultimate risk of asbestos-related disease (i.e., asbestosis, lung cancer, and mesothelioma). To understand the potential health risk of these mineral fibers, researchers generally refer to the three "Ds" of fiber toxicology: dimension, durability, and dose. The range of chemical formulations, physicochemical properties, and industrial and consumer applications currently used or anticipated for nanomaterials, though, add multiple layers of complexity not previously observed in the historical experience with asbestos. Only additional research will reveal which characteristics of exposure, dosimetry, and particle physicochemistry will most correlate with health effects. Instead of the three Ds, toxicologists may refer to the seven Ss of nanoparticle toxicology: size, shape, surface chemistry and reactivity, solubility, source, stability, and surface area; only time will tell.

15.5   CONSIDERATIONS OF PRODUCT SAFETY OF NANOMATERIALS

The days of hypothetical or futuristic visions of how nanomaterials might improve technology are over; nanotechnology has now powerfully taken over society's mindset. In the United States, both industry and government are devoting significant funds to research and development of nanotechnologies, with combined annual investments upwards of $4 to $10 billion, and venture capitalist start-up company investments of $650 million in 2006 (180–182). In addition, to date, there have been over 75 000 nanotechnology patents filed worldwide, the majority for chemistry and molecular biology, drug development, semiconductor devices, and radiant energy applications (183). The question of when the wave of nanotechnology will arrive no longer exists; it is now here.

The new questions now facing researchers, regulators, industry, and insurers are how nanomaterials can be developed and used to maximally benefit our society, but minimally impact human and environmental health. An important part of society's assessment is to not only understand the characteristics that make these materials toxic but to also evaluate how these parameters might be altered to be compatible with human health and the environment. A few organizations, including the Environmental Defense, DuPont, and the International Life Sciences Institute Research Foundation/ Risk Science Institute (ILSI RF/RSI), have proposed a framework to evaluate the potential toxicity and safety of nanomaterials. Table 15.3 summarizes approaches that

TABLE 15.3   Proposed Frameworks for Evaluating the Potential Toxicity and Health Impact of Different Inhaled Nanoparticles

Assays	End points
ILSI RF/RSI (2005)	
Material characterization	
General physicochemical characteristics	Size distribution, agglomeration state, shape, crystal structure, chemical composition, surface area, surface chemistry, surface charge, and porosity
Exposure Monitoring	
General	Particle mass, number and surface area concentration, and size distribution
In vitro testing	
Bronchial and/or alveolar epithelial cells	LDH release, cytokine expression (IL-8, MCP-1) activation of transcription factors (NF-κB AP-1) oxidative and nitrosative stress, upregulation of antioxidant gene (SOD, GPX), cell proliferation, genotoxicity (COMET assay. 8-hydroxy-deoxyguanosine). Include appropriate positive and negative controls (e.g., silica, TiO_2), protein adsorption
Macrophages	Cytokines (TNFα, IL-6), nuclear transcription factors (NF-κB, AP-1), oxidative burst and nitric oxide production, phagocytosis, cytoskeletal arrangement
Other cell types (endothelial cells fibroblasts) and organ systems (lung slices and whole perfused lung)	Similar end point as noted above
In vivo testing	
Tier I – Intratracheal instillation of pharyngeal aspiration or inhalation (2 wk exposure, observation 24 h to 28 days)	Bronochoalveolar lavage (cell profile, protein, LDH, and alkaline phosphatase), oxidative stress markers (DNA damage, cytokines, glutathione, total antioxidants, nitrate/nitrite, lipid peroxidation), histopathology, cell proliferation (BRDU and PCNA). Consider effects in remote organs (liver, spleen, bone marrow, heart, kidney, and CNS). Include benchmark materials (TiO_2, carbon black and crystalline silica)
Tier II – Inhalation (preferred) (4-wk exposure, observation up to 2–3 months)	Deposition, translocation and biopersistence (radiolabeled, fluorescent or magnetically tagged particles), chemical analysis, genomics and proteomics (e. g., mechanisms of oxidative stress, reproductive and development effects (OECD Guideline 422)
Environmental defense/DuPont (2007)	
Material characterization	
General physicochemical characteristics	Flammability explosivity incompatibility reactivity corrosivity, stability, stability, decomposition, polymerization, and photoactivity

(continued)

TABLE 15.3   (*Continued*)

Assays	End points
Exposure monitoring	
Workplace	Particle mass, number size distribution, and surface area concentration
Environmental	Model and monitor routine and non-routine emissions
Consumer	Maximal and time-weighted average particle mass and number concentrations
In vitro and *in vivo* testing	
Inhalation (28-days exposure, 90-day observation)	Histopathology
Instillation (single dose. 90-day observation)	Histopathology
Chronic inhalation (>1 yr) as needed	Histopathology

Note: ILSIRF/RSI, International Life Sciences Institute Research Foundation Risk Science Institute; OECD, Organisation for Economic Co-operation and Development; LDH, lactate dehydrogense; IL-8/IL-6, interleukin; NF-κB, nuclear transcription factor-kappa B; MCP-1, monocyte chemotactic protein-1; AP-1, activator protein-1; SOD, superoxide dismutase; GPX, glutatione peroxidase; TNFα, tumor necrosis factor alpha; PCNA, proliferating cell nuclear antigen; COMET, single cell get electrophoresis assay.

these organizations have developed for understanding the potential impacts of inhaled nanoparticles on human health. As also illustrated in Figure 15.4, a key component of this testing strategy is documenting the physical and chemical characteristics of nanoparticles being generated or released in the setting of interest. While intratracheal instillation studies are useful for investigating the potential effects of nanomaterials, inhalation studies will remain the gold standard for understanding the patterns of particle deposition, translocation, and induction of disease in pulmonary and extrapulmonary organs.

15.6   CONCLUSION

Many nanomaterials exhibit unique electronic, optical, mechanical, or chemical properties. Current research needs to include a better understanding of how these various particle characteristics (e.g., size, agglomeration, morphology, solubility, and surface chemistry) or exposure/dose metrics (e.g., mass, size and surface area) will influence their biological fate and toxicity and elicit biological responses distinct from the parent material. Newly developed engineered nanoparticles (i.e., quantum dots and carbon nanotubes) or model particles (i.e., polystyrene, iridium, gold, and manganese oxide), as well as extensively studied ultrafine particles produced through manufacturing (i.e., ultrafine TiO_2 or carbon black) or combustion processes (i.e., vehicle exhaust and air pollution) are being investigated to understand how surface properties of nanosized particles delivered via the respiratory tract may influence the disposition, fate and transport, and biological responses in pulmonary and extrapulmonary tissues.

Ongoing and future research efforts will benefit from investigations relating exposure and dose metrics administered in experimental studies to those observed in real-world settings. Although the broad range of nanomaterial applications could lead to an incredible array of potential human exposure scenarios, worker health and the industrial environment are still likely to remain the driving force behind research, risk assessment, and future regulation of inhalable nanomaterials. Achieving a better understanding of the dynamics at play between particle physicochemistry and transport patterns and cellular responses in the lungs and other organs (i.e., brain, liver, spleen, and heart) will provide a future basis for establishing predictive measures of toxicity or biocompatibility, for developing testing strategies to evaluate the nanomaterial safety through all pertinent routes of exposure, and for creating frameworks to assess overall potential human health risk.

ACKNOWLEDGMENT

Research supported in part by US EPA STAR Grant #83171401 and University of California Toxic Substances Research & Teaching Program.

REFERENCES

1. Alivisatos, P. Colloidal quantum dots. From scaling laws to biological applications. *Pure Appl Chem* 2002;72(1–2):3–9.

2. Drexler, K.E. *Nanosystems: Molecular, Machinery, Manufacturing, and Composition.* New York, NY: Wiley-Interscience Publication; 1992.

3. Hu, J.T., Odom, T.W., Lieber, C.M. Chemistry and Physics in one dimension: synthesis and properties of nanowires and nanotubes. *Accounts Chem Res* 1999;32(5):435–445.

4. Lieber, C. Nanoscale science and technology: building a big future from small things. *MRS Bull* 2003;28(77):486–491.

5. Navrotsky, A. Thermochemistry of nanomaterials. *Nanoparticles Environ* 2001;44:73–103.

6. Smalley, R. Wires of wonder. *Technol Rev* 2001;104(2):86–91.

7. West, J.L., Halas, N.J. Engineered nanomaterials for biophotonics applications: improving sensing, imaging, and therapeutics. *Ann Rev Biomed Eng* 2003;5:285–292.

8. Hawxhurst, D. Project on emerging nanotechnologies: a nanotechnology consumer products inventory. Washington DC: Woodrow Wilson International Center for Scholars. http://www.nanotechproject.org/44.

9. ASTM International Standard terminology relating to nanotechnology, West Conshohocken, PA: American Society for Testing and Materials (ASTM International); 2006.

10. ICRP. Human respiratory tract model for radiological protection. A report of a Task Group of the International Commission on Radiological Protection. *Ann ICRP* 1994;24(1–3): 1–482.

11. Ferin, J. Pulmonary tissue access of ultrafine particles. *J Aerosol Med* 1991;4(1):57–68.

12. Oberdorster, G., Oberdorster, E., Oberdorster, J. Nanotoxicology: an emerging discipline evolving from studies of ultrafine particles. *Environ Health Perspect* 2005;113(7): 823–839.

13. Roth, C. Clearance measurements with radioactively labelled ultrafine particles. *Ann Occup Hyg* 1994;38:101–106.

14. Roth, C. Deposition and clearance of fine particles in the human respiratory tract. *Ann Occup Hyg* 1997;40:503–508.

15. Roth, C. Clearance of the human lungs for ultrafine partcles. *J Aerosol Sci* 1993;24: S95–S96.

16. Kreyling, W.G., Semmler, M., Moller W. Dosimetry and toxicology of ultrafine particles. *J Aerosol Med* 2004;17(2):140–152.

17. Semmler-Behnke, M., Takenaka, S., Fertsch, S., Wenk, A., Seitz, J., Mayer, P., Oberdorster, G., Kreyling, W.G. Efficient elimination of inhaled nanoparticles from the alveolar region: evidence for interstitial uptake and subsequent reentrainment onto airways epithelium. *Environ Health Perspect* 2007;115(5):728–733.

18. Leong, B.K., Coombs, J.K., Sabaitis, C.P., Rop, D.A., Aaron, C.S. Quantitative morphometric analysis of pulmonary deposition of aerosol particles inhaled via intratracheal nebulization, intratracheal instillation or nose-only inhalation in rats. *J Appl Toxicol* 1998;18(2):149–160.

19. Osier, M., Oberdorster, G. Intratracheal inhalation vs intratracheal instillation: differences in particle effects. *Fundam Appl Toxicol* 1997;40(2):220–227.

20. Rao, G.V., Tinkle, S., Weissman, D.N., Antonini, J.M., Kashon, M.L., Salmen, R., Battelli, L.A., Willard, P.A., Hoover, M.D., Hubbs, A.F. Efficacy of a technique for exposing the mouse lung to particles aspirated from the pharynx. *J Toxicol Environ Health A* 2003;66(15):1441–1452.

21. Brain, J.D., Knudson, D.E., Sorokin, S.P., Davis MA. Pulmonary distribution of particles given by intratracheal instillation or by aerosol inhalation. *Environ Res* 1976;11(1): 13–33.

22. Driscoll, K.E., Costa, D., Hatch, G., Henderson, R., Oberdorster, G., Salem, H., Schlesinger, R.B. Intratracheal instillation as an exposure technique for the evaluation of respiratory tract toxicity; uses and limitations. *Toxicol Sci* 2000;55:24–35.

23. Oberdorster, G., Sharp, Z., Atudorei, V., Elder, A., Gelein, R., Lunts, A., Kreyling, W., Cox, C. Extrapulmonary translocation of ultrafine carbon particles following whole-body inhalation exposure of rats. *J Toxicol Environ Health A* 2002;65(20):1531–1543.

24. Henderson, R.F., Driscoll, K.E., Harkema, J.R., Lindenschmidt, R.C., Chang, I.Y., Maples, K.R., Barr, E.B. A comparison of the inflammatory response of the lung to inhaled versus instilled particles in F344 rats. *Fundam Appl Toxicol* 1995;24(2): 183–197.

25. Osier, M., Baggs, R.B., Oberdorster, G. Intratracheal instillation versus intratracheal inhalation: influence of cytokines on inflammatory response. *Environ Health Perspect* 1997;105 (Suppl 5):1265–1271.

26. Finlayson-Pitts, B. Chemistry of the upper and lower atmosphere: theory, experiments, and applications. San Diego, CA: Academic Press; 2000.

27. Oberdorster, G., Utell, M.J. Ultrafine particles in the urban air: to the respiratory tract, and beyond? *Environ Health Perspect* 2002;110(8):A440–A441.

28. Whitby, K.T., Sverdrup, G.M. California aerosols: their physical and chemical characteristics. *Adv Environ Sci Technol* 1980;8:477–525.

29. Aitken, R.J., Chaudhry, M.Q., Boxall, A.B.,Hull, M. Manufacture and use of nanomaterials: current status in the UK and global trends. *Occup Med (Lond)* 2006;56(5):300–306.

30. Brouwer, D.H., Gijsbers, J.H., Lurvink, M.W. Personal exposure to ultrafine particles in the workplace: exploring sampling techniques and strategies. *Ann Occup Hyg* 2004;48 (5):439–453.

31. Ku, B.K., Maynard, A.D. Comparing aerosol surface-area measurements of monodisperse ultrafine silver agglomerates by mobility analysis, transmission electron microscopy and diffusion charging. *J Aerosol Sci* 2005;36:1108–1124.

32. Ku, B.K., Maynard, A.D. Generation and investigation of airborne silver nanoparticles with specific size and morphology by homogeneous nucleation, coagulation and sintering. *J Aerosol Sci* 2006;37:452–470.

33. Maynard, A.D., Kuempel, E.D. Airborne nanostructured particles and occupational health. *J Nanopart Res* 2005;7:587–614.

34. Powers, K.W. Research strategies for safety evaluation of nanomaterials part VI. Characterization of nanoscale particles for toxicological evaluation. *Toxicol Sci* 2006;90(2):296–303.

35. Baker, S.E., Cai, W., Lasseter, T.L., Weidkamp, K.P.,Hamers, R.J. Covalently bonded adducts of deoxyribonucleic acid (DNA) oligonucleotides with single-wall carbon nanotubes: synthesis and hybridization. *Nano Letters* 2002;2(12):1413–1417.

36. Dresselhaus, M.S., Dresselhaus, G. Carbon nanotubes: synthesis, structure, properties, and applications. New York, NY: Phaedon Avouris and LINK, Springer; 2001.

37. Williams, K.A., Veenhuizen, P.T.M., de la Torre, B.G., Eritja, R., Dekker, C. Nanotechnology—carbon nanotubes with DNA recognition. *Nature* 2002;420(6917):761.

38. De Villiers, M.M., Lvov, Y. Nanoshells for drug delivery. In: Kumar, C., editor. *Nanomaterials for Medical Diagnosis and Therapy.* Weinheim: Wiley-VCH Verlag GmbH & Co.; 2007. pp. 527–556.

39. Han, S.K., Kim, R.S., Lee, J.H., Tae, G., Cho, S.H., Yuk, S.H. Core-shell nanoparticles for drug delivery and molecular imaging. In: Kumar, C., editor. *Nanomaterials for Medical Diagnosis and Therapy.* Weinheim: Wiley-VCH Verlag GmbH & Co.; 2007. pp. 143–188.

40. Idee, J.M., Port, M., Raynal, I., Schaefer, M., Bonnemain, B., Prigent, P., Robert, P., Robic, C.,Corot, C. Superparamagnetic nanoparticles of iron oxides for magnetic resonance imaging applications. In: Kumar, C., editor. *Nanomaterials for Medical Diagnosis and Therapy.* Weinheim: Wiley-VCH Verlag GmbH & Co.; 2007. pp. 51–84.

41. Rieger, J., Jerome, C., Jerome, R., Auzely-Velty, R. Polymeric nanomaterials—synthesis, functionalization and applications in diagnosis and therapy. In: Kumar, C., editor. *Nanomaterials for Medical Diagnosis and Therapy.* Weinheim: Wiley-VCH Verlag GmbH & Co.; 2007. pp. 342–408.

42. Yamaguchi, Y.,Igarashi, R. NANOEGG technology for drug delivery. In: Kumar, C., editor. *Nanomaterials for Medical Diagnosis and Therapy,* Weinheim: Wiley-VCH Verlag GmbH & Co.; 2007. pp. 310–341.

43. Cullen, R.T., Tran, C.L., Buchanan, D., Davis, J.M., Searl, A., Jones, A.D., Donaldson, K. Inhalation of poorly soluble particles. I. Differences in inflammatory response and clearance during exposure. *Inhal Toxicol* 2000;12(12):1089–1111.

44. Renwick, L.C., Donaldson, K., Clouter, A. Impairment of alveolar macrophage phagocytosis by ultrafine particles. *Toxicol Appl Pharmacol* 2001;172(2):119–127.

45. Tran, C.L., Buchanan, D., Cullen, R.T., Searl, A., Jones, A.D., Donaldson, K. Inhalation of poorly soluble particles. II. Influence of particle surface area on inflammation and clearance. *Inhal Toxicol* 2000;12(12):1113–1126.

46. Sayes, C.M., Wahi, R., Kurian, P.A., Liu, Y., West, J.L., Ausman, K.D., Warheit, D.B., Colvin, V.L. Correlating nanoscale titania structure with toxicity: a cytotoxicity and inflammatory response study with human dermal fibroblasts and human lung epithelial cells. *Toxicol Sci* 2006;92(1):174–185.

47. Warheit, D.B., Brock, W.J., Lee, K.P., Webb, T.R.,Reed, K.L. Comparative pulmonary toxicity inhalation and instillation studies with different TiO_2 particle formulations: impact of surface treatments on particle toxicity. *Toxicol Sci* 2005;88(2):514–524.

48. Warheit, D.B., Webb, T.R., Reed, K.L. Pulmonary toxicity screening studies in male rats with TiO_2 particulates substantially encapsulated with pyrogenically deposited, amorphous silica. *Part Fibre Toxicol* 2006a:3, 3–12.

49. Warheit, D.B., Webb, T.R., Reed, K.L., Frerichs, S., Sayes, C.M. Pulmonary toxicity study in rats with three forms of ultrafine-TiO_2 particles: differential responses related to surface properties. *Toxicology* 2007;230(1):90–104.

50. Warheit, D.B., Webb, T.R., Sayes, C.M., Colvin, V.L.,Reed, K.L. Pulmonary instillation studies with nanoscale TiO_2 rods and dots in rats: Toxicity is not dependent upon particle size and surface area. *Toxicol Sci* 2006a;91(1):227–236.

51. Bermudez, E., Mangum, J.B., Asgharian, B., Wong, B.A., Reverdy, E.E., Janszen, D.B., Hext, P.M., Warheit, D.B.,Everitt, J.I. Long-term pulmonary responses of three laboratory rodent species to subchronic inhalation of pigmentary titanium dioxide particles. *Toxicol Sci* 2006;70(1):86–97.

52. Bermudez, E., Mangum, J.B., Wong, B.A., Asgharian, B., Hext, P.M., Warheit, D.B., Everitt, J.I. Pulmonary responses of mice, rats, and hamsters to subchronic inhalation of ultrafine titanium dioxide particles. *Toxicol Sci* 2004;77(2):347–357.

53. Donaldson, K., Li, X.Y., MacNee, W. Ultrafine (nanometre) particle mediated lung injury. *J Aerosol Sci* 1998;29:553–560.

54. Gilmour, P. Surface free radical activity of PM10 and ultrafine titanium dioxide: a unifying factor in their toxicity? *Ann Occup Hyg* 1997;41:32–38.

55. Oberdorster, G. Significance of particle parameters in the evaluation of exposure/dose–response relationships of inhaled particles. *Inhal Toxicol* 1996;8(Suppl):73–89.

56. Oberdorster, G., Ferin, J., Gelein, R., Soderholm, S.C., Finkelstein, J. Role of the alveolar macrophage in lung injury: studies with ultrafine particles. *Environ Health Perspect* 1992;97:193–199.

57. Oberdorster, G., Gelein, R.M., Ferin, J.,Weiss, B. Association of particulate air pollution and acute mortality: involvement of ultrafine particles? *Inhal Toxicol* 1995;7(1):111–124.

58. NIOSH. Current intelligence bulletin: evaluation of health hazard and recommendations for occupational exposure to titanium dioxide (draft). Atlanta, GA: National Institute for Occupational Safety and Health (NIOSH), Centers for Disease Control & Prevention, Department of Human and Health Services; 2005.

59. Adachi, M., Murata, Y., Takao, J., Jiu, J., Sakamoto, M., Wang, F. Highly efficient dye-sensitized solar cells with a titania thin-film electrode composed of a network structure of

single-crystal-like TiO$_2$ nanowires made by the "oriented attachment" mechanism. *J Am Chem Soc* 2004;126(45):14943-14949.

60. Hext, P.M., Tomenson, J.A.,Thompson, P. Titanium dioxide: inhalation toxicology and epidemiology. *Ann Occup Hyg* 2005;49(6):461–472.

61. Hext, P.M. Current perspectives on particulate induced pulmonary tumours. *Hum Exp Toxicol* 1994;13(10):700–715.

62. ILSI. The relevance of the rat lung response to particle overload for human risk assessment: a workshop consensus report. ILSI Risk Science Institute Workshop Participants. *Inhal Toxicol* 2000;12(1–2):1–17.

63. Ferin, J. Translocation of particles from the pulmonary alveoli into the interstitium. *J Aerosol Med* 1992;5(3):179–187.

64. Ferin, J., Oberdorster, G., Penney D.P. Pulmonary retention of ultrafine and fine particles in rats. *Am J Respir Cell Mol Biol* 1992;6(5):535–542.

65. Oberdorster, G., Ferin, J., Lehnert B.E. Correlation between particle size, *in vivo* particle persistence, and lung injury. *Environ Health Perspect* 1994;102(Suppl 5):173–179.

66. Hu, H., Haddon, R.C., Ni, Y., Montana, V., Parpura, V. Chemically functionalized carbon nanotubes as substrates for neuronal growth. *Nano Lett* 2004;4:507–511.

67. Lovat, V., Pantarotto, D., Lagostena, L., Cacciari, B., Grandolfo, M., Righi, M., Spalluto, G., Prato, M., Ballerini, L. Carbon nanotube substrates boost neuronal electrical signaling. *Nano Lett* 2005;5(6):1107–1110.

68. Mattson, M.P., Haddon, R.C., Rao, A.M. Molecular functionalization of carbon nanotubes and use as substrates for neuronal growth. *J Mol Neurosci* 2000;14(3):175–182.

69. Chen, X., Lee, G.S., Zettl, A., Bertozzi, C.R. Biomimetic engineering of carbon nanotubes by using cell surface mucin mimics. *Angew Chem Int Ed Engl* 2004; 43 (45):6111–6116.

70. Park, E.J., Brasuel, M., Behrend, C., Philbert, M.A., Kopelman, R. Ratiometric optical PEBBLE nanosensors for real-time magnesium ion concentrations inside viable cells. *Anal Chem* 2003a;75(15):3784–3791.

71. Park, K.H., Chhowalla, M., Iqbal, Z., Sesti, F. Single-walled carbon nanotubes are a new class of ion channel blockers. *J Biol Chem* 2003b;278(5a):50212–50216.

72. Bianco, A., Hoebeke, J., Kostarelos, K., Prato, M., Partidos, C.D. Carbon nanotubes: on the road to deliver. *Curr Drug Deliv* 2005a;2(3):253–259.

73. Bianco, A., Kostarelos, K., Partidos, C.D., Prato, M. Biomedical applications of functionalised carbon nanotubes. *Chem Commun (Camb)*2005b;5:571–577.

74. Bianco, A., Kostarelos, K.,Prato, M. Applications of carbon nanotubes in drug delivery. *Curr Opin Chem Biol* 2005c;9(6):674–679.

75. Bianco, A., Prato M. Can carbon nanotubes be considered useful tools for biological applications? *Adv Mater* 2003;15:1765–1768.

76. Gao, B., Bower, C., Lorentzen, J.D., Fleming, L., Kleinhammes, A., Tang, X.P., McNeil, L.E., Wu, Y., Zhou, O. Enhanced saturation lithium composition in ball-milled single-walled carbon nanotubes. *Chem Phys Lett* 2000;327(1–2):69–75.

77. Walters, D.A., Casavant, M.J., Qin, X.C., Huffman, C.B., Boul, P.J., Ericson, L.M., Haroz, E.H., O'Connell, M.J., Smith, K., Colbert, D.T., Smalley, R.E. In-plane-aligned membranes of carbon nanotubes. *Chem Phys Lett* 2001;338(1):14–20.

78. Baughman, R.H. Putting a new spin on carbon nanotubes. *Science* 2000;290(5495):1310–1311.

79. Lin, Y., Taylor, S., Li, H., Shiral Fernando, K.A., Qu, L., Wang, W., Gu, L., Zhou, B., Sun, Y. P. Advances toward bioapplications of carbon nanotubes. *J Mater Chem* 2004;14:527–541.

80. Kostarelos, K., Lacerda, L., Partidos, C.D., Prato, M.,Bianco, A. Carbon nanotubeme-diated delivery of peptides and genes to cells: translating nanobiotechnology to thera-peutics. *J Drug Deliv Sci Technol* 2005;15(1):41–47.

81. Katz, E,Willner, I. Biomolecule-functionalized carbon nanotubes: applications in nano-bioelectronics. *Chem Phys Chem* 2004;5(8):1084–1104.

82. Fortina, P., Kricka, L.J., Surrey, S.,Grodzinski, P. Nanobiotechnology: the promise and reality of new approaches to molecular recognition. *Trends Biotechnol* 2005;23(4): 168–173.

83. Maynard, A.D., Baron, P.A., Foley, M., Shvedova, A.A., Kisin, E.R., Castranova, V. Exposure to carbon nanotube material: aerosol release during the handling of unrefined single-walled carbon nanotube material. *J Toxicol Environ Health A* 2004;67 (1): 87–107.

84. Huczko, A., Lange, H. Carbon nanotubes: experimental evidence for a null risk of skin irritation and allergy. *Fullerene Sci Tech* 2001;9(2):247–250.

85. Huczko, A., Lange, H., Calko, E. Fullerenes: experimental evidence for a null risk of skin irritation and allergy. *Fullerene Sci Tech* 1999;7(5):935–939.

86. Huczko, A., Lange, H., Calko, E., Grubek-Jaworska, H.,Droszcz, P. Physiological testing of carbon nanotubes: are they asbestos-like? *Fullerene Sci Tech* 2001;9(2):251–254.

87. Lam, C.W., James, J.T., McCluskey, R., Hunter, R.L. Pulmonary toxicity of single-wall carbon nanotubes in mice 7 and 90 days after intratracheal instillation. *Toxicol Sci* 2004;77(1):126–134.

88. Muller, J., Huaux, F., Moreau, N., Misson, P., Heilier, J.F., Delos, M., Arras, M., Fonseca, A., Nagy, J.B., Lison, D. Respiratory toxicity of multi-wall carbon nanotubes. *Toxicol Appl Pharmacol* 2005;207(3):221–231.

89. Shvedova, A.A., Castranova, V., Kisin, E.R., Schwegler-Berry, D., Murray, A.R., Gandelsman, V.Z., Maynard, A., Baron, P. Exposure to carbon nanotube material: assessment of nanotube cytotoxicity using human keratinocyte cells. *J Toxicol Environ Health A* 2003;66(20):1909–1926.

90. Shvedova, A.A., Kisin, E.R., Mercer, R., Murray, A.R., Johnson, V.J., Potapovich, A.I., Tyurina, Y.Y., Gorelik, O., Arepalli, S., Schwegler-Berry, D., Hubbs, A.F., Antonini, J., Evans, D.E., Ku, B.K., Ramsey, D., Maynard, A., Kagan, V.E., Castranova, V., Baron, P. Unusual inflammatory and fibrogenic pulmonary responses to single-walled carbon nanotubes in mice. *Am J Physiol Lung Cell Mol Physiol* 2005;289(5):L698–L708.

91. Warheit, D.B., Laurence, B.R., Reed, K.L., Roach, D.H., Reynolds, G.A.,Webb, T.R. Comparative pulmonary toxicity assessment of single-wall carbon nanotubes in rats. *Toxicol Sci* 2004;77(1):117–125.

92. Mitchell, L.A., Gao, J., Vander Wal, R., Gigliotti, A., Burchiel, S.W., McDonald, J.D. Pulmonary and systemic immune response to inhaled multiwalled carbon nanotubes. *Tox Sciences* 2007;100:203–214.

93. Dumortier, H., Lacotte, S., Pastorin, G., Marega, R., Wu, W., Bonifazi, D., Briand, J.P., Prato, M., Muller, S.,Bianco, A. Functionalized carbon nanotubes are non-cytotoxic and preserve the functionality of primary immune cells. *Nano Lett* 2006;6(7):1522–1528.

94. Magrez, A., Kasas, S., Salicio, V., Pasquier, N., Seo, J.W., Celio, M., Catsicas, S., Schwaller, B., Forro, L. Cellular toxicity of carbon-based nanomaterials. *Nano Lett* 2006;6(6):1121–1125.

95. Sayes, C.M., Marchione, A.A., Reed, K.L.,Warheit, D.B. Comparative pulmonarytoxicity assessments of C_{60} water suspensions in rats: few differences in fullerenetoxicity *in vivo* in contrast to *in vitro* profiles. *Nano Lett* 2007;7(8):2399–23406.

96. Tian, F., Cui, D., Schwarz, H., Estrada, G.G., Kobayashi, H. Cytotoxicity of single-wall carbon nanotubes on human fibroblasts. *Toxicol In Vitro* 2006;20(7):1202–1212.

97. Bottini, M., Bruckner, S., Nika, K., Bottini, N., Bellucci, S., Magrini, A., Bergamaschi, A., Mustelin, T. Multi-walled carbon nanotubes induce T lymphocyte apoptosis. *Toxicol Lett* 2006;160(2):121–126.

98. Cui, D. Effect of single wall carbon nanotubes on human HEK293 cells. *Toxicol Lett* 2005;155:73–85.

99. Ding, L., Stilwell, J., Zhang, T., Elboudwarej, O., Jiang, H., Selegue, J.P., Cooke, P.A., Gray, J.W.,Chen, F.F. Molecular characterization of the cytotoxic mechanism of multi-wall carbon nanotubes and nano-onions on human skin fibroblast. *Nano Lett* 2005;5 (12):2448–2464.

100. Jia, G., Wang, H., Yan, L., Wang, X., Pei, R., Yan, T., Zhao, Y., Guo, X. Cytotoxicity of carbon nanomaterials: Single-wall nanotube, multi-wall nanotube, and fullerene. *Environ Sci Technol* 2005;39(5):1378–1383.

101. Manna, S.K., Sarkar, S., Barr, J., Wise, K., Barrera, E.V., Jejelowo, O., Rice-Ficht, A.C., Ramesh, G.T. Single-walled carbon nanotube induces oxidative stress and activates nuclear transcription factor-kappaB in human keratinocytes. *Nano Lett* 2005;5(9): 1676–1684.

102. Monteiro-Riviere, N.A., Nemanich, R.J., Inman, A.O., Wang, Y.Y., Riviere, J.E. Multi-walled carbon nanotube interactions with human epidermal keratinocytes. *Toxicol Lett* 2005;155(3):377–384.

103. Sayes, C.M., Fortner, J.D., Guo, W., Lyon, D., Boyd, A.M., Ausman, K.D., Tao, Y.J., Sitharaman, B., Wilson, L.J., Hughes, J.B., West, J., Colvin, V.L. The differential cytotoxicity of water-soluble fullerenes. *Nano Lett* 2004;4(10):1881–1887.

104. Shvedova, A.A., Kisin, E.R., Murray, A.R., Gorelik, O., Arepalli, S., Castranova, V., Young, S.H., Gao, F., Tyurina, Y.Y., Oury, T.D., Kagan, V.E. Vitamin E deficiency enhances pulmonary inflammatory response and oxidative stress induced by single-walled carbon nanotubes in C57BL/6 mice. *Toxicol Appl Pharmacol* 2007;221(3):339–348.

105. Li, Z., Hulderman, T., Salmen, R., Chapman, R., Leonard, S.S., Young, S.H., Shvedova, S., Luster, M.I., Simeonova, P.P. Cardiovascular effects of pulmonary exposure to single-wall carbon nanotubes. *Environ Health Perspect* 2007;115:377–382.

106. Li, Z.J. Pulmonary exposure to carbon nanotubes induces vascular toxicity. *Toxicologist* 2005;84(1):213.

107. Li, Z.J. Relationship between pulmonary exposure to multiple doses of single wall carbon nanotubes and atherosclerosis in ApoE−/− mouse model. *Toxicologist* 2006; 90 (1):318.

108. Bell, D.C., Wu, Y., Barrelet, C.J., Gradecak, S., Xiang, J., Timko, B.P., Lieber, C.M. Imaging and analysis of nanowires. *Microsc Res Tech* 2004;64(5–6):373–389.

109. Patolsky, F. Nanowire nanosensors. *Mater Today* 2005;8(4):20–28.

110. Patolsky, F., Zheng, G., Lieber, C.M. Nanowire sensors for medicine and the life sciences. *Nanomed* 2006;1(1):51–65.

111. Mor, G., Shankar, K., Paulose, M., Varghese, O.,Grimes, C. Use of highly-ordered TiO_2 nanotube arrays in dye-sensitized solar cells. *Nano Lett* 2006;6:215–218.

112. Huang, Y. Integrated nanoscale electronics and optoelectronics: exploring nanoscale science and technology through semiconductor nanowires. *Pure Appl Chem* 2004; 76 (12):2051–2068.

113. Carter, J.D., Qu, Y.R.P., Hoang, L., Masiel, D.J., Guo, T. Silicon-based nanowires from silicon wafers catalyzed by cobalt nanoparticles in hydrogen environment. *Chem Commun* 2005;0:2274–2276.

114. Qu, Y., Porter, R., Shan, F., Carter, J.,Guo, T. Synthesis of tubular gold and silvernanoshells using silica nanowire core templates. *Langmuir* 2006;22:6367–6374.

115. Ekimov, A.I.,Onushchenko, A.A. Quantum size effect in the optical spectra of semiconductor microcrystals. *Sov Phys Semicond* 1982;16:775–777.

116. Agrawal, A., Xing, Y., Gao, X., Nie S. Quantum dots. In: Vo-Dinh, T., editor. *Nanotechnology in Biology and Medicine: Methods, Devices, and Applications*. Boca Rotan, FL: CRC Press, Taylor & Francis Group; 2007.

117. Akerman, M.E., Chan, W.C., Laakkonen, P., Bhatia, S.N., Ruoslahti E. Nanocrystal targeting *in vivo*. *Proc Natl Acad Sci U S A* 2002;99(20):12617–12621.

118. Chen, Y.F., Ji, T.H., Rosenzweig, Z. Synthesis of glyconanospheres containing luminescent CdSe–ZnS quantum dots. *Nano Lett* 2003;3:581–584.

119. Goldman, E.R., Balighian, E.D., Mattoussi, H., Kuno, M.K., Mauro, J.M., Tran, P.T., Anderson, G.P. Avidin: a natural bridge for quantum dot-antibody conjugates. *J Am Chem Soc* 2002;124:6378–6382.

120. Hanaki, K., Momo, A., Oku, T., Komoto, A., Maenosono, S., Yamaguchi, Y., Yamamoto, K. Semiconductor quantum dot/albumin complex is a long-life and highly photostable endosome marker. *Biochem Biophys Res Commun* 2003;302: 496–501.

121. Hardman R. A toxicologic review of quantum dots: toxicity depends on physicochemical and environmental factors. *Environ Health Perspect* 2006;114(2):165–172.

122. Lingerfelt, B.M., Mattoussi, H., Goldman, E.R., Mauro, J.M., Anderson, G.P. Preparation of quantum dot-biotin conjugates and their use in immunochromatography assays. *Anal Chem* 2003;75:4043–4048.

123. Mahtab, R., Rogers, J.P., Murphy, C.J. Protein-sized quantum-dot luminescence can distinguish between straight, bent, and kinked oligonucleotides. *J Am Chem Soc* 1995;117:9099–9100.

124. Osaki, F., Kanamori, T., Sando, S., Sera, T., Aoyama, Y. A quantum dot conjugated sugar ball and its cellular uptake on the size effects of endocytosis in the subviral region. *J Am Chem Soc* 2004;126:6520–6521.

125. Pellegrino, T., Manna, L., Kudera, S., Liedl, T., Koktysh, D., Rogach, A.L., Keller, S., Radler, J., Natile, G.,Parak, W.J. Hydrophobic nanocrystals coated with amphiphilic polymer shell: a general route to water soluble nanocrystals. *Nano Lett* 2004;4: 703–707.

126. Michalet, X., Pinaud, F.F., Bentolila, L.A., Tsay, J.M., Doose, S., Li, J.J., Sundaresan, G., Wu, A.M., Gambhir, S.S.,Weiss, S. Quantum dots for live cells, *in vivo* imaging, and diagnostics. *Science* 2005;307(5709):538–544.

127. Beaurepaire, E., Buissette, V., Sauviat, M.P., Giaume, D., Lahlil, K., Mercuri, A., Casanova, D., Huignard, A., Martin, J.L., Gacoin, T., Boilot, J.-P., Alexandrou, A. Functionalized fluorescent oxide nanoparticles: artificial toxins for sodium channel targeting and imaging at the single-molecule level. *Nano Lett* 2004;4(11):2079–2083.

128. Colton, H.M., Falls, J.G., Ni, H., Kwanyuen, P., Creech, D., McNeil, E., Casey, W.M., Hamilton, G., Cariello, N.F. Visualization and quantitation of peroxisomes using fluorescent nanocrystals: treatment of rats and monkeys with fibrates and detection in the liver. *Toxicol Sci* 2004;80(1):183–192.

129. Dubertret, B., Skourides, P., Norris, D.J., Noireaux, V., Brivanlou, A.H.,Libchaber, A. *In vivo* imaging of quantum dots encapsulated in phospholipid micelles. *Science* 2002;298:1759–1762.

130. Gao, X., Cui, Y., Levenson, R.M., Chung, L.W., Nie, S. *In vivo* cancer targeting and imaging with semiconductor quantum dots. *Nat Biotechnol* 2004;22(8):969–976.

131. Lidke, D.S., Nagy, P., Heintzmann, R., Arndt-Jovin, D.J., Post, J.N., Grecco, H.E., Jares-Erijman, E.A., Jovin, T.M. Quantum dot ligands provide new insights into erbB/HER receptormediated signal transduction. *Nat Biotechnol* 2004;22(2):198–203.

132. Wu, X., Liu, H., Liu, J., Haley, K.N., Treadway, J.A., Larson, J.P., Ge, N., Peale, F., Bruchez, M.P. Immunofluorescent labeling of cancer marker Her2 and other cellular targets with semiconductor quantum dots. *Nat Biotechnol* 2003;21(1):41–46.

133. Hopkins, L., Pinkerton, K.E. Nose-to-brain translocation of inhaled nanoparticles. *Toxicologist* 2007;96(1):234.

134. Ryman-Rasmussen, J.P., Riviere, J.E.,Monteiro-Riviere, N.A. Penetration of intact skin by quantum dots with diverse physicochemical properties. *Toxicol Sci* 2006;91(1):159–165.

135. Johnson, L.N., Cashman, S.M., Kumar-Singh, R. Cell-penetrating peptide for enhanced delivery of nucleic acids and drugs to ocular tissues including retina and cornea. *Mol Ther* 2007;16:107–114.

136. Kreyling, W.G., Semmler, M., Erbe, F., Mayer, P., Takenaka, S., Schulz, H., Oberdorster, G., Ziesenis A. Translocation of ultrafine insoluble iridium particles from lungepithelium to extrapulmonary organs is size dependent but very low. *J Toxicol Environ Health A* 2002;65(20):1513–1530.

137. Oberdorster, G., Sharp, Z., Atudorei, V., Elder, A., Gelein, R., Kreyling, W.,Cox, C. Translocation of inhaled ultrafine particles to the brain. *Inhal Toxicol* 2004;16(6–7):437–445.

138. Derfus, A.M., Chan, W.C.W., Bhatia, S.N. Probing the cytotoxicity of semiconductor quantum dots. *Nano Letters* 2004;4(1):11–18.

139. Tsay, J.M., Michalet, X. New light on quantum dot cytotoxicity. *Chem Biol* 2005; 12 (11):1159–1161.

140. Zhang, T., Stilwell, J.L., Gerion, D., Ding, L., Elboudwarej, O., Cooke, P.A., Gray, J.W., Alivisatos, A.P., Chen, F.F. Cellular effect of high doses of silica-coated quantum dot profiled with high throughput gene expression analysis and high content cellomics measurements. *Nano Lett* 2006;6(4):800–808.

141. Ballou, B., Lagerholm, B.C., Ernst, L.A., Bruchez, M.P.,Waggoner, A.S. Noninvasive imaging of quantum dots in mice. *Bioconjug Chem* 2004;15(1):79–86.

142. Hoshino, A. Physicochemical properties and cellular toxicity of nanocrystal quantum dots depend on their surface modification. *Nano Lett* 2004;4(11):2163–2169.

143. Brown, J.S., Zeman, K.L., Bennett, W.D. Ultrafine particle deposition and clearance in the healthy and obstructed lung. *Am J Respir Crit Care Med* 2002;166(9):1240–1247.

144. Kreyling, W.G., Semmler-Behnke, M., Moller, W. Ultrafine particle-lung interactions: does size matter? *J Aerosol Med* 2006;19(1):74–83.

145. Semmler, M., Seitz, J., Erbe, F., Mayer, P., Heyder, J., Oberdorster, G., Kreyling, W.G. Long-term clearance kinetics of inhaled ultrafine insoluble iridium particles from the rat lung, including transient translocation into secondary organs. *Inhal Toxicol* 2004; 16 (6–7):453–459.

146. Wiebert, P., Sanchez-Crespo, A., Seitz, J., Falk, R., Philipson, K., Kreyling, W.G., Moller, W., Sommerer, K., Larsson, S., Svartengren, M. Negligible clearance of ultrafine particles retained in healthy and affected human lungs. *Eur Respir J* 2006;28(2):286–290.

147. Green, G.M. Alveolobronchiolar transport mechanisms. *Arch Intern Med* 1973;131 (1):109–114.

148. Takenaka, S., Karg, E., Kreyling, W.G., Lentner, B., Moller, W., Behnke-Semmler, M., Jennen, L., Walch, A., Michalke, B., Schramel, P., Heyder J., Schulz, H. Distribution pattern of inhaled ultrafine gold particles in the rat lung. *Inhal Toxicol* 2006;18(10): 733–740.

149. Rinderknecht, A., Elder, A., Prud'homme, R., Gindy, M., Oberdörster, G. Are poly (ethylene glycol)-functionalized nanoparticles biocompatible? *Toxicologist* 2007; 96(1):232.

150. Inoue, K., Takano, H., Yanagisawa, R., Sakurai, M., Ichinose, T., Sadakane, K., Yoshikawa, T. Effects of nano particles on antigen-related airway inflammation in mice. *Respir Res* 2005;6:106–118.

151. Yamamoto, A., Honma, R., Sumita, M., Hanawa, T. Cytotoxicity evaluation of ceramic particles of different sizes and shapes. *J Biomed Mater Res A* 2003;68(2):244–256.

152. Elder, A., Gelein, R., Silva, V., Feikert, T., Opanashuk, L., Carter, J., Potter, R., Maynard, A., Ito, Y., Finkelstein, J., Oberdorster, G. Translocation of inhaled ultrafine manganese oxide particles to the central nervous system. *Environ Health Perspect* 2006;114(8):1172–1178.

153. Elder, A., Oberdorster, G. Translocation and effects of ultrafine particles outside of the lung. *Clin Occup Environ Med* 2006;5(4):785–796.

154. Geldenhuys, W.J., Lockman, P.R., McAfee, J.H., Fitzpatrick, K.T., Van der Schyf, C.J., Allen, D.D. Molecular modeling studies on the active binding site of the blood–brain barrier choline transporter. *Bioorg Med Chem Lett* 2004;14(12):3085–3092.

155. Henriksson, J., Tallkvist, J., Tjalve, H. Transport of manganese via the olfactory pathway in rats: dosage dependency of the uptake and subcellular distribution of the metal in the olfactory epithelium and the brain. *Toxicol Appl Pharmacol* 1999;156 (2):119–128.

156. Kreuter, J. Nanoparticulate systems for brain delivery of drugs. *Adv Drug Deliv Rev* 2001;47(1):65–81.

157. Lockman, P.R., Koziara, J.M., Mumper, R.J., Allen, D.D. Nanoparticle surface charges alter blood–brain barrier integrity and permeability. *J Drug Target* 2004a;12(9–10): 635–641.

158. Lockman, P.R., McAfee, J.H., Geldenhuys, W.J., Allen, D.D. Cation transport specificity at the blood–brain barrier. *Neurochem Res* 2004b;29(12):2245–2250.

159. Persson, E., Henriksson, J., Tallkvist, J., Rouleau, C., Tjalve, H. Transport and subcellular distribution of intranasally administered zinc in the olfactory system of rats and pikes. *Toxicology* 2003;191(2–3):97–108.

160. Nemmar, A. Passage of inhaled particles into the blood circulation in humans. *Circulation* 2002;105:411–414.

161. Kato, T., Yashiro, T., Murata, Y., Herbert, D.C., Oshikawa, K., Bando, M., Ohno, S., Sugiyama, Y. Evidence that exogenous substances can be phagocytized by alveolar epithelial cells and transported into blood capillaries. *Cell Tissue Res* 2003;311:47–51.

162. Barbeau, A. Manganese and extrapyramidal disorders. *Neurotoxicology* 1984;5:13–36.

163. Donaldson, J. The physiopathologic significance of manganese in brain: its relation to schizophrenia and neurodegenerative disorders. *Neurotoxicology* 1987;8(3):451–462.

164. Fechter, L.D., Johnson, D.L.,Lynch, R.A. The relationship of particle size to olfactory nerve uptake of a non-soluble form of manganese into brain. *Neurotoxicology* 2002; 23(2):177–183.

165. Tjalve, H., Henriksson, J., Tallkvist, J., Larsson, B.S., Lindquist, N.G. Uptake of manganese and cadmium from the nasal mucosa into the central nervous system via olfactory pathways in rats. *Pharmacol Toxicol* 1996;79(6):347–356.

166. Tjalve, H., Mejare, C., Borg-Neczak, K. Uptake and transport of manganese in primary and secondary olfactory neurones in pike. *Pharmacol Toxicol* 1995;77(1):23–31.

167. Hunter, D.D., Undem, B.J. Identification and Substance P content of vagal afferent neurons innervating the epithelium of the guinea pig trachea. *Am J Respir Crit Care Med* 1999;159(6):1943–1948.

168. Bodian, D. Experimental studies on intraneural spread of poliomyelitis virus. *Bull Johns Hopkins Hosp* 1941a;69:248–267.

169. Bodian, D. The rate of progression of poliomyelitis virus in nerves. *Bull Johns Hopkins Hosp* 1941b, 69:79–85.

170. DeLorenzo, A. The old factory neuron and the blood–brain barrier. In: Wolstenholme G., Knight J., editors. *Taste and Smell in Vertebrates*, London: Churchill; 1970. pp. 151–176.

171. Hunter, D.D.,Dey, R.D. Identification and neuropeptide content of trigeminal neurons innervating the rat nasal epithelium. *Neuroscience* 1998;83(2):591–599.

172. Frampton, M.W., Stewart, J.C., Oberdorster, G., Morrow, P.E., Chalupa, D., Pietropaoli, A.P., Frasier, L.M., Speers, D.M., Cox, C.,Huang, L.S., Utell, M.J. Inhalation of ultrafine particles alters blood leukocyte expression of adhesion molecules in humans. *Environ Health Perspect* 2006;114(1):51–58.

173. Frampton, M.W., Utell, M.J., Zareba, W., Oberdorster, G., Cox, C., Huang, L.S., Morrow, P.E., Lee, F.E., Chalupa, D.,Frasier, L.M., Speer, D.M., Stewart, J. Effects of exposure to ultrafine carbon particles in healthy subjects and subjects with asthma. *Res Rep Health Eff Inst.* 2004;(126):1–47.

174. Chalupa, D.C., Morrow, P.E., Oberdorster, G., Utell, M.J.,Frampton, M.W. Ultrafine particle deposition in subjects with asthma. *Environ Health Perspect* 2004;112(8):879–882.

175. Daigle, C.C., Chalupa, D.C., Gibb, F.R., Morrow, P.E., Oberdorster, G., Utell, M.J., Frampton, M.W. Ultrafine particle deposition in humans during rest and exercise. *Inhal Toxicol* 2003;15(6):539–552.

176. Mills, N.L. Do inhaled carbon nanoparticles translocate directly into the circulation in humans? *Am J Respir Crit Care Med* 2005;173:426–431.

177. Moller, W., Seitz, J., Sommerer, K., Heyder, J.,Kreyling, W.G. Deposition and clearance of radiolabelled ultrafine carbon particles in the human airways after bolus inhalation. *J Aerosol Med* 2003;16(2):201.

178. Pietropaoli, A.P., Frampton, M.W., Hyde, R.W., Morrow, P.E., Oberdorster, G., Cox, C., Speers, D.M., Frasier, L.M., Chalupa, D.C., Huang, L.S., Utell, M.J. Pulmonary function, diffusing capacity, and inflammation in healthy Utell, M.J. subjects exposed to ultrafine particles. *Inhal Toxicol* 2004;16(Suppl 1):59–72.

179. Donaldson, K., Stone, V., Clouter, A., Renwick, L.,MacNee, W. Ultrafine particles. *Occup Environ Med* 2001;58(3):211–216.

180. Hebert, P. Nanotech venture capital to exceed $650 million in 2006. New York: Lux Research, Inc.; 2006. http://www.luxresearchinc.com/press/RELEASE_VCreport.pdf.

181. Magrez, A., Kasas, S., Salicio, V., Pasquier, N., Seo, J.W., Celio, M., Catsicas, S., Schwaller, B., Forro, L. Cellular toxicity of carbon-based nanomaterials. *Nano Lett* 2006;6(6):1121–1125.

182. NNI. Funding opportunities. National Nanotechnology Initiative, Washington D.C., 2007. http://www.nano.gov/html/funding/home_funding.html.

183. Huang, Z. Longitudinal patent analysis for nanoscale science and engineering: country, institution and technology field. *J Nanoparticle Res* 2003;5:333–363.

CHAPTER 16

Neurotoxicity of Manufactured Nanoparticles

JAIME M. HATCHER[1,2], DEAN P. JONES[3], GARY W. MILLER[1,2], and KURT D. PENNELL[1,4]

[1]Center for Neurodegenerative Disease, Department of Neurology, Emory University School of Medicine, 615 Michael Street, Atlanta, GA 30322, USA

[2]Department of Environmental and Occupational Health, Rollins School of Public Health, Emory University, Atlanta, GA 30332, USA

[3]Department of Medicine, Emory University School of Medicine, 615 Michael Street, Atlanta, GA 30322, USA

[4]School of Civil and Environmental Engineering, Georgia Institute of Technology, 311 Ferst Drive, Atlanta, GA 30332-0512, USA

16.1 INTRODUCTION

The neurotoxicity of manufactured nanoparticles is a relatively new field and few studies have examined the potential neurologic damage that may result from nano-particle exposure. The central nervous system (CNS) is susceptible to many insults, and damage can lead to both neurological and psychiatric impairment. Because of the limited amount of information available, it is necessary to draw from other exposure paradigms, such as inhalational exposures, to anticipate or reasonably estimate the potential neurotoxicity of manufactured nanoparticles.

Engineered nanoscale materials or "manufactured nanomaterials" are defined as synthesized materials with at least one dimension in the range of 1–100 nm, which possess unique properties based upon their size (1). It has been estimated that there are approximately 300 consumer products on the market that contain nanoscale components, including metal oxides (e.g., manganese and titanium dioxides) and carbon (C_{60})-based structures (e.g., single-walled nanotubes and fullerenes), and that over 600 nanotechnology-based products such as films, coatings, and filters are currently used in electronics production, drug delivery, and research instrumentation (2, 3). In addition to the dramatic increases in manufactured nanomaterial production, natural

Nanoscience and Nanotechnology Edited by Vicki H. Grassian
Copyright © 2008 John Wiley & Sons, Inc.

airborne or "ambient" nanoscale particles can be generated from forest fires, volcanic eruptions, and dust storms. Anthropogenic sources of airborne nanoparticles, which increased dramatically after the industrial revolution in the early 1800s, include internal combustion engines, coal-fired power plants, incinerators, prescribed burning, and land development. Despite the abundance of nanoparticles in the environment, only limited data are available regarding the environmental fate, human uptake and translocation, and toxicity of manufactured and natural nanoparticles (2, 4–6). This lack of information is reflected in material data safety sheets (MSDSs), which employ a classification system based on chemical name and formula, but typically do not report potential or known differences in toxicity as a function of particle size or specific surface area.

The very properties that render nanomaterials superior to conventional materials, including small size, large surface area, and chemical reactivity, make the assessment of nanoparticle toxicity particularly complex. For example, the same base nanomaterial (e.g., C_{60} fullerene) with different physicochemical properties or surface functional groups can elicit conflicting toxicological responses in organisms. In one of the first manufactured nanomaterial toxicity studies, Oberdörster (7) reported lipid peroxidation and depletion of glutathione (GSH), a free radical scavenger, in the brain of juvenile largemouth bass after a 48 h exposure to a 0.5 mg/L aqueous solution of C_{60} fullerene. However, C-3 and C-5 carboxylic fullerene derivatives have been shown to function as antioxidants against iron-induced oxidative stress, exhibiting protective responses that were similar to those observed with vitamin E (as Trolox) or GSH treatments (8). A second, and often underappreciated factor, is the potential for nanoparticles to aggregate or agglomerate, particularly when they are delivered in aqueous solutions containing electrolytes (9–11). Thus, particles that were initially nanoscale in size may aggregate to form micron-size particles when used in experimental systems (e.g., cell culture) (12, 13), which can dramatically alter toxicologic responses. The strong dependence of toxicological response on the particle size and surface area of nanoparticles (6) highlights the need for careful *in situ* quantification of the mean particle size and distribution to ensure that the delivered particles are in fact nanoscale. In addition, the solutions used to prepare nanoparticle suspensions may contain additional compounds, such as surfactants and solvents, that can elicit toxicological responses (12, 14).

A growing body of literature has shown that the inhalation of ultrafine particles and manufactured nanomaterials can induce inflammation of the upper and lower respiratory tract, which depends on particle size, surface area, and chemical composition (2, 15, 16). Furthermore, the integrity of both the air–blood barrier (ABB) and blood–brain barrier (BBB) may be compromised as a result of exposure to nanoparticles or other toxic agents. For example, chronic exposure of canines to air pollution-derived particulate matter (PM) in Mexico City resulted in persistent pulmonary inflammation, deterioration of olfactory and respiratory barriers, and brain pathology associated with neurodegenerative disease (17, 18). Recent studies demonstrate that exposure to nanoscale diesel exhaust particles (DEP) and manufactured nanomaterials can induce generalized oxidative stress (19), microglial activation, and selective toxicity toward dopaminergic neurons (20). These findings provide some of the first evidence of a direct link between nanoparticle exposure and neurologic damage. In this chapter, we explore the potential neurotoxicity of nanoparticles, with particular emphasis on translocation to the CNS and subsequent neuronal responses including oxidative stress, inflammation, and other factors that are associated with neurodegenerative disease. Much of the early work in the field of nanotoxicology focused on exposure and subsequent toxicological responses to airborne particulate matter, and therefore, we consider relevant studies involving both ambient and manufactured nanoparticles. Although still a relatively novel area of research, many of the toxicity studies conducted to date indicate that increased oxidative stress and inflammation are frequent outcomes of nanoparticle exposure. The potential implications of these immune responses are discussed within the context of neurodegenerative diseases.

16.1.1 Nanoparticle Inhalation and Pulmonary Response

Much of the current information on nanoparticle uptake and translocation is based on ultrafine ($<$0.1 μm diameter) particulate matter (PM0.1) exposure studies (21, 22). These studies provide a sound foundation from which to gain useful insights into potential mechanisms of nanoparticle uptake, deposition, and translocation. Human uptake of nanoscale particles can occur via several pathways including inhalation, ingestion, dermal uptake, and direct injection. Of these pathways, the most well characterized, and arguably the most relevant for nanoparticle uptake, is inhalation.

Respiratory deposition of micron-size particles is governed by inertial impaction in the upper airways, with gravitational settling and interception controlling micron-size particle deposition in smaller conducting airways and gas-exchange regions of the lung (23). Owing to their smaller size, diffusion dominates nanoparticle deposition in the respiratory tract. Numerical simulations of nanoparticle deposition, based on a bifurcation airway model and advection-diffusion mass transport, indicate that the deposition distribution is relatively uniform, with local enhancement factors ranging from 2 to 11 versus 40 to 2400 for micronsize particles (24). Similarly, simulated deposition of 5-nm diameter particles in the human respiratory tract is relatively uniform, with the inhaled mass distributed evenly among the nasopharyngeal, tracheobronchial, and alveolar regions of the respiratory system (25).

The results of several studies indicate that nanosize particles elicit far greater pulmonary toxicity (e.g., inflammatory response) than micron-size particles of the same composition. When rat lung tissue was exposed to varying sizes of titanium dioxide (TiO_2) and carbon black particles, a 10-fold increase in the inflammatory response was observed for 64-nm diameter particles compared to 535-nm diameter particles (26). Oberdörster (27) found that intratracheal instillation of 20-nm diameter TiO_2 particles in rats and mice resulted in a substantially larger pulmonary neutrophil response relative to 250-nm diameter TiO_2 particles when expressed on a mass basis. However, when the nanoparticle dose was expressed on a surface area basis, the dose–response curves were nearly identical, demonstrating the importance of surface area in nanoparticle toxicity. Warheit et al. (16) reported that rats exposed to single-walled carbon nanotubes (SWCNTs) via intratracheal instillation exhibited a non dose-dependent response characterized by multifocal granulomas, whereas rats exposed to a positive control (quartz particles) exhibited a dose-dependent inflammatory response. The highest SWCNT treatment (5 mg/kg) resulted in 15% mortality within 24 h, which was attributed to suffocation resulting from mechanical blockage of the upper airways by agglomerated nanotubes. These findings can be explained by the fact that SWCNTs, due to their strong electrostatic properties, readily aggregate to form micron-size or larger particles that elicit a weaker immune response but have the capacity to obstruct air passageways. In summary, the enhanced pulmonary toxicity of nanoparticles relative to micron-sized particles or aggregated nanoparticles can be attributed to several factors including, greater available surface area per unit mass (i.e., specific surface area), prolonged retention time and lower clearance rates, and potential for deposition in the alveolar region where surfactant films may enhance particle uptake (5, 23).

16.1.2   Nanoparticle Translocation to the Central Nervous System

Following deposition in the lung, one of the most common means of particle clearance (disposition) is macrophage phagocytosis (i.e., engulfment and digestion by immune cells). However, the ability of macrophages to sequester particles has been shown to strongly depend on particle size. For example, less than 25% of several types of 20- and 80-nm diameter nanoparticles were retained by alveolar macrophages compared to 80% retention for 500-nm diameter particles (6). The remaining nanoparticle mass was recovered from lavaged lung tissue, indicating that approximately 80% of the 20- and 80-nm diameter particles were present in epithelial cells or interstitium. Upon reaching interstitial sites, uptake of nanoparticles into the blood or lymphatic pathways can occur (28). Translocation of nanoparticles from deposition sites in the lung to systemic circulation was demonstrated by Kreyling et al. (29), who ventilated rats with airborne insoluble ^{192}iridium particles (15 and 80 nm diameter) for 1 h and measured particle distribution and excretion up to 7 days after exposure (Fig. 16.1). Although most of the particles were cleared via the gastrointestinal tract and feces, a small fraction of the particles were translocated by systemic circulation to extrapulmonary organs, including the liver, spleen, kidneys, heart, and brain. Shimada et al. (30) investigated nanoparticle transport from the lung into the systemic system, and concluded that ultrafine carbon black particles pass the air–blood barrier through gaps between alveolar epithelial cells. In general, nanomaterials with effective diameters of less than approximately 100 nm have a much greater tendency to exhibit alveolar–capillary translocation and accumulation in the CNS (6).

In two related studies of ^{13}C-labeled graphite particle (36 nm diameter) uptake and translocation, rats received two whole-body inhalation exposures of 160 μg/m^3 for 6 h (21, 31). Effective translocation of ^{13}C-labeled graphite nanoparticles from the lung to the liver was demonstrated at 18 and 24 h after exposure (21). In the second study, significant and persistent increases of ^{13}C levels in the olfactory bulb (0.35 μg/g) were observed 1 day after exposure, which increased to a concentration

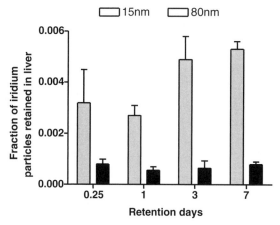

FIGURE 16.1 Mass fraction of 15- and 80-nm insoluble iridium (^{192}Ir) particles taken up and translocated to the liver of rats up to 7 days after inhalation exposure (adapted from (29), reproduced with permission from *J Toxicol Environ Health A*, Copyright 2002, Taylor & Francis).

of 0.43 μg/g 7 days after exposure. Significant enhancements in the concentration of ^{13}C in the cerebrum and cerebellum were also detected 1 day after exposure and increased thereafter, possibly due to selective transport across the BBB in different regions of the brain (31).

The observed accumulation of nanoparticles in the olfactory bulb and detection in the brain is of particular interest from a neurotoxicology perspective. A recent study by Elder et al. (32) examined the olfactory translocation of manganese dioxide (MnO_2) particles to the CNS in rats exposed to 500 μg/m^3 of MnO_2, (30 nm diameter) for up to 12 days. Significant increases in manganese concentration were observed in the olfactory bulb more so than the lungs and brain regions such as the striatum, cerebellum, and frontal cortex (Fig. 16.2). Although blood-borne manganese could

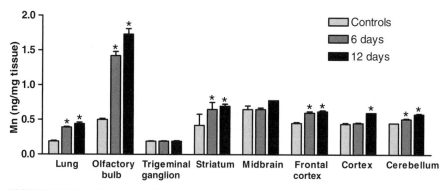

FIGURE 16.2 Manganese (Mn) content in the lung and brain tissue of rats exposed to 500 μg/m^3 of MnO_2 particles (30 nm diameter) for up to 12 days (adapted from (32), reproduced with permission from *Environmental Health Perspectives*).

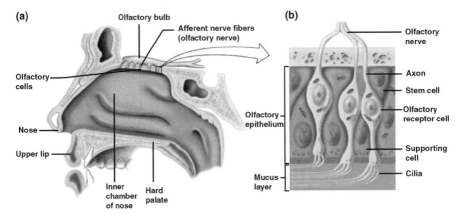

FIGURE 16.3   Schematic diagram of the nasal airway (a) and olfactory nerve (b). Note the close proximity of the nasal airway and the olfactory bulbs and the interconnectivity provided by the olfactory nerves. (From Vander et al., (134). Reproduced with permission from The McGraw-Hill Companies, Inc.).

have contributed to the increased levels in some brain regions, the observed increase in olfactory bulb concentrations indicate that the insoluble nanoparticles were translocated along axons (either within or on the outer surface) of the olfactory nerve into the CNS. The anatomy of the upper respiratory pathway is shown in Fig. 16.3 to illustrate the close proximity of the nasal airway and the olfactory bulb, and the interconnectivity provided by olfactory nerves. Thus, inhalation studies indicate three potential pathways for translocation of nanoparticles from the respiratory tract to the CNS: (a) deposition of nanoparticles along the respiratory tract and entry into systemic circulatory system, (b) deposition of nanoparticles in the nasal olfactory mucosa and migration along the olfactory nerve, and (c) transport of nanoparticles along the perineural (around the nerve) pathway across the olfactory mucosa and ethmoid bone into the cerebrospinal fluid.

For nanoparticles to enter the brain from a systemic source, they must cross the BBB, which serves as a physical barrier that allows controlled exchange of substances and cells between blood and the nervous system (33). The brain is often referred to as being "immunoprivileged" because it is relatively inaccessible to circulating immune cells (34), and cells that do gain access often circulate without any apparent pathological response (35). However, long-term exposure of canines to ambient air pollution resulted in deterioration of olfactory and respiratory barriers that rendered the nervous system vulnerable to damage (17, 18). In addition, nanoparticle surface chemistry strongly influences the ability of nanoparticles to enter the brain (32, 36). Modification of nanoparticle surfaces with surfactants, in particular the Tween® series (ethoxylated sorbitans), has been shown to greatly enhance transport across the BBB (37–39). Such surfactant-modified nanoparticles appear to mimic lipoprotein particles, which can be taken up by capillary endothelial cells via receptor-mediated endocytosis (40).

16.2 NEUROTOXICITY

While health concerns related to nanoparticle exposure initially focused on pulmonary toxicity resulting from the inhalation of airborne particles, the potential for nano-particles to induce neurotoxic responses was recognized (41). In addition to demon-strating that nanomaterials induce inflammation of the pulmonary system (16, 42–47), results of several recent studies indicate that nanomaterials may activate and recruit several proinflammatory cytokines and chemokines into the brain, leading to neuroin-flammation and pathologic damage (15, 17, 18, 32, 48). The immune response within the brain and subsequent inflammatory responses are directly related to the production of prooxidative molecules (49–51). In the following sections, fundamental aspects inflammatory response and oxidative stress in the brain are discussed within the framework of nanoparticle exposure.

16.2.1 Neuroinflammation

Inflammation is one of the primary responses of the CNS to insult or injury, and has been linked to both ambient and manufactured nanoparticle exposures (17, 18, 48). One of the first cell types to respond to injury or immunological stimuli are microglia, which function as the immune cells, or macrophages, of the CNS (52). Microglia are one of the three types of glial cells, including oligodendrocytes and astrocytes (Fig. 16.4).

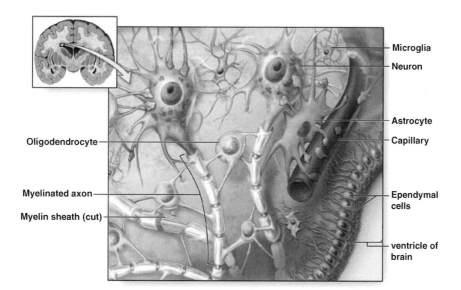

FIGURE 16.4 Three types of glial cells are found in the CNS, microglia, astrocytes, and oligodendrocytes, which provide support and nutrition, maintain homeostasis, form myelin, and participate in signal transmission in the nervous system. Epdendymal cells line the ventricles of the brain and secrete cerebrospinal fluid (CSF). (From McKinley and O'Loughlin, (135). Reproduced with permission from The McGraw-Hill Companies, Inc.).

Oligodendrocytes are specialized cells that form the myelin sheath, an electrical insulator around nerve fibers that enables rapid electrical signal transmission. Astrocytes are involved in forming the BBB, modulating signal transmission, and maintaining neuronal homeostasis (53).

In their normal resting state, microglia are characterized by a ramified morphology consisting of motile processes and protrusions that continuously monitor the extracellular space of the CNS (54). Microglia cells may be activated by pathogens or foreign material directly or as a secondary response to proinflammatory signals secreted from other cells (49). Upon activation, microglia transform into a reactive phenotype with an amoeboid morphology that is characterized by a more spherical cell body, hypertrophy of the nuclei, and elongated processes. In this activated state, microglia phagocytose (i.e., engulf and digest) cellular debris, foreign material, and pathogens release proinflammatory cytokines, reactive nitrogen and oxygen species (discussed below), and upregulate expression of many surface membrane receptors, such as major histocompatibility complexes (MHC) I and II (54).

Although activated microglia play an essential role in the development and homeostasis of the CNS, overactivation and dysregulation of microglia can exacerbate inflammation and have been associated with neurodegenerative disorders (55). Microglial release of cytokines, a diverse group of polypeptides including interleukins (ILs), interferons (IFNs), and tumor necrosis factors (TNFs), is strongly associated with inflammation and immune activation (56, 57). Several recent studies indicate that nanoparticle exposure can induce an inflammatory response in the CNS. For example, exposure of mice to concentrated airborne particulate matter resulted in elevated levels of the proinflammatory cytokines TNF-α and IL-1α, as well as the related transcription factor, nuclear factor-kappa beta (NF-κB) (48). TNF-α is a prototypical inflammatory cytokine involved in regulation of the immune response, cellular effects such as apoptosis and necrosis, and promotion of growth throughout the body (58, 59). Within the CNS, TNF-α is active in thermoregulation, cell fate, and neuroinflammation (59). While TNF-α can be neuroprotective or neuromodulatory under certain conditions (58, 59), results of numerous studies suggest that in pathological conditions TNF-α can be neurotoxic (59–63). IL-1α is produced by macrophages and monocytes, and is responsible for activating a variety of genes associated with inflammation and immune responses (64). In particular, IL-1 initiates cyclooxygenase (COX)-2, phospholipase A$_2$, and inducible nitric oxide synthase (iNOS) activity, and is also partially responsible for increasing permeability of the BBB allowing the transmigration of white blood cells into the brain parenchyma (65, 66). NF-κB is a transcription factor that plays a central role in many immunological processes (67, 68). Activation of NF-κB results in its translocation from the cytoplasm to the nucleus, where it induces the expression of many immune and inflammatory response genes including iNOS and COX-2 (69). Oberdörster (7) reported increased expression of several COX genes in largemouth bass following exposure to C$_{60}$ fullerene in aqueous solution. In addition, this exposure led to significant upregulation of inflammatory response genes such as macrophage stimulating factor, immunosuppressive proteins such as lipocalins, and a decrease in inflammatory cytokines (6), indicative of an immune response.

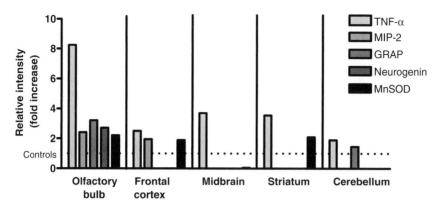

FIGURE 16.5 Comparison of TNF-α, macrophage inflammatory protein (MIP)-2, GFAP, neurogin, and manganese superoxide dismutase (MnSOD) expression in different brain regions of rats exposed to 500 μg/m³ of MnO₂ particles (30 nm diameter) (adapted from (32), reproduced with permission from *Environmental Health Perspectives*).

In rats exposed to MnO₂ nanoparticles via inhalation, inflammatory and stress response factors were elevated over controls (32). Specifically, expression of TNF-α, macrophage inflammatory protein (MIP)-2, glial fibrillary acidic protein (GFAP), and neuronal cell adhesion molecules increased in several regions of the brain (Fig. 16.5). The observed increase in GFAP expression in the brain following exposure to MnO₂ nanoparticles is indicative of astrocyte activation. Astrocytes are macroglial cells that contribute to the structure and integrity of the BBB, generate metabolic substrates for neurons, and assist in maintaining ionic and neurotransmitter homeostasis in the extracellular space (see Fig. 16.4) (52, 70). Astrocytes are known to respond, in parti-cular, to proinflammatory cytokines such as IL-1β and TNF-α. In contrast to microglia, activated astrocytes are thought to play a protective role in the brain (71, 72). Following activation, astrocytes have been shown to create a physical barrier between healthy and injured neural cells and tissue through the deposition of proteoglycans, and may assist in repairing damage to the BBB (52, 73). In addition, astrocytes can scavenge and metabolize compounds released by dying neurons that may contribute to oxidative stress (72). A summary of relevant cytokines secreted by activated microglia in response to nanoparticle exposures is provided in Table 16.1.

16.2.2 Oxidative Stress

Oxidation–reduction (redox) reactions play a central role in aerobic respiration, where oxygen serves as an oxidant or oxidizing agent that is reduced (accepts electrons) to form water, resulting in the release of energy that is subsequently stored as adenosine triphosphate (ATP). Because oxygen and its associated intermediates can be toxic, aerobic organisms possess antioxidant systems that act to minimize the production of and scavenge toxic intermediates, and can repair resulting damage.

TABLE 16.1 **Reported Mediators of Nanoparticle-induced Neurotoxicity**

Compound Name	Abbreviation	Nanoparticle	Putative Function
Hydrogen peroxide	H_2O_2	TiO_2[a]	RS[a]
Superoxide dismutase	SOD	MnO_2[b]	RS, antioxidant[b]
Inducible nitric oxide synthase	iNOS	PM[c, d]	RS, NO$^{\bullet}$ synthesis[e]
Superoxide	$O_2^{\bullet -}$	DEP[f], TiO_2[a]	RS, respiratory burst[g]
Nicotinamide adenine dinucleotide phosphate-oxidase	NADPH-oxidase or NOX	DEP[f]	Respiratory burst[f]
Nuclear factor-κB	NF-κB	PM[c, d, h]	Transcription factor[i]
Glial fibrillary acidic protein	GFAP	PM[c, d], MnO_2[b]	Marker of astrocyte activation[j]
Macrophage inflammatory protein 2	MIP2	MnO_2[b]	Cytokine/chemokine[k]
Tumor necrosis factor-α	TNF-α	MnO_2[b], PM[h]	Cytokine/chemokine[l]
Interleukin-1α	IL-1α	PM[h]	Cytokine/chemokine[m]
Cyclooxygenase-2	COX-2	PM[c, d], Fullerenes[n, o]	Prostaglandin synthesis
Macrophage stimulating factor	MCSF	Fullerenes[o]	Inflammation response gene[o]
Lipocalins	LCN	Fullerenes[o]	Immunosuppressive protein[o]
Neuronal cell adhesion molecule	NCAM	MnO_2[b]	Cell–Cell adhesion, synaptic plasticity[p]

DEP, diesel exhaust particles; MnO_2, manganese dioxide; PM, particulate matter; RS, reactive species; TiO_2, titanium dioxide.

[a]Long et al. (13);
[b]Elder et al. (32);
[c]Calderón-Garcidueñas et al. (17);
[d]Calderón-Garcidueñas et al. (18);
[e]Beal (81);
[f]Block et al. (20);
[g]Colton et al. (129);
[h]Campbell et al. (48);
[i]Baldwin (67);
[j]Hamill et al. (52);
[k]Tomita et al. (130).
[l]Pan et al. (58);
[m]Gibson et al. (64);
[n]Oberdörster (7);
[o]Oberdörster et al. (6);
[p]Doherty et al. (131).

However, a substantial body of scientific literature has established that an individual's oxidation state gradually increases with age, and that oxidation reactions contribute to major diseases, including pulmonary disease, cancer, and neurodegenerative disease (74–77). These underlying principles led to the concept of oxidative stress, which was initially defined as "a disturbance in the prooxidant and antioxidant balance in favor of the former, leading to potential damage" (78). Based on a wealth of studies performed on this topic over the past 20 years, a more comprehensive definition has been proposed: "a disruption of redox signaling and control" (79). The latter definition accounts for the complexity of biologic redox systems, and the fact that reactive species, although potentially harmful, play an essential role in signaling, synthesis, and detoxification.

Owing to their high electronegativity (i.e., tendency to accept electrons), oxygen and nitrogen have the capacity to form highly reactive oxidants known as reactive oxygen species (ROS) and reactive nitrogen species (RNS), respectively. ROS include superoxide ($O_2^{\bullet-}$), hydrogen peroxide (H_2O_2), hydroxyl radicals ($OH^{\bullet-}$), and singlet oxygen (O_2). However, other classes of reactive species may exist and hence the more general term reactive species (RS) is often used (80). In summary, RS are capable of reacting with a wide range of biological molecules such as DNA, proteins, and lipids, compromising the functions of these molecules and often creating more highly reactive species (81).

Reactive species are present throughout the body due to normal cellular function and are products of a variety of metabolic processes. RS are utilized in many processes such as the respiratory burst used by macrophages and the controlled shuttling of electrons in the mitochondria (82). While the production of RS in healthy individuals is balanced by molecules and systems that detoxify them, imbalances in this relationship can lead to cellular injury, dysfunction, and disease (80). Despite the fact that these compounds are present systemically, the brain is especially sensitive to their damaging effects. While many factors contribute to the brain's inherent susceptibility to RS, the most important attributes include (80, 83)

1. High oxygen consumption due to the large amounts of ATP required to maintain ionic and neurotransmitter homeostasis in both the intra- and extracellular space.

2. Significant generation of superoxide radicals, which can be formed by the electron transport chain in mitochondria and the endoplasmic reticulum from which electrons can "leak" and react with oxygen. Superoxide can also form secondary to autooxidation of neurotransmitters such as dopamine (84–86).

3. Several metabolic reactions in the brain generate H_2O_2 as a metabolic product including the superoxide dismutase (SOD) reaction. SOD quenches two superoxides by reducing one to hydrogen peroxide and oxidizing the other to molecular oxygen. Additionally, monoamine oxidases, enzymes responsible for the breakdown of dopamine and norepinephrine, among others, generate hydrogen peroxide as a product of normal metabolism (80).

4. There are several autooxidizable compounds within the brain. Many of the monoaminergic neurotransmitters ($R-CH_2NH_2$), such as dopamine, as well as the dopamine precursor, L-DOPA, can autooxidize and form superoxide as well as reactive quinones and semiquinones that can adduct protein sulfhydryl groups (87, 88).

5. Transition metals present in the brain, such as iron and copper, can participate in Fenton's chemistry and catalyze the formation of free radicals. In the Fenton reaction, ferrous (II) iron reacts with hydrogen peroxide to yield ferric (III) iron, a hydroxyl anion (OH^-), and a hydroxyl radical (89). In addition, the cerebrospinal fluid (CSF) has significantly lower levels of transferrin, an iron binding protein, than plasma. This reduced capacity to bind released iron may result in the CSF normally existing at, or near, iron saturation (83, 90).

6. High calcium flux across neuronal membranes. Any disruption of ion transport and rapid increases in intracellular free calcium can lead to oxidative stress.

7. Lipid membranes within the brain are often rich in polyunsaturated fatty acid side chains (80). Products of lipid peroxidation are known to be injurious. One such product, 4-hydroxynonenal, is particularly toxic secondary to increasing intracellular calcium levels, inactivating glutamate transporters, thereby preventing clearance of the excitotoxic neurotransmitter, and damage to neurofilaments (91, 92).

8. Coupled with prooxidative processes, the brain possesses a relatively weak antioxidant defense systems (93, 94). The modest buffering capacity between RS production and antioxidant levels results in a precarious oxidative balance in the brain that can easily be disrupted.

As noted above, superoxide and hydrogen peroxide are common ROS species produced in the brain. In addition, nitric oxide can be generated from arginine by one of the three forms of nitric oxide synthase (NOS): endothelial (eNOS), iNOS, and neuronal (nNOS) (81). Superoxide can react with nitric oxide to form peroxynitrite ($ONOO^-$) that, at physiological pH (7.4), can rapidly protonate to form peroxynitrous acid (ONOOH), which can undergo hemolytic fission to create hydroxyl radicals (95). The hydroxyl radical is another highly reactive species produced by a variety of reactions. The short half-life and high reactivity of the hydroxyl radical make quenching of the radical and prevention of its attack on numerous cellular components difficult. While many of the RS are nonradicals (no free electrons), they are still capable of oxidizing and nitrating DNA, proteins, and lipids (95). Relevant redox reactions occurring in the brain are summarized in Table 16.2.

The brain is sensitive to changes in the oxidative environment and it is especially susceptible to any compounds that are capable of causing oxidative stress. Many nanoparticles have been shown to cause oxidative stress secondary to inflammation in the brain (7, 15, 17, 18, 20, 32, 48, 96). The role of inflammation as both a precursor and sequela of neurological damage and disease has been well documented (35, 59, 62, 97, 99–101). Disorders such as multiple sclerosis and infectious conditions are traditionally considered to be primary inflammatory conditions, while stroke and Parkinson's disease (PD) are thought to demonstrate inflammation secondary to their ischemic and neurodegenerative etiologies, respectively (35). In general, any insult to the brain can lead to a complex cascade of events that ultimately leads to inflammation.

Microglia use an oxidative or respiratory burst (rapid release of ROS) to digest or destroy ingested material. Microglia are capable of initiating neuronal damage after

TABLE 16.2   Key Reactions Relevant to Oxidative Stress in the Brain

Reaction	Description
$2O_2^{\bullet-} + 2H^+ \rightarrow H_2O_2 + O_2$	Formation of hydrogen peroxide by SOD scavenging of superoxide radical[a]
$C_8H_{11}NO_2$ (dopamine) $+ O_2 + H_2O \rightarrow H_2O_2 + NH_3 + C_8H_8O_4$ (DOPAC)	Production of hydrogen peroxide by MAO metabolism of dopamine to DOPAC[b]
RCH_2NH_2 (monoamine) $+ O_2 + H_2O \rightarrow$ RCHO (aldehyde) $+ H_2O_2 + NH_3$	Autooxidation of monoaminergic neurotransmitter[c, d]
$Fe^{2+} + H_2O_2 \rightarrow Fe^{3+} + HO^- + HO^{\bullet}$	Fenton reaction resulting in the formation of hydroxyl radical[e]
$NO^{\bullet} + O_2^{\bullet-} \rightarrow ONOO^-$	Formation of peroxynitrite by reaction of nitric oxide with superoxide[f]
$ONOO^- \rightarrow ONOOH$	Protonation of peroxynitrite to form peroxynitrous acid[f]
$ONOOH \rightarrow NO_2^{\bullet} + HO^{\bullet}$	Hemolytic fission to form hydroxyl radical[f]
2GSH (reduced glutathione) $+ H_2O_2 \rightarrow$ GSSG (oxidized glutathione) $+ 2H_2O$	Scavenging of hydrogen peroxide by glutathione[a, g]

SOD, superoxide dismutase; MAO, monoamine oxidase; $O_2^{\bullet-}$, superoxide; H_2O_2, hydrogen peroxide; GSH, reduced glutathione; GSSG, oxidized glutathione.

[a]Halliwell (80);
[b]Spina et al. (132);
[c]Spencer et al. (87);
[d]Spencer et al. (88);
[e]Wardman and Candeias (89);
[f]Alvarez and Radi (95);
[g]Maker et al. (133).

activating and producing neurotoxic proinflammatory factors (55), generating RS, and recruiting white blood cells (52). This neuronal damage can then lead to secondary activation of microglia (reactive microgliosis) and the production of more toxic factors and more damage resulting in a self-propelling cycle (Fig. 16.6) (20, 49). Thus, exposure to nanoparticles such as diesel exhaust particles can lead to a lethal cycle of microglial activation, oxidative stress, and neuronal damage (19, 20). Additional concerns about inflammation and damage in the brain exist because of the low regenerative capacity of neurons in the brain (33). Most neurons are postmitotic and, therefore, do not readily divide and reproduce in an effort to repair the region after injury (33).

Exposure of canines to air-pollution-derived PM resulted in expression of neuronal NF-κB and iNOS and disruption of the BBB (17). The expression of iNOS is enhanced at sites of chronic inflammation and is partially responsible for generating high localized concentrations of nitric oxide (83) that are capable of permeabilizing the BBB. Additional neuropathology observed in PM-exposed canines included degeneration of cortical neurons, apoptotic glial cells, reactive astrocytes, deposition of apolipoprotein E (ApoE) lipid droplets in smooth muscle cells and pericytes, and occasional nonneuritic plaques and neurofibrillary tangles (NFTs) (17). Damage to the

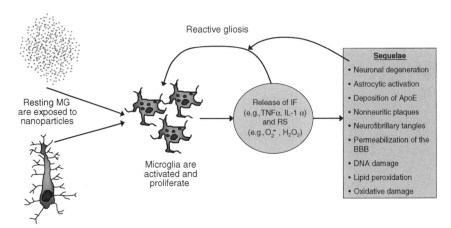

FIGURE 16.6 Schematic diagram of reactive microgliosis. Activated microglia can damage neurons by releasing proinflammatory factors (IF) and generating RS. This neuronal damage leads to secondary activation of microglia, release of more IF and RS, resulting in a self-propelling cycle.

DNA of PM-exposed canines was characterized by analyzing apurinic/apyrimidinic (AP) sites in the nasal and brain genomic DNA (18). Oxidized DNA bases are often repaired via a base excision repair pathway (18, 101) during which an AP site is formed. AP sites result in a number of sequelae including base substitution mutations and inhibition of DNA replication (18, 102). In addition to increased NF-κB, iNOS, and GFAP expression previously reported in exposed canines, this study also reported increased expression of endothelial and glial cyclooxygenase-2 (COX-2). COX-2 is an inducible, rate-limiting enzyme involved in the synthesis of several inflammatory mediators known as prostanoids (99) and is involved in several neurodegenerative disorders (97, 99, 100, 103). Finally, the presence of amyloid precursor protein (APP) and amyloid beta (Aβ) was noted in neurons, diffuse plaques, and in subarachnoid blood vessels (18). Cleavage of APP by the enzyme gamma-secretase produces Aβ, the main constituent of the amyloid plaques characteristic of Alzheimer's disease (AD) (104–107). The pathology reported in the exposed canines is indicative of the activation and inflammatory response of endothelial, neuronal, and microglial cells that can lead to the production of free radicals and subsequent oxidative damage to proteins, lipids, and DNA (74, 81). In human samples collected from the same exposure regions as the canine studies, highly exposed individuals exhibited elevated COX-2 mRNA and protein expression in both the frontal cortex and hippocampus. There was also nuclear localization of NF-κB expression, indicating activation, in the frontal cortex and glial cells, as well as Aβ accumulation in pyramidal frontal neurons, cortical and white matter astrocytes, and subarachnoid and cortical blood vessels (18).

Primary mixed cultures of neurons and glial cells treated with diesel exhaust particles (DEP, size 0.22 μm) showed PD-like pathology, including a dose-dependent decrease in the ability to take up dopamine, the neurotransmitter that is depleted in PD (20). However, DEP exposure had no effect on the uptake of GABA, another neurotransmitter used by cells in some of the same regions involved in PD. Immuno-histochemical analysis revealed activation of microglia in DEP-treated cultures. Interestingly, removing the microglia from the cultures prevented the decrease in dopamine uptake and reintroduction of increasing densities of microglia resulted in dose-dependent decreases in dopamine uptake, suggesting that DEP were not directly toxic to the dopaminergic neurons, but rather exert toxic effects via microglial activation. Intracellular ROS, including superoxide, were produced in microglia-enriched cultures exposed to DEP, but supernatant from DEP-exposed cultures showed no TNF-α, nitrite (an indicator of nitric oxide production), or prostaglandin-E_2 (a prostanoid), which may represent a decreased ability to recruit inflammatory cytokines and chemokines in a culture system or demonstrate that not all nanomaterials induce toxicity via the same mechanisms. In addition, inhibition of microglial superoxide production by genetically depleting NADPH oxidase, the enzyme respon-sible for the respiratory burst, and treatment with cytochalasin D, which inhibits the ability of microglia to phagocytize DEP, prevented the neuronal toxicity associated with DEP exposure. These findings suggest that NADPH oxidase was the source of DEP-induced microglial ROS and that phagocytosis of DEP was required for microglial superoxide production (20). The interrelationships between inflammatory and oxidative stress responses to nanoparticle exposure are summarized in Fig. 16.7.

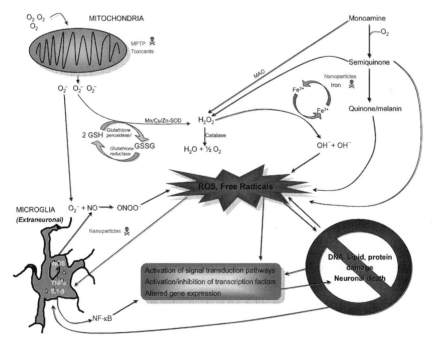

FIGURE 16.7 Interplay of nanoparticles with oxidative stress and inflammation in the neuron and microglia. Nanoparticle-induced release of microglial inflammatory factors can lead to oxidative damage within the neuron.

16.3   IMPLICATIONS FOR NEURODEGENERATION

Many of the oxidative and inflammatory sequelae observed in the brain following exposure to various nanoparticles mimic pathological responses seen in several neurological disorders, most notably the neurodegenerative diseases (35, 81, 98, 108). Oxidative stress and inflammation are key pathologic features in diseases such as PD, AD, and amyotrophic lateral sclerosis (ALS) (109–112). Consequences of nanoparticle exposure such as neuronal degeneration, NF-κB activation, and increased expression of proinflammatory cytokines are seen in a variety of neurodegenerative diseases (69, 113). A-beta accumulation, ApoE deposition, and formation of nonneuritic plaques and neurofibrillary tangles seen in PM-exposed canines and humans recapitulate pathologic features of AD. AD, the leading cause of dementia in the elderly, is the most common neurodegenerative disease and is characterized by cognitive and behavioral dysfunction (114). The neuroendocrine cell line, PC-12, has been used to study the interaction of MnO_2 nanoparticles with the neuronal system involved in PD (12). PD is a neurodegenerative disorder clinically characterized by tremor, slowness of movement, imbalance, and rigidity. Pathologically, there is death of cells in an area of the brain known as the substantia nigra pars compacta resulting in the loss of the neurotransmitter, dopamine (115). MnO_2 exposure caused depletion of dopamine, generated intracellular oxidative stress, and caused mitochondrial dysfunction (12). While genetic causes have been identified in early-onset PD (<50 years of age) (116), environmental factors including exposure to polychlorinated biphenyls (117), pesticides (118–120), and metal species (121–123) have been linked to sporadic cases of PD. Many of these environmental toxicants have been found to induce the same oxidative and inflammatory neurological changes observed following exposure to manufactured nanoparticles.

16.4   SUMMARY AND CONCLUSIONS

The presence of nanoparticles in the environment is of obvious concern. While many neurodegenerative diseases have genetic components, the role of environmental factors in accentuating the risk or accelerating the pathology seen in these diseases, especially PD, has been extensively examined (123–125). The development of medicinal imaging and drug delivery methods based on direct injection of nanomaterials into the CNS or modified nanoparticles that can cross the BBB could prove to be even more neurotoxic. Nanoparticles are now being modified to serve as intracellular biomarkers and biosensors that can directly target specific cell types (122, 126–128). While the application of these technologies may prove to be very useful for advanced imaging techniques, concerns must be raised over delivering compounds such as transition metal and paramagnetic nanoparticles directly to neurons that are extremely vulnerable to oxidative stress and where dangerous reactions such Fenten chemistry can produce hydroxyl and superoxide radicals (90). At present, it is unclear what risks nanoparticle exposures pose to the human nervous system. Currently available data suggest that manufactured nanoparticles represent an emerging class of environmental neurotoxicants and further studies will be necessary to evaluate their potential to negatively impact human, and in particular, neurodegenerative diseases.

REFERENCES

1. NNCO. *The National Nanotechnology Initiative: Environmental, Health, and Safety Research Needs for Engineered Nanoscale Materials.* Arlington, VA, 2006. p. 62.

2. Maynard, A.D., Aitken, R.J., Butz, T., Colvin, V., Donaldson, K., Oberdörster, G., Philbert, M.A., Ryan, J., Seaton, A., Stone, V., Tinkle, S.S., Tran, L., Walker, N.J., Warheit, D.B. Safe handling of nanotechnology. *Nature* 2006;444:267–269.

3. Nel, A., Xia, T., Madler, L., Li, N. Toxic potential of materials at the nanolevel. *Science* 2006;311:622–627.

4. Holsapple, M.P., Farland, W.H., Landry, T.D., Monteiro-Riviere, N.A., Carter, J.M., Walker, N.J., Thomas, K.V. Research strategies for safety evaluation of nanomaterials, part II: toxicological and safety evaluation of nanomaterials, current challenges and data needs. *Toxicol Sci* 2005;88:12–17.

5. Oberdörster, G., Maynard, A., Donaldson, K., Castranova, V., Fitzpatrick, J., Ausman, K., Carter, J., Karn, B., Kreyling, W., Lai, D., Olin, S., Monteiro-Riviere, N., Warheit, D., Yang, H. Principles for characterizing the potential human health effects from exposure to nanomaterials: elements of a screening strategy. *Part Fibre Toxicol* 2005a;2:8.

6. Oberdörster, G., Oberdörster, E., Oberdörster, J. Nanotoxicology: an emerging discipline evolving from studies of ultrafine particles. *Environ Health Perspect* 2005b;113: 823–839.

7. Oberdörster, E. Manufactured nanomaterials (fullerenes, C60) induce oxidative stress in the brain of juvenile largemouth bass. *Environ Health Perspect* 2004;112:1058–1062.

8. Lin, A.M., Ho, L.T. Melatonin suppresses iron-induced neurodegeneration in rat brain. *Free Radic Biol Med* 2000;28:904–911.

9. Chen, K.L., Elimelech, M. Aggregation and deposition kinetics of fullerene (C-60) nanoparticles. *Langmuir* 2006;22:10994–11001.

10. Deguchi, S., Alargova, R.G., Tsujii, K. Stable dispersions of fullerenes, C60 and C70, in water: preparation and characterization. *Langmuir* 2001;17:6013–6017.

11. Fortner, J.D., Lyon, D.Y., Sayes, C.M., Boyd, A.M., Falkner, J.C., Hotze, E.M., Alemany, L.B., Tao, Y.J., Guo, W., Ausman, K.D., Colvin, V.L., Hughes, J.B. C60 in water: nanocrystal formation and microbial response. *Environ Sci Technol* 2005; 39:4307–4316.

12. Hussain, S.M., Javorina, A.K., Schrand, A.M., Duhart, H.M., Ali, S.F., Schlager, J.J. The interaction of manganese nanoparticles with PC-12 cells induces dopamine depletion. *Toxicol Sci* 2006;92:456–463.

13. Long, T.C., Saleh, N., Tilton, R.D., Lowry, G.V., Veronesi, B., Titanium dioxide (P 25) produces reactive oxygen species in immortalized brain microglia (BV2): implications for nanoparticle neurotoxicity. *Environ Sci Technol* 2006;40:4346–4352.

14. Henry, T.B., Menn, F.M., Fleming, J.T., Wilgus, J., Compton, R.N., Sayler, G.S. Attributing effects of aqueous C60 nano-aggregates to tetrahydrofuran decomposition products in larval zebrafish by assessment of gene expression. *Environ Health Perspect* 2007;115:1059–1065.

15. Veronesi, B., Oortgiesen, M. Neurogenic inflammation and particulate matter (PM) air pollutants. *Neurotoxicology* 2001;22:795–810.

16. Warheit, D.B., Laurence, B.R., Reed, K.L., Roach, D.H., Reynolds, G.A.M., Webb, T.R. Comparative pulmonary toxicity assessment of single-wall carbon nanotubes in rats. *Toxicol Sci* 2004;77:117–125.

17. Calderón-Garcidueñas, L., Azzarelli, B., Acuna, H., Garcia, R., Gambling, T.M., Osnaya, N., Monroy, S., Tizapartzi, M.D.R., Carson, J.L., Villarreal-Calderon, A., Rewcastle, B. Air pollution and brain damage. *Toxicol Pathol* 2002;30:373–389.

18. Calderón-Garcidueñas, L., Maronpot, R.R., Torres-Jardon, R., Henriquez-Roldan, C., Schoonhoven, R., Acuna-Ayala, H., Villarreal-Calderon, A., Nakamura, J., Fernando, R., Reed, W., Azzarelli, B., Swenberg, J.A. DNA damage in nasal and brain tissues of canines exposed to air pollutants is associated with evidence of chronic brain inflammation and neurodegeneration. *Toxicol Pathol* 2003;31:524–538.

19. Xia, T., Kovochich, M., Brant, J., Hotze, M., Sempf, J., Oberley, T., Sioutas, C., Yeh, J.I., Wiesner, M.R., Nel, A.E. Comparison of the abilities of ambient and manufactured nanoparticles to induce cellular toxicity according to an oxidative stress paradigm. *Nano Lett* 2006;6:1794–1807.

20. Block, M.L., Wu, X., Pei, Z., Li, G., Wang, T., Qin, L., Wilson, B., Yang, J., Hong, J.S., Veronesi, B. Nanometer size diesel exhaust particles are selectively toxic to dopaminergic neurons: the role of microglia, phagocytosis, and NADPH oxidase. *Faseb J* 2004; 18:1618–1620.

21. Oberdörster, G., Sharp, Z., Atudorei, V., Elder, A., Gelein, R., Lunts, A., Kreyling, W., Cox, C. Extrapulmonary translocation of ultrafine carbon particles following whole-body inhalation exposure of rats. *J Toxicol Environ Health A* 2002;65:1531–1543.

22. Peters, A., Wichmann, H.E., Tuch, T., Heinrich, J., Heyder, J. Respiratory effects are associated with the number of ultrafine particles. *Am J Respir Crit Care Med* 1997;155:1376–1383.

23. Geiser, M., Schurch, S., Im Hof, V., Gehr, P. Retention of particles: structural and interfacial aspects. In: Gehr, P., Heyder, J., editors. Particle-lung interactions. New York: Marcel Dekker; 2000. pp. 291–322.

24. Zhang, Z., Kleinstreuer, C., Donohue, J.F., Kim, C.S. Comparison of micro- and nanosize particle depositions in a human upper airway model. *J Aerosol Sci* 2005;36:211–233.

25. ICRP Human respiratory model for radiological protection. *Ann ICRP* 1994; 24: 1–300.

26. Donaldson, K., Stone, V., Gilmour, P.S., Brown, D.M., MacNee, W. Ultrafine particles: mechanisms of lung injury. *Philosophical Transactions of the Royal Society A: Mathematical, Physical and Engineering Sciences* 2000;358:2741–2749.

27. Oberdörster, G. Toxicology of ultrafine particles: *in vivo* studies. *Philos Trans Royal Soc A* 2000;358:2719–2740.

28. Berry, J.P., Arnoux, B., Stanislas, G., Galle, P., Chretien, J. A microanalytic study of particles transport across the alveoli: role of blood platelets. *Biomedicine* 1977;27: 354–357.

29. Kreyling, W.G., Semmler, M., Erbe, F., Mayer, P., Takenaka, S., Schulz, H., Oberdörster, G., Ziesenis, A. Translocation of ultrafine insoluble iridium particles from lung epithelium to extrapulmonary organs is size dependent but very low. *J Toxicol Environ Health A* 2002;65:1513–1530.

30. Shimada, A., Kawamura, N., Okajima, M., Kaewamatawong, T., Inoue, H., Morita, T. Translocation pathway of the Intratracheally instilled ultrafine particles from the lung into the blood circulation in the mouse. *Toxicol Pathol* 2006;34:949–957.

31. Oberdörster, G., Sharp, Z., Atudorei, V., Elder, A., Gelein, R., Kreyling, W., Cox, C. Translocation of inhaled ultrafine particles to the brain. *Inhal Toxicol* 2004;16:437–445.

32. Elder, A., Gelein, R., Silva, V., Feikert, T., Opanashuk, L., Carter, J., Potter, R., Maynard, A., Ito, Y., Finkelstein, J., Oberdörster, G. Translocation of inhaled ultrafine manganese oxide particles to the central nervous system. *Environ Health Perspect* 2006;114:1172–1178.

33. Wekerle, H., Linington, C., Lassmann, H., Meyermann, R. Cellular immune reactivity within the CNS. *Trends Neurosci* 1986;9:271–277.

34. Becher, B., Prat, A., Antel, J.P. Brain-immune connection: immuno-regulatory properties of CNS-resident cells. *Glia* 2000;29:293–304.

35. Zipp, F., Aktas, O. The brain as a target of inflammation: common pathways link inflammatory and neurodegenerative diseases. *Trends Neurosci* 2006;29:518–527.

36. Peters, A., Veronesi, B., Calderon-Garciduenas, L., Gehr, P., Chen, L.C., Geiser, M., Reed, W., Rothen-Rutishauser, B., Schurch, S., Schulz, H. Translocation and potential neurological effects of fine and ultrafine particles a critical update. *Part Fibre Toxicol* 2006;3:13.

37. Kreuter, J. Influence of the surface properties on nanoparticle-mediated transport of drugs to the brain. *J Nanosci Nanotechnol* 2004;4:484–488.

38. Lockman, P.R., Koziara, J.M., Mumper, R.J., Allen, D.D. Nanoparticle surface charges alter blood-brain barrier integrity and permeability. *J Drug Target* 2004;12:635–641.

39. Muller, R.H., Keck, C.M. Drug delivery to the brain—realization by novel drug carriers. *J Nanosci Nanotechnol* 2004;4:471–483.

40. Kreuter, J., Shamenkov, D., Petrov, V., Ramge, P., Cychutek, K., Koch-Brandt, C., Alyautdin, R. Apolipoprotein-mediated transport of nanoparticle-bound drugs across the blood-brain barrier. *J Drug Target* 2002;10:317–325.

41. Oberdörster, G., Utell, M.J. Ultrafine particles in the urban air: to the respiratory tract and beyond? *Environ Health Perspect* 2002;110:A440–A441.

42. Bergamaschi, E., Bussolati, O., Magrini, A., Bottini, M., Migliore, L., Bellucci, S., Iavicoli, I., Bergamaschi, A. Nanomaterials and lung toxicity: interactions with airways cells and relevance for occupational health risk assessment. *Int J Immunopathol Pharmacol* 2006;19:3–10.

43. Chen, H.W., Su, S.F., Chien, C.T., Lin, W.H., Yu, S.L., Chou, C.C., Chen, J.J., Yang, P.C. Titanium dioxide nanoparticles induce emphysema-like lung injury in mice. *Faseb J* 2006;20:2393–2395.

44. Evans, S.A., Al-Mosawi, A., Adams, R.A., Berube, K.A. Inflammation, edema, and peripheral blood changes in lung-compromised rats after instillation with combustion-derived and manufactured nanoparticles. *Exp Lung Res* 2006;32:363–378.

45. Lam, C.-W., James, J.T., McCluskey, R., Hunter, R.L. Pulmonary toxicity of single-wall carbon nanotubes in mice 7 and 90 days after intratracheal instillation. *Toxicol Sci* 2004;77:126–134.

46. Renwick, L.C., Brown, D., Clouter, A., Donaldson, K. Increased inflammation and altered macrophage chemotactic responses caused by two ultrafine particle types. *Occup Environ Med* 2004;61:442–447.

47. Shvedova, A.A., Kisin, E.R., Mercer, R., Murray, A.R., Johnson, V.J., Potapovich, A.I., Tyurina, Y.Y., Gorelik, O., Arepalli, S., Schwegler-Berry, D., Hubbs, A.F., Antonini, J., Evans, D.E., Ku, B.-K., Ramsey, D., Maynard, A., Kagan, V.E., Castranova, V., Baron, P. Unusual inflammatory and fibrogenic pulmonary responses to single-walled carbon nanotubes in mice. *Am J Physiol Lung Cell Mol Physiol* 2005;289:L698–L708.

48. Campbell, A., Oldham, M., Becaria, A., Bondy, S.C., Meacher, D., Sioutas, C., Misra, C., Mendez, L.B., Kleinman, M. Particulate matter in polluted air may increase biomarkers of inflammation in mouse brain. *Neurotoxicology* 2005;26:133–140.

49. Liu, B., Gao, H.M., Hong, J.S. Parkinson's disease and exposure to infectious agents and pesticides and the occurrence of brain injuries: role of neuroinflammation. *Environ Health Perspect* 2003;111:1065–1073.

50. Milatovic, D., Zaja-Milatovic, S., Montine, K.S., Horner, P.J., Montine, T.J. Pharmacologic suppression of neuronal oxidative damage and dendritic degeneration following direct activation of glial innate immunity in mouse cerebrum. *J Neurochem* 2003; 87:1518–1526.

51. Montine, T.J., Milatovic, D., Gupta, R.C., Valyi-Nagy, T., Morrow, J.D., Breyer, R.M. Neuronal oxidative damage from activated innate immunity is EP2 receptor-dependent. *J Neurochem* 2002;83:463–470.

52. Hamill, C.E., Goldshmidt, A., Nicole, O., McKeon, R.J., Brat, D.J., Traynelis, S.F. Special lecture: glial reactivity after damage: implications for scar formation and neuronal recovery. *Clin Neurosurg* 2005;52:29–44.

53. Jessen, K.R. Glial cells. *Int J Biochem Cell Biol* 2004;36:1861–1867.

54. Nimmerjahn, A., Kirchhoff, F., Helmchen, F. Resting microglial cells are highly dynamic surveillants of brain parenchyma *in vivo*. *Science* 2005;308:1314–1318.

55. Block, M.L., Zecca, L., Hong, J.-S. Microglia-mediated neurotoxicity: uncovering the molecular mechanisms. *Nat Rev Neurosci* 2007;8:57–69.

56. Allan, S.M., Rothwell, N.J. Cytokines and acute neurodegeneration. *Nat Rev Neurosci* 2001;2:734–744.

57. Rock, R.B., Gekker, G., Hu, S., Sheng, W.S., Cheeran, M., Lokensgard, J.R., Peterson, P.K. Role of microglia in central nervous system infections. *Clin Microbiol Rev* 2004;17: 942–964.

58. Pan, W., Zadina, J.E., Harlan, R.E., Weber, J.T., Banks, W.A., Kastin, A.J. Tumor necrosis factor-alpha: a neuromodulator in the CNS. *Neurosci Biobehav Rev* 1997;21:603–613.

59. Perry, S.W., Dewhurst, S., Bellizzi, M.J., Gelbard, H.A. Tumor necrosis factor-alpha in normal and diseased brain: conflicting effects via intraneuronal receptor crosstalk? *J Neurovirol* 2002;8:611–624.

60. Kim, Y.S., Joh, T.H. Microglia, major player in the brain inflammation: their roles in the pathogenesis of Parkinson's disease. *Exp Mol Med* 2006;38:333–347.

61. Minghetti, L., Ajmone-Cat, M.A., De Berardinis, M.A., De Simone, R. Microglial activation in chronic neurodegenerative diseases: roles of apoptotic neurons and chronic stimulation. *Brain Res Brain Res Rev* 2005;48:251–256.

62. Rosenberg, P.B. Clinical aspects of inflammation in Alzheimer's disease. *Int Rev Psychiatry* 2005;17:503–514.

63. Woodroofe, M.N. Cytokine production in the central nervous system. *Neurology* 1995;45: S6–S10.

64. Gibson, R.M., Rothwell, N.J., Le Feuvre, R.A. CNS injury: the role of the cytokine IL-1. *Vet J* 2004;168:230–237.

65. Dinarello, C.A. The interleukin-1 family: 10 years of discovery. *Faseb J* 1994;8: 1314–1325.

66. Dinarello, C.A. The IL-1 family and inflammatory diseases. *Clin Exp Rheumatol* 2002;20:S1–S13.

67. Baldwin, A.S., Jr. The NF-kappa B and I kappa B proteins: new discoveries and insights. *Annu Rev Immunol* 1996;14:649–683.

68. Townsend, K.P., Pratico, D. Novel therapeutic opportunities for Alzheimer's disease: focus on nonsteroidal anti-inflammatory drugs. *FASEB J* 2005;19:1592–1601.

69. Memet, S. NF-kappaB functions in the nervous system: from development to disease. *Biochem Pharmacol* 2006;72:1180–1195.

70. Nedergaard, M., Ransom, B., Goldman, S.A. New roles for astrocytes: redefining the functional architecture of the brain. *Trends Neurosci* 2003;26:523–530.

71. Moore, A.H., O'Banion, M.K. Neuroinflammation and anti-inflammatory therapy for Alzheimer's disease. *Adv Drug Deliv Rev* 2002;54:1627–1656.

72. Teismann, P., Schulz, J.B. Cellular pathology of Parkinson's disease: astrocytes, microglia and inflammation. *Cell Tissue Res* 2004;318:149–161.

73. Faulkner, J.R., Herrmann, J.E., Woo, M.J., Tansey, K.E., Doan, N.B., Sofroniew, M.V. Reactive astrocytes protect tissue and preserve function after spinal cord injury. *J Neurosci* 2004;24:2143–2155.

74. Jenner, P. Oxidative stress in Parkinson's disease. *Ann Neurol* 2003;53 (Suppl 3): S26–S36; discussion S36-28.

75. Luppi, F., Hiemstra, P.S. Epithelial responses to oxidative stress in chronic obstructive pulmonary disease—Lessons from expression profiling. *Am J Respir Crit Care Med* 2007;175:527–528.

76. Ragu, S., Faye, G., Iraqui, I., Masurel-Heneman, A., Kolodner, R.D., Huang, M.E. Oxygen metabolism and reactive oxygen species cause chromosomal rearrangements and cell death. *Proc Nat Acad Sci USA* 2007;104:9747–9752.

77. Repine, J.E., Bast, A., Lankhorst, I., deBacker, W., Dekhuijzen, R., Demedts, M., vanHerwaarden, C., vanKlaveren, R., Lammers, J.W., Larsson, S., Lundback, B., Petruzelli, S., Postma, D., Riise, G., Vermeire, P., Wouters, E., Yernault, J.C., van Zandwijk, N. Oxidative stress in chronic obstructive pulmonary disease. *Am J Resp Crit Care Med* 1997;156:341–357.

78. Sies, H. Oxidative stress: from basic research to clinical application. *Am J Med* 1991;91:31–38.

79. Jones, D.P. Redefining oxidative stress. *Antioxid Redox Signal* 2006;8:1865–1879.

80. Halliwell, B. Oxidative stress and neurodegeneration: where are we now? *J Neurochem* 2006;97:1634–1658.

81. Beal, M.F. Oxidatively modified proteins in aging and disease. *Free Radic Biol Med* 2002;32:797–803.

82. Beal, M.F. Mitochondria take center stage in aging and neurodegeneration. *Ann Neurol* 2005;58:495–505.

83. Halliwell, B., Gutteridge, J.M.C. *Free Radicals in Biology and Medicine*, New York: Oxford University Press, Inc.; 1999.

84. Fornstedt, B., Brun, A., Rosengren, E., Carlsson, A. The apparent autooxidation rate of catechols in dopamine-rich regions of human brains increases with the degree of depigmentation of substantia nigra. *J Neural Transm Park Dis Dement Sect* 1989; 1:279–295.

85. Fornstedt, B. Role of catechol autooxidation in the degeneration of dopamine neurons. *Acta Neurol Scand Suppl* 1990;129:12–14.

86. Fornstedt, B., Pileblad, E., Carlsson, A. *In vivo* autoxidation of dopamine in guinea pig striatum increases with age. *J Neurochem* 1990;55:655–659.

87. Spencer, J.P., Jenner, P., Daniel, S.E., Lees, A.J., Marsden, D.C., Halliwell, B. Conjugates of catecholamines with cysteine and GSH in Parkinson's disease: possible mechanisms of formation involving reactive oxygen species. *J Neurochem* 1998;71: 2112–2122.

88. Spencer, J.P., Whiteman, M., Jenner, P., Halliwell, B. 5-s-Cysteinyl-conjugates of catecholamines induce cell damage, extensive DNA base modification and increases in caspase-3 activity in neurons. *J Neurochem* 2002;81:122–129.

89. Wardman, P., Candeias, L.P. Fenton chemistry: an introduction. *Radiat Res* 1996;145: 523–531.

90. Halliwell, B., Gutteridge, J.M. Oxygen toxicity, oxygen radicals, transition metals and disease. *Biochem J* 1984;219:1–14.

91. Farooqui, A.A., Horrocks, L.A. Phospholipase A2-generated lipid mediators in the brain: the good, the bad, and the ugly. *Neuroscientist* 2006;12:245–260.

92. Mark, R.J., Lovell, M.A., Markesbery, W.R., Uchida, K., Mattson, M.P. A role for 4-hydroxynonenal, an aldehydic product of lipid peroxidation, in disruption of ion homeostasis and neuronal death induced by amyloid beta-peptide. *J Neurochem* 1997; 68:255–264.

93. Cardozo-Pelaez, F., Brooks, P.J., Stedeford, T., Song, S., Sanchez-Ramos, J. DNA damage, repair, and antioxidant systems in brain regions: a correlative study. *Free Radic Biol Med* 2000;28:779–785.

94. Harman, D. Free radical involvement in aging. Pathophysiology and therapeutic implications. *Drugs Aging* 1993;3:60–80.

95. Alvarez, B., Radi, R. Peroxynitrite reactivity with amino acids and proteins. *Amino Acids* 2003;25:295–311.

96. Calderón-Garcidueñas, L., Reed, W., Maronpot, R.R., Henriquez-Roldan, C., Delgado-Chavez, R., Calderon-Garciduenas, A., Dragustinovis, I., Franco-Lira, M., Aragon-Flores, M., Solt, A.C., Altenburg, M., Torres-Jardon, R., Swenberg, J.A. Brain inflammation and Alzheimer's-like pathology in individuals exposed to severe air pollution. *Toxicol Pathol* 2004;32:650–658.

97. Hoozemans, J.J., O'Banion, M.K. The role of COX-1 and COX-2 in Alzheimer's disease pathology and the therapeutic potentials of non-steroidal anti-inflammatory drugs. *Curr Drug Targets CNS Neurol Disord* 2005;4:307–315.

98. Ringheim, G.E., Conant, K. Neurodegenerative disease and the neuroimmune axis (Alzheimer's and Parkinson's disease, and viral infections) *J Neuroimmunol* 2004; 147:43–49.

99. Teismann, P., Tieu, K., Choi, D.K., Wu, D.C., Naini, A., Hunot, S., Vila, M., Jackson-Lewis, V., Przedborski, S. Cyclooxygenase-2 is instrumental in Parkinson's disease neurodegeneration. *Proc Natl Acad Sci USA* 2003;100:5473–5478.

100. Tzeng, S.F., Hsiao, H.Y., Mak, O.T. Prostaglandins and cyclooxygenases in glial cells during brain inflammation. *Curr Drug Targets Inflamm Allergy* 2005;4: 335–340.

101. Demple, B., Harrison, L. Repair of oxidative damage to DNA: enzymology and biology. *Annu Rev Biochem* 1994;63:915–948.

102. Loeb, L.A., Preston, B.D. Mutagenesis by apurinic/apyrimidinic sites. *Annu Rev Genet* 1986;20:201–230.

103. McGeer, E.G., McGeer, P.L. Pharmacologic approaches to the treatment of amyotrophic lateral sclerosis. *BioDrugs* 2005;19:31–37.

104. Beyreuther, K., Bush, A.I., Dyrks, T., Hilbich, C., Konig, G., Monning, U., Multhaup, G., Prior, R., Rumble, B., Schubert, W., et al. Mechanisms of amyloid deposition in Alzheimer's disease. *Ann NY Acad Sci* 1991;640:129–139.

105. Hardy, J. Amyloid, the presenilins and Alzheimer's disease. *Trends Neurosci* 1997;20:154–159.

106. Selkoe, D.J. The genetics and molecular pathology of Alzheimer's disease: roles of amyloid and the presenilins. *Neurol Clin* 2000;18:903–922.

107. Selkoe, D.J. Alzheimer's disease results from the cerebral accumulation and cytotoxicity of amyloid beta-protein. *J Alzheimers Dis* 2001;3:75–80.

108. Jenner, P. Oxidative stress in Parkinson's disease and other neurodegenerative disorders. *Pathol Biol (Paris)* 1996;44:57–64.

109. Christen, Y. Oxidative stress and Alzheimer disease. *Am J Clin Nutr* 2000;71: 621S–629S.

110. Hensley, K., Mhatre, M., Mou, S., Pye, Q.N., Stewart, C., West, M., Williamson, K.S. On the relation of oxidative stress to neuroinflammation: lessons learned from the G93A-SOD1 mouse model of amyotrophic lateral sclerosis. *Antioxid Redox Signal* 2006;8: 2075–2087.

111. Jenner, P. Oxidative mechanisms in nigral cell death in Parkinson's disease. *Mov Disord* 13 (Suppl 1): 1998; 24–34.

112. Owen, A.D., Schapira, A.H., Jenner, P., Marsden, C.D. Oxidative stress and Parkinson's disease. *Ann NY Acad Sci* 1996;786:217–223.

113. Pizzi, M., Spano, P. Distinct roles of diverse nuclear factor-kappa B complexes in neuropathological mechanisms. *Eur J Pharmacol* 2006;545:22–28.

114. Selkoe, D.J. Alzheimer's disease: genotypes, phenotypes, and treatments. *Science* 1997;275:630–631.

115. Marsden, C.D. The pathophysiology of movement disorders. *Neurol Clin* 1984;2: 435–459.

116. Tanner, C.M. Is the cause of Parkinson's disease environmental or hereditary? Evidence from twin studies. *Adv Neurol* 2003;91:133–142.

117. Caudle, W.M., Richardson, J.R., Delea, K.C., Guillot, T.S., Wang, M., Pennell, K.D., Miller, G.W. Polychlorinated biphenyl-induced reduction of dopamine transporter expression as a precursor to Parkinson's disease-associated dopamine toxicity. *Toxicol Sci* 2006;92:490–499.

118. Ascherio, A., Chen, H., Weisskopf, M.G., O'Reilly, E., McCullough, M.L., Calle, E.E., Schwarzschild, M.A., Thun, M.J. Pesticide exposure and risk for Parkinson's disease. *Ann Neurol* 2006;60:197–203.

119. Hatcher, J.M., Richardson, J.R., Guillot, T.S., McCormack, A.I., Di Monte, D.A., Jones, D.P., Pennell, K.D., Miller, G.W. Dieldrin exposure induces oxidative damage in the mouse nigrostriatal dopamine system. *Exp Neurol* 2007;204:619–630.

120. Overstreet, D.H. Organophosphate pesticides, cholinergic function and cognitive performance in advanced age. *Neurotoxicology* 2000;21:75–81.

121. Aremu, D.A., Meshitsuka, S. Some aspects of astroglial functions and aluminum implications for neurodegeneration. *Brain Res Brain Res Rev* 2006;52:193–200.

122. Berg, D., Youdim, M.B. Role of iron in neurodegenerative disorders. *Top Magn Reson Imaging* 2006;17:5–17.

123. Zatta, P., Lucchini, R., van Rensburg, S.J., Taylor, A. The role of metals in neurodegenerative processes: aluminum, manganese, and zinc. *Brain Res Bull* 2003;62: 15–28.

124. Di Monte, D.A., Lavasani, M., Manning-Bog, A.B. Environmental factors in Parkinson's disease. *Neurotoxicology* 2002;23:487–502.

125. Wicklund, M.P. Amyotrophic lateral sclerosis: possible role of environmental influences. *Neurol Clin* 2005;23:461–484.

126. Zawia, N.H., Basha, M.R. Environmental risk factors and the developmental basis for Alzheimer's disease. *Rev Neurosci* 2005;16:325–337.

127. Peira, E., Marzola, P., Podio, V., Aime, S., Sbarbati, A., Gasco, M.R. *In vitro* and *in vivo* study of solid lipid nanoparticles loaded with superparamagnetic iron oxide. *J Drug Target* 2003;11:19–24.

128. Sykova, E., Jendelova, P. Magnetic resonance tracking of implanted adult and embryonic stem cells in injured brain and spinal cord. *Ann NY Acad Sci* 2005;1049:146–160.

129. Colton, C.A., Gilbert, D.L. Production of superoxide anions by a CNS macrophage, the microglia. *FEBS Lett* 1987;223:284–288.

130. Tomita, M., Holman, B.J., Santoro, C.P., Santoro, T.J. Astrocyte production of the chemokine macrophage inflammatory protein-2 is inhibited by the spice principle curcumin at the level of gene transcription. *J Neuroinflammation* 2005;2: Doi 10.1186/1742-2094-2-8.

131. Doherty, P., Fazeli, M.S., Walsh, F.S. The neural cell adhesion molecule and synaptic plasticity. *J Neurobiol* 1995;26:437–446.

132. Spina, M.B., Cohen, G. Dopamine turnover and glutathione oxidation: implications for Parkinson disease. *Proc Natl Acad Sci* 1989;86:1398–1400.

133. Maker, H.S., Weiss, C., Silides, D.J., Cohen, G. Coupling of dopamine oxidation (monoamine oxidase activity) to glutathione oxidation via the generation of hydrogen peroxide in rat brain homogenates. *J Neurochem* 1981;36:589–593.

134. Vander, A.J., Sherman, J.H., Luciano, D.S. *Human Physiology: The Mechanisms of Body Function*, 8th edition, New York: McGraw-Hill Companies, Inc.; 2001.

135. McKinley, M., O'Loughlin, V.D. *Human Anatomy*, 1st edition, New York: McGraw-Hill Companies, Inc.; 2001.

■■■■■■ CHAPTER 17

Occupational Health Hazards of Nanoparticles

PATRICK T. O'SHAUGHNESSY

Oakdale Campus, University of Iowa, Iowa City, IA 52242-5000, USA

17.1 INTRODUCTION

Nanotechnology encompasses a broad array of methods and products. The current rapid expansion of industries that produce and utilize nanoparticles, defined as particles having diameters less than 100 nm, has created a need to thoroughly determine how the production of these particles will pose a health or safety threat to the workers involved. Although the number of industries associated with nanoparticle creation or use is still relatively small, the industry is expected to grow rapidly and therefore putting a significant number of workers at risk of exposure to these particles. An informative document published in the United Kingdom estimates that 2000 people in the United Kingdom were working with nanoparticles in 2004, of which over half were associated with universities and research centers (1) while over 2 million people will be needed to support nanotechnology companies worldwide (2). Investment in nanotechnology is also growing rapidly with a worldwide investment of $4.5 billion in 2006, an 18% increase from 2004. Furthermore, investments have increased in North America by 42% since 2004 to $1.9 billion (3).

As detailed in previous chapters of this book, toxicological research efforts on the effects of ultrafine particles, which, like nanoparticles, have a diameter of less than $0.1 \mu m$ (100 nm), have involved *in vivo* exposures of animal subjects by inhalation (4–7) and intratracheal instillation (8–12), as well as *in vitro* studies involving cell culture exposures to nanoparticles (13–15). In general, these studies suggest that (a) smaller particles can be more toxic than larger particles on a per mass basis, (b) for some nanoparticles reactivity is related to particle surface area rather than count or mass concentration, and (c) inhalation of nanoparticles can lead to oxidative stress and a subsequent chain of events that produce adverse health effects. Studies specifically

Nanoscience and Nanotechnology Edited by Vicki H. Grassian
Copyright © 2008 John Wiley & Sons, Inc.

related to the toxicity of manufactured nanoparticles (16–18) have found similar toxicological properties. Recent research also indicates that nanoparticles can translocate to the brain via the olfactory nerves (19–21). These studies suggest that concern for the health effects of nanoparticles to workers as an inhalation hazard is justified. Nanoparticles may also pose a dermal hazard given that many commercial products such as sunscreens and lotions are being developed with nanoparticles as additives; however, research in this area is not well developed. The work of Monteiro-Riviere and collaborators on dermal exposures to nanoparticles constitutes the principal research conducted in this area to date (22, 23), although Shvedova et al. (24) have also performed preliminary work demonstrating that dermal exposure to carbon nanotubes may promote oxidative stress and subsequent dermal toxicity. Despite their potential as a dermal exposure hazard, this chapter will focus exclusively on nanoparticles as an inhalation hazard.

Given the long-standing exposure scenarios to ultrafine particles in both the workplace and the ambient environment, the exposure of workers and the general population to particles with diameters less than 100 nm is not new to the advent of nanotechnology. However, what may be new with regard to the assessment of the risk of workers to nanoparticles is the proper evaluation and control of a submicrometer airborne hazard that may emanate from various sources throughout a production facility rather than localized areas such as when welding. This chapter will address the occupational safety and health issues of nanoparticle production using the industrial hygiene paradigm of recognition, evaluation, and control of a workplace hazard by discussing the following:

- the probable hazards and their characteristics,
- available exposure assessment and control methods, and
- risk management options.

17.2 NANOPARTICLE HAZARDS

17.2.1 Historical Perspective

The study of adverse health effects caused by inhaled particles has focused on progressively smaller aerosol size fractions, ranging from micrometer-sized particles to submicrometer-sized particles, and now to nanosized particles. This shift in research focus has largely been influenced by an increased awareness of the harmful effects of smaller particles and the development of instruments capable of detecting submicrometer-sized particles. From an occupational standpoint, original concern centered on dust created in mines (25), especially those used to extract coal, asbestos, and silica. Although producing a small fraction of particles in the submicrometer range, particles in mines tended to be larger than those presently of concern in association with nanotechnology production facilities. However, workplace exposures to "ultrafine" particles have also been commonplace for certain occupations (26, 27). Welding fumes, which typically exist as particles less than 1 μm, have caused both chronic and acute respiratory effects among welders (28–30). Grinding operations also produce ultrafine particles (31, 32). Workers are also at risk from the fine particulate produced by diesel emissions at work sites (33). Likewise, carbon black ("lamp black") one of the intentional products of combustion, has been an occupational health problem for centuries (34). In addition to the presence of submicrometer particles in the workplace, their existence in the ambient atmosphere is also a well-known health concern (35–39). As explained by Oberdörster et al. (40), human exposure to ultrafines has also occurred historically via anthropogenic sources such as internal combustion engines, power plants, and other combustion sources. The U.S. Environmental Protection Agency (41) has not only set regulations for "fine" particles, those less than 2.5 μm, but is also aware that ultrafine particles are a subset of all particles in the atmosphere that require special attention because of their potential to cause adverse health effects (42, 43).

The primary reason for distinguishing between various size fractions of particles in the atmosphere is related to their formation (Fig. 17.1). Whereas most particles greater than 1 μm are derived from mechanical action on a parent material (grinding, crushing), those in the ultrafine region result from the nucleation of low vapor pressure pollutant gases emitted from sources or formed by chemical reactions in the atmosphere from other pollutants. For example, as a

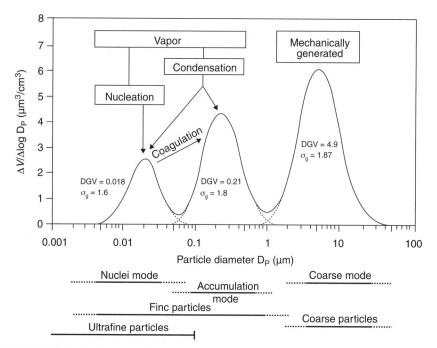

FIGURE 17.1 Typical size distribution of particles existing in an urban area with traffic showing the geometric mean diameter by volume (DGV) and geometric standard deviation (σ_g) for each of the three primary size modes. Adapted from Reference 41.

consequence of motor vehicle combustion as a source of ultrafine particles, researchers have measured a 25-fold increase in exposure levels over background for people living adjacent to highways in California (44). As shown in Fig. 17.1, particles may also be formed by condensation of gas-phase materials on existing particles. Likewise, particles in the nanometer range can grow as a result of the coagulation of two particles into one.

Exposures to both ultrafines and nanoparticles may occur as a consequence of an unintentional release into a workplace atmosphere. The primary difference between the ultrafine particles encountered in workplaces and the ambient environment relative to the current concern for exposure to airborne nanoparticles is that ultrafines are unintentionally created as a consequence of some process, whereas nanoparticles are intentionally created. However, both may have similar properties in terms of how they behave as an aerosol. The most pertinent physical properties of nanoparticles are described in the following section to better understand how they can ultimately be sampled and controlled.

17.2.2 Nanoparticle Properties

Recognition of a workplace hazard also involves anticipating the hazard before workers are adversely affected (45). This anticipation process is at the forefront of current efforts to minimize adverse health effects caused by nanoparticle exposure. As of the date of this publication, no worker has been identified who has suffered deleterious health effects as a consequence of nanoparticle exposure. It should be the goal of all occupational health professionals to maintain this perfect record in the future. As previously mentioned, there are a variety of health effects, both acute and chronic, that may be potentially induced by airborne nanoparticle exposure. Exposure of this type implies that nanoparticles have the potential to be generated as an uncontrolled fugitive emission from some source in a workplace and exist as an aerosol long enough to be inhaled. It is, therefore, worth considering the behavior of a nanoparticle in the atmosphere as well as its potential to be deposited in the lungs.

17.2.2.1 Movement by Diffusion   Although gravity affects all objects, particles in air with a diameter less than about 1 μm will not settle in an appreciable amount of time primarily because their path of descent is strongly influenced by friction with air molecules (46). For example, a 10-μm particle with unit density will fall 1 m in 5.4 min, but a 100 nm particle will require over 13 days to fall the same distance. Therefore, the motion of particles in the nanometer size range is not influenced by gravity (or inertia) but by diffusion—the net transport of particles in a concentration gradient from high to low concentration caused by the Brownian (irregular) motion of particles as they are bombarded by gas molecules (46). As will be explained later, diffusion is an important concept relative to the measurement of nanoparticles, and it also influences how well a nanoparticle is deposited in the lung.

17.2.2.2 *Agglomeration by Coagulation*

In addition to diffusion as the primary transport mechanism for nanoparticles, the tendency for individual nanoparticles to coagulate into multiparticle agglomerates is also fundamental to their behavior in the atmosphere (47). As given by Hinds (46), the rate of change of the number concentration (N) of a particle is proportional to the square of the number concentration:

$$dN/dt = -K_0 N^2$$

where K_0 is the coagulation coefficient that is directly proportional to the diffusion coefficient and particle size and therefore decreases with particle size. An indication of the rate of coagulation can be stated in terms of the half-life of an individual particle suddenly introduced into an atmosphere containing a certain concentration of such particles. This half-life will therefore depend on the particle's size and the number concentration of other particles in that atmosphere. For example, a 200-nm particle has a half-life of 20 s in a high number concentration of $10^{14}\,m^{-3}$ and 231 days if in a low concentration of $10^8\,m^{-3}$. The corresponding half-lives for a 10-nm particle are approximately half those for a 200-nm particle, therefore half-lives largely depend on concentration rather than particle size (Fig. 17.2a). However, when related on a mass concentration basis, there is greater disparity between particle sizes because smaller particles contribute much less to the mass per volume of air (Fig. 17.2b). As shown in this figure, the half-life of a 10-nm particle in a $1\,mg\,m^{-3}$ aerosol cloud is 0.5 s compared to 730 and 8400 s for 100 and 200 nm particles, respectively.

Historical concerns with aerosol exposures typically focused on micrometer-sized particles. As stated by Hinds (46), coagulation could be neglected for laboratory experiments and occupational hygiene work if concentrations were less than $10^{12}\,m^{-3}$ because concentrations less than this resulted in half-lives of more than an hour. Aerosols could therefore be considered as discrete particles in these cases. Most manufactured nanoparticle powders contain primary particles (the individual particles that together comprise the bulk powder) that have diameters ranging from 5 to 30 nm. These particles can therefore coagulate readily to form agglomerates of various sizes. Figure 17.3a and b shows agglomerates formed from titanium dioxide particles with a primary diameter of 20 nm. These were formed by dispersion of the bulk powder in a small chamber and collected for microscopic analysis. Because nanoparticles readily agglomerate by coagulation, the term "agglomerate" was recently formally defined relative to the term "aggregate" by the American Society of Testing and Materials (ASTM) (48), where an agglomerate is

> a group of particles held together by relatively weak forces (for example, Van der Waals or capillary), that may break apart into smaller particles upon processing, for example

and an aggregate is

> a discrete group of particles in which the various individual components are not easily broken apart, such as in the case of primary particles that are strongly bonded together (for example, fused, sintered, or metallically bonded particles).

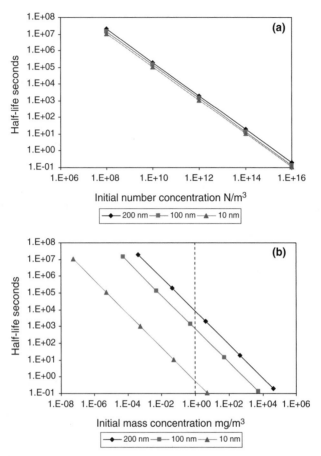

FIGURE 17.2   The half-life of monodisperse spherical particles relative to number concentration (a) and mass concentration (b). Mass concentration determined for particles with a density of $1000 \, \text{kg/m}^3$.

An important caveat to the ready agglomeration of aerosolized nanoparticles is the use of compounds to coat nanoparticles, in some cases specifically to reduce agglomeration (49). Likewise, the agglomeration properties of many engineered particles, such as single- or multiwalled carbon nanotubes (SWCNTs, MWCNTs) and metallic oxides, are currently unknown, which necessarily limits our ability to predict the resulting size distribution of a fugitive nanoparticle aerosol in the workplace and therefore our ability to predict the degree of deposition of these particles in the lung (49).

17.2.2.3   Lung Deposition   The deposition of particles in various regions of the lung has been well researched. The International Commission on Radiological Protection (ICRP) has published a well-known, generalized association between particle size and deposition (Fig. 17.4) that is worth considering in the context of

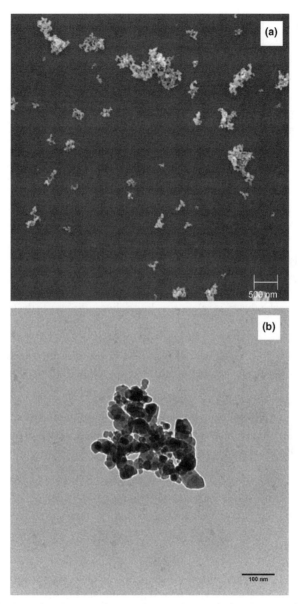

FIGURE 17.3   Scanning electron micrograph (a) and transmission electron micrograph (b) of titanium dioxide nanoparticles with a primary diameter of 20 nm.) (Images taken by Hyun Ju Park, laboratory of Dr. Patrick O'Shaughnessy, The University of Iowa.)

nanoparticle inhalation hazards (50). Historically, aerosol sizes of concern were those greater than the nanometer size range (51, 52) and for which the deposition curve has been well developed. However, less work has been performed to verify the deposition of nanoparticles in the human lungs (53–55), although it is an active research area in

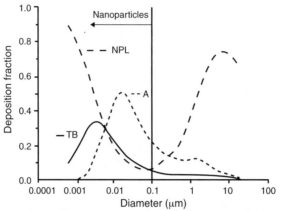

A = Alveolar; TB = Tracheobronchial; NPL = Nasal, pharyngeal, laryngeal

FIGURE 17.4 Lung deposition efficiencies for various regions of the lung and particle sizes. Adapted from Reference 50.

the pharmaceutical industry in relation to drug delivery directly to the lungs (56) and in the area of mathematical modeling using computational fluid dynamics (CFD) methods (57, 58).

Using the ICRP curves shown in Fig. 17.4 to estimate the deposition of nano-particles, it is evident that nanoparticles will readily deposit in all regions of the lung. Maximal deposition for particles in the alveolar region occurs for particles with diameters of approximately 20 nm (0.02 μm), whereas particles less than 10 nm (0.01 μm) are more efficiently deposited in the higher regions of the lung with diffusion of the particles to the inner surface of the lung being the primary mechanism of deposition. However, as shown in Fig. 17.4, any growth of nanoparticles (≤100 nm) via agglomeration will decrease their deposition in the alveolar and tracheobronchial regions of the lung but increase their deposition in the nasopharyngeal region. Therefore, the potential for a nanoparticle to deposit in the lung is not necessarily influenced by its stated primary particle size but by the distribution of agglomerate sizes that exist after the nanoparticles have been dispersed by some means as a fugitive emission in a workplace.

17.2.2.4 Nanoparticle Shape Depending on the fabrication process and type, nanoparticles can have a wide variety of shapes such as fibrous single-walled carbon nanotubes, spherical cerium oxide particles, and platelike nanoclays. In terms of aerosol behavior, shape has been known to affect settling velocity for which a "dynamic shape factor" has been developed to compensate for this effect (46). However, such a correction may be irrelevant to particles in the nanometer range, which are affected more by diffusion than gravity settling as previously mentioned. Current research related to particle shape and toxicity is most importantly associated with how previous work with asbestos and other fibrous aerosols can be translated to the behavior of nanoparticles, especially SWCNTs and MWCNTs (59, 60). A large

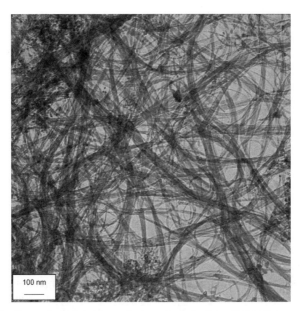

FIGURE 17.5 Transmission electron micrograph of single-walled carbon nanotubes. (TEM image taken by Sherrie Elzey, laboratory of Dr. Vicki Grassian, The University of Iowa.)

body of work has been developed to understand the toxic effects of asbestos. This research has resulted in the commonly understood concept that asbestos fibers travel lengthwise through the airways and can therefore penetrate deep into the lung based on the size of their diameter, whereas they become difficult to clear because of their long length. However, an analogous behavior for SWCNTs, for example, may not prove to be valid because they are at least an order of magnitude smaller than asbestos fibers and tend to agglomerate into a mesh-like clump rather than retain themselves as individual fibers (Fig. 17.5).

Another aspect of nanoparticle shape has been raised by Balbus et al. (61)—their homogeneity in shape and size relative to ultrafine particles derived from combustion sources. Whereas exposure to combustion aerosols will involve a complex mixture of chemicals adhered to a carbon core, exposure to fugitive nanoparticles will present a worker with a much narrower range of shape and size distribution. This type of exposure could lead to something similar to a targeted delivery of the nanoparticles to certain areas of the lung and/or compartments within individual cells.

17.2.2.5 Dustiness If there is potential for nanoparticles to be released in a production facility, then their ability to adversely affect workers is related to not only their toxicity but also their ability to disperse as an inhalable aerosol in the workplace atmosphere. This concept is taken for granted in mines, for example, but should not necessarily be assumed for all nanoparticles. The potential for nanoparticles to be dispersed or resuspended into a workplace atmosphere is related to their "dustiness," a concept that has gained renewed interest with the advent of this new class of aerosol-forming particles.

Maynard discussed the use of dustiness as a nanoparticle quality that can be used to assess their exposure potential (62, 63). However, Aitken et al. (1) suggest that most nanoparticles may not be easily resuspended once they have been collected into a bulk material because of their propensity to agglomerate. This assessment was based on the study by Maynard et al. (64) who used a benchtop shaking method to investigate the propensity for SWCNTs to be resuspended. However, Aitken et al. concede that information on the dustiness of nanoparticles would be useful. Lidén (65) recently not only reviewed the concept of dustiness as a dust quality needing further evaluation with regard to worker exposure but also stated the need to standardize testing procedures (66). Examples of techniques used in the past to assess the dustiness of powders are given by Hietbrink (67) and Brouwer (68).

17.2.2.6 Chemical and Physical Properties   In addition to the nanoparticle properties described above, there are other attributes of importance to a full characterization of the particles in a workplace setting. Although, as mentioned, an exposure to nanoparticles in a workplace may involve a rather homogeneous assortment of particles that make up the aerosol, by nature of their genesis as part of a controlled fabrication process, even nanoparticles of the same general construct can have very different surface chemistries and impurities. Other physical properties such as porosity, solubility, and crystal structure may also be important in terms of the potential toxicity of a nanoparticle (69). For example, an enhanced pulmonary response was found in rats exposed to titanium dioxide (TiO_2) nanoparticles with a crystal structure consisting primarily of the anatase structure of TiO_2 compared to a similar exposure to primarily rutile TiO_2 (70). Likewise, the carbon nanotube synthesis process can result in a very high percentage of impurities in the nanotube matrix (71). A thorough review of carbon nanotube toxicity, including the influence of impurities on toxicity, is given by Lam et al. (72). These two examples illustrate the need to fully characterize chemical and physical properties of nanoparticles during an exposure assessment—a topic further discussed in Section 3.2.

17.3 NANOPARTICLE DETECTION INSTRUMENTS AND ASSESSMENT STRATEGIES

17.3.1 Overview of Available Instruments and Methods

Studies to determine the most relevant dose metric (mass, count, and surface area) for predicting lung inflammatory response as a result of nanoparticle exposure is currently under investigation (73). However, instruments and procedures have been developed to properly evaluate aerosol exposures in the workplace that can be used in the interim to quantify exposure risk from nanoparticles. These instruments fall into two general categories: "time integrated" and "direct reading." Time-integrated measurements involve those that require the completion of a sampling duration after which an analysis is made to determine aerosol concentration, whereas direct-reading instruments provide concentration values in "real time" and typically employ a digital memory device to store the measurements taken for subsequent display and mathematical analysis. The time-integrated devices—primarily filter collection methods—have been used for decades to determine the threat caused by dusts containing, for example, asbestos and silica, whereas direct-reading devices have become more accurate and therefore more often used over the past 20 years.

17.3.1.1 Time-Integrated Measurements Originally, time-integrated aerosol sampling was divided into two general classes by the National Institute for Occupational Safety and Health (NIOSH): a "total" dust sample (74) and "respirable" (75). In 1994, an international agreement was reached on the nature of the efficiency curves that define the ability of particles of varying sizes to infiltrate the human mouth (inhalable), penetrate below the larynx and into the tracheobronchial region (thoracic), and penetrate down to the alveolar region (respirable). The mathematical models associated with each of these three size ranges (Fig. 17.6) is published each year by the American Conference of Governmental Hygienists (ACGIH) (76). The curves shown in Fig. 17.6 represent the expected collection efficiency of the human respiratory system as well as that of the filter-based samplers developed to collect particles with the same efficiency. For example, "respirable" particles constitute the subset of all particles in the atmosphere such that 50% are below 4 μm (77); therefore, a sampler with the same collection efficiency will collect 4 μm particles with 50% efficiency. In the context of the sampling for nanoparticles, however, Fig. 17.6 demonstrates that the three curves converge for particles less than 1 μm with a sampler collection efficiency of 97.1%, which increases to 99.7% for 0.1 μm (100 nm) particles. Therefore, as mentioned above, size-selective sampling via gravimetric analysis of samplers designed with any one of these efficiency curves is an irrelevant concept when attempting to sample for nanoparticles exclusively since all will be collected at the

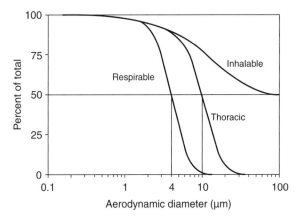

FIGURE 17.6 Conventional curves for the respirable, thoracic, and inhalable fractions of the total dust in a workplace.

same rate. Furthermore, given their extremely small size, nanoparticles will not contribute significantly to the mass loading of a collection filter, therefore, other means of sampling for nanoparticles is necessary. For example, the exposure study completed by Maynard et al. (64) in a nanoparticle production facility determined the airborne concentration of carbon nanotubes to be less than 53 $\mu g/m^3$.

In the past, an indication of aerosol concentration by means of a passive aerosol monitor has not been utilized because of the inherent difficulties associated with accounting for large particles that do not diffuse to the sampler. However, this method may prove to be useful for assessing nanoparticle exposure levels. Wagner and Leith (78) developed a technique to estimate ambient particulate matter (PM) concentration from particles collected passively, which has been expanded upon by Peters et al. (79). To establish a mass concentration using this device, first airborne number concentration N by particle size i is calculated as

$$N_i = \mathrm{SL}_i / V_{\mathrm{dep},\,i} t$$

where SL is surface loading determined by a scanning electron microscope (SEM), V_{dep} is deposition velocity, and t is sampling time. This number distribution is then converted to mass distribution and summed to obtain PM2.5 and PM10. This technique depends on a wind-dependent, semiempirical model to calculate deposition velocity (80). In small research studies of occupational and indoor settings where wind speed is negligible, Wagner and coworkers found good correlation ($r_2 > 0.73$) between PM measured passively and that measured with filter-based samplers for concentrations ranging from 10 to 1000 $\mu g/m^3$ (81, 82). A passive aerosol monitor was also developed by Vinzents (83) and validated by Schneider et al. (84). This device utilizes opaque foils from which the projected area equivalent diameter is used to enumerate particles.

Cascade impactors have been developed to determine the size distribution of an aerosol based on gravimetric sampling. However, historically these devices were designed for use in conventional atmospheres and did not have the ability to distinguish between size ranges much less than 1 μm. For example, the impactor developed by Marple and his collaborators (85) and further characterized by Rader et al. (86) has cut diameter for the lowest stage of 0.52 μm. However, the microorifice uniform deposit impactor (MOUDI) (87) was designed with a lower cutoff of 56 nm and a subsequent version of the instrument, the nano-MOUDI, can distinguish particles down to 10 nm. This instrument may therefore be useful in some applications in which a nanoparticle aerosol size distribution needs to be characterized, although it has primarily been used in ambient air (88) and diesel-exhaust (89)studies. A direct-reading version of the impactor, the electrical low pressure impactor (ELPI, Dekati, Tampere, Finland), can size separate particles ranging from 7 nm to 10 μm by charging the particles landing on each stage and reading the resulting change in current as an indication of the mass buildup on each stage.

17.3.1.2 Direct-Reading Methods
Excellent descriptions of direct-reading instruments for aerosol detection and measurement are given by Pui (90) as well as those in the text by Willeke and Baron (91–93). The reader is referred to these sources

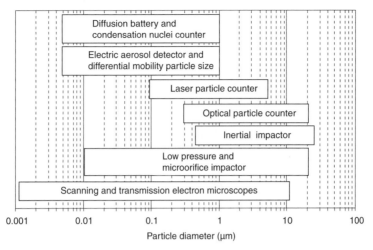

FIGURE 17.7 Typical measurement range of real-time aerosol detection instruments. Adapted from Pui (90), Fig. 17.

for the operational details of these instruments. In general, direct reading instruments fall into two general types: those that provide a measure of aerosol concentration, either mass or number based, and those that determine the aerosol size distribution (90). Available instruments and their respective measurement size range are given in Fig. 17.7. Not shown in this figure are direct-reading aerosol nephelometers (or "photometers"), which have been used to indicate the mass concentration of aerosols in occupational settings (94, 95). However, the use of these instruments to indicate exposures to nanoparticles may be limited because their sensitivity declines exponentially with a decrease in particle size below 1 µm (96).

Optical particle counters (OPCs) have been available for several decades. These instruments both size and count particles to allow for the determination of a number concentration and particle size distribution. As a particle passes through an illuminated viewing volume of the detector, it scatters the light, which is then detected by a photodetector. Particle size is then based on the amplitude of a voltage pulse generated by the photodetector. These instruments vary with regard to the light source used (laser, white light) and the number of size bins, or "channels," into which the sized particles are grouped. Another class of particle counter utilizes the inertial properties of the particles to differentiate their size. These "time-of-flight" instruments accelerate the particles through a nozzle. The larger particles take longer to accelerate and therefore travel for a longer period between two lasers that are interrupted by the particle triggering a set of photosensors to record the travel time. Although these instruments have been frequently used in the past for aerosol assessment and research studies, their lower limit in the range of 300 nm makes them less desirable for assessing nanoparticle exposures except in cases where large agglomerates are expected.

Of the commonly available direct-reading instruments shown in Fig. 17.7, only the condensation nuclei counter (CNC) and the scanning mobility particle sizer

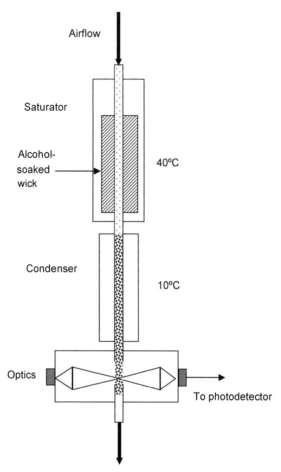

FIGURE 17.8   Schematic representation of the operational aspects of a condensation particle counter.

(SMPS) can accurately measure particles in the size range below 100 nm (0.1 μm). The fundamental difference between these two instruments is that the CNC can only provide a total count whereas the SMPS can also distinguish particle sizes. A CNC (now also referred to as a condensation particle counter, CPC) utilizes the light scattered by a particle to detect and count its presence in the gas stream pumped through the instrument (Fig. 17.8). However, the signal received from a nanoparticle is too small to be accurately detected. To compensate for this problem, the aerosol stream is first passed through a volume containing an alcohol-soaked wick held near 40°C to saturate the stream with alcohol vapors. The stream then passes through a second chamber held near 10°C, which condenses the alcohol on the particles promoting their growth to a size that can be easily detected. Handheld CNCs respond to particles greater than ~10 nm and can measure up to 10^5 particles/cc, whereas benchtop versions can detect up to 10^7 particles/cc.

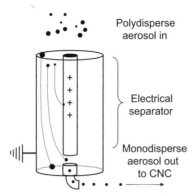

FIGURE 17.9   Schematic diagram of the operational aspects of a differential mobility analyzer.

A CNC is also used as part of an SMPS system. An SMPS combines the size-separating properties of a differential mobility analyzer (DMA) with the counting capabilities of CNC to obtain size distribution information for particles in the nanometer range. For example, under one flow configuration, a commercially available SMPS (TSI Inc., St. Paul, MN, USA) scans 105 size channels between 7.4 and 311 nm over a 3-min period. A diagram of the working parts of a DMA is given in Fig. 17.9. After flowing through an impactor to separate out particles bigger than ~0.5 μm, the charges on a polydispersed aerosol stream are first neutralized and then given a predictable charge distribution based on their size before entering the size-separation section. This "classifier" section consists of an annular space through which the aerosol travels with a laminar flow of filtered air of known flow rate. The annular space is created by a grounded coaxial tube surrounding an inner rod that is charged between 20 and 10,000 V. The voltage potential created along the rod affects the "electrical mobility" of an aerosol, which causes it to travel toward the rod as it moves down along its side. (For this reason, the resulting diameter is referred to as "mobility diameter.") For a set flow rate and voltage, only one particle diameter will have a trajectory that causes it to travel to the bottom of the inner rod, where an annular slit removes them for counting by the CNC. By varying the voltage (only), different particle sizes will reach the slit for counting. Equations are available to relate voltage applied with particle diameter. These equations are used to apply a stepped voltage signal to the DMA to produce a series of aerosols of known size to the CNC which then sends counts to a computer for storage and display. Although very accurate, an SMPS is a benchtop device costing between U.S. $50,000 and $70,000.

17.3.1.3 *Surface Area Measurement*
Given the research that supports the increased reactivity of small particles relative to their surface area, an electronic instrument that gives an indication of particle surface area relative to particle diameter may be the optimal detection device when measuring nanoparticles (26). An instrument designed to measure particle surface area directly (Matter Engineering AG, Wohlen, Switzerland) utilizes one of the principles associated with the operation of a DMA. As with the DMA, particles flowing into the instrument are charged by

bombardment with unipolar ions in a process known as diffusion charging. The number of ions a particle can carry depends on its surface area. Therefore, a measure of the total charge applied to an aerosol stream is directly related to the total surface area of all particles in the stream. Bon and Maynard (97) compared a commercially available diffusion charger with estimates of surface area made with an SMPS and by transmission electron microscopy (TEM) and found good agreement for particles smaller than 100 nm, where the diffusion charger response was proportional to the mobility diameter squared. However, for larger particles the diffusion charger response varied as diameter to the power of 1.5. Another surface area analyzer developed by TSI Inc. (St. Paul, MN, USA) was found to be proportional to the mobility diameter to the 1.16 power but correlated well if related to typical lung deposition curves (98, 99), and therefore provides an indication of the potential surface area dose received.

17.3.2 Nanoparticle Characterization

The ability to fully characterize the chemical and physical properties of nanoparticle surfaces is extremely important for detailing the potential adverse health effects they may cause. At the most basic level, a morphological and size assessment can be obtained via microscopic methods. However, these assessment methods involve both an accurate particle capture method and the methods needed to adequately characterize particle size and morphology with the microscope. One technique of collecting nanoparticles for analysis by TEM is to utilize electrostatic precipitation onto a TEM grid (100). A device for this purpose was developed by Morrow (101) and characterized by Cheng et al. (102), and is fabricated by Intox Products of Moriarty, NM, USA. Enumeration methods such as those prescribed for the analysis of asbestos by TEM can be used to determine a count distribution of nanoparticles (103). Another collection method utilizes diffusion rather than electromotive forces to collect particles in the nanometer range. These "diffusion batteries" have become somewhat obsolete since the development of the SMPS but may now be more widely used for enumerating even sub-10 nm particles (69).

Once particles have been collected on a TEM stub, they can be photographed by the microscope and the resulting digital image analyzed to determine particle size and shape (Fig. 17.10). A shareware image processing software package, ImageJ (among other image analysis software packages), is available from the National Institute of Health (NIH) that is capable of analyzing TEM images with a variety of options. For example, as shown in Fig. 17.10, a digital image of Fe_2O_3 nanoparticle agglomerates was taken via TEM and processed with ImageJ to delineate the borders of each agglomerate. The software can then be used to determine a variety of diameter types, for example, the diameter of a sphere with an equivalent projected surface area as the agglomerate. As with the standard method for determining the size distribution of asbestos particles by phase-contrast microscopy, it is important to obtain an image of the particles that provides a representative sample of the particles. This may require taking multiple samples, each having different sample times, to obtain one with the best array of particles. Unlike asbestos counting,

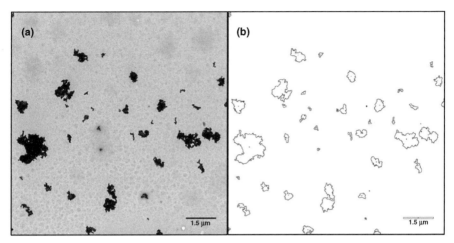

FIGURE 17.10   TEM image of Fe_2O_3 nanoparticles (a) and corresponding area delineation of the particles (b). (Courtesy of Patrick O'Shaughnessy lab.)

however, the presence of agglomerates with a wide range of geometries poses a new problem when using this analysis technique. For example, the imaging software may not properly distinguish between several nearby agglomerates composed of loose associations of primary particles or, more typically, may separate particles in one agglomerate into many individual particles. Some manual oversight of the software results may be necessary to correct these potential errors.

In addition to the size of an agglomerate, its "state" may also be important. In its most basic form, this concept involves the relative porosity of an agglomerate, which indicates how tightly the primary particles are bound within an agglomerate. An agglomeration state is not well defined but is best discussed by Powers et al. (104, 105), who suggest that an "average agglomeration number" can be derived from the ratio of the volume-based median particle size to the average equivalent spherical volume derived from BET gas adsorption (where "BET" is the first initial of the family names of the three scientists who developed the method). Agglomerates have also been characterized in terms of their "fractal" geometry. For example, Virtanen et al. (106) provide a method for determining an agglomerate's "effective density," related to particle mobility and mass, and a "fractal dimension," related to the number of primary particles relative to the size of the agglomerate, based on measurements made with an SMPS and ELPI.

In addition to determining the morphological characteristics of nanoparticles, a wide variety of methods are available to characterize their surface properties and chemical composition. Some of these methods are summarized by Maynard (100) and the ISO (107). If a sample is available after collection of a nanoparticle aerosol onto a TEM grid, then characterization methods associated with electron microscopy such as electron energy loss spectroscopy (EELS) and energy dispersive X-ray analysis (EDX) can be used to quantify particle elemental species. As mentioned in

Section 2.2, knowing the crystallinity of nanopowders such as titanium dioxide is important because it can affect their toxicity. A common method for determining crystallinity is by X-ray powder diffraction (XRD), but it requires a sample of the bulk powder (108). If a bulk sample is available, the surface area relative to the mass of the sample can also be determined by BET analysis. Likewise, surface chemical properties from a bulk sample can be determined by X-ray photoelectron spectroscopy (XPS), and the identification of surface functional groups can be obtained via attenuated total reflection Fourier transform infrared (ATR FTIR) spectroscopy (109, 110). Due to the sophistication of these instruments and their expense, utilizing these methods for characterizing nanoparticles will necessarily involve the aid of a research facility or company specializing in powder characterization techniques.

17.3.3 Site Assessments

An excellent overview of nanoparticle production methods is given by Aitken et al. (1). These methods fall into three main groups: gas-phase processes such as flame pyrolysis; vapor deposition synthesis; and colloidal or liquid-phase methods, which lead to the formation of colloids. For example, carbon nanotubes are produced via chemical vapor deposition (CVD). Aitken et al. also add a fourth method, mechanical processes such as grinding, milling, and alloying, but admit that these attrition methods contrast with the other three groups because they constitute a "top-down" approach rather than the "bottom-up" methods more typically associated with modern nanoparticle production. Of the three main bottom-up methods, only a single workplace assessment specifically related to the evaluation of nanoparticle exposures has been conducted to date. A study by Maynard et al. (64) investigated a carbon nanotube processing facility that found airborne concentrations less than 53 $\mu m/m^3$. These investigators also measured nanotube deposits on gloves, which ranged from 0.2 to 6 mg per hand. Furthermore, they discovered that the bulk powder was difficult to aerosolize and deagglomerate when brought back to a laboratory for investigation.

Other recent studies have focused on all ultrafines in a workplace atmosphere rather than nanoparticles in particular. For example, Thomassen et al. (111) evaluated the number concentration and size distribution of ultrafine particles in an aluminum smelter by TEM analysis and SMPS (111). These researchers were able to identify a particular process (anode changing) that caused the highest concentrations. Likewise, Lee et al. (112) documented a reduction in nanoparticle exposure levels caused by welding aerosols after modification of the workplace ventilation system.

Recently, efforts have been made to determine the spatial or temporal variation of nanosized particles in workplace atmospheres. Peters et al. (113) utilized an aerosol mapping method similar to that described by Heitbrink et al. (114) and Dasch et al. (115) to characterize the spatial distribution of ultrafine particles in a workplace during different seasons. This work involved capturing particle counts with a handheld CPC at a large number of locations throughout the plant and then integrating those measurements into a concentration profile map linked to the dimensions of the plant. Interestingly, they found that the largest number concentration of ultrafine particles was attributed to the exhaust of the direct-fire, natural gas burners used to heat the supply air. Rather than focusing on the spatial distribution of ultrafine particles, Kuhlbush et al. (116) effectively enumerated carbon black particles during the packaging stage in bag filling areas of three carbon black plants over time with the use of an SMPS. The SMPS captured size distribution and total count information every 10 min at each operation location, which enabled the investigators to determine processes that produced the highest fugitive particle emissions. Although these spatial and temporal investigations were not conducted specifically in nanoparticle production companies, the methods used are certainly applicable to the evaluation of nanoparticles exposures.

Although performed in a laboratory, a study by Hsu and Chein (117) is worth mentioning here because it is the only study in which a nanoparticle process was duplicated in a laboratory with the intent of determining which aspects of the production process caused the greatest concentration of aerosolized nanoparticles. In this case, the researchers evaluated titanium dioxide nanoparticle emissions when the bulk powder was coated on different substrates including wood, polymer, and tile while subjected to UV light, a fan, and a rubber knife to simulate the effects of sunlight, wind, and human contact conditions, respectively. An SMPS was used to enumerate the resulting aerosol. Results indicated that the highest particle emissions developed from coating the tile, and UV light increased the release of particles below 200 nm. More studies of this type would be very helpful in determining the potential for the production of a fugitive nanoparticle aerosol in a workplace.

17.4 NANOPARTICLE CONTROL

In addition to the need to assess the effectiveness of exposure evaluation methods for the determination of nanoparticle concentrations in the workplace, there is also a need to study the relative effectiveness of personal protective equipment, notably respirators, used to mitigate aerosol exposures. Current regulations governing respirators (42 CFR Part 84) require testing with a 0.3-μm (300 nm mass median diameter) aerosol to establish filter efficiency ratings for particles of that size. A test particle size of 300 nm was chosen because this was considered to be the diameter of a particle that most efficiently penetrates through a filter (118, 119). Larger particles are more readily captured by impaction and interception, whereas smaller particles are expected to be more efficiently captured because they are greatly influenced by Brownian diffusion and eventual transport to the filter fibers (Fig. 17.11) (118). Therefore, the particles with diameters where these two capture mechanisms are jointly minimized have the lowest capture efficiency.

Previous work discussing the effectiveness of filter media when challenged specifically with ultrafine powders was conducted to determine the effect of flow rate (120) and humidity (121) on capture efficiency. More recent studies have focused specifically on nanoparticle penetration through filters (122–129) as well as the possibility of a "thermal rebound" effect (130). Thermal rebound is thought to occur for particles whose diameters approach that of air molecules (<10 nm) and therefore

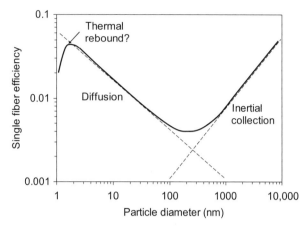

FIGURE 17.11 Idealized fiber efficiency curve demonstrating minimum efficiency where both inertial collection and diffusion are jointly minimized.

literally bounce through a filter media with the effect of reducing the filter's capture efficiency for these particles. Kim et al. (125) reported just such a drop in collection efficiency for particles ≤ 2 nm (125) whereas Heim et al. (128) could not detect a thermal rebound for particles as small as 2.5 nm. Therefore, the thermal rebound effect may only occur for extremely small nanoparticles.

17.4.1 Filter Media Tests

Tests of filter media are typically conducted with instruments specifically designed for this purpose, such as the model 8160 automated filter tester (TSI, Inc., Shoreview, MN, USA). This instrument, and one developed with the use of an SMPS, is described by Japuntich et al. (131). Essentially, these instruments incorporate an aerosol generation system to produce either dioctyl phthalate (DOP) or sodium chloride particles, which are then used to challenge a filter media specimen. A condensation particle counter (CPC) measures the challenge concentration and another CPC measures the air downstream of the filter. The ratio of downstream to upstream concentration is then calculated to determine the fraction or percent penetration of particles through the filter. Because the CPC cannot size-separate particles, the generated aerosol must first be sent to an electrostatic classifier (the same device, a DMA, as that used in an SMPS system) to provide a monodisperse aerosol of known size. However, this step was not necessary when utilizing two SMPS systems in place of the CPCs as the SMPS can both count and size aerosols. Japuntich et al. emphasize the need to adequately purge sampling lines to the counters before taking another sample as residual particles will elevate the percent penetration detected especially for the lower particle sizes (<100 nm).

Kim et al. (126) developed a filter-testing apparatus to evaluate a variety of different filter media types challenged with silver nanoparticles. The filters tested consisted of thin sections (0.053–0.074 mm) of electret filter media as well as an e-PTFE membrane filter commonly used to fabricate respirator filters. This study focused on particles between 3 and 20 nm and therefore represents particles captured by the diffusion mechanism, which, by theory, should result in less efficiency (or greater penetration) with increasing size (see Fig. 17.11). Although penetration values differed for the different media types, all penetration curves declined as expected with a decrease in particle size. Furthermore, no subsequent increase in penetration for the smallest particles as a consequence of thermal rebound was measured.

17.4.2 Respirator Tests

Only one article to date was located that was designed to evaluate nanoparticle penetration through a respirator (129). In this study, a mannequin head equipped with a face-sealed N95 filtering facepiece was placed in a dome into which a challenge aerosol of sodium chloride particles (10–600 nm) was administered. Particles were measured outside and inside the mask with a wide-range particle spectrometer (WPS) that not only incorporates a DMA and CPC but also includes a laser particle spectrometer to increase the measurement range up to 10 μm. The flow rate through the filter was varied between either 30 (light workload) or 85 L/min (heavy workload and the flow rate, adopted for filter testing). Results indicated that at the high flow rate, some filters were not able to maintain a maximum penetration less than 5% (the peak penetration was not reported but a graph shows one curve with a peak at 6%). Furthermore, the most penetrating particles were found to be in the range of 40–50 nm (mobility diameter). Particles <20 and >100 nm had almost no penetration. These results suggest that nanosized particles challenging an electret filter facepiece exhibit a penetration curve with a peak shifted to much smaller particles than the 300 nm size evaluated during standard filter tests.

17.5 NANOPARTICLE RISK MANAGEMENT

Concern for the health and safety of workers in nanotechnology industries was originally addressed by the Institute of Occupational Medicine in 2004 (1). In that publication, Aitken and his collaborators reviewed nanoparticle characteristics, exposure, and control, as well as provided predictions of the number of workers with the potential to be exposed to nanoparticles. This document was also one of the first to address issues associated with developing a proper risk assessment for exposure to nanoparticles with the primary conclusion that "knowledge concerning nanoparticle risks is inadequate for risk assessments." This statement may still be considered accurate at the present time although information is now accruing at a fast pace with a number of investigators recently publishing reports on this issue (49, 61, 132–135).

As described by Tsuji et al. (135), a framework for the risk assessment of nanomaterials involves four interrelated concepts: hazard identification, exposure assessment, toxicity assessment, and risk characterization (Fig. 17.12). Hazard identification includes a physical characterization of the nanomaterial, which may be less predictable

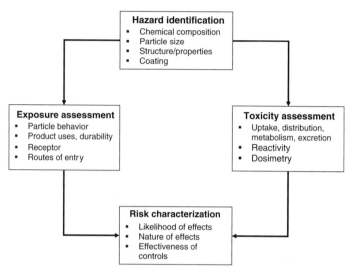

FIGURE 17.12   Nanomaterial risk assessment framework. Adapted from Reference 135.

if the material is novel. Likewise, a proper exposure assessment needs to incorporate the entire life cycle of the material. Certainly, workers in manufacturing facilities will be at risk of exposure to a nanomaterial, but an analysis of the fate of the material over time, its consumer uses, and potential for degradation in the environment may also reveal potential human and ecological exposure routes. Two fundamental criteria for identifying nanomaterials that may have the potential to cause a health risk are provided by Maynard and Kuempel: the material must be biologically accessible to the body after exposure and the material should have the potential to elicit an enhanced biological response because of its nanostructure (63).

An example of the application of hazard identification toward a risk assessment of nanoparticles is given by Maynard and Aitken (69) who investigated the application of different physical structure metrics to a range of particle combinations as a first step in developing a proper assessment of nanoparticles in the workplace. Their approach suggests that investigators first attempt to classify the nanoparticles produced in a workplace by morphology with a classification scheme that ranges from simple spherical particles to particles with increasing structural complexity such as heterogeneous aggregates. They then provide a table of nanostructured particle attributes having the potential to cause adverse health impacts after inhalation such as shape, composition, and crystal structure. With these two qualities in mind, Maynard and Aitken discuss the need to identify the relevant physical exposure metrics that are most important to measure to best determine the health risk to workers. Charts are provided to link morphological characteristics and particle attributes with their expected relevance to workers' health relative to the four metrics of number, length, surface area, and mass concentration. Application of this framework would provide an investigator with a logical approach to identify the most relevant assessment (Section 3.1) and characterization (Section 3.2) method for the particular nanoparticle exposures expected in a workplace.

As considered in this chapter, the primary route of entry for many nanoparticles is via the pulmonary system. Therefore, the toxicity assessment needs to focus on health effects from inhalation. Oberdörster et al. (136) provide a screening strategy for the hazard identification of nanomaterials. In the context of this report, the hazard identification also includes toxicity assessment. The authors concisely outline this process in three general categories involving physicochemical characterization, *in vitro* testing methods, and *in vivo* assays. The authors admit that in many cases still limited data are available and therefore present data-gathering strategies for each category as well as an assessment of current research gaps.

17.5.1 Risk Assessment Methods

The upper three areas of the risk assessment framework shown in Fig. 17.12 are now the subject of intense research efforts throughout the world. However, the fourth aspect of the framework, risk characterization, is more difficult to determine largely because of the current nascent level of nanomaterial production. To emphasize this point, Renn and Roco (133) suggest that, at present, innovations in nanotechnology are outpacing the policy and regulatory environment. One approach to determining adequate controls for minimizing exposures in the absence of sufficient information is by applying the concept of control banding (62). This process allows for an assessment of appropriate control strategies based on products and processes without complete information on hazard and exposure. An example is given by Maynard in which a nanomaterial is related on a scale associated with its "impact" (bulk hazard, surface area, size, etc.) and a scale associated with its "exposure" (amount used, dustiness, etc.) (69). Nanomaterials placed high on both scales may then be associated with an action such as "containment" versus materials lower on the scales, which may only require "general ventilation" as a control solution.

Another approach to establishing a nanomaterial risk framework without the sufficient information and the associated uncertainty is described by Kandlikar et al. (49). Their approach is to study the degree of consensus and/or disagreement between experts familiar with the various exposure/response aspects of nanoparticle exposure—particle characteristics, exposure characteristics, and adverse health effects. This process begins with a set of qualitative questions applied in an interview format, which then leads to the development of an assessment protocol administered to a larger group of experts. An evaluation of answers obtained is then used to create probabilistic information assigned to the risk causal chain to determine the most reasonable assessment of risk. Another sophisticated risk assessment method, multi-criteria decision analysis (MCDA), is proposed by Linkov et al. (132) that also takes into account a scarcity of data available for conventional risk assessment approaches. An example is provided that demonstrates how the application of MCDA can balance societal benefits against unintended side effects and risks posed by the nanomaterial. Similar to Kanlikar's method, MCDA also utilizes expert judgment from scientists and managers to provide decision criteria and weightings used to scale the relative effects of various aspects of the assessment paradigm.

17.6 SUMMARY

This chapter summarized the primary aspects of nanoparticles as an occupational health issue by reviewing the literature associated with the characterization of nanoparticles, relevant detection instruments, filtration of nanoparticles as a control device, and risk assessment issues. This review demonstrates that there is currently a large body of knowledge that can be applied to nanoparticle assessment and control in general but that a significant amount of research is needed to analyze nanoparticle exposure threats in the workplace in particular. Without a single case of morbidity or mortality due to exposure from nanoparticles in the workplace documented to date, the literature demonstrates not only a current proactive concern for potentially exposed workers that is commendable but also a large degree of uncertainty that can stymie efforts to adequately protect affected workers. However, the knowledge of occupational health issues, and the proper protection of workers, has advanced significantly since, for example, asbestos was manufactured and applied without proper protection. Despite the lack of current nanoparticle exposure limits, or even an established assessment strategy to evaluate their levels in a workplace, the accrued knowledge of sound industrial hygiene practice should be at the forefront of any efforts to protect nanoparticle workers while research continues to detail and elucidate specific approaches to best guarantee their health and safety.

REFERENCES

1. Aitken, R.J., Creely, K.S., Tran, C.S. 2004. Nanoparticles: an occupational hygiene review. Institute of Occupational Medicine. Available at: http://www.hse.gov.uk/research/rrpdf/rr274.pdf. Accessed 2004 Nov 12.

2. National Nanotechnology Initiative (NNI). 2004. Available at http://www.nano.gov/html/res/faqs.html#1020.

3. Luxresearch. The Nanotech Report: Investment Overview and Market Research for Nanotechnology, 4th ed. New York: Lux Research Inc.; 2006.

4. Bermudez, E., Mangum, J.B., Wing, B.A., Asgharian, B., Hext, P.M., Warheit, D.B., Everitt, J.L. Pulmonary responses of mice, rats, and hamsters to subchronic inhalation of ultrafine titanium dioxide particles. *Toxicol Sci* 2004;77(2):347–357.

5. Oberdörster, G., Sharp, Z., Atudorei, V., Elder, A., Gelein, R., Lunts, A., Kreyling, W., Cox, C. Acute pulmonary effects of ultrafine particles in rats and mice. *Res Rep Health Eff Inst* 2000;96:5–74, disc. 75–86.

6. Strom, K.A., Johnson, J.T., Chan, T.L. Retention and clearance of inhaled submicron carbon black particles. *J Toxicol Environ Health* 1989;26(2):183–202.

7. Zhou, Y.M., Zhong, C.Y., Kennedy, I.M., Pinkerton, K.E. Pulmonary responses of acute exposure to ultrafine iron particles in healthy adult rats. *Environ Toxicol* 2003;18(4):227–235.

8. Benson, J.M., Holmes, A.M., Barr, E.B., Nikula, K.J., March, T.H. Particle clearance and histopathology in lungs of C3H/HeJ mice administered beryllium/copper alloy by intratracheal instillation. *Inhal Toxicol* 2000;12(8):733–749.

9. Bowden, D.H., Adamson, I.Y. Response of pulmonary macrophages to unilateral instillation of carbon. *Am J Pathol* 1984;115(2):151–155.

10. Ernst, H., Rittinghausen, S., Bartsch, W., Creutzenberg, O., Dasenbrock, C., Görlitz, B. D., Hecht, M., Kairies, U., Muhle, H., Müller, M., Heinrich, U., Pott, F. Pulmonary inflammation in rats after intratracheal instillation of quartz, amorphous SiO_2, carbon black, and coal dust and the influence of poly-2-vinylpyridine- N-oxide (PVNO) *Exp Toxicol Pathol* 2002;54(2):109–126.

11. Li, X.Y., Brown, D., Smith, S., MacNee, W., Donaldson, K. Short-term inflammatory responses following intratracheal instillation of fine and ultrafine carbon black in rats. *Inhal Toxicol* 1999;11(8):709–731.

12. Zhang, Q., Kusaka, Y., Sato, K., Nakakuki, K., Kohyama, N., Donaldson, K. Differences in the extent of inflammation caused by intratracheal exposure to three ultrafine metals: role of free radicals. *J Toxicol Environ Health A* 1998;53(6):423–438.

13. Lundborg, M., Johard, U., Lastbom, L., Gerde, P., Camner, P. Human alveolar macrophage phagocytic function is impaired by aggregates of ultrafine carbon particles. *Environ Res* 2001;86(3):244–253.

14. Renwick, L.C., Donaldson, K., Clouter, A. Impairment of alveolar macrophage phagocytosis by ultrafine particles. *Toxicol Appl Pharmacol* 2001;172(2):119–127.

15. Monteiller, C., Tran, L., Macnee, W., Faux, S.P., Jones, A.D., Miller, B.,G., Donaldson, K. The pro-inflammatory effects of low solubility low toxicity particles, nanoparticles and fine particles, on epithelial cells *in vitro*: the role of surface area. *Occup Environ Med* 2007;64:609–615.

16. Lam, C.W., James, J.T., McCluskey, R., Hunter, R.L. Pulmonary toxicity of single-wall carbon nanotubes in mice 7 and 90 days after intratracheal instillation. *Toxicol Sci* 2004;77(1):126–134.

17. Shvedova, A.A., Castranova, V., Kisin, E.R., Schwegler-Berry, D., Murray, A.R., Gandelsman, V.Z., Maynard, A., Baron, P. Exposure to carbon nanotube material: assessment of nanotube cytotoxicity using human keratinocyte cells. *J Toxicol Environ Health A* 2003;66(20):1909–1926.

18. Warheit, D.B., Laurence, B.R., Reed, K.L., Roach, D.H., Reynolds, G.A.M., Webb, T.R. Comparative pulmonary toxicity assessment of single-wall carbon nanotubes in rats. *Toxicol Sci* 2004;77(1):117–125.

19. Elder, A., Gelein, R., Silva, V., Feikert, T., Opanashuk, L., Carter, J., Potter, R., Maynard, A., Ito, Y., Finkelstein, J., Oberdorster, G. Translocation of inhaled ultrafine manganese oxide particles to the central nervous system. *Environ Health Perspect* 2006;114 (8):1172–1178.

20. Oberdörster, G., Sharp, Z., Kreyling, W., Cox, C., Atudorei, V., Elder, A., Gelein, R. Translocation of inhaled ultrafine particles to the brain. *Inhal Toxicol* 2004;16(6–7): 437–445.

21. Oberdörster, G., Sharp, Z., Atudorei, V., Elder, A., Gelein, R., Kreyling, W., Cox, C. Extrapulmonary translocation of ultrafine carbon particles following whole-body inhalation exposure of rats. *J Toxicol Environ Health A* 2002;65(20):1531–1543.

22. Xia, X.R., Baynes, R.E., Monteiro-Riviere, N.A., Riviere, J.E. An experimentally based approach for predicting skin permeability of chemicals and drugs using a membrane-coated fiber array. *Toxicol Appl Pharmacol* 2007;221(3):320–328.

23. Ryman-Rasmussen, J.P., Riviere, J.E., Monteiro-Riviere, N.A. Variables influencing interactions of untargeted quantum dot nanoparticles with skin cells and identification of biochemical modulators. *Nano Lett* 2007;7(5):1344–1348.

24. Shvedova, A.A., Castranova, V., Kisin, E.R., Schwegler-Berry, D., Murray, A.R., Gandelsman, V.Z., Maynard, A., Baron, P. Exposure to carbon nanotube material: assessment of nanotube cytotoxicity using human keratinocyte cells. *J Toxicol Environ Health A* 2003;66(20):1909–1926.

25. Borm, P.J., Paul, J.A. Particle toxicology: from coal mining to nanotechnology. *Inhal Toxicol* 2002;14(3):311–324.

26. Brouwer, D.H., Gijsbers, J.H., Lurvink, M.W. Personal exposure to ultrafine particles in the workplace: exploring sampling techniques and strategies. *Ann Occup Hyg* 2004; 48(5):439–453.

27. Vincent, J., Clement, C. Ultrafine particles in workplace atmospheres. *Royal Soc* 2000;358:2673–2682.

28. Sobaszek, A., Boulenguez, C., Frimat, P., Robin, H., Haguenoer, J. M., Edme, J.L. Acute respiratory effects of exposure to stainless steel and mild steel welding fumes. *J Occup Environ Med* 2000;42(9):923–931.

29. El-Zein, M., Malo, J.L., Infante-Rivard, C., Gautrin, D. Prevalence and association of welding related systemic and respiratory symptoms in welders. *Occup Environ Med* 2003;60(9):655–661.

30. Stephenson, D., Seshadri, G., Veranth, J.M. Workplace exposure to submicronparticle mass and number concentrations from manual arc welding of carbon steel. *AIHAJ* 2003;64(4):516–521.

31. O'Brien, D., Baron, P., Willeke, K. Size and concentration measurement of an industrial aerosol. *Am Ind Hyg Assoc J* 1986;47(7):386–392.

32. O'Brien, D., Froehlich, P.A, Gressel, M.G., Hall, R.M., Clark, N.J., Bost, P., Fischbach, T. Silica exposure in hand grinding steel castings. *Am Ind Hyg Assoc J* 1992;53(1):42–48.

33. Ono-Ogasawara, M., Smith, T.J. Diesel exhaust particles in the work environment and their analysis. *Ind Health* 2004;42(4):389–399.

34. Dell, L.D., Mundt, K.A., Luippold, R.S., Nunes, A.P., Cohen, L., Burch, M.T., Heidenreich, M.J., Bachand, A.M. A cohort mortality study of employees in the U.S. carbon black industry. *J Occup Environ Med* 2006;48(12):1219–1229.

35. Englert, N. Fine particles and human health: a review of epidemiological studies. *Toxicol Lett* 2004;149(1–3):235–242.

36. Ibald-Mulli, A., Wichmann, H.E., Kreyling, W., Peters, A. Epidemiological evidence of the effects of ultrafine particles exposure. *J Aerosol Med* 2002;15(2):189–201.

37. Donaldson, K., Brown, D., Clouter, A., Duffin, R., MacNee, W., Renwick, L., Tran, L., Stone, V. The pulmonary toxicology of ultrafine particles. *J Aerosol Med* 2002;15 (2):213–220.

38. Pope, C.A. 3rd, Burnett, R.T., Thun, M.J., Calle, E.,E., Krewski, D., Ito, K., Thurston, G. D. Lung cancer, cardiopulmonary mortality, and long-term exposure to fine particulate air pollution. *JAMA* 2002;287(9):1132–1141.

39. Kreyling, W.G., Semmler-Behnke, M., Moller, W. Ultrafine particle-lung interactions: does size matter? *J Aerosol Med* 2006;19(1):74–83.

40. Oberdorster, G., Oberdorster, E., Oberdorster, J. Nanotoxicology: an emerging discipline evolving from studies of ultrafine particles. *Environ Health Perspect* 2005;113(7):823–839.

41. USEPA. editor. *Air Quality Criteria for Particulate Matter.* Office of Research and Development, United State, Environmental Protection Agency, Washington DC; 2004.

42. Oberdörster, G.G. Pulmonary effects of inhaled ultrafine particles. *Int Arch Occup Environ Health* 2001;74(1):1–8.

43. Frampton, M.W. Systemic and cardiovascular effects of airway injury and inflammation: ultrafine particle exposure in humans. *Environ Health Perspect* 2001;109 (Suppl 4):529–532.

44. Zhu, Y., Hinds, W.C., Kim, S., Sioutas, C. Concentration and size distribution of ultrafine particles near a major highway. *J Air Waste Manag Assoc* 2002;52(9):1032–1042.

45. Rose, V. History and philosophy of industrial hygiene. In: Dinardi, S., editor. *The Occupational Environment: Its Evaluation, Control, and Management.* Fairfax, VA: American Industrial Hygiene Association; 2003. pp. 3–18.

46. Hinds, W.C. *Aerosol Technology: Properties, Behavior, and Measurement of Airborne Particles,* 2nd ed. New York: John Wiley & Sons, Inc.; 1999.

47. Preining, O. The physical nature of very, very small particles and its impact on their behaviour. *J Aerosol Sci* 1998;29(5/6):481–495.

48. E 2456-06, ASTM's Terminology for Nanotechnology. West Conshohocken, PA: American Society for Testing and Materials; 2006.

49. Kandlikar, M., Ramachandran, G., Maynard, A., Murdock, B., Toscano, W.A. Health risk assessment for nanoparticles: a case for using expert judgment. *J Nanopart Res* 2007;9 (1):137.

50. International Commission on Radiological Protection (ICRP), Human respiratory tract model for radiological protection, Annals of the *ICRP, Publication 66,* Volume 24(1–3), Elsevier Science Inc., Tarrytown, NY; 1994.

51. Lippmann, M. Sampling criteria for fine fractions of ambient air. In: Vincent, J., editor. *Particle Size-Selective Sampling for Particulate Air Contaminants.* Cincinnati, OH: American Conference of Governmental Industrial Hygienists (ACGIH); 1999.

52. Phalen, R. Airway anatomy and physiology. In: Vincent, J., editor. *Particle Size-Selective Sampling for Particulate Air Contaminants.* Cincinnati, OH: American Conference of Governmental Industrial Hygienists (ACGIH); 1999.

53. Gradon, L., Orlicki, D., Podgorski, A. Deposition and retention of ultrafine aerosol particles in the human respiratory system. *Normal and pathological cases. Int J Occup Saf Ergon* 2000;6(2):189–207.

54. Jaques, P.A., Kim, C.S. Measurement of total lung deposition of inhaled ultrafine particles in healthy men and women. *Inhal Toxicol* 2000;12(8):715–731.

55. Kuempel, E.D., Tran, C.L., Castranova, V., Bailer, A.J. Lung dosimetry and risk assessment of nanoparticles: evaluating and extending current models in rats and humans. *Inhal Toxicol* 2006;18(10):717–724.

56. Dickinson, P.A., Howells, S.W., Kellaway, I.W. Novel nanoparticles for pulmonary drug administration. *J Drug Target* 2001;9(4):295–302.

57. Asgharian, B., Price, O.T. Airflow distribution in the human lung and its influence on particle deposition. *Inhal Toxicol* 2006;18(10):795–801.

58. Nowak, N., Kakade, P.P., Annapragada, A.V. Computational fluid dynamics simulation of airflow and aerosol deposition in human lungs. *Ann Biomed Eng* 2003;31(4):374–390.

59. Powers, K.W. *Characterization issues in evaluating toxicity of nanoparticles: size and shape.* Littleton, CO: Society for Mining, Metallurgy and Exploration; 2006.

60. Mossman, B.T., Borm, P.J., Castranova, V., Costa, D.L., Donaldson, K., Kleeberger, S.R. Mechanisms of action of inhaled fibers, particles and nanoparticles in lung and cardio-vascular diseases. *Part Fibre Toxicol* 2007;4:4.

61. Balbus, J.M., Florini, K., Denison, R.A., Walsh, S.A. Protecting workers and the environment: an environmental NGO's perspective on nanotechnology. *J Nanopart Res* 2007;9(1):11.

62. Maynard, A.D. Nanotechnology: the next big thing, or much ado about nothing? *Ann Occup Hyg* 2007;51(1):1–12.

63. Maynard, A.D., Kuempel, E.D. Airborne nanostructured particles and occupational health. *J Nanopart Res* 2005;7(6):587.

64. Maynard, A.D., Baron, P.A., Foley, M., Shvedova, A.A., Kisin, E.R., Castranova, V. Exposure to carbon nanotube material: aerosol release during the handling of unrefined single walled carbon nanotube material. *J Toxicol Environ Health* 2004;67(1):87–107.

65. Liden, G. Dustiness testing of materials handled at workplaces. *Ann Occup Hyg* 2006; 50(5):437.

66. Hamelmann, F., Schmidt, E. Methods for characterizing the dustiness estimation of powders. *Chem Eng Technol* 2004;27(8):844.

67. Heitbrink, W.A., Todd, W.F., Cooper, T.C., O'Brien, D.M. The application of dustiness tests to the prediction of worker dust exposure. *Am Ind Hyg Assoc J* 1990;51(4):217–223.

68. Brouwer, D.H., Links, I.H.M., De Vreede, S.A.F., Christopher, Y. Size selective dustiness and exposure; simulated workplace comparisons. *Ann Occup Hyg* 2006;50(5):445.

69. Maynard, A.D., Aitken, R.J. Assessing exposure to airborne nanomaterials: current abilities and future requirements. *Nanotoxicology* 2007;1(1):26–41.

70. Warheit, D.B., Webb, T.R., Reed, K.L., Frerichs, S., Sayes, C.M. Pulmonary toxicity study in rats with three forms of ultrafine-TiO_2 particles: differential responses related to surface properties. *Toxicology* 2007;230(1): 90–104.

71. Landi, B.J., Ruf, H.J., Evans, C.M., Cress, C.D., Raffaelle, R.P. Purity assessment of single-wall carbon nanotubes, using optical absorption spectroscopy. *J Phys Chem B* 2005;109(20):9952–9965.

72. Lam, C.W., James, J.T., McCluskey, R., Arepalli, S., Hunter, R.L. A review of carbon nanotube toxicity and assessment of potential occupational and environmental health risks. *Crit Rev Toxicol* 2006;36(3):189–217.

73. Wittmaack, K. In search of the most relevant parameter for quantifying lung inflammatory response to nanoparticle exposure: particle number, surface area, or what? *Environ Health Perspect* 2007;115(2):187–194.

74. NIOSH, editor. Particulates not otherwise regulated, total (0500). *NIOSH Manual of Analytical Methods*, 4th ed. Cincinnati, OH: National Institute for Occupational Safety and Health; 1994.

75. NIOSH, editor. Particulates not otherwise regulated, respirable (0600). *NIOSH Manual of Analytical Methods*, 4th ed. Cincinnati, OH: National Institute for Occupational Safety and Health; 1994.

76. ACGIH. Threshold Limit Values and Biological Exposure Indices. Cincinnati, OH: American Conference of Governmental Industrial Hygienists; 2006.

77. Vincent, J.H. *Aerosol Science for Industrial Hygienists*. Tarrytown, NY: Elsevier Science, Inc.; 1995. pp. 278–279.

78. Wagner, J., Leith, D. Passive aerosol sampler. Part I: principle of operation. *Aerosol Sci Technol* 2001;34(2):186–192.

79. Peters, T., Leith, D., Rappaport, S. Developing a passive sampler for ultrafine particles. *23rd Annual American Association of Aerosol Researchers Conference. Atlanta, GA*; 2004.

80. Wagner, J., Leith, D. Passive aerosol sampler. Part II: wind tunnel experiments. *Aerosol Sci Technol* 2001;34(2):193–201.

81. Wagner, J., Leith, D. Field tests of a passive aerosol sampler. *J Aerosol Sci* 2001;32(1):33.

82. Wagner, J., Macher, J. Comparison of a passive aerosol sampler to size-selective pump samplers in indoor environments. *AIHAJ* 2003;64:630–639.

83. Vinzents, P.S. A passive personal dust monitor. *Ann Occup Hyg* 1996;40(3):261–280.

84. Schneider, T., Schlunssen, V., Vinzents, P.S., Kildeso, J. Passive sampler used for simultaneous measurement of breathing zone size distribution, inhalable dust concentration and other size fractions involving large particles. *Ann Occup Hyg* 2002;46(2):187.

85. Rubow, K.L., Marple, V.A., Olin, J., McCawley, M.A. A personal cascade impactor: design, evaluation and calibration. *Am Ind Hyg Assoc J* 1987;48(6):532–538.

86. Rader, D.J., Mondy, L.A., Brockmann, J.E., Lucero, D.A., Rubow, K.L. Stage response calibration of the Mark III and Marple personal cascade impactors. *Aerosol Sci Technol* 1991;14:365–379.

87. Marple, V.A., Rubow, K.I., Behm, S.M. Microorifice uniform deposit impactor (MOUDI): description, calibration, and use. *Aerosol Sci Technol* 1991;14(4):434.

88. Lin, C.-C., Chen, S.-J., Huang, Kuo-Lin., Hwang, Wen-Ing., Chang-Chien, Guo-Ping., Lin, Wen-Yinn. Characteristics of metals in nano/ultrafine/fine/coarse particles collected beside a heavily trafficked road. *Environ Sci Technol* 2005;39(21):8113.

89. Kubo, S., Chatani, S., Kondoh, T., Yamamoto, M., Inoue, M. Detailed properties of diesel volatile nanoparticles. *Tran Jpn Soc Mech Eng B* 2006;72(11):2619.

90. Pui, D.Y.H. Direct-reading instrumentation for workplace aerosol measurements. *Analyst* 1996;121:1215–1224.

91. Baron, P.A., Mazumder, M.K., Cheng, Y.-S. Direct-reading techniques using particle motion and optical detection. In: Willeke, K., Baron, P., editors. *Aerosol Measurement: Principles, Techniques, and Applications*. New York: Van Nostrand Reinhold; 2001. pp. 495–536.

92. Flagan, R.C. Electrical techniques. In: Willeke, K., Baron, P., editors. *Aerosol Measurement: Principles, Techniques, and Applications*. New York: Van Nostrand Reinhold; 2001. pp. 537–568.

93. Cheng, Y.-S. Condensation detection and diffusion size separation techniques. In: Willeke, K., Baron, P., editors. *Aerosol Measurement: Principles, Techniques, and Applications*. New York: Van Nostrand Reinhold; 2001. pp. 569–602.

94. Smith, J.P. Use of scattered light particulate monitors with a foundry air recirculation system. *Appl Ind Hyg* 1987;2(2):74–78.

95. Page, S., Jankowski, R. Correlations between measurements with RAM-1 and gravimetric samplers on longwall shearer faces. *Am Ind Hyg Assoc J* 1984;45(9):610–616.

96. O'Shaughnessy, P.T., Slagley, J.M. Photometer response determination based on aerosol physical characteristics. *AIHAJ* 2002;63(5):578–585.

97. Bon, K.K., Maynard, A.D. Comparing aerosol surface-area measurements of monodisperse ultrafine silver agglomerates by mobility analysis, transmission electron microscopy and diffusion charging. *J Aerosol Sci* 2005;36(9):1108.

98. Wilson, W.E., Stanek, J., Han, H-S., Johnson, T., Sakurai, H., Pui, D.Y.H., Turner, J., Chen, D-R., Duthie, S. Use of the electrical aerosol detector as an indicator of the surface area of fine particles deposited in the lung. *J Air Waste Manag Assoc* 2007;57(2):211.

99. Fissan, H., Neumann, S., Trampe, A., Pui, D.Y.H., Shin, W.G. Rationale and principle of an instrument measuring lung deposited nanoparticle surface area. *J Nanopart Res* 2007;9(1):53.

100. Maynard, A.D. Overview of methods for analysing single ultrafine particles. *Philos Trans R Soc Lond A* 2000;358(1775):2593.

101. Morrow, P.E., Mercer, T.T. A point-to-plane electrostatic precipitator for particle size sampling. *Am Ind Hyg Assoc J* 1964;25:8–14.

102. Cheng, Y.S., Yeh, H.C., Kanapilly, G.M. Collection efficiencies of a point-to-plane electrostatic precipitator. *Am Ind Hyg Assoc J* 1981;42:605–610.

103. NIOSH, editor. Asbestos by TEM (7402). *NIOSH Manual of Analytical Methods*, 4th ed. Cincinnati, OH: National Institute for Occupational Safety and Health; 1994.

104. Powers, K.W., Palazuelos, M., Moudgil, B.M., Roberts, S.M. Characterization of the size, shape, and state of dispersion of nanoparticles for toxicological studies. *Nanotoxicology* 2007;1(1):42–51.

105. Powers, K.W., Brown, S.C., Krishna, V.B., Wasdo, S.C., Moudgil, B.M., Roberts, S.M. Research strategies for safety evaluation of nanomaterials. Part VI. Characterization of nanoscale particles for toxicological evaluation. *Toxicol Sci* 2006;90(2):296–303.

106. Virtanen, A., Ristimäki, J., Keskinen, J. Method for measuring effective density and fractal dimension of aerosol agglomerates. *Aerosol Sci Technol* 2004;38(5):437–446.

107. International Organization for Standardization, Workplace Atmospheres - Ultrafine, nanoparticle and nano-structured aerosols - Inhalation exposure characterization and assessment. Geneva, Switzerland. ISO/TR 27628:2007.

108. Narkiewicz, U., Guskos, N., Arabczyk, W., Typek, J., Bodziony, T., Konicki, W., Gasiorek, G., Kucharewicz, I., Anagnostakis, E.A. XRD, TEM and magnetic resonance studies of iron carbide nanoparticle agglomerates in a carbon matrix. Strasbourg: Elsevier; 2004.

109. Tunc, I., Suzer, S., Correa-Duarte, M.A., Liz-Marzan, L.M. XPS characterization of Au (core)/SiO$_2$ (shell) nanoparticles. *J Phys Chem B* 2005;109(16):7597.

110. Park, Y.-J., Kim, J.H. The use of ATR-FTIR to determine the effects of functional groups on interfacial crosslinking of reactive nanoparticles. *J Disper Sci Technol* 2003;24(3–4):537.

111. Thomassen, Y., Koch, W., Dunkhorst, W., Ellingsen, D.G., Skaugset, N.P., Jordbekken, L., Arne Drablos, P., Weinbruch, S. Ultrafine particles at workplaces of a primary aluminium smelter. *J Environ Monit* 2006;8(1):127–133.

112. Lee, M.H., McClellan, W.H., Candela, J., Andrews, D., Biswas, P. Reduction of nanoparticle exposure to welding aerosols by modification of the ventilation system in a workplace. *J Nanopart Res* 2007;9(1):127.

113. Peters, T.M., Heitbrink, W.A., Evans, D.E., Slavin, T.J., Maynard, A.D. The mapping of fine and ultrafine particle concentrations in an engine machining and assembly facility. *Ann Occup Hyg* 2006;50(3):249–257.

114. Heitbrink, W.A., Evans, D.E., Peters, T.M., Slavin, T.J. Characterization and mapping of very fine particles in an engine machining and assembly facility. *J Occup Environ Hyg* 2007;4(5):341–351.

115. Dasch, J., D'Arcy, J., Gundrum, A., Sutherland, J., Johnson, J., Carlson, D. Characterization of fine particles from machining in automotive plants. *J Occup Environ Hyg* 2005;2 (12):609–625.

116. Kuhlbusch, T.A., Fissan, H. Particle characteristics in the reactor and palletizing areas of carbon black production. *J Occup Environ Hyg* 2006;3(10):558–567.

117. Li-Yeh, H., Hung-Min, C. Evaluation of nanoparticle emission for TiO_2 nanopowder coating materials. *J Nanopart Res* 2007;9(1):157.

118. Lee, K.W., Liu, B.Y.H. On the minimum efficiency and most penetrating particle size for fibrous filters. *J Air Poll Control Assoc* 1980;30:377–381.

119. Colton, C., Nelson, T. Respirator protection. In: Dinardi, S., editor. *The Occupational Environment: Its Evaluation, Control, and Management.* Fairfax, VA: American Industrial Hygiene Association; 2003. pp. 931–953.

120. Stevens, G.A., Moyer, E.S. "Worst case" aerosol testing parameters: I. Sodium chloride and dioctyl phthalate aerosol filter efficiency as a function of particle size and flow rate. *Am Ind Hyg Assoc J* 1989;50(5):257–264.

121. Moyer, E.S., Stevens, G.A. "Worst case" aerosol testing parameters: II. Efficiency dependence of commercial respirator filters on humidity pretreatment. *Am Ind Hyg Assoc J* 1989;50(5):265–270.

122. Otani, Y., Emi, H., Cho, S-J., Namiki, N. Generation of nanometer size particles and their removal from air. *Adv Powder Technol* 1995;6(4):271–281.

123. Ichitsubo, H., Hashimoto, T., Alonso, M., Kousaka, Y. Penetration of ultrafine particles and ion clusters through wire screens. *Aerosol Sci Technol* 1996;24:119–127.

124. Alonso, M., Kousaka, Y., Hashimoto, T., Hashimoto, N. Penetration of nanometer-sized aerosol particles through wire screen and laminar flow tube. *Aerosol Sci Technol* 1997;27:471–480.

125. Kim, C.S., Bao, L., Okuyama, K., Shimada, M., Niinuma, H. Filtration efficiency of a fibrous filter for nanoparticles. *J Nanopart Res* 2006;8(2):215.

126. Kim, S.C., Harrington, M.S., Pui, D.Y.H. Experimental study of nanoparticles penetration through commercial filter media. *J Nanopart Res* 2007;9(1):117.

127. Wang, J., Chen, D.R., Pui, D.Y.H. Modeling of filtration efficiency of nanoparticles in standard filter media. *J Nanopart Res* 2007;9(1):109.

128. Heim, M., Mullins, B.J., Wild, M., Meyer, J., Kasper, G. Filtration efficiency of aerosol particles below 20 nanometers. *Aerosol Sci Technol* 2005;39(8):782.

129. Balazy, A., Toivola, M., Reponen, T., Podgorski, A., Zimmer, A., Grinshpun, S.A. Manikin-based performance evaluation of N95 filtering-facepiece respirators challenged with nanoparticles. *Ann Occup Hyg* 2006;50(3):259.

130. Wang, H.-C. Comparison of thermal rebound theory with penetration measurements of nanometer particles through wire screens. *Aerosol Sci Technol* 1996;24:129–134.

131. Japuntich, D.A., Franklin, L.M., Pui, D.Y., Kuehn, T.H., Kim, S.C., Viner, A.S. A comparison of two nano-sized particle air filtration tests in the diameter range of 10 to 400 nanometers. *J Nanopart Res* 2007;9(1):93.

132. Linkov, I., Satterstrom, F.K., Steevens, J., Ferguson, E., Pleus, R.C. Multi-criteria decision analysis and environmental risk assessment for nanomaterials. *J Nanopart Res* 2007;9(4):543.

133. Renn, O., Roco, M.C. Nanotechnology and the need for risk governance. *J Nanopart Res* 2006;8(2):153.

134. Mowat, F.S., Hartzell, A.L., Da Silva, M.G., Tsuji, J.S. *Health risk assessment of products containing nano-engineered materials*. Cambridge, MA: Nano Science and Technology Institute; 2007.

135. Tsuji, J.S., Maynard, A.D., Howard, P.C., James, J.T., Lam, C.W., Warheit, D.B., Santamaria, A.B. Research strategies for safety evaluation of nanomaterials, part IV: risk assessment of nanoparticles. *Toxicol Sci* 2006;89(1):42–50.

136. Oberdorster, G., Maynard, A., Donaldson, K., Castranova, V., Fitzpatrick, J., Ausman, K., Carter, J., Karn, B., Kreyling, W., Lai, D., Olin, S., Monteiro-Riviere, N., Warheit, D., Yang, H. Principles for characterizing the potential human health effects from exposure to nanomaterials: elements of a screening strategy. *Part Fibre Toxicol* 2005;2:8.

Nanoscience and Nanotechnology Edited by Vicki H. Grassian
Copyright © 2008 John Wiley & Sons, Inc.

Water-saturated porous medium, 95, 103
Water-soluble fullerene derivatives, 171
  fullerols, 171
Water-soluble groups, 380
Water-soluble nanomaterials, 6
Water treatment plants (WTPs), 76
Wet-packed cell, 121
White flocs, 84
Woodrow Wilson International Center, 46

X-ray absorption near-edge spectroscopy
  (XANES), 210, 353
X-ray diffraction (XRD), 48

patterns, 48
techniques, 47
X-ray photoelectron spectroscopy (XPS),
  52, 141
Xenobiotoics, 352
XPS measurements, 52, 53, 61
XRD diffraction pattern, 48, 53

Years lived disabled (YLD), 22
Years of life lost (YLL), 22
Young–Laplace equation of capillarity, 115

Zeolites, characteristics of, 289, 296, 300